Communications
in Computer and Information Science

49

Miltiadis D. Lytras
Patricia Ordonez de Pablos
Ernesto Damiani David Avison
Ambjörn Naeve David G. Horner (Eds.)

Best Practices for the Knowledge Society

Knowledge, Learning, Development and Technology for All

Second World Summit
on the Knowledge Society, WSKS 2009
Chania, Crete, Greece, September 16-18, 2009
Proceedings

Volume Editors

Miltiadis D. Lytras, lytras@acgmail.gr
Patricia Ordonez de Pablos, patriop@uniovi.es
Ernesto Damiani, damiani@dti.unimi.it
David Avison, avison@essec.fr
Ambjörn Naeve, amb@nada.kth.se
David G. Horner, president@acg.edu

Library of Congress Control Number: 2009935331

CR Subject Classification (1998): I.2.6, I.2.4, K.4, K.3, J.1, H.2.8

ISSN 1865-0929
ISBN-10 3-642-04756-4 Springer Berlin Heidelberg New York
ISBN-13 978-3-642-04756-5 Springer Berlin Heidelberg New York

springer.com

Printed in Germany

Typesetting: Camera-ready by author, data conversion by Scientific Publishing Services, Chennai, India
Printed on acid-free paper SPIN: 12768554 06/3180 5 4 3 2 1 0

Preface

It is a great pleasure to share with you the Springer LNCS proceedings of the Second World Summit on the Knowledge Society, WSKS 2009, organized by the Open Research Society, Ngo, http://www.open-knowledge-society.org, and held in Samaria Hotel, in the beautiful city of Chania in Crete, Greece, September 16–18, 2009.

The 2nd World Summit on the Knowledge Society (WSKS 2009) was an international scientific event devoted to promoting dialogue on the main aspects of the knowledge society towards a better world for all. The multidimensional economic and social crisis of the last couple of years has brought to the fore the need to discuss in depth new policies and strategies for a human centric developmental processes in the global context.

This annual summit brings together key stakeholders involved in the worldwide development of the knowledge society, from academia, industry, and government, including policy makers and active citizens, to look at the impact and prospects of information technology, and the knowledge-based era it is creating, on key facets of living, working, learning, innovating, and collaborating in today's hyper-complex world.

The summit provides a distinct, unique forum for cross-disciplinary fertilization of research, favoring the dissemination of research on new scientific ideas relevant to international research agendas such as the EU (FP7), OECD, or UNESCO. We focus on the key aspects of a new sustainable deal for a bold response to the multidimensional crisis of our times.

Eleven general pillars provide the constitutional elements of the summit:

Pillar 1. Information Technologies – Knowledge Management Systems – E-business and Business, and Organizational and Inter-organizational Information Systems for the Knowledge Society
Pillar 2. Knowledge, Learning, Education, Learning Technologies, and E-learning for the Knowledge Society
Pillar 3. Social and Humanistic Computing for the Knowledge Society – Emerging Technologies for Society and Humanity
Pillar 4. Culture and Cultural Heritage – Technology for Culture Management – Management of Tourism and Entertainment – Tourism Networks in the Knowledge Society
Pillar 5. E-government and E-democracy in the Knowledge Society
Pillar 6. Innovation, Sustainable Development, and Strategic Management for the Knowledge Society
Pillar 7. Service Science, Management, Engineering, and Technology
Pillar 8. Intellectual and Human Capital Development in the Knowledge Society
Pillar 9. Advanced Applications for Environmental Protection and Green Economy Management
Pillar 10. Future Prospects for the Knowledge Society: from Foresight Studies to Projects and Public Policies
Pillar 11. Technologies and Business Models for the Creative Industries

In the 2nd World Summit on the Knowledge Society, six main tracks and three workshops were organized. The CCIS 49 Volume, summarizes 61 full research articles that were selected after a double blind review process from 256 submissions, contributed by 480 co-authors.

We are very happy, because in this volume of CCIS you will find excellent quality research giving sound propositions for advanced systems towards the knowledge society.

In the next figure we summarize the context of the research contributions presented at WSKS 2009.

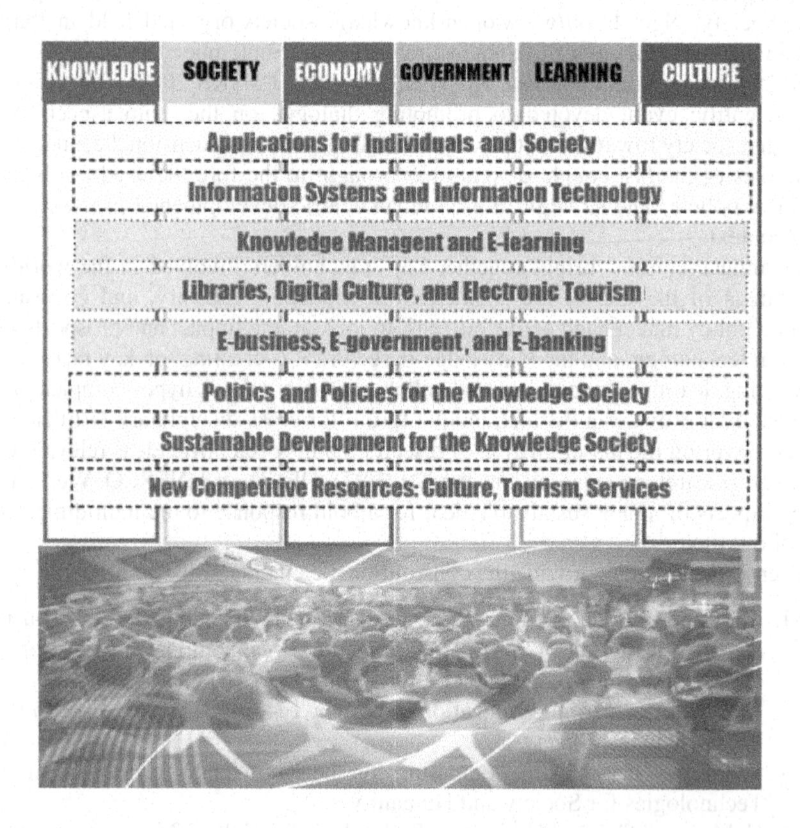

I would like to thank the more than 480 co-authors from 59 countries for their submissions, the Program Committee members and their subreviewers for the thoroughness of their reviews and the colleagues in the Open Research Society for the great support they offered during the organization of the event in Chania.

We are honored by the support and encouragement of the Editors-in-Chief of the five ISI SCI/SSCI listed journals that agreed to publish special issues from extended versions of papers presented at the summit:

- Robert Tennyson, Editor-in-Chief of Computers in Human Behaviour
- Amit Sheth, Editor-in-Chief of the International Journal of Semantic Web and Information Systems
- Adrian Dale, Editor-in-Chief of the Journal of Information Science

- Felix Tan, Editor-in-Chief of the Journal of Global Information Management
- Janice Sipior, Editor-in-Chief of Information Systems Management

A great thank you also to Alfred Hofmann from Springer and his staff for the excellent support during the publication of LNCS/LNAI 5736 and CCIS 49.

Last but not least, I would like to express my gratitude to the staff and members of the Open Research Society for their hard work throughout the organization of the summit and their efforts to promote a better world for all based on knowledge and learning.

We need a better world. We contribute with our sound voices to the agenda, policies, and actions. We invite you to join your voice with ours and all together to shape a new deal for our world: education, sustainable development, health, opportunities for well being, culture, collaboration, peace, democracy, and technology for all.

Looking forward to seeing you at the third event in the series, about which you can find more information at: http://www.open-knowledge-society.org/summit.htm.

With 30 special issues already agreed for WSKS 2010, and 6 main tracks planned, we would like to ask for your involvement, and we would be happy to see you joining us.

THANK YOU – Efharisto Poli!!

July 2009 Miltiadis D. Lytras

Organization

WSKS 2009 was organized by the International Scientific Council for the Knowledge Society and supported by the Open Research Society, Ngo, http://www.open-knowledge-society.org and the International Journal of the Knowledge Society Research, http://www.igi-global.com/ijksr

Executive Committee

General Chair of WSKS 2009

Professor Miltiadis D. Lytras

President, Open Research Society, Ngo

Miltiadis D. Lytras is the President and Founder of the Open Research Society, NGO. His research focuses on semantic web, knowledge management, and e-learning, with more than 100 publications in these areas. He has co-edited / co-edits 25 special issues in International Journals (e.g., IEEE Transaction on Knowledge and Data Engineering, IEEE Internet Computing, IEEE Transactions on Education, Computers in Human Behavior, Interactive Learning Environments, the Journal of Knowledge Management, the Journal of Computer Assisted Learning, etc.) and has authored/ (co-)edited 25 books (e.g., Open Source for Knowledge and Learning Management, Ubiquitous and Pervasive Knowledge Management, Intelligent Learning Infrastructures for Knowledge Intensive Organizations, Semantic Web Based Information Systems, China Information Technology Hanbook, Real World Applications of Semantic Web and Ontologies, Web 2.0: The Business Model, etc.). He is the founder and officer of the Semantic Web and Information Systems Special Interest Group of the Association for Information Systems (http://www.sigsemis.org). He serves as the (Co-) Editor-in-Chief of 12 international journals (e.g., the International Journal of Knowledge and Learning, the International Journal of Technology Enhanced Learning, the International Journal on Social and Humanistic Computing, the International Journal on Semantic Web and Information Systems, the International Journal on Digital Culture and Electronic Tourism, the International Journal of Electronic Democracy, the International Journal of Electronic Banking, and the International Journal of Electronic Trade, etc.) while he is associate editor or editorial board member in seven more.

WSKS 2009 Co-chairs

Professor Ernesto Damiani

University of Milan, Italy

Ernesto Damiani is a professor at the Dept. of Information Technology, University of Milan, where he leads the Software Architectures Lab. Prof. Damiani holds/has held visiting positions at several international institutions, including George Mason University (Fairfax, VA, USA) and LaTrobe University (Melbourne, Australia). Prof. Damiani is an Adjunct Professor at the Sydney University of Technology (Australia). He has written several books and filed international patents; also, he has co-authored more than two hundred research papers on advanced secure service-oriented architectures, open source software and business process design, software reuse, and Web data semantics. Prof. Damiani is the Vice Chair of IFIP WG 2.12 on Web Data Semantics and the secretary of IFIP WG 2.13 on Open Source Software Development. He coordinates several research projects funded by the Italian Ministry of Research and by private companies, including Siemens Mobile, Cisco Systems, ST Microelectronics, BT Exact, Engineering, Telecom Italy, and others.

Professor John M. Carroll

The Pennsylvania State University, USA

John M. Carroll was a founder of human-computer interaction, the youngest of the 9 core areas of computer science identified by the Association for Computing Machinery (ACM). He served on the program committee of the 1982 Bureau of Standards Conference on the Human Factors of Computing Systems that, in effect, inaugurated the field, and was the direct predecessor of the field's flagship conference series, the ACM CHI conferences. Through the past two decades, Carroll has been a leader in the development of the field of Human-Computer Interaction. In 1984 he founded the User Interface Institute at the IBM Thomas J. Watson Research Center, the most influential corporate research laboratory during the latter 1980s. In 1994, he joined Virginia Tech as Department Head of Computer Science where he established an internationally renowned HCI focus in research and teaching. Carroll has served on the editorial boards of every major HCI journal – the International Journal of Human-Computer Interaction, the International Journal of Human-Computer Studies, Human-Computer Interaction, Transactions on Computer-Human Interaction, Transactions on Information Systems, Interacting with Computers, and Behavior and Information Technology. He was a founding associate editor of the field's premier journal, ACM Transactions on Computer-Human Interaction, and a founding member of the editorial boards of Transactions on Information Systems, Behavior and Information Technology, and the International Journal of Human-Computer Interaction. He is currently on the Senior Editorial Advisory Board for the field's oldest journal, the International Journal of Human-Computer Systems. He served on the editorial board of all three editions of the Handbook of Human-Computer Interaction, and was associate editor for the section on Human-Computer Interaction in the Handbook of Computer Science and Engineering. He has served on more than 50 program committees for international HCI conferences, serving as chair or associate chair 12 times. He has been nominated to become the next Editor-in-Chief of the ACM Transactions on Computer-Human Interaction. He is currently serving his second term on the National Research Council's Committee on Human Factors. Carroll has published 13 books and more than 250 technical papers, and has produced more than 70 miscellaneous reports (videotapes, workshops, tutorials, conference demonstrations and discussant talks). He has presented more than 30 plenary or distinguished lectures.

Professor Robert Tennyson

University of Minnesota, USA

Robert Tennyson is currently a professor of educational psychology and technology in learning and cognition. In addition to his faculty position, he is a program coordinator for the psychological foundations of education. His published works range from basic theoretical articles on human learning to applied books on instructional design and technology. He serves as the Editor-in-Chief of the scientific journal, Computers in Human Behavior, published by Elsevier Science and now in its 17th year, as well as serving on several editorial boards for professional journals. His research and publications include topics such as cognitive learning and complex cognitive processes, intelligent systems, complex-dynamic simulations, testing and measurement, instructional design, and advanced learning technologies. His international activities include directing a NATO-sponsored advanced research workshop in Barcelona and a NATO advanced study institute in Grimstad, Norway – both on the topic of automated instructional design and delivery. He has recently directed an institute on technology in Athens and Kuala Lumpur. His other international activities include twice receiving a Fulbright Research Award in Germany and once in Russia. His teaching interests include psychology of learning, technology-based systems design, evaluation, and management systems.

Professor David Avison

ESSEC Business School, France

David Avison is Distinguished Professor of Information Systems at ESSEC Business School, near Paris, France, after being Professor at the School of Management at Southampton University for nine years. He has also held posts at Brunel and Aston Universities in England, and the University of Technology Sydney and University of New South Wales in Australia, and elsewhere. He is President-elect of the Association of Information Systems (AIS). He is joint editor of Blackwell Science's Information Systems Journal now in its eighteenth volume, rated as a 'core' international journal. So far, 25 books are to his credit including the fourth edition of the well-used text Information Systems Development: Methodologies, Techniques and Tools (jointly authored with Guy Fitzgerald). He has published a large number of research papers in learned journals, and has edited texts and conference papers. He was Chair of the International Federation of Information Processing (IFIP) 8.2 group on the impact of IS/IT on organizations and society and is now vice chair of IFIP technical committee 8. He was past President of the UK Academy for Information Systems and also chair of the UK Heads and Professors of IS and is presently member of the IS

Senior Scholars Forum. He was joint program chair of the International Conference in Information Systems (ICIS) in Las Vegas (previously also research program stream chair at ICIS Atlanta), joint program chair of IFIP TC8 conference at Santiago Chile, program chair of the IFIP WG8.2 conference in Amsterdam, panels chair for the European Conference in Information Systems at Copenhagen and publicity chair for the entity-relationship conference in Paris, and chair of several other UK and European conferences. He will be joint program chair of the IFIP TC8 conference in Milan, Italy in 2008. He also acts as consultant and has most recently worked with a leading manufacturer developing their IT/IS strategy. He researches in the area of information systems development and more generally on information systems in their natural organizational setting, in particular using action research, though he has also used a number of other qualitative research approaches.

Dr. Ambjorn Naeve

KTH-Royal Institute of Technology, Sweden

Ambjörn Naeve (www.nada.kth.se/~amb) has a background in mathematics and computer science and received his Ph.D. in computer science from KTH in 1993. He is presently coordinator of research on Interactive Learning Environments and the Semantic Web at the Centre for user-oriented Information technology Design (CID: http://cid.nada.kth.se) at the Royal Institute of Technology (KTH: www.kth.se) in Stockholm, where he heads the Knowledge Management Research group (KMR: http://kmr.nada.kth.se).

Professor Adrian Dale

Creatifica Associates, UK

Adrian Dale is renowned as a radical thinker in the information and knowledge management fields. He has managed the information, records, and knowledge functions of several UK organizations in the public and private sector, driving the shift from the paper to electronic worlds. He coaches the knowledge management teams of a number of public and private sector organizations, helping them to create the radical change necessary today. Current clients include the Cabinet Office; Shell Exploration and Production; CBI; the Department for Children, Schools and Families; Scottish Enterprise; Health Scotland; NICE; and the National Library for Health. Past clients have included Health London, HM Treasury, the Ministry of Defence, the House of Commons, DeFRA, Environment Agency, the Competition Commission, KPMG, Christian Aid, Motor Neurone Disease Associate, the Learning and Skills Council, and Bedford Hospital NHS Trust. He has 21 years of experience in the fields of IT, Knowledge, and Information Management. Before becoming independent, he was Chief Knowledge Officer for Unilever Research with responsibility for IT & Knowledge Management in their Research and Innovation Programs. Adrian is Chairman of the International Online Information Conference, Editor of the Journal of Information Science, and a fellow of the Chartered Institute of Library and Information Professionals.

Professor Paul Lefrere

University of Tampere, Finland

Paul Lefrere is a professor of eLearning at the University of Tampere, Finland (Hypermedia Lab; Vocational Learning and e-skills Centre). Before that, he was Senior Lecturer at the UK Open University's Institute of Educational Technology. From 2003-2005, he was on leave of absence as Microsoft's Executive Director for eLearning, in which role he served on various European and national advisory groups concerned with professional learning and related topics, and also on Europe's e-learning Industry Group, eLIG. Until 2003 he was Policy Adviser at the Open University, where he was also Academic Director of a number of international multi-university projects concerned with e-skills and knowledge management. Since he returned to the OU he has been engaged in a number of development and consultancy activities in many countries of the world including the Middle East and Pakistan.

Professor Felix Tan

Auckland University of Technology, New Zealand

Dr. Felix B. Tan is Professor of Information Systems and Chair of the Business Information Systems discipline. He is also Director of the Centre for Research on Information Systems Management (CRISM). He serves as the Editor-in-Chief of the Journal of Global Information Management. He is a Fellow of the Information Resources Management Association as well as the New Zealand Computer Society. He also served on the Council of the Association for Information Systems from 2003-2005. He has held visiting positions with the National University of Singapore, The University of Western Ontario, Canada, and was a visiting professor at Georgia State University, USA, in May/June 2005 and the University of Hawaii at Manoa in January 2006. Dr. Tan is internationally known for his work in the global IT field. Dr. Tan's current research interests are in electronic commerce, global information management, business-IT alignment, and the management of IT. He actively uses the repertory grid and narrative inquiry methods in his research. Dr. Tan has published in MIS Quarterly, Information Management, the Journal of Information Technology, the Information Systems Journal, IEEE Transactions on Engineering Management, IEEE Transactions on Personal Communications, the International Journal of HCI, and the International Journal of Electronic Commerce, as well as other journals, and has refereed a number of conference proceedings.

Professor Janice Sipior

School of Business, Villanova University, USA

Janice C. Sipior is Associate Professor of Management Information Systems at Villanova University, an Augustinian university located in Pennsylvania, USA. Her academic experience also includes faculty positions at Canisius College, USA; the University of North Carolina, USA; Moscow State Linguistic University, Russia; and the University of Warsaw, Poland. She was previously employed in computer planning at HSBC (Hong Kong-Shanghai Bank Corporation). Her research interests include ethical and legal aspects of information technology, system development strategies, and knowledge management. Her research has been published in over 65 refereed journals, international conference proceedings, and books. She is Chair of the Association for Computing Machinery, Special Interest Group on Management Information Systems (ACM-SIGMIS), and serves as a Senior Editor of Data Base, an Associate Editor of the Information Resources Management Journal, and Editorial Board Member of Information Systems Management.

Professor Gottfried Vossen

University of Muenster, Germany

Gottfried Vossen is Professor of Computer Science in the Department of Information Systems at the University of Muenster in Germany. He is the European Editor-in-Chief of Elsevier's Information Systems and Director of the European Research Center for Information Systems (ERCIS) in Muenster. His research interests include conceptual as well as application-oriented problems concerning databases, information systems, electronic learning, and the Web.

Program Chairs

Miltiadis D. Lytras	American College of Greece, Greece
Patricia Ordonez De Pablos	University of Oviedo, Spain
Miguel Angel Sicilia	University of Alcala, Spain

Knowledge Management and E-Learning Symposium Chairs

Ambjorn Naeve	Royal Institute of Technology, Sweden
Miguel Angel Sicilia	University of Alcala, Spain

Publicity Chair

Ekaterini Pitsa	Open Research Society, Greece

Exhibition Chair

Efstathia Pitsa	University of Cambridge, UK

Sponsoring Organizations

Gold

Inderscience Publishers, http://www.inderscience.com

Program and Scientific Committee Members (Serving also as Reviewers)

Adrian Paschke	Technical University Dresden, Germany
Adriana Schiopoiu Burlea	University of Craiova, Romania
Agnes Kukulska-Hulme	The Open University, UK
Ahmad Syamil	Arkansas State University, USA
Aimé Lay-Ekuakille	University of Salento, Italy
Alan Chamberlain	University of Nottingham, UK
Alejandro Diaz-Morcillo	University of Cartagena, Spain
Alok Mishra	Atilim University, Turkey
Alyson Gill	Arkansas State University, USA
Ambjörn Naeve	Royal Institute of Technology, Sweden
Ana Isabel Jiménez-Zarco	Open University of Catalonia, Spain
Anas Tawileh	Cardiff University, UK
Anastasia Petrou	University of Peloponnese, Greece
Anastasios A. Economides	University of Macedonia, Greece
Andreas Holzinger	Medical University Graz, Austria
Andy Dearden	Sheffield Hallam University, UK
Ane Troger	Aston Business School, UK
Angela J. Daniels	Arkansas State University, USA
Anna Lisa Guido	University of Salento, Italy
Anna Maddalena	DISI, University of Genoa, Italy
Anna Maria Tammaro	University of Parma, Italy
Ansgar Scherp	OFFIS - Multimedia and Internet Information Services, Germany
Antonio Cartelli	University of Cassino, Italy
Antonio Tomeu	University of Cadiz, Spain
Riccardo Lancellotti	University of Modena and Reggio Emilia, Italy
Apostolos Gkamas	University of Peloponnese, Greece
Arianna D'Ulizia	National Research Council, Italy
Aristomenis Macris	University of Piraeus, Greece
Badr Al-Daihani	Cardiff University, UK
Beatriz Fariña	University of Valladolid, Spain
Berardina Nadja De Carolis	University of Bari, Italy
Bob Folden	Texas A&M University-Commerce, USA
Bodil Nistrup Madsen	Copenhagen Business School, Denmark
Bradley Moore	University of West Alabama, USA
Campbell R. Harvey	Duke University, USA
Carla Limongelli	Università "Roma Tre", Italy
Carlos Bobed	University of Zaragoza, Spain
Carlos Ferran	Penn State Great Valley University, USA
Carmen Costilla	Technical University of Madrid, Spain
Carolina Lopez Nicolas	University of Murcia, Spain
Charles A. Morrissey	Pepperdine University, USA
Chengbo Wang	Glasgow Caledonian University, UK
Chengcui Zhang	University of Alabama at Birmingham, USA

Christian Wagner	City University of Hong Kong, Hong Kong (China)
Christos Bouras	University of Patras, Greece
Chunzhao Liu	Chinese Academy of Sciences, China
Claire Dormann	Carleton University, Canada
Claus Pahl	Dublin City University, Ireland
Cui Tao	Brigham Young University, USA
Damaris Fuentes-Lorenzo	IMDEA Networks, Spain
Daniel R. Fesenmaier	Temple University, USA
Daniela Leal Musa	Federal University of Sao Paulo, Brazil
Daniela Tsaneva	Cardiff School of Computer Science,UK
Darijus Strasunskas	Norwegian University of Science and Technology (NTNU), Norway
David O'Donnell	Intellectual Capital Research Institute of Ireland, Ireland
David R. Harding	Jr. Arkansas State University, USA
Dawn Jutla	Saint Mary's University, Canada
Denis Gillet	Swiss Federal Institute of Technology in Lausanne (EPFL), Switzerland
Diane H. Sonnenwald	Göteborg University and University College of Borås, Sweden
Dimitri Konstantas	University of Geneva, Switzerland
Dimitris N. Chryssochoou	University of Crete, Greece
Douglas L. Micklich	Illinois State University, USA
Dusica Novakovic	London Metropolitan University, USA
Edward Dieterle	Harvard Graduate School of Education, USA
Ejub Kajan	High School of Applied Studies, Serbia
Elena García-Barriocanal	University of Alcalá, Spain
Emma O'Brien	University of Limerick, Ireland
Eric Tsui	The Hong Kong Polytechnic University, Hong Kong (China)
Eva Rimbau-Gilabert	Open University of Catalonia, Spain
Evanegelia Karagiannopoulou	University of Ioannina, Greece
Evangelos Sakkopoulos	University of Patras, Greece
Fernanda Lima	Universidade Catolica de Brasilia, Brazil
Filippo Sciarrone	Università "Roma Tre", Italy
Francesc Burrull	Polytechnic University of Cartagena, Spain
Francesca Lonetti	ISTI - Area della Ricerca CNR, Italy
Francisco Palomo Lozano	University of Cádiz, Spain
Gang Wu	Tsinghua University, China
Gavin McArdle	University College Dublin, Ireland
George A. Jacinto	Arkansas State University, USA
Georgios Skoulas	University of Macedonia, Greece
Gianluca Elia	University of Salento, Italy
Gianluigi Viscusi	University of Milano-Bicocca, Italy
Giovanni Vincenti	S.r.l. Rome, Italy
Giuseppe Pirrò	University of Calabria, Italy
Giuseppe Vendramin	University of Salento, Italy

Vincenza Pelillo	University of Salento, Italy
Gregg Janie	University of West Alabama, USA
Guillermo Ibañez	Universidad de Alcalá, Spain
Hai Jiang	Arkansas State University, USA
Haim Kilov	Stevens Institute of Technology, USA
Hanh H. Hoang	Hue University, Vietnam
Hanne Erdman Thomsen	Copenhagen Business School, Denmark
Hanno Schauer	Universität Duisburg-Essen, Germany
Heinz V. Dreher	Curtin University of Technology, Australia
Helena Corrales Herrero	University of Valladolid, Spain
Helena Villarejo	University of Valladolid. Spain
Hyggo Almeida	University of Campina Grande, Brazil
Inma Rodríguez-Ardura	Open University of Catalonia, Spain
Ino Martínez León	Universidad Politécnica de Cartagena, Spain
Ioan Marius Bilasco	Laboratoire Informatique de Grenoble (LIG), France
Ioanna Constantiou	Copenhagen Business School, Denmark
Ioannis Papadakis	Ionian University, Greece
Ioannis Stamelos	AUTH, Greece
Irene Daskalopoulou	University of Peloponnese, Greece
Isabel Ramos	University of Minho, Portugal
James Braman	Towson University, USA
Jan-Willem Strijbos	Leiden University, The Netherlands
Javier De Andrés	University of Oviedo, Spain
Javier Fabra	University of Zaragoza, Spain
Jeanne D. Maes	University of South Alabama, USA
Jens O. Meissner	Lucerne School of Business, Switzerland
Jerome Darmont	University of Lyon (ERIC Lyon 2), France
Jesus Contreras	ISOCO, Spain
Jesús Ibáñez	University Pompeu Fabra, Spain
Jianhan Zhu	The Open University, UK
Johann Gamper	Free University of Bozen-Bolzano, Italy
Jon A. Preston	Clayton State University, USA
Jorge Gracia	University of Zaragoza, Spain
Jose Jesus García Rueda	Carlos III University of Madrid, Spain
José Luis García-Lapresta	Universidad de Valladolid, Spain
Jose Luis Isla Montes	University of Cadiz, Spain
Josemaria Maalgosa Sanahuja	Polytechnic University of Cartagena, Spain
Joseph C. Paradi	University of Toronto, Canada
Joseph Feller	University College Cork, Ireland
Joseph Hardin	University of Michigan, USA
Joze Gricar	University of Maribor, Slovenia
Juan Gabriel Cegarra Navarro	Universidad Politécnica de Cartagena, Spain
Juan Manuel Dodero	University of Cádiz, Spain
Juan Miguel Gómez Berbís	Univesidad Carlos III de Madrid, Spain
Juan Pablo de Castro Fernández	University of Valladolid, Spain
Juan Vicente Perdiz	University of Valladolid, Spain

Juan Ye	University College Dublin, Ireland
Julià Minguillón	Universitat Oberta de Catalunya (UOC), Spain
Jyotishman Pathak	Mayo Clinic College of Medicine, USA
Karim Mohammed Rezaul	University of Wales, UK
Karl-Heinz Pognmer	Copenhagen Business School, Denmark
Katerina Pastra	Institute for Language and Speech Processing, Greece
Ken Fisher	London Metropolitan University, UK
Kleanthis Thramboulidis	University of Patras, Greece
Konstantinos Tarabanis	University of Macedonia, Greece
Kylie Hsu	California State University at Los Angeles, USA
Laura Papaleo	DISI, Italy
Laura Sanchez Garcia	Universidade Federal do Parana, Brazil
Laurel D. Riek	Cambridge University, UK
Lazar Rusu	Royal Institute of Technology (KTH), Sweden
Leonel Morgado	University of Trás-os-Montes e Alto Douro, Portugal
Leyla Zhuhadar	Western Kentucky University, USA
Lily Diaz-Kommonen	University of Art and Design Helsinki, Finland
Linda A. Jackson	Michigan State University, USA
Liqiong Deng	University of West Georgia, USA
Lori L. Scarlatos	Stony Brook University, USA
Lubomir Stanchev	Indiana University - Purdue University Fort Wayne, USA
Luis Angel Galindo Sanchez	Universidad Carlos III de Madrid, Spain
Luis Iribarne	Departamento de Lenguajes y Computacion, Universidad de Almeria, Spain
Luke Tredinnick	London Metropolitan University, UK
Lynne Nikolychuk	King's College London, UK
M. Carmen de Castro	Universidad de Cádiz, Spain
M. Latif	Manchester Metropolitan University, UK
Mahmoud Youssef	Arab Academy for Science and Technology, Egypt
Maiga Chang	Athabasca University, Canada
Manolis Vavalis	University of Thessaly, Greece
Manuel Rubio-Sanchez	Rey Juan Carlos University, Spain
Marco Temperini	Università La Sapienza, Italy
Marcos Castilho	Departamento de Informática da Universidade Federal Paraná, Brazil
Marcos Martin-Fernandez	Associate Valladolid University, Spain
Maria Chiara Caschera	IRPPS-CNR, Rome, Italy
Maria Grazia Gnoni	University of Salento, Lecce, Italy
Maria Helena Braz	Technical University of Lisbon, Portugal
Maria Jesús Martinez-Argüelles	Open University of Catalonia, Spain
María Jesús Segovia Vargas	Universidad Complutense de Madrid, Spain
María Jesús Verdú Pérez	University of Valladolid, Spain
Maria Joao Ferreira	Universidade Portucalense, Portugal
Maria Papadaki	University of Ioannina, Greece

Paola Mello	University of Bologna, Italy
Paolo Toth	University of Bologna, Italy
Patricia A. Walls	Arkansas State University, USA
Paul A. Kirschner	Open University of the Netherlands, The Netherlands
Paul G. Mezey	Memorial University, Canada
Pedro J. Muñoz Merino	Universidad Carlos III de Madrid, Spain
Pedro Soto-Acosta	University of Murcia, Spain
Peisheng Zhao	George Mason University, USA
Pekka Muukkonen	University of Turku, Finland
Per Anker Jensen	Copenhagen Business School, Denmark
Peter Gomber	Johann Wolfgang Goethe-Universität Frankfurt, Germany
Phil Fitzsimmons	University of Wollongong, Australia
Pierre Deransart	INRIA-Rocquencourt, France
Pilar Manzanares-Lopez	Technical University of Cartagena, Spain
Pirkko Walden	Abo Akademi University, Finland
Ralf Klamma	RWTH Aachen University, Germany
Raquel Hijón Neira	Universidad Rey Juan Carlos, Spain
Raymond Y.K. Lau	City University of Hong Kong, Hong Kong SAR
Razaq Raj	Leeds Metropolitan University, UK
Razvan Daniel Zota	Academy of Economic Studies Bucharest, Romania
Ricardo Colomo Palacios	Universidad Carlos III de Madrid, Spain
Ricardo Lopez-Ruiz	University of Zaragoza, Spain.
Rob Potharst	Erasmus University, The Netherlands
Robert Fullér	Åbo Akademi University, Finland
Roberto García	Universitat de Lleida, Spain
Roberto Paiano	University of Salento, Italy
Roman Povalej	University of Karlsruhe, Germany
Rushed Kanawati	LIPN – CNRS, France
Russel Pears	Auckland University of Technology, New Zealand
Ryan Robeson	Arkansas State University, USA
Sabine H. Hoffmann	Macquarie University, Australia
Sadat Shami	Cornell University, USA
Salam Abdallah	Abu Dhabi University, United Arab Emirates
Samiaji Sarosa	Atma Jaya Yogyakarta University, Indonesia
Sean Mehan	University of the Highlands and Islands, Scotland, UK
Sean Wolfgand M. Siqueira	Federal University of the State of Rio de Janeiro (UNIRIO), Brazil
Sebastian Matyas	Otto-Friedrich-Universität Bamberg, Germany
Sergio Ilarri	University of Zaragoza, Spain
Shantha Liyanage	Macquarie University, Australia
Shaoyi He	California State University, USA
She-I Chang	National Chung Cheng University, Taiwan (China)
Sherif Sakr	University of New South Wales, Australia
Sijung Hu	Loughborough University, UK
Silvia Rita Viola	Università Politecnica delle Marche - Ancona, Italy

Silvia Schiaffino	Univ. Nac. del Centro de la Provincia de Buenos Aives, Argentina
Sima Yazdani	Cisco Systems, USA
Sinuhe Arroyo	University of Alcalá de Henares, Spain
Sonia Berman	University of Cape Town, South Africa
Soror Sahri	Université Paris-Dauphine, France
Spyros D. Arsenis	University of Thessaly, Greece
Staffan Elgelid	Arizona State University, USA
Stefan Hrastinski	Uppsala University, Sweden
Stephan Lukosch	FernUniversität in Hagen, Germany
Steve Barker	King's College London, UK
Steve Creason	Metropolitan State University, USA
Stone Schmidt	Arkansas State University, USA
Suneeti Rekhari	University of Wollongong, Australia
Taha Osman	Nottingham Trent University, UK
Takaharu Kameoka	Mie Univeristy, Japan
Teck Yong Eng	University of London, UK
Terrill L. Frantz	Carnegie Mellon University, USA
Thanos C. Papadopoulos	The University of Warwick, UK
Tobias Ley	Know-Center, Austria
Toyohide Watanabe	Nagoya University, Japan
	Université Libre de Bruxelles , Belgium
Upasana Singh	University of KwaZulu, South Africa
Vaclav Snasel	VSB-Technical University of Ostrava, Czech Republic
Vagan Terziyan	University of Jyvaskyla , Finland
Val Clulow	Monash University, Australia
Véronique Heiwy	IUT de Paris, France
Vincenzo Ciancia	University of Pisa, Italy
Violeta Damjanovic	Salzburg Research, Austria
Virginie Sans	Université de Cergy-Pontoise, France
Vlasios Voudouris	London Metropolitan University, UK
Walt Scacchi	University of California, USA
Weili Wu	University of Texas at Dallas, USA
Xavier Calbet	EUMETSAT, Germany
Xiaohua (Tony) Hu	Drexel University, USA
Xihui (Paul) Zhang	University of Memphis, USA
Yihong Ding	Brigham Young University, USA
Yolaine Bourda	SUPELEC, France
Yuan An	Drexel University, USA
Yun Shen	University of Bristol, UK
Yvon Kermarrec	Institut Télécom / Télécom Bretagne, France
Ziqi Liao	Hong Kong Baptist University, Hong Kong (China)

Table of Contents

Knowledge, Learning, Education, Learning Technologies and E-Learning for the Knowledge Society

Information Technologies - Knowledge Management Systems - E-Business and Business, Organizational and Inter-organizational Information Systems for the Knowledge Society

Engineering the Knowledge Society through Web Science: Advanced Systems, Semantics, and Social Networks

Language Micro-gaming: Fun and Informal Microblogging Activities for Language Learning

Maria A. Perifanou

Department of Italian and Spanish Language and Literature, University of Athens, Ilisia, 15784 Greece
mariaperif@gmail.com

Abstract. 'Learning is an active process of constructing rather than acquiring knowledge and instruction is a process of supporting that construction rather than communicating knowledge' [1]. Can this process of learning be fun for the learner? Successful learning involves a mixture of work and fun. One of the recent web 2.0 services that can offer great possibilities for learning is Microblogging [2]. This kind of motivation can raise students' natural curiosity and interest which promotes learning. Play can also promote excitement, enjoyment, and a relaxing atmosphere. As Vygotsky (1933) [3] advocates, play creates a zone of proximal development (ZDP) in children. According to Vygotsky, the ZDP is the distance between one's actual developmental level and one's potential developmental level when interacting with someone and/or something in the social environment [4]. Play can be highly influential in learning. What happens when play becomes informal learning supported by web 2.0 technologies? Practical ideas applied in an Italian foreign language classroom using microblogging to promote fun and informal learning showed that microblogging can enhance motivation, participation, collaboration and practice in basic language skills.

Keywords: Knowledge Society, Web 2.0, Microblogging, Language Learning, Play, Games, Informal Learning.

1 Introduction

Our society which has now become what is best described by the term 'Knowledge Society' is undergoing tremendous changes and this has a great impact on business and industry field but even more on education. The arrival of internet and especially of the new era of web 2.0 social technologies have changed dramatically the way we communicate and we get the information. Nowadays, language competencies and intercultural skills will more than ever be a part of the key qualifications needed to live and work in this new reality. The integration of new technologies into language learning is a necessary step to ensure the acquisition of this kind of skills and competencies. New methodological approaches are needed in order to make the best use of the potentials of the new technologies.

'Learning is an active process of constructing rather than acquiring knowledge and instruction is a process of supporting that construction rather than communicating knowledge' [1]. Successful learning involves a mixture of work and fun. One of the

M.D. Lytras et al. (Eds.): WSKS 2009, CCIS 49, pp. 1–14, 2009.

recent web 2.0 services that can offer great possibilities for learning is microblogging [2].This kind of motivation can raise students' natural curiosity and interest which promotes learning. Play can also promote excitement, enjoyment, a relaxing atmosphere and can be highly influential in learning.

In this paper first will be discussed some previous theories of language play in order to understand the important influence of play in learning. Then it will be presented how web 2.0 technologies have been used to promote learning through play. With the analysis of the microblogging technology and its use in language learning the theoretical part will be concluded. It follows the description of a series of microgaming activities that were applied in an Italian foreign language classroom using microblogging in order to promote fun and informal learning. Finally, there is a discussion of the results of this experience that aims to give an answer to the research question whether the proposed activities can enhance motivation, participation, collaboration in the participants and practice of the basic language skills.

In the first section the author will introduce at first place the basic theories regarding the topic of Play in general and then will analyze the importance of Play in Language Learning according to researches' findings. How the web 2.0 tools can enhance Language Learning in a playful and entertaining way will be discussed in the end of this section.

1.1 Introduction to Play

There is extensive amount of literature on the topic of play [3], [5], [6], [7], [8]. Play has several forms like fantasy, story telling, music, movement, games, etc. Play almost always promotes excitement, enjoyment, a relaxing atmosphere but also organization, problem solving and effective learning.

Sylva, (1974) [9] and Lantolf, (2001) [8] support the first research on play that was conducted by Vygotsky in 1933. This research is probably one of the most influential research on play and shows that play creates a zone of proximal development (ZDP) in children. According to Vygotsky, the ZDP is the distance between one's actual developmental level and one's potential developmental level when interacting with someone and/or something in the social environment [4]. Lantolf (2001) [8] explains that children engage in activities that are not just about enjoyment, but they go ahead of their actual development level getting prepared for the future.

Some other researchers such as Huizinga (1955) [5] sustains that play is difficult to be defined even though is a part of our life as everyone plays and has the ability to play. Play depends on the age of participants, of the activities, and the social contexts. It can be based on reality but also can be a fruit of imagination. It might involves a game but not all the times. Cook (2000) [7] distinguishes play among adults and children in a discussion about fantasy. He notes that children are more likely to take part in make-believe and adults in fantasy. Also, Bateson (1972) [6] tries to give his own definition for play. He explains that when two individuals are playing, there is an intuitive sense that the current activity being engaged in is "play." "Play" looks something like, "the actions which are now being engaged in do not denote what these actions, for which they stand, would traditionally denote." For example, two children involved in playing "house" are playing because the activities carried out in playing house (pretending to prepare lunch, etc.) do not stand for what "house" denotes in an adult society. Butzkamm (1980) [10] argues that verbal play can provide children with the training phases necessary for developing new verbal skills.

Types of language play vary widely, from vocabulary games to talking one's self. Language play incorporates word play, unregulated rules, and enjoyment. But how can play influence Language Learning and how this can be introduced in the Language classroom?

1.2 Play in Language Learning

Play is also highly influential in language learning. With the term play are intended all the fun activities in which the learners can be engaged in such as story telling, fantasy, music, movement, role-plays, language games, problem solving etc as it is mentioned earlier. The terms "games", "problem solving" and "tasks" have surfaced continually throughout the literature on language pedagogy over the years. Well designed language learning games should include all the benefits of a Task Based Language Teaching (TBLT) assignment since from a functional perspective they are simply a meaningful concatenation of tasks [11]. With the current emphasis on extending communicative language teaching methodology to center around task-based instruction, many of the tasks people engage in outside of the classroom are actually the central core of a foreign language classroom. Foreign language educators already have a strong history of incorporating everyday activities into their lessons and in this way they can introduce fun activities removing at the same time this barrier between formal education and entertainment.

Some researchers and teachers emphasize that play should not simply be secondary to the focus in our classroom, but should be the main activity [12]. Guy Cook (2000) [7] takes an extensive look at the role of language play in language learning in his book *Language Play, Language Learning* and he also insists that play should not be thought of as peripheral activity or a trivial, but as central to learning, creativity, to human thought and culture and intellectual enquiry. It fulfils a major function of language, underpinning the human capacity to adapt: as individuals, as societies, and as a species. A play-centered approach to curricular design encourages the extensive learning that comes with play, reduces the affective barrier of failure and provides a robust framework for non-linear sequencing. The lower of the affective filter in classroom is also supported by the Natural Approach [13]. Buckby (1984) [14] explores the benefits of using games in the ESL classroom and how they might be used to motivate students. He maintains that games can be a main part of an ESL curriculum and do not just have to be introduced as fun extracurricular activities. Also he insists that games have to be meaningful and relevant to the material of the curriculum. Finally, he adds that games can offer a "plausible incentive" for students to use the target language and they can encourage competition.

Web 2.0 tools though they are not designed specially for learning; and it's getting more popular everyday among our students

1.3 Web 2.0 and Play

Web changed from a medium to a platform, from a read-web to a read-write-web, thereby fulfilling Berners-Lee's original vision of the Web [15]. We are more connected than ever and anyone can create, upload and share information. What it's important to point out is that three crucial factors are responsible for the success of this "social revolution" of the WorldWideWeb [16], [17]: a) *accessibility:* according to different statistical data nearly the whole population of the European Union has access to the WorldWideWeb; b) *usability*: in other words the increasing "ease of use" [18] especially amongst younger generations called among other names as Digital Natives [19] c) *device:* that are getting improved by the companies as more people use them, such as new generations of mobile devices like iPhone.

These new social software applications like wikis, blogs, social bookmarking, podcasting etc. enhance creativity, communication, sharing between users and collaboration; the words we use in education. These principles are in line with modern educational theories such as constructivism and connectionism and thus make Web 2.0 applications very attractive for teachers and learners. Many web 2.0 tools are now commonly used in learning [20]. Alm (2006) [21] maintains that language learners have two communities of learning: the learning community in the classroom, and the target language community. Web 2.0 applications can be used as a mean that can bring these two learning communities together. In this way language learners can have the opportunity to combine formal and informal learning. As Özge Karaoğlu (2009) [22] maintains Web 2.0 platform can open up the classroom walls, creating freedom and independence in learning, and provide a broader range of opportunities for authentic and varied language interaction. Students can get involved in the writing process by posting blog entries, editing to other pages, creating their own e-portfolios. Not only it increases the creativity, as any student can write, film and publish a video or an audio but it's also collaborative as students can easily create social networks and communities of interests. Additionally, it supports student centered learning as it allows users to become the producers of the knowledge and to share their work with a wider audience. Finally, he adds that web 2.0 promotes motivation and participation as it lowers the affective filter even to shy students.

According to the British Council, nowadays only for the English language there are 750 million language learners who are using the masses of web 2.0 services to learn one of the 5,000 active languages of the world. guest author On one hand, there are plenty of new Online Services that Offer Web 2.0 Language Instruction such as Livemocha, (http://www.livemocha.com), Palabea, (http://www.palabea.net) Italki.com (http://www.italki.com) Soziety.com (http://www.soziety.com/LanguageExchange.do) and many companies that offer language tutoring using collaborative, real-time software as well as virtual worlds and on the other hand, there are great possibilities to choose on their own the tools that they want to introduce in their language classroom. [23]. The possibilities of fun language learning that can practice all the basic language abilities with really interesting activities based on the idea of play and entertainment are enormous.Web 2.0 services such as *Flickr* for sharing photos, *Blip.fm* for personalized radio and sharing music, *Utube* for sharing videos, *Diigo* for sharing bookmarks, blogs and wikis for sharing news, knowledge and multimedia material can be used in multiple ways to introduce language learners to a new culture with interesting fun activities. Additionally, another great idea for creating fun web 2.0 activities can be to use mobile Internet, Google Maps and Geotagging. In this way, the learner is guided and guides himself around in real-life locations where the target language is used. With tools like *Google Earth* [Visit: http://earth.google.com/] language teacher can create fun activities like games to familiarize students with foreign geography, city landmarks and navigation in the target language. Meyer B. et al, (2008) [24] have developed a mobile educational simulation for foreign and second language learning for adults.

Other Web 2.0 applications that can support fun language activities offering enormous promise to people who want to learn or perfect languages are the virtual worlds applications like this of *Second Life* [Visit: http://secondlife.com/]. Second Life is a 3D virtual game that can support playing 'yourself' outside the formal educational context with authentic communicative tasks can be a new challenge for language teachers. For instance, learners can visit a virtual representation of the Basilica of Saint Francis of Assisi to find native speakers. Also, this can be a venue for language classes like Language Lab [Visit: http://www.languagelab.com/en-gb/]. or a networking opportunity for educationalists (EduNation, created by Consultants-E Visit: http://www.theconsultants-e.com/edunation/edunation.asp) [25].

Games and simulations are not introduced now with web 2.0 applications in Language Learning but they have been part of second and foreign language education for decades [26], [27] as it has been proved that these techniques can be extremely powerful means of helping people to acquire foreign or second language skills [28], [29], [27], [30].

These were some of the web 2.0 services that can support fun language learning activities. In the next section, the author will introduce a web 2.0 application called microblogging that has been chosen for the creation of fun language activities for the Italian classroom and the case study will be presented in the fourth section of this paper.

2 Microblogging in Language Learning

2.1 Microblogging Services

A faster and easier mode of communication compared to the well known blogging is microblogging [31] and it enables users to post short messages that are distributed within their community [32].This is a real time publishing and it's a unique way in which people can connect and relate to one another. Users can post messages from their mobile devices, a Web page, and from Instant Messengers [30]. Constructing a meaningful environment around the individual, based on the people he/she follows and who follow them back is the great value of this web 2.0 system. It can be used to ask questions, exchange ideas and useful links, or provide pieces of advice and support [2]. There are many well known microblogging systems such as are *Pownce, Tumblr, Utterli, Plurk, Jaiku. Yappd, Edmodo, Cirip.ro, Logoree.ro, Zazazoo* but they are still less prominent in the micro-blogging sphere [33), [34]. Despite the potentials of these microbbloging platforms, none of these have such a powerful network as twitter does [35]. *Twitter* is the most popular microblogging services currently available, which can generically be described as a social networking site hosting interaction in 140 character messages [34]. Twitter has a very powerful network with millions of energetic participants connected to each other and it is global and multilingual [36].

Language educators should check each microblogging service before introducing them in class [2] or choosing them as part of their Personal/Professional Learning Networks (PLNs), [37] because each of them displays particular features. A microblogging service that is ideal for Language Learning is *Edmodo*. This is the service that has been used in the fun micro-language activities that will be presented in the fourth section of this paper. *Edmodo* [Visit: www.edmodo.com]. is a private communication platform built for teachers and students. It provides a way for teachers and students to share notes, links, and files. Teachers can have a closed group collaboration, customise and manage their groups and obtain the groups code. Teachers have also the ability to send alerts, events, and assignments to students. It's worth notice that Edmodo has also a public component, which allows teachers to post any privately shared item to a public timeline and RSS feed. Furthermore, it offers students the option of communicating in whichever language they feel confortable. The calendar and the email notifications are two options that allow teachers and students to control better their work and communication. [2]

After the quick presentation of the most popular microblogging services existing at the moment, the next paragraph will focus on some of the first Language Learning experiences with microblogging.

2.2 Microblogging Language Learning Experiences

Micro-blogging is a communicative approach to the teaching and learning of a foreign language for a linguistic point of view. In fact it gives the opportunity to language students to communicate n their 'virtual classroom' at any time they like about their daily activities, current events and also to share material, to exchange opinions or to ask for any help. The important thing is that they can actually practice their language communication skills while they are having the sense that they are relaxing at the same time. Another key element is that they can not only use micro-blogging among each other, but also with native speakers, which will further improve their communication skills concerning "real-use" of language. There is also a possibility of privacy, as the micro-technology allows all the users to send direct messages and in this way only two parties can view the message. This is particularly helpful when a student wants to communicate directly with his teacher or with another student.

Sharon Scinicariello of the University of Richmond and George Mayo of the Silver International Middle School in Montgomery Country (Maryland) are two of the first language teachers who tried to introduce the microblogging technology in their language classrooms. Additionally, Carsten Ullrich and his collaborators of the University Jiao Tong of Shanghai were the first who made a research study in microblogging for English Language Learning within the context of vocational English language learning. But which were the findings of these first microblogging experiences?

Sharon Scinicariello made some experiences regarding the application of twitter-based activities in the French language classroom [36]. Some of the language activities that she applied were question answer activities in order to help her students to practice verbal constructions and digital story telling that could encourage his/her engagement in an active and enthusiastic way. The most important positive outcome of these practices is that microblogging technology can offer the feeling of 'real learning' and give the opportunity to students to practice and use language in an easy, fun, and interactive way connecting simultaneously language learning with the target culture. On the other hand, she points out that users should remember two things; first, the cost when they try to tweet through their mobile phones using SMS and second, that they should protect their updates when in k-12 level.

"Twitter was incredibly simple and really amazing…safe for kids... with limitations for classroom use" This is how George Mayo, an English teacher, [38], explained his first microblogging experience with his students in an online-collaborative project. His students practiced their writing skills through a unique collaborating, interactive and enjoyable experience, which connected their classroom to a wider world. One hundred students from six different countries participated in a storytelling project that his class has started and what they composed in collaboration was a suspenseful science-fiction narrative about a mermaid-turned-human.

Their research aimed at observing how students' active participation in conversational courses for English as a second language could be increased. Students didn't have time to practice the foreign language outside the classroom hours and were also very shy in face to face environments, allegedly due to their limited vocabulary.

In the University of Shanghai, Jiao Tong, where it was carried out the first microblogging study for English Language Learning [32] results showed that the use of microblogging through public and private messages, increased very much students' participation in the conversational courses for English as a second language. Students had to publish a minimum of seven micro-messages a week, and read the incoming messages of their fellow students using the web and their mobile phones. In this way all the learners, including those who were very shy or they had difficulties with vocabulary, participated actively forming a sense of community and practicing the language in a informal, spontaneous and relaxing way. The contact with native speakers of English played also a key role in supporting learners to achieve a real and authentic communication. The only problem occurred was the fact that the students didn't use the target language during the whole project.

In this section, the author has succinctly introduced some of the first microblogging experiences in Language Learning classroom. In the next section, she will try to answer the question that has motivated her research: if microblogging can enhance motivation, participation, collaboration and practice in basic language skills.

3 Case Study: Italian Language Micro-gaming

Regardless the other experiences and the study reported above, the adoption of microblogging in teaching and learning of a language, and especially of Italian as a foreign language, is a particular area that has not been sufficiently explored yet. This happens probably, due to the fact that microblogging in the classroom is a fairly new adopted approach. There is also a lack of research, and practice, on the integration of microblogging in language courses, or even as part of studies focusing on learning technologies and innovative teaching and learning practices.

This section will describe some practical ideas that are applied in an Italian foreign language classroom using microblogging to promote fun and informal learning. These micro-gaming language activities are a part of a long term research project that has been conducted in a well known private language institution in Thessaloniki, Greece during the academic year 2008-2009 from October to May. This research explored the use of wiki, blog and microblogging in combination within a very specific context: an Italian as a foreign language classroom. Despite the fact that the teacher applied these three web 2.0 online approaches to support the face to face lessons beyond class and time borders, the focus of this paper will be exclusively on the use of a microblogging service as a tool that can promote an enjoyable way to learn Italian language through play. Additionally, the main goal of this research presented here is to answer to the basic research question whether these micro-gaming language activities can enhance motivation, participation, collaboration in the participants and practice of the basic language skills.

3.1 Methodology

The participants of this case study were 10 students attending the second year of Italian studies and regarding their gender, the group was composed of three women and seven men, being their average age twenty-two. All students had a good level of competence in Italian language (B1-B2) and with the exception of three of the students, all the other individuals, who took part in this study, had no or very limited experience with Information and Communication Technologies prior to this experience. The microblogging tool chosen for this research project was Edmodo, a tool especially developed for educational use. For the creation of the microblogging group, the teacher took a leading role, signing up for it and creating the class group. The microblogging group was attributed a code which was given out by the teacher to the students. To join the group students had to use the code, after signing up for an *Edmodo* account. *Edmodo* is a microblogging tool which does not require an email address, making it easy to set up and use by the different educational stakeholders.

The methodology for collecting the first samples of qualitative data included the analysis of students' messages and activities, questionnaires, and informal group interviews. The analysis regarding the frequency of the students' registered messages in the microblogging's log can also generate concrete quantitative data in relation to the frequency of their participation. As it is described in a previous paper regarding this research [2] even though the teacher's initial intention was to use *Edmodo* as a private microblogging platform that would work as the class' 'virtual announcement board' with alerts for events, assignments and exchange of notes, links, and files between the participants, it evolved in a more meaningful way. Students made this communication channel a part of their daily life and they used it to express feelings and ideas about class' subjects, to share materials and information, to ask for help regarding their homework or to discuss social problems. This back-up communication class channel succeeded to bridge the face to face lessons with the web environment supporting the language learning of the students inside and outside the classroom. In this way, informal and formal learning coexisted and were acknowledged as part of students' learning. The sense of being part of a community and also able to reach beyond the classroom's walls, while in school, has given the students a feeling of freedom and tranquility where their learning was concerned. The students' participation and satisfaction have increased as they practiced the Italian language in a rather natural and more fluent way and this has also influenced their levels of learning outcomes very positively.

Due to the fact that the students embraced so quickly and in such an enthusiastic way this communication channel and bearing in mind the research's first positive outcomes, the researcher wanted to apply this tool also in a different way in order to support the learning process. She created a series of fun micro-gaming language activities that aimed to enhance motivation, participation, collaboration in the participants and practice of their basic language skills.

This time the students didn't face any problem in the use of the tool despite their lack of technology literacy as they have been using it for several months. The class was divided in groups of two students, in total five groups. Each group had a different *edmodo* account and a password for this created by the teacher. In the class microblogging space *'incontritaliani 2'* the students could find the task to do with all the instructions.

a) First micro-activity example: *'digital stories'*
Each group had to visit utube and flickr to make a little research and find one viseo and one photo related to the Italian culture. The teacher proposed some keywords such as *cucina italiana, feste italiane*, etc to enhance students' research. Based on the material found each group had to create a story. What was a really interesting and suspenseful side in this activity was the fact that each group after has found the video and the photo of its preference had to start a story and post everything to the group indicated by the teacher, in order to continue and finish the story in its turn. The group that could publish first in the classes' microblogging space *'incontritaliani 2'* would be the winner of the game.

b) Second micro-activity example: *'Film scripts'*
Each group had to choose a scene between several scenes of five famous Italian films proposed and published by the teacher at the wikis' class. The purpose of the activity was to write the subtitles of the scene chosen. After the creation of these imaginary dialogues using the microblogging tool the teacher published the dialogues at the class microblogging space and then made the corrections needed giving explanations to the students in private or public messages. The activity was concluded with the performance of the scenes by the students and then followed, first, the screening of the original film scenes and, second, the votes for the best script creation by the students. The goal of the activity was to practice basically the oral skill through the creation of these script-dialogues in a spontaneous way as if they were 'acting' in real life.

c) Third example: *'Italian music'*
Each group had to visit *Utube* and find the video of one of the five Italian songs proposed by the teacher and then post it to the class microblogging space. Every group had to watch the film that has chosen and then write and post the verses of the song. The group that would have finished first the activity would be the winner. After this students could visit the class blog and find the music site *italianissima.net* [Visit: http://www.italianissima.net/] where they could explore a database of many famous Italian songs and find the verses of their songs and correct them. The activity finished with the students singing the three songs. In this case two of the students were musicians and played also the guitar. The goal of this activity was to practice mostly listening and comprehension skills.

d) Fourth example: *'cartoon based story'*
In this activity students worked all together and not in groups. The teacher gave out to them a story based on photos of cartoons designs and each of them had to write the dialogue of each photo of the 'cartoon heroes' every time it was his/her turn. The story was corrected and posted in the class microblogging space. The goal of this activity

was to practice mostly oral skills and the common goal of all these activities was the enhancement of motivation, participation and collaboration in the participants.

Before referring to results of the study regarding the micro-gaming language activities, it's worth mentioning that students were very positive when they had been asked in previous questionnaires if they would like to use *Edmodo* as microblogging service for participating in collaborative fun language activities. They also gave to the researcher many ideas and based on their personal preferences the researcher created the activities.

3.2 Results

According to the data collected in the questionnaires and their micro-posts all the students were very enthusiastic knowing that they would participate in these fun activities that were based on topics they have chosen. The 90% of the students stated that they loved the experience of being involved in these activities and only the 10% of the students have expressed that was a little bit hesitant since have continued to affront some problems with the use of technology in general.

When they were asked if they liked participating in teams or if they would have preferred to be involved in the activities on their own all of them chose the first option. In a similar question at the group interviews students stated that they felt free, more confident in themselves knowing that could be supported by their group companions. In fact, they commented that they enriched their knowledge a lot by their collaboration with their classmates. The 90% also commented that teacher's continuous support with her immediate feedback also encouraged them to continue their effort with no stress as she resolved any problem they affronted at once. They also stated that it was very important to them that the task was clearly explained and they understood every step of the process. As it is reported by the 90% of the students being engaged in these activities felt that they were having such a great fun that the time seemed to pass so quickly and so they were looking forward to repeating such an experience again very soon.

The 100% of the participants agreed that this was definitely not a lost of time as they were also preparing hard for the certification exams in that period. In contrary, they stated that not only they have learned a lot of things about the Italian culture, and they have enriched their vocabulary but they had also the possibility to use the language in real situations and be really creative.

The majority of them also agreed that they had the chance to practice oral, writing, listening and comprehension skills. To the question which one was their favourite activity the 80% of the students answered that even though it was difficult to decide they chose the activity named *'Film scripts'* and especially the part in which they had to perform their roles reading their dialogues. All of the students agreed that the competition provided by the game motivated them very much and gave them the feeling of learning in a more 'natural' and not very formal way. To the question if their expectations were less than what they expected the 100% of the students said that what they had experienced was really beyond their thoughts as they hadn't imagined that they could have such a fun, playing and learning at the same time. What has been characterized by the 80% as a barrier to the successful conclusion of the activities in time was the slow internet connection sometimes and some other technical problems that have been occurred. Despite these problems all the students succeed to complete their activities even after the face to face meetings. The 80% stated that this time and space flexibility was a really advantageous point of the tool whereas the rest 20% said that they would prefer to have finished everything at the lab as they didn't have much free time at this period. From the group interviews the researcher also found out that the 80% agreed that this kind of fun activities should not be secondary in the focus of the curriculum that would like to have in future but it should be the main type of activities. It is also worth notice that the 90% of the participants stated that having become familiar with micro-technology by engaging in with it for a long time, gave them confidence and freedom during the process.

Finally to the question if they would visit again the sites proposed during the fun activities after the lessons the 60% of the participants answered yes while the 40% said that they will do that in future when they will have more time to discover more material.

4 Conclusions

This paper attempted, first, to show how a series of micro-gaming language activities that were applied in an Italian foreign language classroom promoted fun and informal learning and, second, tried to give an answer to the research question whether the microblogging technology could enhance motivation, participation, collaboration and practice in basic language skills. Despite some technical problems that have been occurred, the students' feedback from this experience was very good and beyond any expectations. In fact, micro-gaming language activities influenced their levels of participation, collaboration and learning outcomes positively. Playing and practicing the Italian language at the same time outside the formal educational context but with authentic communicative tasks, appeared to have significant impact on the way students were learning the target language. Learning became a more fluent and natural action. They were really motivated and full of enthusiasm and excitement. The dynamism that Edmodo had brought into this particular Italian language classroom after the first month of use has been strengthen even more now after these activities. Further analysis of the results of this project will be made in a future paper.

Authentic, meaningful, interactive, student-centered, fun, Web-based learning activities can improve student performance in much the same manner as learning the language and culture while studying abroad. These tremendous changes initiated by the new technologies necessitate a re-thinking of the way we teach and learn. It is no longer a question of whether to take advantage of these electronic technologies in foreign language instruction, but of how to harness them and guide our students in their use. It is necessary to consider that complete new didactical approaches and teaching challenges have to be evolved by teachers. Experimenting new ways of combining play and new web 2.0 technologies in the foreign language classroom it is what the author of this paper aims to do in future and invites other language teachers to follow her example.

References

1. Duffy, T.M., Cunningham, D.J.: Constructivism: Implications for the design and delivery of instruction. In: Jonassen, D.H. (ed.) Handbook of Research for Educational Communications and Technology, pp. 170–198. Simon & Shuster Macmillan, New York (1996)
2. Perifanou, M., Costa, C.: Microblogging in language learning - analysis of experiences and suggestions of practices for the 21st century classroom. In: Conference Proceedings INTED, March, Valencia, Spain, March 9-11 (2009)
3. Vygotsky, L.: Play and its role in the mental development of the child. In: Bruner, J., Jolly, A., Sylva, K. (eds.): Play: It's role in development and evolution, pp. 461–463. Penguin Books, New York (1933/1976)
4. Vygotsky, L.: Mind in Society: The Development of Higher Psychological Processes. In: Cole, M. (ed.). Harvard University Press, MA (1978)
5. Huizinga, J.: Homo ludens: A study of the play-element in culture. Beacon Press, Boston (1955); Cited in: Joel Bacha, Play and Affect in Language Learning, http://genkienglish.net/playandaffect.htm (last accessed: 09.07/09)
6. Bateson, G.: Steps to an ecology of mind. Granada, London (1972); Cited in: Joel Bacha, Play and Affect in Language Learning, http://genkienglish.net/playandaffect.htm (last accessed 9/07/09)
7. Cook, G.: Language play, language learning. Oxford University Press, New York (2000), http://www.oup.com/elt/catalogue/isbn/0-19-442153-8?cc=global
8. Lantolf, J.: Sociocultural theory and second language learning. Oxford University Press, Oxford (2001)
9. Sylva, K., Bruner, J., Genova, P.: The role of play in problem-solving of children 3-5 years old. In: Bruner, J., Jolly, A., Sylva, K. (eds.) Play: It's role in development and evolution, pp. 244–257. Penguin Books, New York (1974/1976); Cited in: Joel Bacha, Play and Affect in Language Learning, http://genkienglish.net/playandaffect.htm
10. Butzkamm, W.: Verbal play and pattern practice. In: Felix, S. (ed.) Second language development: Trends and issues, pp. 233–248. Gunther Narr., Tubingen (1980)
11. Language Learning with New Media and Video Games, http://www.lingualgamers.com/thesis/
12. Kopecky, A.: Using Games to Motivate Your Adult ESL Students (2009), http://www.eslfocus.com/articles/using_games_to_motivate_your_adult_esl_students-400.html (last accessed: 10/07/09)
13. Stephen, K.D.: The Natural Approach: Language Acquisition in the Classroom (Language Teaching Methodology) (1996)
14. Buckby, Buckby, Why Use Games to Teach English, from Games for Language Learning. Cambridge University Press, Cambridge (1984)
15. Berners-Lee, T.: Weaving the Web. Orion Business Books (1999)
16. Ebner, M., Holzinger, A., Maurer, H.: Web 2.0 Technology: Future Interfaces for Technology Enhanced Learning? In: HCI 2007, Bejing (2007)
17. Downes, S.: E-learning 2.0. eLearn Magazine, October 17 (2005), http://www.elearnmag.org/subpage.cfm?section=articles&article=29-1 (last accessed 10/07/09)
18. Nielsen- Egenfeldt, S.: Exploration in computer games - a new starting point. In: Digra - Level up conference 2003 Electronic Proceedings (2003)

19. Prensky, M.: Digital natives, digital immigrants. From on the Horizon 9(5) (2001), http://www.marcprensky.com/writing/Prensky%20-%20Digital%20Natives,%20Digital%20Immigrants%20-%20Part1.pdf (last accessed 10/07/09)
20. Alexander, B.: Web 2.0: A new wave of innovation for teaching and learning? Educause Review 41(2), 32–44 (2006)
21. Alm, 2006. Guest Author on August 25 (2008), http://uk.techcrunch.com/2008/08/25/learning-a-language-the-web-20-way/ (last accessed 10/07/09)
22. Karaoğlu, Ö.: The Future of Education: Is it Web 2.0 or not? January 28 (2009), http://www.teachingenglish.org.uk/blogs/%C3%B6zge-karao%C4%9Flu/future-education-it-web-20-or-not?page=1 (last accessed 10/07/09)
23. Guest author: Learning a language the Web 2.0 way. TechCrunch UK, August 25 (2008), http://uk.techcrunch.com/2008/08/25/learning-a-language-the-web-20-way/
24. Meyer, B., Kristensen, M.: Bo: Designing location aware games for mobile language learning (2008), http://209.85.129.132/search?q=cache:KfnwISVIqUcJ:www.formatex.org/micte2009/book/1086-1090.pdf+Bo-Kristensen+et+al+2008&cd=4&hl=el&ct=clnk&gl=gr&client=firefox-a (last accessed 10/07/09)
25. Helen Newstead: Web 2.0 / Language Learning (December 2007), http://web20andlanguagelearning.wikidot.com/second-life (last accessed 10/07/09)
26. Crookall, D.: Second language acquisition and simulation. Simulation & Gaming 38, 6 (2007)
27. Li, R.-C., Topolewski, D.: ZIP & TERRY: a new attempt at designing language learning simulation. Simulation and gaming 33, 2 (2002), Cited in Bente Meyer: Designing serious games for foreign language education in a global Perspective Retrieved from, http://209.85.129.132/search?q=cache:jaJTUsk47lUJ:www.formatex.org/micte2009/book/715-719.pdf+Language+Learning+but+they+have+been+part+of+second+and+foreign+language+education+for+decades+Crookall+2007,+Li+%26+Topolewski+2002&cd=1&hl=el&ct=clnk&gl=gr&client=firefox-a (last accessed 10/07/09)
28. Crookall, D., Oxford, R.L.: Simulation, Gaming, and Language Learning. Newbury House Publishers (1990)
29. Baltra, A.: Language Learning Through Computer Adventure Games. Simulation & Gaming 21, 4 (1990)
30. Garcia-Carbonell, A., Rising, B., Montero, B., Watts, F.: Simulation/gaming and the acquisition of communicative competence in another language. Simulation and gaming 32, 4 (2001)
31. Java, A., Song, X., Finin, T., Tseng, B.: Why we twitter: understanding microblogging usage and communities. In: Procedings of the Joint 9th WEBKDD and 1st SNA-KDD Workshop, August 12 (2007), http://ebiquity.umbc.edu/paper/html/id/367/Why-We-Twitter-Understanding-Microblogging-Usage-and-Communities (last accessed 10/07/09)

32. Ullrich, C., Borau, K., Luo, H., Tan, X., Shen, L., Shen, R.: Why Web 2.0 is Good for Learning and for Research: Principles and Prototypes. In: World Wide Web Conference 2008, Beijing, China, pp. 705–714. International World WideWeb Committee, IW3C2 (2008), http://www2008.org/papers/pdf/p705-ullrichA.pdf (last accessed 10/01/09)
33. Pegrum, Mark: Microblogging, E-language wiki, December 20 (2008), https://e-language.wikispaces.com/microblogging (last accessed at 18/07/09)
34. Holotesca, C., Grossek, G.: Using microblogging in education. Case Study: Cirip.ro to run online courses and to enhance traditional courses. In: 6th Conference on e-Learning Applications, Cairo (January 2009), http://www.scribd.com/doc/8551345/Using-microblogging-in-education-Case-Study-Ciripro (last accessed 15/07/09)
35. Stevens, V.: Trial by Twitter: The rise and slide of the years most viral microbogging platform (2008), http://prosites-vstevens.homestead.com/files/efi/papers/tesl-ej/08june/twitter.htm (last accessed 10/07/09)
36. Scinicariello Sharon,: Que fais-tu? Twitter for Language and Culture. The University of Richmond (2008), http://www.frenchteachers.org/technology/Twitter.pdf (last accessed 15/07/09)
37. Sue, W.: Here Are The Results From My PLN Survey! Blog: Mobile Technology in TAFE (2009), http://aquaculturepda.edublogs.org/2008/12/04/here-are-the-results-from-my-pln-survey/ (last accessed 28/07/09)
38. Katie, A.: Educators Test the Limits of Twitter Microblogging Tool Published Online at Digital directions, June 25 (2008), http://www.edweek.org/dd/articles/2008/06/24/01twitter_web.h02.html (last accessed: 15/07/09)

Appropriating Technologies for Contextual Knowledge: Mobile Personal Learning Environments

Graham Attwell[1], John Cook[2], and Andrew Ravenscroft[3]

[1] Pontydysgu, Pontypridd, Wales, UK
graham10@mac.com
[2,3] Learning Technology Research Institute, London Metropolitan University, UK
john.cook@londonmet.ac.uk, a.ravenscroft@londonmet.ac.uk

Abstract. The development of Technology Enhanced Learning has been dominated by the education paradigm. However social software and new forms of knowledge development and collaborative meaning making are challenging such domination. Technology is increasingly being used to mediate the development of work process knowledge and these processes are leading to the evolution of rhizomatic forms of community based knowledge development. Technologies can support different forms of contextual knowledge development through Personal Learning Environments. The appropriation or shaping of technologies to develop Personal Learning Environments may be seen as an outcome of learning in itself. Mobile devices have the potential to support situated and context based learning, as exemplified in projects undertaken at London Metropolitan University. This work provides the basis for the development of a Work Orientated MoBile Learning Environment (WOMBLE).

Keywords: technology enhanced learning, mobile learning, media, context-aware, personal learning environment.

1 Introduction

The development and implementation of Technology Enhanced Learning has not been unproblematic. Despite optimistic industry predictions and sustained governmental support, instances of effective practice in using technology for learning often remain isolated, and initiatives fail to prove sustainable with one generation of technology replacing another and one project following on another in rapid succession.

The reasons put forward for this are numerous – poor interface design, the failures to adapt appropriate pedagogic approaches, the lack of confidence of teachers an trainers with new technologies, the digital divide and the lack of digital literacies often being cited.

However the difficulties and challenges of new technologies for learning may lay deeper, resting in the relations of technologies and media to learning and knowledge development within societies and to the organisation and purposes of institutions and education systems.

It may not be going too far to say that the rapid development and use of new internet based technologies and media, particularly social software and the 'read-write

M.D. Lytras et al. (Eds.): WSKS 2009, CCIS 49, pp. 15–25, 2009.

web' are challenging the present structures and purpose of educational institutions with learning increasingly taking place outside formal pedagogic structures. Equally the use of social media for information exchange, social networking and knowledge development challenges the relevance of traditional purpose-built education technology and the idea of instructional design.

Given the above context, in this paper we explore the development of a Work Oriented Mobile MoBile Learning Environment, with a view to enabling us to overcome a number of contradictions and dilemmas in the development of technology for learning.

2 The Education Paradigm and Technology Development

It has been suggested [1] suggested that the adoption of educational technology and instructional design has been shaped by the educational paradigm. Thus successive waves of technology have been developed to support the management of existing schooling systems with the development of Virtual Management Systems and Management Information Systems. Even in the pedagogic sphere the development of 'virtual classrooms' has tended towards supporting and even reinforcing traditional pedagogic approaches and teacher-learner relationships – albeit whilst extending the potential for access to classroom based education. It is little surprise that the leading commercial educational technology platform is named 'Blackboard'.

The adoption and effective use of any innovation is not solely a function of the affordances of new tools [2-4]. Innovation is socially constructed and constrained by the social context and the individuals who are offered opportunity to use the innovation [5]. In seeking to explain the relationships between learning and technology and the social construction and context of educational innovation it is necessary to explore further the relationship between the educational paradigms and other social and economic paradigms.

The potential for shaping of new technologies of learning may be more easily understood in the changes in economies and wider social structures and in particular in the ways in which technology is being used outside education than at looking at technology form within the educational paradigm.

2.1 The Industrial Model of Schooling

The present 'industrial' model of schooling evolved to meet the needs and forms of a particular phase of industrial development [1].

The industrial revolution imposed new requirements in terms of skills and knowledge – in particular the need to extend general education to wider layers of society. In 1893 the Elementary Education (School Attendance) Act raised the school leaving age to 11 and in 1902 the state took over education, through the organisation of Local Education Authorities and the provision of funding for schools from taxation. These reforms were based on a perceived need for Britain to remain competitive in the world by being at the forefront of manufacture and improvement. The form of organisation of schooling and the predominant pedagogy were based on the forms of production developed through the industrial revolution. Schools resembled large scale factories for knowledge [6].

The curriculum was closely tied to the needs of industry. In the early years of the 19th century the major emphasis was on basic skills and literacy. In 1872 a Revised code of Regulations laid down six levels of standards for reading, writing and arithmetic.

Despite a series of reforms throughout the 20th Century, the paradigmatic forms of organisation and delivery of education, the institutional form of schooling, the development of curriculum and approaches to pedagogy continue to be based on the Taylorist organisation of production stemming from the industrial revolution and from the economic and social needs of society to reproduce the workforce.

Of course there are different interpretations of the rise of the schooling system. Whilst acknowledging the importance of how social, technological, economic, and political forces influenced the evolution of schools, Arenas [7] explores the epistemological origin of schools. Arenas refers to the "artificial and ritualistic nature of classroom learning coupled by a virtual monopoly of book and abstract/fragmented knowledge". "Children had been given the role of 'students' and had to learn the necessary skills and behaviors to become honorable citizens and, eventually, also effective workers in the emerging industrial-bureaucratic world."

2.2 Education, Media and Technology

Friesen and Hug [8] argues that "the practices and institutions of education need to be understood in a frame of reference that is mediatic: "as a part of a media-ecological configuration of technologies specific to a particular age or era." This configuration, they say, is one in which print has been dominant. They quote McLuhan [9] who has described the role of the school specifically as the "custodian of print culture" (1962). It provides, he says, a socially sanctioned "civil defense against media fallout" [10] — against threatening changes in the mediatic environs.

Neil Postman [11] says that "school was an invention of the printing press and must stand or fall on the issue of how much importance the printed word will have in the future". Schooling and education, by extension, appear as the formal setting that is the necessary institutional correlative to this conception of development. As the "custodian of print culture," it is the task of education to provide students with a structured, controlled environment that is conducive to the quiet repose that print media demand of their audiences. This further positions the school as a kind of separate, reflective, critical pedagogical "space," isolated from the multiple sources of informational "noise" in an otherwise media-saturated lifeworld.

However Friesen and Hug [8] consider that: "any simple binary opposition between logical, hierarchical print culture on the one hand, and the visceral visual and audio flows on the other, has seriously undermined by the eclectic mix of media available via the internet, Web, and mobile communications. This is illustrated with special clarity in the case of textual forms of communication that have been collectively labeled "Web 2.0" or "the read-write Web."... Education, fighting on the side of literacy, no longer needs to fend off the attacks of "mass media" on the one hand while wielding instructional media from its curricular arsenal on the other. Instead of working against media and insulating its use of media against the mass mediatic environment, education now has the opportunity of working *with* media, in greater consonance with the larger mediatic ecology."

Thus the opportunities and challenges posed for the development of Technology Enhanced Learning may be seen in moving beyond the industrial model of schooling and the artificial and ritualistic nature of classroom learning to embracing new forms of learning based on a wider mediatic ecology, with learning embedded in forms of media and mediated experience outside the classroom paradigm.

2.3 Social Learning and Knowledge Development

However it is not only the social form of the classroom and the media which have defined educational practices and are being challenged by social media based on new technologies. It is also the nature and content of the curricula. Cormier [12] has written that the present speed of information based on new technologies has undermined traditional expert driven processes of knowledge development and dissemination. The explosion of freely available sources of information has helped drive rapid expansion in the accessibility of the canon and in the range of knowledge available to learners. We are being forced to reexamine what constitutes knowledge and are moving from expert developed and sanctioned knowledge to collaborative forms of knowledge construction. Social learning practices are leading to new forms of knowledge discovery. "Social learning is the practice of working in groups, not only to explore an established canon but also to negotiate what qualifies as knowledge." Cormier [12] cites Brown and Adler [13] who say: "The most profound impact of the Internet, an impact that has yet to be fully realized, is its ability to support and expand the various aspects of social learning." Ravenscroft [14] has argued that one of the reasons why this hasn't happened is due to the conceptual misalignment of (old) pedagogy and (new) communicative practices. He has pointed out that, typically, educationalists attempt to shoe-horn the radical possibilities offered by social software into, relatively, traditional pedagogical approaches, which unsurprisingly gives rise to mixed results and limited success. He argues that instead, we need to embrace new and emerging literacies through devising new, and typically more 'ambient' pedagogies [15] that embrace the new possibilities for engaging social learning and highly communicative interaction. Along these lines, some new pedagogical approaches to learning through dialogue have been proposed [16], that have also been realized and tested through the development and deployment of dialogue game technologies that have been used with hundreds of users. Crucial to these approaches is the setting up of a dialogic space [17] that reconciles new linguistic forms, such as instant messaging and immediate textual communication with tested pedagogical frameworks (e.g. for reflective thinking and reasoned discussion). So, this sort of technology provides affordances and structure that can be deployed in various, different and evolving ways, and so can harmonise with what Cormier [12] proposes a "rhizomatic" model of learning in which "a community can construct a model of education flexible enough for the way knowledge develops and changes today by producing a map of contextual knowledge."

2.4 Using Social Software for Learning

The central point of our argument is that educational technology has been developed within a particular educational paradigm to support the diffusion of expert based knowledge. Tools have been designed to support particular forms of knowledge within the particular environment of instructional design in school settings. Given the domination of universities and schools as the context for the implementation of educational technology, the focus has been on academic and disciplinary knowledge.

It is particularly important to note that the forms of technology and the media have been separated from the contents or subjects of learning. Knowledge has been embedded within the idea of learning objects, to be consumed through technology players, thus separating technology from content and process. Within the formal educational institutions, social practices have been limited to a critique (at best) of an established canon of disciplinary knowledge, within groups defined by age and relation to progression within that canon (and by extension by class and gender etc.).

However outside the classroom social software is increasingly being expropriated for learning. Young people are increasingly using technology for creating and sharing multi media objects and for social networking. A Pew Research study [18] found that 56 per cent of young people in America were using computers for 'creative activities, writing and posting of the internet, mixing and constructing multimedia and developing their own content. Twelve to 17-year-olds look to web tools to share what they think and do online. One in five who use the net said they used other people's images, audio or text to help make their own creations. According to Raine [19], "These teens were born into a digital world where they expect to be able to create, consume, remix, and share material with each other and lots of strangers."

But it is not just young people who are using social media. A survey into the use of technology for learning in Small and Medium Enterprises found few instances of the use of formal educational technologies [20]. But the study found the widespread everyday use of internet technologies for informal learning, utilizing a wide range of business and social software applications. This finding is confirmed by a recent study on the adoption of social networking in the workplace and Enterprise 2.0 [21]. The study found almost two-thirds of those responding (65%) said that social networks had increased either their efficiency at work, or the efficiency of their colleagues. 63% of respondents who said that using them had enabled them to do something that they hadn't been able to do before.

Of course such studies beg the question of the nature and purpose of the use of social software in the workplace. The findings of the ICT and SME project, which was based on 106 case studies in six European countries focused on the use of technologies for informal learning. The study suggested that although social software was used for information seeking and for social and communication purposes it was also being widely used for informal learning. In such a context:

- Learning takes place in response to problems or issues or is driven by the interests of the learner
- Learning is sequenced by the learner
- Learning is episodic
- Learning is controlled by the learner in terms of pace and time
- Learning is heavily contextual in terms of time, place and use
- Learning is cross disciplinary or cross subject
- Learning is interactive with practice
- Learning builds on often idiosyncratic and personal knowledge bases
- Learning takes place in communities of practice

However, it is important to note that the technology was not being used for formal learning, nor in the most part was it for following a traditionally curriculum or academic body of knowledge. Instead business applications and social and networking software were being used to develop what has been described as Work Process Knowledge [22]. The concept of Work Process Knowledge emphasises the relevance of practice in the workplace and is related to concepts of competence and qualification that stress the idea that learning processes not only include cognitive, but also affective, personal and social factors. They include the relevance of such non-cognitive and affective-social factors for the acquisition and use of work process knowledge in practical action. Work often takes place, and is carried out, in different circumstances and contexts. Therefore, it is necessary for the individual to acquire and demonstrate a certain capacity to reflect and act on the task (system) and the wider work environment in order to adapt, act and shape it. Such competence is captured in the notion of "developmental competence" [23] and includes 'the idea of social shaping of work and technology as a principle of vocational education and training' [24]. Work process knowledge embraces 'developmental competence', the developmental perspective emphasising that individuals have the capacity to reflect and act upon the environment and thereby forming or shaping it.

In using technologies to develop such work process knowledge, individuals are also shaping or appropriating technologies, often developed or designed for different purposes, for social learning.

3 Developing a Mobile Personal Learning Environment

The findings of the different studies cited above suggest dichotomies and dilemmas in developing and implementing technologies for learning.

Educational technology has been developed within the paradigm of educational systems and institutions and is primarily based on acquiring formal academic and expert sanctioned knowledge.

However business applications and social software have been widely appropriated outside the education systems for informal learning and for knowledge development, through social learning in communities of practice.

Is it possible to reconcile these two different worlds and to develop or facilitate the mediation of technologies for investigative and learning and developing developmental competence and the ability to reflect and act on the environment?

Based on the ideas of collaborative learning and social networks within communities of practice, the notion of Personal Learning Environments is being put forward as a new approach to the development of e-learning tools [25,26]. In contrast to Virtual Learning environments, PLEs are made-up of a collection of loosely coupled tools, including Web 2.0 technologies, used for working, learning, reflection and collaboration with others. PLEs can be seen as the spaces in which people interact and communicate and whose ultimate result is learning and the development of collective know-how. A PLE can use social software for informal learning which is learner driven, problem-based and motivated by interest – not as a process triggered by a single learning provider, but as a continuing activity. The 'Learning in Process' project [27] and the APOSDLE project [28] have attempted to develop embedded, or work-integrated, learning support where learning opportunities (learning objects, documents, checklists and also colleagues) are recommended based on a virtual understanding of the learner's context. While these development activities acknowledge the importance of collaboration, community engagement and of embedding learning into working and living processes, they have not so far addressed the linkage of individual learning processes and the further development of both individual and collective understanding as the knowledge and learning processes mature [29]. In order to achieve that transition (to what we term a 'community of innovation'), processes of reflection and formative assessment have a critical role to play.

John Cook [30] has suggested that Work Orientated MoBile Learning Environments (Womble) could play a key role in such a process. He points out "around 4 billion users around the world are already appropriating mobile devices in their every day lives, sometimes with increasingly sophisticated practices, spawned through their own agency and personal/collective interests."

However, in line with Jenkins at al [31] it is not just the material and functional character of the technologies which is important but the potential of the use of mobile devices to contribute to a new "participatory culture." They define such a culture as one "with relatively low barriers to artistic expression and civic engagement, strong support for creating and sharing one's creations, and some type of informal mentorship whereby what is known by the most experienced is passed along to novices... Participatory culture is emerging as the culture absorbs and responds to the explosion of new media technologies that make it possible for average consumers to archive, annotate, appropriate, and recirculate media content in powerful new ways."

The specific skills that Jenkins and his coauthors describe as arising through involvement of "average consumers" in this "participatory culture" include ludic forms of problem solving, identity construction, multitasking, "distributed cognition," and "transmedial navigation."

Importantly modern mobile devices can easily be user customized, including the appearance, operation and applications. Wild, Mödritscher and Sigurdarson [32] suggest that "establishing a learning environment, i.e. a network of people, artefacts, and tools (consciously or unconsciously) involved in learning activities, is part of the learning outcomes, not an instructional condition." They go on to say: "Considering the learning environment not only a condition for but also an outcome of learning, moves the learning environment further away from being a monolithic platform which is personalisable or customisable by learners ('easy to use') and heading towards providing an open set of learning tools, an unrestricted number of actors, and an open corpus of artefacts, either pre-existing or created by the learning process – freely combinable and utilisable by learners within their learning activities ('easy to develop'). "

Critically, mobile devices can facilitate the recognition of context as a key factor in work related and social learning processes. Cook [33] proposes that new digital media can be regarded as cultural resources for learning and can enable the bringing together of the informal learning contexts in the world outside the institution with those processes and contexts that are valued inside the intuitions.

He suggests that informal learning in social networks is not enabling the "critical, creative and reflective learning that we value in formal education."

Instead he argues for the scaffolding of learning in a new context for learning through learning activities that take place outside formal institutions and on platforms that are selected by learners.

Cook [30] describes two experimental learning activities for mobile devices developed through projects at London Metropolitan University. In the first, targeted at trainee teachers an urban area close to London Metropolitan University, from 1850 to the present day, is being used to explore how schools are signifiers of both urban change and continuity of educational policy and practice.

The aim of this project is to provide a contextualised, social and historical account of urban education, focusing on systems and beliefs that contribute to the construction of the surrounding discourses. A second aim is to scaffold the trainee teachers' understanding of what is possible with mobile learning in terms of field trips. In an evaluation of the project, 91% of participants thought the mobile device enhanced the learning experience. Furthermore, they considered the information easy to assimilate allowing more time to concentrate on tasks and said the application allowed instant reflection in situ and promoted "active learning" through triggering their own thoughts and encouraging them to think more about the area.

In the second project, archaeology students were provided with a tour of context aware objects triggered by different artifacts in the remains of a Cistercian abbey in Yorkshire. The objects allowed learners to expire not only the physical entity of the reconstructed abbey through the virtual representation, but also to examine different aspects including social and cultural history and the construction methods deployed. According to Cook [30] "the gap between physical world (what is left of Cistercian), virtual world on mobile is inhabited by the shared cognition of the students for deep learning."

The use of the mobile technology allowed the development and exploration of boundary objects transcending the physical and virtual worlds. Boundary objects have been defined as "objects which are both plastic enough to adapt to local needs and constraints of the several parties employing them, yet robust enough to maintain a common identity across sites. They are weakly structured in common use, and become strongly structured in individual-site use. They may be abstract or concrete. They have different meanings in different social worlds but their structure is common enough to more than one world to make them recognizable means of translation. The creation and management of boundary objects is key in developing and maintaining coherence across intersecting social worlds." [33]. The creation and management of boundary objects which can be explored through mobile devices can allow the inter-linking of formal and academic knowledge to practical and work process knowledge.

Practically, if we consider models for personalized and highly communicative learning interaction in concert with mobile devices, whilst employing context aware techniques, startling possibilities can arise. For example, we can combine the immediacy of mobile interaction with an emergent need for a collaborative problem solving dialogue, in vivo, during everyday working practices, where the contextual dimensions can constrain and structure (through semantic operations) the choices about a suitable problem solving partner or the type of contextualised knowledge that will support the problem solving. In brief, combining dialogue design, social software techniques, mobility and context sensitivity means we have greater opportunities for learning rich dialogues in situations where they are needed - to address concrete and emergent problems or opportunities at work.

Such approaches to work oriented mobile learning also supports Levi Strauss's idea of bricolage [34]. The concept of bricolage refers to the rearrangement and juxtaposition of previously unconnected signifying objects to produce new meanings in fresh contexts. Bricolage involves a process of resignification by which cultural signs with established meanings are re-organised into new codes of meaning. In such a pedagogic approach the task of educators is to help co-shape the learning environment.

Of course, such approaches are possible using social software on desktop and lap top computers. The key to the mobile environment is in facilitating the use of context. This is particularly important as traditional elearning, focused on academic learning, has failed to support the context based learning inherent in informal and work based environments.

Whilst the use of context is limited in the experiments undertaken by London Metropolitan University, being mainly based on location specific and temporal factors, it is not difficult to imagine that applications could be developed which seek to build on wider contextual factors. These might include tasks being undertaken, the nature of any given social network, competences being deployed, individual learner preferences and identities and of course the semantic relation involved.

4 Conclusions

The development of a Work Oriented MoBile Learning Environment is seen as potentially allowing us to overcome a number of contradictions and dilemmas in the development of technology for learning as outlined in the first part of this paper.

These include:

- Classroom learning versus 'real life' informal or work based learning
- Academic knowledge as opposed to work process knowledge
- Individual learning versus social learning
- Educational technology versus social software
- Decontextualised knowledge acquisition versus context knowledge development
- Information acquisition versus reflective learning

At the same time the Work Oriented MoBile Learning environment can be developed by learners themselves through the appropriation or shaping of technologies, thus allowing learners to co-design their own learning environments through the process of learning.

The technologies to develop such environments exist. The challenge is to further explore the nature of context and to co-develop or appropriate with learners, applications which can facilitate such social learning processes. At the same time though the process of interacting with the wider environment learners themselves can construct and shape the corpus of knowledge itself within the community.

References

1. Attwell, G.: The Social Impact of Personal Learning Environments in Connected Minds, Emerging Cultures: Cybercultures in Online Learning. In: Wheeler, S. (ed.) Information Age Publishing, Charlotte (2008)
2. Rogers, E.: Diffusion of Innovations. Free Press, New York (2003)
3. Bijker, W.: Of Bicycles, Bakelites and Bulbs: Towards a Theory of Sociotechnical Change. MIT Press, Cambridge (1999)

4. Anderson, T.: Open Educational Resources Plus Social Software: Threat or opportunity for Canadian Higher Education? (2008), http://auspace.athabascau.ca:8080/dspace/bitstream/2149/1609/1/Open+Educational+Resources+Plus+Social+Software+to+CSSHE.doc (September 1, 2008)
5. Fulk, J.: Social construction of communication technology. Academy of Management Review 36(5), 921–950 (1993)
6. Steeves, K.A.: Schools like Factories: Exploring Connections between Philosophy and Structure. Paper presented at the annual meeting of the American Historical Association, http://www.allacademic.com/meta/p26104_index.html (accessed September 2, 2008)
7. Arenas: The Intellectual Development of Modern Schooling: An Epistemological Analysis, universitas humanística no. 64 julio-diciembre de, Bogotá, Columbia, pp. 165–192 (2007)
8. Friesen, N., Hug, T.: The Mediatic Turn: Exploring Concepts for Media Pedagogy. In: Lundby, K. (ed.) Mediatization: Concept, Changes, Consequences. Peter Lang, New York (2009)
9. McLuhan, M.: The Gutenberg Galaxy: The Making of Typographic Man. University of Toronto Press, Toronto (1962)
10. McLuhan, M.: Understanding Media: The Extensions of Man. McGraw-Hill, New York (1964)
11. Postman, N.: The Disappearance of Childhood. Random House, New York (1982)
12. Cormier, D.: Rhizomatic education: Community as curriculum. Innovate 4(5) (2008), http://www.innovateonline.info/index.php?view=article&id=550 (accessed May 31, 2008)
13. Brown, J.S., Adler, R.P.: Minds on fire: Open education, the long tail, and Learning 2.0. Educause Review 43(1), 16–32 (2008), http://connect.educause.edu/Library/EDUCAUSE+Review/MindsonFireOpenEducationt/45823 (accessed August 13, 2008)
14. Ravenscroft, A.: Social Software, Web 2.0 and Learning: Status and implications of an evolving paradigm. Guest Editorial for Special Issue of Journal of Computer Assisted Learning (JCAL) 21(1), 1–5 (2008/2009)
15. Ravenscroft, A.: Promoting Thinking and Conceptual Change with Digital Dialogue Games. Journal of Computer Assisted Learning 23(6), 453–465 (2007)
16. Ravenscroft, A., Wegerif, R.B., Hartley, J.R.: Reclaiming thinking: dialectic, dialogic and learning in the digital age. Underwood. J., Dockrell, J. (Guest eds.) British Journal of Educational Psychology Monograph Series, Learning through Digital Technologies, Underwood, Series II (5), 39–57 (2007)
17. Wegerif, R.: Dialogic education and technology: Expanding the space of learning. Springer, New York (2007)
18. Lenhart, A., Madden, M.: Teen Content Creators and Consumers, Pew Internet (2005), http://www.pewinternet.org/pdfs/PIP_Teens_Content_Creation.pdf (accessed 21 April 2007)
19. Raine, L.: BBC, US youths use internet to create, November 4 (2005), http://news.bbc.co.uk/2/hi/technology/4403574.stm (accessed 20 April 2007)
20. Attwell, G. (ed.): Searching, Lurking and the Zone of Proximal Development, e-learning in Small and Medium enterprises in Europe, Vienna, Navreme (2007)

21. Oliver Young, G.: Global Enterprise Web 2.0 Market Forecast: 2007 to 2013, Forrester (2009)
22. Boreham, N., Samurçay, R., Fischer, M.: Work Process Knowledge. Routledge, New York (2002)
23. Elltrom, P.E.: The many meanings of occupational competence and qualifications. In: Brown, A. (ed.) Promoting Vocational Education and Training: European Perspectives. University of Tampere Press, Tampere (1997)
24. Heidegger, G., Rauner, F.: Vocational Education in Need of Reform, Institut Technik und Bildung, Bremen (1997)
25. Attwell, G.: The personal learning environments – The future of eLearning? eLearning Papers 2(1) (2007), http://www.elearningeuropa.info/files/media/media11561.pdf (accessed 11 August 2008)
26. Wilson, S., Liber, O., Johnson, M., Beauvoir, P., Sharples, P., Milligan, C.: Personal learning environments challenging the dominant design of educational systems. In: Paper presented at the ECTEL Workshops 2006, Heraklion, Crete, October 1-4 (2006)
27. Schmidt, A.: Knowledge Maturing and the Continuity of Context as a Unifying Concept for Knowledge Management and E-Learning. In: Paper presented at the I-KNOW 2005, Graz (2005)
28. Lindstaedt, S., Mayer, H.: A storyboard of the APOSDLE vision. In: Paper presented at the 1st European Conference on Technology-Enhanced Learning, Crete, October 1-4 (2006)
29. Attwell, G., Barnes, S.A., Bimrose, J., Brown, A.: Maturing Learning: Mashup Personal Learning Environments. In: CEUR Workshops proceedings, Aachen, Germany (2008)
30. Cook, J.: Scaffolding the Mobile wave, Presentation at the Jisc Institutional Impact programme on-line meeting (July 9, 2009), http://www.slideshare.net/johnnigelcook/cook-1697245?src=embed (accessed July 10, 2009)
31. Jenkins, H., Purushotoma, R., Clinton, K.A., Weigel, M., Robison, A.J.: Confronting the Challenges of Participatory Culture: Media Education for the 21st Century. White paper co-written for the MacArthur Foundation (2006), http://www.projectnml.org/files/working/NMLWhitePaper.pdf (Accessed July 14, 2008)
32. Wild, F., Mödritscher, F., Sigurdarson, S.: Designing for Change: Mash-Up Personal Learning Environments, elearning papers (2008), http://www.elearningeuropa.info/out/?doc_id=15055&rsr_id=15972 (accessed 2 September, 2008)
33. Star, S.L., Griesemer, J.R.: Institutional Ecology, Translations and Boundary Objects: Amateurs and Professionals in Berkeley's Museum of Vertebrate Zoology, 1907-39. Social Studies of Science 19(4), 387–420 (1989)
34. Levi Strauss, C.: The savage mind. University of Chicago Press, Chicago (1966) [first published in 1962]

Teachers Professional Development through Web 2.0 Environments

Cristina Costa

University of Salford,
Salford, Greater Manchester, M5 4WT, UK
cristainacost@gmail.com

Abstract. Teacher professional development is no longer synonymous with acquiring new teaching techniques, it is rather about starting new processes as to engage with new forms of learning, reflected in the practice of teaching. With easy access to the panoply of online communications tools, new opportunities for further development have been enabled. Learning within a wider community has not only become a possibility, but rather a reality accessible to a larger number of individuals interested in pursuing their learning path both in a personalised and networked way. The web provides the space for learning, but the learning environment is decidedly dependent on the interrelationships that are established amongst individuals. The effectiveness of the web is reflected in the unconventional opportunities it offers for people to emerge as knowledge producers rather than information collectors. Hence, it is not the tools that most matter to develop a learning environment where more personalized learning opportunities and collective intelligence prospers as the result of personal and collaborative effort. Although web tools provide the space for interaction, it is the enhancement of a meaningful learning atmosphere, resulting in a joint enterprise to learn and excel in their practice, which will transform a space for learning into an effective, interactive learning environment. The paper will examine learning and training experiences in informal web environments as the basis for an open discussion about professional development in web 2.0 environments.

Keywords: Web 2.0, Learning Environment, Informal Learning, Teacher Professional Development.

1 The Challenges of Our Era

In the last decades we have witnessed a massive change in our society. We have not only become technologically more advanced, we have also become inherently dependent on the latest technologies. From the simplest household appliances to the most sophisticated machines, the fact is that technology has been embedded in one's existence in such a way we cannot ignore it nor can we reject it if we want to keep up with the pace of this Era. The truth is that the job market is more than ever reliant on such approaches to prosper, be competitive and thus stand a better chance in the reality ahead. Hence, the demands of the real world do include first hand knowledge on the latest technologies. Web technologies are no exception. In fact, they have

M.D. Lytras et al. (Eds.): WSKS 2009, CCIS 49, pp. 26–32, 2009.

instigated change in our society. Web technologies have come to revolutionize the way we access, broadcast and interact with information. They have also enabled new forms of working (tele-work), networking and socializing. Consequently, new forms of learning, and teaching, have also emerged as an answer to this phenomenon. Informal Learning is being given special attention as an important part of one's learning experience. On the web learning in community, networking and socializing at a larger scale, independently of time or space has not only become a possibility but also a reality. Regardless of the potential of this new modality to establish contact and learning relationships with individuals interested in the same areas, and develop shared understanding, and even practice, about topics of interest, online learning is still far from being mainstream.

Although learning online is not the only route individuals should be seeking as to update or acquire new skills, the fact is that web environments do prove to be a rich source for continuous development, free of scheduled programmes, contrived curricula or restricted to a specific location or time zone. We can thus say that Web 2.0 Learning Environments have become quite liberating in the way they allow individuals interested in pursuing their own learning, their own way, at their own pace and with their own adopted communities and/or networks.

2 Virtual Learning Environments: A Replication of the Classroom or an Emancipation of the Traditional Learning Spaces?

The Virtual Learning Environment, commonly referred to as VLE, has been a topic of lively debates in the recent years. In the English speaking reality, the VLE is often synonym of the well-known Learning Management Systems (LMS) or Course Management Systems (CMS), which many Educational Institutions have adopted as part of their strategy to introduce e-learning or blended learning in their teaching curriculum. The adoption of such system has been, in many cases, used to store content, provide students with reading lists and deal with class work. Only in few cases have these management tools been used to introduce interactivity to the learning activity or provide the learner with the scope for creating and participating in a less contrived environment. These tools, as crucial as they may have been in the first phase of introducing e-learning across the educational sector, are not proving efficient when compared with the autonomy the second generation of web tools are allowing learners to do and become. The 21st century learner has the choice of becoming more autonomous in his/her learning process. The 21st century learner no longer is restricted to the institutional provision of tools or learning resources and spaces. The 21st century learner has the power to plot his/her own learning trajectory autonomously and in company (through relevant communities and networks) in more dynamic environments. He/she is in charge of his/her learning curriculum.

In this sense we believe that the VLE of the 21st century is in no way constrained to a management system. We would even like to advance the idea that **no** Virtual Learning Environment is synonym of the provision of any tool or set of tools. The concept of a Virtual Learning Environment has to be understood as something that is pervasive beyond a technological platform people can join or use. Important as they may be - and indeed they are essential - the tools do provide the virtual space for connection, we would like to argue that it is the atmosphere that might emerge in that virtual space that will affect one's personal and collective learning experiences. Hence, through the observations we have made online, we have noticed the learning environment is not as much bond to the features the tool might display as it is dependent on learners' engagement with one another and the learning interrelationships they are able to develop as to create a trustworthy environment, in which they feel comfortable in threading their uncertain learning path. Learning can and is most likely to happen in environments which display a friendly and reliable atmosphere, and where everyone feels welcome to take part in. Although the tools are provided from the very beginning to 'house' a given learning activity, the learning environment, just as it happens in analogy with a 'home', often takes longer to be created. It is a process and relies heavily on how the environment is nurtured from the very beginning and how much effort learners are willing to put in it, especially in terms of a culture of sharing and ongoing support. A house in which there is 'giving and taking' can more easily be transformed into a 'home' people are willing to return to. On the opposite, places free of such close learning relationships are often most likely not to persist in the long run, independently how sophisticated they might be.

In this sense strict, planning and constricted structures provide a more artificial scenario for learning , leaving little scope for the learner to emerge and feel part of that same environment as someone who also has something to contribute to and not only take from. That is also where the power of web 2.0 environments lie. The participatory web is indeed effective and appealing to those who see the benefit of taking ownership of their own learning. It becomes a meaningful alternative to formal learning and training. Implicit to learning 2.0 is also a curriculum that evolves naturally, mirroring the emergent requirements of its participants. A curriculum 2.0 also helps reshape the learning environment in a continuum effort to be in synch with the pace of the real world. In this sense, online Learning Environments add a new dimension to learning, in the way the curriculum is designed and meets the needs of its target audience. Such learning approach still seems rather impracticable in a formal educational learning framework, where the curricula are previously planned and intended to be meticulously applied. This leaves little scope for alterations, changes and adaptations to a learning environment in progress. However, shifts in plans are fairly common in the real world. Being able to be responsive and pro-active to new daily challenges is a skill the 21st century worker needs to be used to. Similarly, informal learning environments are often associated with flexible learning trajectories, being its configuration characteristically more loosen and focused on both the expectations of the group and on the advancements of the surrounding reality, rather than on a static structure. This seems to be particularly useful when addressing the new requirements for lifelong learning and updated know-how in a society in constant demand for new skills. Teachers, as 'actors of change', specialized in preparing the younger generations to the new job market should see in informal forms of learning, mediated by web 2.0 environments, a potential alternative to standard, traditional teaching and teacher training.

In the next section we will provide an overview of a study we conducted on informal teacher professional development through web environments, with a special focus on how the learning practice orients the curriculum. We will also draw some reflections on the role of online environments in pursuing shared teacher professional development.

3 Informal Teacher Professional Development Online

It is argued that most of the skills acquired and knowledge developed happens informally [1], through belonging to multiple communities [2]. Autonomy, networking, communal activity and collaboration are also concepts directly linked to the ways of being, thinking and acting in the digital age [3]. The social web appears to be equally a very productive field in the educational area, as well as an efficient complement to the enrichment of learning opportunities. Accordingly, adjustments to how educators learn and facilitate learning also need to be considered [4]. This is not only a time where the individuals can develop a voice, he/she can create his/her learning environment in collaboration with others, in a way that is more adjusted not only to his/her needs but also to his/her own personality and learning expectations. The informal world online seems to be a productive field to cultivate new learning experiences and develop expertise in collaboration and cooperation. As Prof. Carneiro stated in an interview for the Online Educa Berlin , "We are standing at the threshold of a new era in learning approaches and itineraries where the greatest novelty of ICT resides in the full use of the C: C for community, communication and care." [5]

In this section we will briefly report on a study we have conducted on Informal Teacher Professional Development Online. The target audience was a group of International English as a Foreign Language (EFL) Teachers – The Webheads in Action - known for their innovative teaching and learning practice in online environments. The study focused particularly on the curriculum inherent to their online activity and consequently on the perceived benefits of belonging and taking part in its shared environment and practice.

As to set a background understanding about our study it is useful to provide an insight of the group under study. The Webheads in Action have grown from a group of learners informally organized into a recognized Community of Practice operating online, and independent of any institutional body. They have developed a historical past over the last ten years in terms of informal teacher development through informal training, learning online and in community. They also display a shared repertoire based on their joint enterprise and mutual engagement. All of these are features Wenger et al [2,6] have identified as being part of a Community of Practice (CoP). All of these also contribute to a real sense of identity with the community, thus enabling a more trustworthy and comfortable environment for shared practice, learning and collaboration. Equally, we have noticed that the social dynamics of the group has been crucial to maintain the group active and relevant both to the individual and to the Community. The social affections, the way individuals cater for each other, is a nuance in their daily practice with a crucial impact in the perception of the individual towards the community and vice-versa.

That can be verified in the participants' own words:

We learn a lot about ourselves, how we deal with a virtual group of people from all over the world, how we can live so far but still share the same values, how there are people in the world whom we have never met personally but are still there for us when we need them, when we are in a crisis (be it technical because we cannot get the results we wanted or if it is a personal loss or situation.(…)) (Be)

WiA participants start exploring without fear of making mistakes, because they know that if they do, they will get help from other colleagues just by sending an e-mail, or contacting them in Yahoo Messenger, Skype, or MSN (Da)

The members scaffold each other and the teaching/learning cycle continues. A member can count on the expertise and support of other members. (Rm)

The affective side of the learning interrelationships, as observed during our field work, and attested through our data analysis, show how important it is for the individual to feel valued and acknowledged within the wider group and how that enables them to emerge with their own learning needs, while helping others pursue their learning purposes too. It is in this daily dialectic of giving and taking that the curriculum is designed and constantly re-adjusted to accommodate the individual expectations and the community's activity.

I guess the best of this all is that each of us learns what they need/want to learn (Gl)

This kind of flexibility in education is seldom permitted outside informal approaches. Deprived of any organisational accreditation, the perceived effectiveness of a practice oriented by an open, organic curriculum is probably bound to the informality of the learning process. With the focus on the practice and on one's learning, participants' attention is not shifted to a quantitative evaluation. They are rather focused on a common practice, similar goals and shared experiences, which wrapped by spontaneous feedback provides a more constructive kind of formative evaluation of one's practices.

Participants provide their opinion about it:

How to take turns, (...) , and, develop the critical listening skills to listen to more than our own opinion on what we want to hear or say, and learn to listen and See better, what it it that is actually being said, despite our own prejudices which would like to see things from a personal, and, naive way of looking at the world. I also know that my writing skills have improved, So, in summation, Critical Listening, and Thinking skills, Speech and Communication and Writing Skills across the curriculum are improved. (Rg)

Equally, we have also observed that participation in such web 2.0 environments also trigger self reflection about one's practice as part of one's constant dialog with the other members of the web environments in which they are embedded. All of this seems to have an important impact in one's personal learning as well as on the dynamics of the group's practice. In a way it does determine the way the learning curriculum evolves:

Feedback and reflection is a big part of the learning process.(Lb)

4 The Curriculum within Informal Learning Web Environments

According to our research we have noticed that online learning environments ask for an open, progressive curriculum supported by an equally flexible group of people willing to construct communal meaning in a rather personal, yet not individual, way.

As Oliver and Herrington [7] point out, at the core of such learning model is an authentic learning context, based not only on realistic learning activities, but also anchored on the individuals' own needs, wishes and practice requirements. Hence, in this case, the curriculum contributing to the learning environment is neither formal nor prescribed. It is rather a curriculum that is implicitly co-developed as the learners, engaged in their joint activities, spontaneously build a friendly, relevant learning environment. The learning activity, that ongoing, community tailored-made curriculum, ends up reflecting what the community is in its essence. In it are included its approach to learning, the strong social-affective component, the ongoing negotiation not only of meaning but also personal and collective goals, the collaborative aspect, and of course the drive to constantly keep their practice updated and in touch with the surrounding reality. It is the learning curriculum "as a field of learning resources in everyday resources in everyday practice viewed from the perspective of the learners" [8] that most influences the 'preservation' of a productive, effective and meaningful web 2.0 learning environment. Whereas web 2.0 are useful in bridging the connection, it is the effort learners put in their learning and also in others' practice that most transforms and adds coherence to the experience of the individual and consequently their professional development.

As Wenger suggests [2] "learning cannot be designed", but it can certainly be based on the experience and progressive practice of a group of people operating in a single or multiple communal environments in which the significant construction of knowledge and shared understanding is supported through a "soft approach" – that of interpersonal connections and learning relationships where the individual tends to feel more compelled to emerge as an active contributor.

The participation of educators in such approach is crucial not only as form of updating their own knowledge and be in contact with the wider community, but also as a way of preparing themselves to prepare the current learner generations [9]. By engaging in informal web 2.0 learning environments, educators will be able to connect better to their students. They will also feel more confident to invest a pedagogy of change, more focused on a 'humanized' approach where the individual is not only at the center of the learning process, but is also responsible for pursuing their lifeling learning.

As one of the participants notes:

(…) it [the informal practice] allows us to rotate expertise, to negotiate meaning and to build bonds of trust which enhance and increase our knowledge in a way that would be impossible individually, and in a much shorter period of time. (Te)

5 Some Considerations

In the recent years we have noticed that the web landscape has changed dramatically. An increasing number of educators are seeing the value of informal learning online. Web 2.0 environments are providing a new form for informal lifelong learning and professional development. The offer has increased largely with the widespread of volunteered initiatives by educators for educators in different fields of knowledge. The generosity of teaching and learning with others is impressive, and there are many interesting ideas which are worth exploring and being part of. However, the increasing numbers of individuals seeking informal, free learning opportunities have a new dilemma to face: that of critically selecting and cultivating Web Learning Environment(s), which will add true, personal value to their experience. These days, people co-participate in several environments simultaneously. This is not only a Digital Era we are living in. This is equally the Era of multi-tasking and multi-participating in different environments. The widespread of the self is much more notorious and more common in this first decade of the 21st century than in the decades that introduced learning online. This colossal variety of possibilities is also having enormous implications in the way people congregate and in the way web learning environments are currently fostered and sustained. We have moved from centralised, institutionally controlled system to more disperse web environments. Lifelong learning and professional further development is no longer only bound to institutional provision. The individual now has the autonomy to make his/her own learning choices. There is a strong emphasis on the 'self made learner'. And there is also the 'fashionable' idea that any web space is automatically a learning environment ready to accommodate community activity. But not all of those spaces are what they were intended to, especially because in many of them the core to its success was missing. And that is passionate people committed to share personal narratives which will entice others to do the same. Above all, the web is helping bring the conversational and practical tone of the 'original communities of practice', as researched by Lave and Wenger (1994) into lifelong learning through a more humanized approach, where the technology's role is just one: that of bridging the connection between people and giving them a space to congregate. The environment is up to the individuals to create and maintain through meaningful practices and activities.

References

1. Cross, J.: Informal Learning, Rediscovering the Natural Pathways That Inspire Innovation and Performance, Pfeiffer, an imprint. Wiley, San Franscisco (2007)
2. Wenger, E.: Communities of practice. Learning, meaning and identity. Cambridge University Press, Cambridge (1998)
3. Veen, W., Vrakking, B.: Homo Zappiens, Growing up in the digital age. MPG Books ltd., Bodmin (2006)
4. Bell, F.: Communities of Practice Online? The case for 'going Feral' in Academic Development, in ILIA Innovative Learning in Action, 5: Learning Technologies in the Curriculum, pp. 3–16 (2006)
5. Carneiro, R.: Interview with Prof. Roberto Carneiro, keynote speaker in the Online Educa Berlin Plenary on "The Role of Technology in Supporting Cradle to Grave Learning, from News and Background Information on the Conference, Online Educa Berlin (2007), http://www.icwe.net/oeb_special/news50.php
6. Wenger, E., McDermott, R., Snyder, W.M.: Cultivating communities of practice. HBS press (2002)
7. Oliver, R., Herrington, J.: An instructional design framework for authentic learning environments. Educational Technology Research and Development 48(3), 23–48 (2000)
8. Lave, J., Wenger, E.: Situated learning: legitimate peripheral participation. Cambridge University Press, New York (1991)
9. Wesch, M.: From Knowledgable to Knowledge-able: Learning in New Media Environments (2009), http://www.academiccommons.org/commons/essay/knowledgable-knowledge-able (accessed 13 July 2009)

Using Web 2.0 Applications as Supporting Tools for Personal Learning Environments

Ricardo Torres Kompen[1], Palitha Edirisingha[2], and Josep M. Monguet[3]

[1] Fundació i2Cat, C/Gran Capità 2-4, Edifici Nexus I
Barcelona, Spain
ricardo.torres@i2cat.net
[2] Beyond Distance Research Alliance, Attenborough Tower
University of Leicester, Leicester, UK
palitha.edirisingha@le.ac.uk
[3] Universitat Politècnica de Catalunya, Laboratori d'Aplicacions Multimèdia
Diagonal 647, Barcelona, Spain
jm.monguet@upc.edu

Abstract. This paper shows the results of a pilot study based on a proposed framework for building Personal Learning Environments using Web 2.0 tools. A group of 33 students from a Business Administration program were introduced to Web 2.0 tools in the context of an Information Systems class, during the academic year 2008-2009, and reflected about this experience through essays and interviews. The responses show evidence of learning and acquiring skills, strengthening social interactions and improvement in the organization and management of content and learning resources.

Keywords: personalized learning, personal learning environments, informal learning, lifelong learning, Web 2.0.

1 Personal Learning Environments

Personal Learning Environments, or PLEs, are a relatively new concept [1] that changes the focus of the learning processes from the VLE towards a user-built, personalized set of tools - not necessarily digital ones- that are used to manage content and interactions, and support the learning experience.

There is a growing interest on PLEs from both practitioners and researchers, and from many levels of education. The 2009 Horizon Report [2] mentions the personal web as one of the trends in the second horizon of adoption (Time-to-Adoption: Two to Three Years). Personalization of learning is based on the idea that learning technologies should enable the various aspects of learning (the content, the mode of delivery and Access) to be offered, according to the personal characteristics of the learner, thus providing the learner with greater flexibility and options for learning.

Another factor behind the increased attention on PLEs is the emergence of and widespread access to the so-called Web 2.0 tools: a group of internet-based tools and technologies, with a strong social component. The term Web 2.0 was coined by Tim O'Reilly, and it captures a "trend towards greater creativity, information sharing and

M.D. Lytras et al. (Eds.): WSKS 2009, CCIS 49, pp. 33–40, 2009.

collaboration amongst internet users" [3]. Web 2.0 tools can have a central role in personalizing learning, enabling the learner to take a more active role.

There seems to be a widespread notion that the majority of the current generation of learners – usually referred to as Net Gen, Millennials or Digital Natives - are familiar with computers and internet-based technologies and are capable of using Web 2.0 technologies for learning. Nevertheless, there is no strong evidence that suggests that they are familiar with using Web 2.0 tools for formal learning.

It is a fact that students do use a variety of Web 2.0 tools and applications [4], [5]. However, there is no strong evidence that students use these tools in an integrated manner suited to academic learning [6]. Such integration would follow a constructivist approach, as students would build their own personal learning environment and thus their knowledge. In this sense, the PLE will be the result of using and connecting all these tools and applications.

There are two main approaches to PLEs: PLE as an actual object, a program or platform common to all users (that might be customizable to a certain extent), that allows them to organize, collect, process and share information and knowledge. This approach raises many problems, mostly technical ones, related to interoperability and interfaces between the different applications.

Another approach is to consider the PLE not as a specific tool, but rather as a concept, a way of organizing a variety of Web 2.0 technologies. The PLE would be unique to each user, and would change according to the user's needs and experiences. As Attwell [7] wrote in his blog, "Clearly any PLE application will be a perpetual beta."

We have chosen the second definition of a PLE for the PELICANS (Personal ELearning In Community And Networking Spaces) project based at the University of Leicester, UK and at the Universitat Politècnica de Catalunya, i2Cat Foundation and Citilab, in Catalonia, Spain. In this conceptualization of PLE, each learner chooses their own Web 2.0 tools and connects them to collect, organise, process and share information, and manage their knowledge. Thus the sum effect of the tools, information, connections, storage and resultant knowledge actually creates the PLE. In fact, the PLE would also include the users, their network of contacts, books, any resources they use to learn. What we usually call a PLE is in fact the PLE support tools.

We would like to close this introduction with two quotes:

"The PLE is what happens when we apply Web 2.0 principles to e-learning"
(M. Metcalfe)

"…the ideal PLE will vary from person to person, as each individual will add different elements to his or her Personal Learning Environment. Subsequently I believe that the ideal PLE for an individual should not be created by someone else than this person"
(K. van Westenbrugge)

Defining what a PLE is usually proves a difficult task; but in the end, there seems to be general agreement on the fact that it is something unique to each individual; a set of tools that support that person's learning experience.

2 Methodology: Pilot Study

This section presents the methodology followed during the pilot study conducted between September 2008 and May 2009, with a group of 33 students from the 2nd year of a Business Management program, at the *Escuela Superior de Estudios Internacionales (ESEI)* in Barcelona, Spain.

The study was based on a proposed framework for creating Personal Learning Environments using Web 2.0 applications [8]. The aim of the study was to test the conceptual framework in practice, by guiding the students in the development of their own PLEs, and to gather empirical evidence on students' engagement with PLEs. The study's initial approach had been to present the concept of PLE and the four approaches (wiki-based, social network-based, aggregator page-based and browser-based) to students, and then provide the guidance needed for them to build their own PLEs. Nevertheless, discussions with the students during the planning stage of the pilot study suggested that a bottom-up approach to PLEs would be more efficient. Thus, students were gradually introduced to Web 2.0 tools, usually chosen by them, after they were given an overview of Web 2.0 and the variety of tools available.

As a previous step, the students took a survey, based on the 2007 ECAR study, and focused on students' technology skills before taking the subject, and their experience with ICTs. After that, the study was started, and was divided into four phases:

2.1 1st Phase: Introduction to Web 2.0 Tools and Applications

During the first stage of the pilot study, which spanned 12 weeks, the Web 2.0 concept was presented and discussed; Web 2.0 tools were gradually introduced. The first one was Twitter; it was the only one presented by the teacher, and was suggested as a new channel of communications for the class. The adoption was slow, but after three weeks 90% of students were using Twitter, not only for academic purposes, but also for social affairs and casual chat. The percentage of use varied during the semester, and showed peaks just before exams and school events. The percentage of frequent users (i.e., students that were using Twitter as their main channel of communications) was around 20% of class.

After the introductory session, the Web 2.0 tools to be considered and discussed were suggested by the students themselves, either because they were already using them and thought they could be interesting for their classmates, or because they felt the need to learn a particular tool. Amongst these, Flickr, FriendFeed, Clipperz, Jooce, RSS, Blip.fm, last.fm, MOG, Blogger, and Picasa were discussed, and some of them were used in class activities or online e-tivities. See Table 1 for a description of some of the activities and projects.

As a final activity in this semester, students were asked to think about the way they used Web 2.0 tools, and how did they see them "connect", or how would they like them to interact with each other. This exercise was approached in a more formal way during the second phase of the study.

In summary, during the first semester of the academic year 2008-2009, students were introduced to the concept of Web 2.0, and some of the so-called Web 2.0 applications. They also reflected on the way they used these tools, and the potential ways they could interact. Some other concepts, like Digital ID and issues on digital safety, were also discussed.

2.2 2nd Phase: Web 2.0 Diagrams

At the beginning of the second semester of the academic year 2008–2009, students were asked to continue working on the exercise proposed at the end of the 1st semester, and draw pictures of their "ideal" work environment, based on Web 2.0 tools, or any other tools they used. Thus, this exercise was not limited to the tools and applications that had been used during the previous semester, and it was also emphasized that this environment should not be restricted to tools, platforms or applications that they already knew: it was rather focused on their objectives, the way they used the tools, and their needs.

It was decided that the participation in the study was going to be an optional activity during this semester, which resulted on 10 students dropping out of the study, since this was no longer a graded activity. The group was now down to 21 students (three students from the original group transferred to another school, while an exchange student joined the class). Out of these, 6 submitted very simple diagrams, while eight students went so far as to attempt to establish links among the tools, and even checked which of these links actually existed and which ones were merely "wishful thinking" on their part.

The proposals covered a wide range of approaches. Eight students proposed a platform or web service that would allow them to access their sets of tools, and most of them pointed out that some kind of one-time, safe access should be provided as part of the service. One of the students called the diagram his "personal page of everything"; this diagram matched one of our approaches, the browser-based PLE.

The Start-page/aggregator-based PLE approach was also proposed by one of the students. In her words, "a centralized platform allowing the access of user-selected Web 2.0 applications through a single password from one site". The student even searched for such a tool, and found and set-up a Pageflakes account.

Another student also mentioned the aggregator page, but using Google applications by means of iGoogle. This matches to some degree one of our approaches (the Wiki-based one), in the sense that it relies mostly on Google applications, but the idea of using a Wiki for a single user was not even considered by the students, probably because they already had two Wikis being used for collaborative projects.

The Social-network approach proposed in the framework was not considered by the class. Not a single student thought of using Facebook (or any other social network) as a hub for their PLEs.

A fifth approach was proposed, one that had not been considered in the proposed framework. This involved the use of a virtual desktop utility (Jooce) that allows users to manage multiple desktops from one account, allows them to share desktops and files if they wish to do so, and provides access to multiple working spaces This approach will be included in a revised version of our framework. Interestingly enough, Jooce was presented by one of the students, but she did not use it in her "Web 2.0 diagram" (as this exercise was called – the PLE concept had not been introduced to the class yet).

2.3 3rd Phase: Approaches to PLEs

During the 3rd phase of the study, the PLE concept was introduced and explained, as well as the proposed frameworks. Students were able to compare the "Web 2.0 diagrams" they had drawn with the framework approaches; at this point, they were asked to "build" or structure their PLEs, either around the diagrams they had proposed, or following a particular approach, or combinations of them, something they discussed later in their essays. The fifth approach using a virtual desktop was also presented and discussed; that particular student decided to stop using Jooce as a hub, and focused on finding alternatives and comparing them, and wrote a report on this. At the end of this phase, 17 students had built or developed a PLE, while 4 students reported that they did not see the usefulness and chose to drop out of the study. As was mentioned earlier in the text, 10 students did not participate at all.

2.4 4th Phase: Essays and Interviews

The final phase of the study involved written essays as well as interview conducted face-to-face and through e-mail. The results have been classified in three main categories: Evidence of PLEs as organization and management tools, Evidence of strengthening social interactions, and Evidence of learning and developing skills. Some of the obstacles, criticism and suggestions the students mentioned in their essays and interviews are also included.

3 Study Results

3.1 Evidence of PLEs as Organisation and Management Tools

"this is 'not only my PLE but also my PEE (Personal Entertainment Environment" and PSE (Personal Socialisation Environment)'"

"I really support the use of PLEs, because it can help me to share information and exchange many things through the web 2.0 tools"

"there is such an overload of tools today that we need some kind of organiser for them"

"Flock has taken it to another level for me, by centralising all my different [web] stops that I do in one page"

"[a PLE] is an easy way to manage and organise all the information I get from online sources, and also offline ones"

Most of the students reported a sense of chaos and confusion regarding the wide range of Web 2.0 tools and the need for some way of organising them. The PLE approaches gave them suggestions and ideas, and most of them came up with some way of managing the applications and tools. Flock was repeatedly mentioned as a useful tool for centralizing the applications and offering a one-stop access to them, and at the same time a way of dealing with logins and passwords.

Some of the advantages mentioned are: the ability to organize and manage data and contents they already knew about, as well as the access to new sources of information; the chance to integrate the tools they were more comfortable with and the new tools they were introduced to and that seemed like good resources; filtering information, as the use of the PLE helped them pick out only the most valuable information.

3.2 Evidence of Strengthening Social Interactions

"one thing I started using through Flock is Facebook [...] I started to have conversations with classmates, friends and even teachers"

"what I like the most about all these web 2.0 tools is the ability to get inspiration, knowledge and to be able to interact with other people"

"I am developing a network that most probably will become extremely valuable in the near future"

"the social element has had a large impact in my learning process, helped me to create stronger links with classmates, friends and teacher because you interact more and put your opinions forward"

The social element was one of the most important aspects of Web 2.0 tools, according to the students' opinions. The collaborative approach followed in this class through the use of the wiki and blog, together with microblogging and social bookmarking increased the learning opportunities and the availability of useful resources. Surprisingly, some students did not use Facebook, and only joined the social network after seeing it was one of the tools available through Flock. As mentioned before, Facebook was not considered as an option for building or managing a PLE.

Some of the highlighted benefits were: the ability to share and discuss different points of view, the fact that Web 2.0 tools seem to be ideal tools to collaborate, and share and create knowledge; the exchange of ideas with fellow students and teachers taking place outside of the learning institution, and the discussion of concepts that was taking place online.

3.3 Evidence of Learning and Developing Skills

"PLEs increase my level of learning opportunities, as I don't miss anything in [the] news' perspective"

"a proper working PLE decreases the level of stress and increases the opportunities to learn"

"a well-developed PLE can lead to enhanced (autodidactic) learning"

"[PLE] has changed my personal learning process"

Most of the students mention the new tools they learned and the skills they acquired as the highlights of having participated in the study. Although some of them were familiar with Web 2.0 applications, for the most part they did not realize they could use them in their learning process. A high percentage of them were Facebook users prior to the beginning of the study, but none of them had used Twitter or del.icio.us before, or any other microblogging or social bookmarking tools, for that matter. They did use blogs as a source of news (mostly on entertainment, news or specific interests), but very few of them knew what RSS were. Although the use of Wikipedia was widespread among the class, they said they did not use wikis.

Some of the benefits they reported are: Web 2.0 tools made the learning process more dynamic and interesting; it helped them in transforming information provided in class and course textbooks into knowledge, as discussions forced them to reflect on the concepts covered in class; the fact that they can use their PLE not only in a formal context, but also outside the school, where –in their own words- "a lot of the learning takes place".

3.4 Disadvantages and Recommendations

The users were also asked about any particular problems and obstacles they might have found during the pilot study. What follows is a summary of the main points they mentioned, and their suggestions.

- The activities were sometimes confusing – a “big picture” is required from the start. This is a difficult issue, since the discussion of PLEs as a concept was avoided on purpose, to avoid leading the students in choosing one of the framework approaches, and cutting their creativity and options. More guidance and support might be needed during the initial phase.
- Creating a PLE can be too time-consuming. This depends on the approach and the user experience, but it might indeed be a complex task. It is important to explain the advantages of an organised environment, and that the benefits outweigh the disadvantages.
- Some tools cannot be tailored to the users’ needs. This comment made reference to Flock´s limitation for adding tools other than the ones already provided with the browser, as well as the limited tools found in the Google sites Wiki.
- Adoption problems. These were mostly related to student’s experience with technology and Web 2.0 tools.
- Interoperability. Some tools do not “speak” to each other, which makes it hard to integrate them into one environment.
- PLE might lead to distraction and procrastination. A consequence of mixing academic and “fun” tools, according to some of the students, who still see a marked difference between these two environments. This is reflected in one of the comments: “ the Facebook approach [to PLEs] is ‘too social’”
- Technology problems. Mainly in the case of systems failure (down time) of some of the tools, like Twitter. They also mentioned the lack of support and the differences between the tools – while some of them are easy to learn and adopt, other tools require more time and practice.

4 Conclusions and Lessons Learned from the Study

Although there is a widespread notion that students are familiar with the Internet and Web 2.0 tools in particular, this was not what we found in our study. Few of them had heard the term “Web 2.0” before, or knew what it meant. This is reflected in the following comment: “I was quite surprised that I actually already used some web 2.0 tools without even knowing what they were”. The fact that these tools could be used for learning was even less obvious to them: “In the course of the classes, I realized the possibility of using web 2.0 applications for actual learning. […] blogs, wikis and online communities, provided my with much information, which made studying easier”.

There were divided opinions on whether the PLE concept and tools such as Flock and Netvibes should be presented earlier in the study. While some of them think they should be given a “big picture” and a clear purpose of how can Web 2.0 tools be incorporated and integrated, some of them thought that the hands-on, do-it-yourself approach actually made them become more familiar and knowledgeable with the applications, and that the “chaotic” situation forced them to come up with solutions on their own.

The social element was observed not only in the collaboration online and the increased communication, but also on a "network effect" with some of the tools: adoption of these was in some cases motivated by some of their peers; since they saw their classmates trying some of the applications, some of the students decided to join and try them too. Word of mouth and comments were also effective in bringing some of the students on board. There also some "advocates" for some of the tools, such as in the case of Twitter.

Probably the main conclusion of the study was the fact that, in the end, the choice of hub for the PLE and the specific tools that the student adopts are not the end result nor the most important point in the experience: it is the journey itself, the way they discover and try new tools and applications, how they use them to share and collaborate, and how they take on a "prosumer" approach, learning together, even teaching each other. Teachers are, of course, a very important part of the learning experience: they are the guides in this journey.

Acknowledgements

We would like to thank the Beyond Distance Research Alliance at the University of Leiceser (UK), the Universitat Politècnica de Catalunya, i2cat Foundation and Citilab, and the Escuela Superior de Estudios Internacionales in Spain for supporting the work described in this paper. We are also grateful to all the students that participated in the pilot study, for their enthusiasm and inspiration, and for allowing us to quote their answers and comments in the essays and interviews.

References

1. JISC. A report on the JISC CETIS PLE project (2007), http://wiki.cetis.ac.uk/Ple/Report (accessed February 8, 2008)
2. Johnson, L., Levine, A., Smith, R.: The 2009 Horizon Report. The New Media Consortium, Austin (2009)
3. The Economist. Innovation: Home Invention, The Economist, p. 98 (May 3, 2008)
4. ECAR, The ECAR study of undergraduate students and information technology (2007)
5. Trinder, K., Guiller, J., Margaryan, A.: Learning from digital na-tives:bridging formal and informal learning. The Higher Education Academy, New York (2008)
6. Mcloughlin, C., Lee, M.: Future learning landscapes: Transforming pedagogy through social software. Innovate 4(5) (2008)
7. Atwell, G.: Personal Learning Environments – a position paper (2006), http://www.knownet.com/writing/weblogs/Graham_Attwell/entries/6521819364 (accessed February 11, 2009)
8. Szücs, A., Tait, A., Vidal, M., Bernath, U.: Distance Learning in Transition. In: Putting the Pieces Together: Conceptual Frameworks for Building PLEs with Web 2.0 Tools, ch. 55. Wiley-ISTE, Chichester (2009)

An Infrastructure for Intercommunication between Widgets in Personal Learning Environments

Tobias Nelkner

Computer Science and Education, University of Paderborn,
Fürstenallee 11, 33102 Paderborn, Germany
tobin@upb.de

Abstract. Widget based mashups seem to be a proper approach to realise self-organisable Personal Learning Environments. In comparison to integrated and monolithic pieces of software developed for supporting certain workflows, widgets provide small sets of functionality. The results of one widget can hardly be used in other widgets for further processing. In order to overcome this gap and to provide an environment allowing easily developing PLEs with complex functionality, the based on the TenCompetence Widget Server [1], we developed a server that allows widgets to exchange data. This key functionality allows developers to create synergetic effects with other widgets without increasing the effort of developing widgets nor having to deal with web services or similar techniques. Looking for available data and events of other widgets, developing the own widget and uploading it to the server is an easy way publishing new widgets. With this approach, the knowledge worker is enabled to create a PLE with more sophisticated functionality by choosing the combination of widgets needed for the current task. This paper describes the Widget Server developed within the EU funded IP project Mature, which possibilities it provides and which consequences follow for widget developer.

Keywords: PLE, SOA, mashup, informal learning, personalized learning, widgets.

1 Introduction

The term *Widget* is the short version of "window gadget" and has been applied during Project Athena [2]. A widget is usually a window the user can interact with and provides a small set of functionality. Widgets are nowadays well known and used in very different contexts like in a WeBlog to aggregate a RSS feed or for weather information on the operation system like in Windows Vista or Mac OS X. The last years widgets became very important in the research area of computer supported education and (informal) learning [3]. Because of their small set of functionality the learner can personalise its learning environment and choose exactly the one he or she needs to fulfill the current task. Therefore, many environments have been developed to mashup widgets in order to allow learners creating their Personal Learning Environment (PLE), for example ELGG[1]. A PLE is not a formal educational concept

[1] http://elgg.org/

M.D. Lytras et al. (Eds.): WSKS 2009, CCIS 49, pp. 41–48, 2009.

but one for computer supported work integrated learning [4]. Hence, within working processes also more complex tasks have to be fulfilled by the user and therefore need to be supported by the PLE. As Attwell mentioned, a PLE has to support learning among others by aggregating and scaffolding, sharing, and reflecting knowledge with a key focus on communication and collaboration [5]. Bringing the mashup concept of widgets and the requirements on a work integrated PLE together, intercommunication between widgets is a possible solution for creating PLEs more effecient, interactive and immersive. Exchanging data enables two or more widgets to create more sophisticated functionality by linking the output of one widget to the input of another one.

The widget server has been implemented within the EU funded IP project MATURE and is presented in this paper. It provides communication channels between widgets and allows also the collaboration of users via their widgets. This server allows running widgets almost independent from platform and programming language as long as a broker to the JAVA programming language is available.

The next section provides a look to related work in order to show up similarities but also improvements compared to other projects and work. Section 3 then explains the concept, different components of the server and its special functionalities. Section 4 closes the report with a conclusion and an outlook.

2 Related Work

Well known environments exist allowing users including, running, and configuring a widget in its mashups, like iGoogle or Netvibes. Both of them provide an own widget definition for inlcuding self developed widgets and they provide a huge repository of existing ones. But both environments lack of the possibility of intercommunication between widgets and data persistence and therefore do not allow to combine the functionalities of widgets to more complex ones. This leads to isolated functionality of small widgets without supporting users in more complex tasks in a workflow. Moreover, the missing possibility of saving data and sharing it in the user‘s community is a barrier for creating new knowledge. Nevertheless, some widgets provide possibility of saving data but neither it is transparent where and how it is saved (private/public) nor is it accessible for any kind of analysis for tasks like expert finding, provision of information on demand or reuse of certain information in another context.

A closed systems like WordPress that is set up quickly on an own server, provides transparency and sustainability. It is easy to add widgets on the user interface but they also lack of intercommunication. Yahoo has developed the Yahoo Pipes system that allows mashing up and filtering information streams of RSS feeds. However, intercommunication between widgets means additionally exchanging data and firing events on other widgets for instance to update the user interface or run some computations.

Within the EzWeb/Fast EU funded project [6] a SOA based environment has been created that allows the composition of mashups of widgets by providing them in a special repository. The EzWeb composition environment is a user interface that allows graphically connecting widgets in order to create intercommunication by linking the output of one widget to the input of another one. In the backend, this connection results in an orchestration of webservices. The architecture presented here, bases on the idea of providing widget communication facilities and it does not matter in which environment they run, nor in which programming language they are developed. A minimum of overhead in administration shall be achieved. Moreover, widgets shall not only be visualisations of data but also may realise business logic, depending on what the developer wants to provide to the user. It is not the aim to link the underlying data or logical services they shall remain independent.

Another approach is a result of the TenCompetence partly EU-funded project with the Wookie server as one of the outcomes. The Wookie server allows hosting and instantiation of widgets and provides a proxy server that allows access to different kind of services and therefore also to backend services. However, the Wookie server misses also intercommunication of widgets although it provides a data pool where each widget has access to. This allows the group wide use of data (for example for a chat widget) but no communication in the nearer sense, for example point to point. Nevertheless, this architecture has served for a first version of the server presented in this paper. It has been redesigned and completely new implemented in order to concentrate on the special issue of intercommunication between widgets and between one widget and groups of them within the server. Therefore an independence from the user interface has been achieved. The user can choose in which platform his or her widgets shall run, there is no difference running them in iGoogle, WordPress or an integrated environment as long as they can use the provided API for the server.

An infrastructure of a Widget Server that hosts widgets, provides communication facilities and transparent access to backend systems for persistence and sustainability is a valuable challenge for improving PLE implementations. In the next section the server concept and its different aspects are described.

3 Widget Server Architecture

In order to describe the server structure, the requirements shall be described before explaining the server concept.

3.1 Requirements on the Server

As described before, the most important aspects are the provision of a widget repository, the possibility of intercommunication between widgets and a transparent access to a backend-system for storing, retrieving and analysing data.

The widget repository serves the user as a kind of catalogue to find widgets he or she likes to mashup. Therefore, it provides an upload and download functionality where especially the upload is enriched with validation tests on the widget. Download means, the user can start the widget locally in his client, for instance in case of HTML Widgets, in the web browser. Consequently, a user management is necessary in order to provide user bound persistence services as profile management or configuration settings.

The access to a backend system is necessary to provide persistence services to the widgets for storing or retrieving data. Such a backend system can be a middleware like the Knowledge Bus concept [7] but also a set of fix services provided by a certain system.

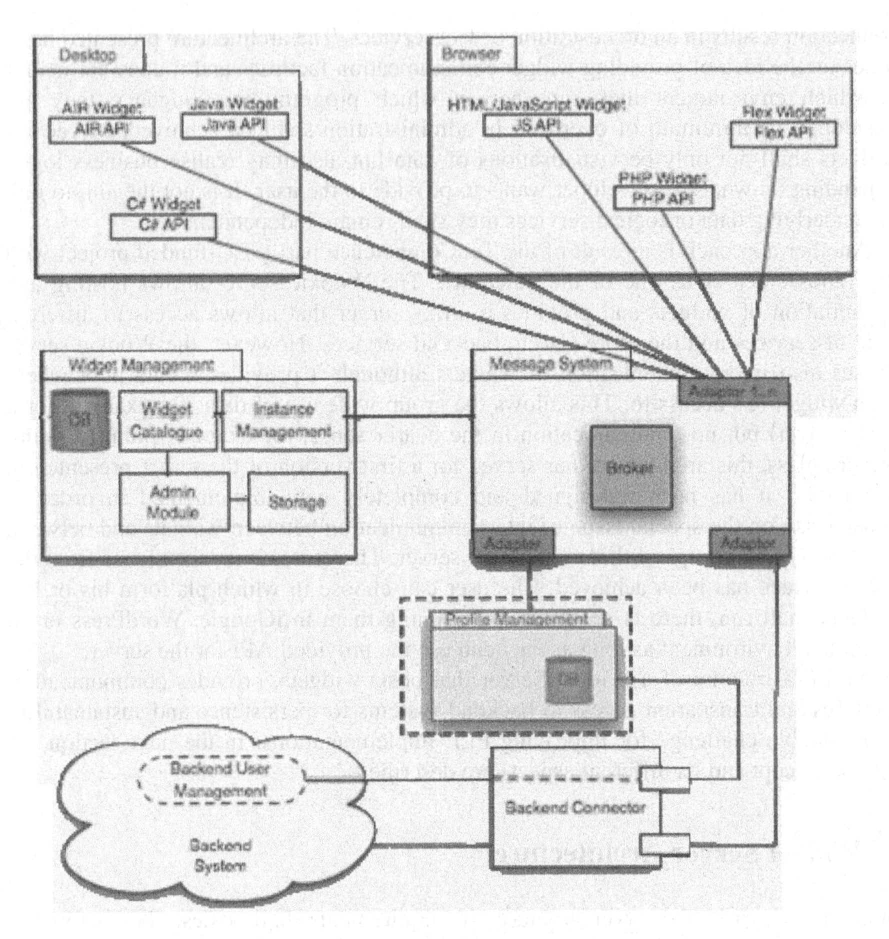

Fig. 1. The Widget Server Architecture

A communication system allows widgets to get data and events from other widgets, this can be on a direct channel but also via broadcast to all widgets that are interested in a special data set of another widget.

3.2 The Architecture

Ongoing from these requirements a first version of the server has been developed based up on the Wookie server of the TenCompetence project. This server already allows the hosting of widgets and undirected information flow. For example, a chat widget is available that allows chatting with all users that opened the widget.

Based on this server, a version was developed that provided a user centric communication channel for all loaded widgets but had also some limitations especially concerning a transparent backend interface and the possiblity of loading widgets developed in different programming languages. For instance, it was not possible to use Flex widgets or JavaFX widgets as the access to the communication channel was not available.

With the experiences of this development, a complete new server has been build that meets our expectations and requirements (see Figure 1).

The server is divided into the four parts of Widget Management, Message System, Profile Management, and Knowledge Bus Connector that will be explained in detail in the following.

3.2.1 Widget Management

The Widget Management package of the server contains all modules dealing with the issues of the widget repository mentioned above. This includes hosting, adding and managing running and inactive widgets. Especially the *Widget Catalogue* needs a database to save metadata about hosted widgets and as a backend system is not necessarily available, this database is not outsourced but under the control of this package.

The *Widget Catalogue* holds a collection of the currently available widget types or templates, for example a Weather Widget, Chat Widget, etc. These are then instantiated and executed in the user's client. In addition to this, the catalogue also keeps track of each widget's preferred run-time environment. It is thought to be queried by users' workspaces and provides a list of available and fitting widgets for that workspace. This is important especially to include mobile devices with their special requirements on performance, display size etc.

The *Instance Management* is the most important component of the widget management package of the server. It keeps track of all running instances of widgets in all connected workspaces and automatically purges widgets that have become inactive. It is always the most up-to-date source on the current global state of the widget server, and messaging and administration functions depend on it. As well as keeping track of running widgets, it also manages the state of their communication, for example which widgets are connected to communication units. The messaging system uses this information to distribute messages correctly.

The *Admin Module* performs management functions for the widget management component, usually working on the catalogue. It provides options for catalogue maintenance, widget upload, widget deletion and other administrative functionality. It is intended to be accessed via internal messaging or an authentication-enabled web frontend.

The *Storage Modul* stores widget files, such as Flash, Flex, HTML and JavaScript files. This is basically a shared and protected directory or drive from which the hosted widgets are being loaded via HTTP. This is transparent to the user.

3.2.2 Message System

The Message System is used for widget-to-widget and widget-to-server communication. The *Message Broker* supports point to point communication as well as group based communication models. Groups are created automatically, either because of the input and output paramters of loaded widgets or according to the user's configuration. They are provided by the *Instance Management* module of the *Widget Management* component. All external entities like widgets, an administration frontend or the backend system accessing the server via adapters. The used adapter concept is flexible and allows the connection of all kinds of systems as long as an adapter can be developed that works as a translator between the certain programming languages and JAVA.

3.2.3 Profile Management

The *Profile Management* module is conceptualised for managing server-bound user data that is strictly necessary for the widgets to work. This includes data that is only used to provide essential functions of the widget system and does not necessarily need to be stored in the backend. It includes for example user authentication, default widget positions, saved environment configurations (if available), possibly third-party logins and others. Especially the user management led to the possibility of intercommunication between widgets of different users. This allows to develop communication and collaboration facilities.

Because of its special relevance this module is part of the server as the backend system is not necessarily available and not important for the server and the hosted widgets to run. As long as a backend system is integrated in the server, it makes sense to use this module as a stub and provide the used database as a backend service. Because of this conceptual duality this package is greyed in the image.

3.2.4 Backend Connector

The *Backend Connector* is responsible for providing the access to appropriate services where it is unimportant if this is a middleware encapsulating several services or only one service. All requests and responses are transferred through this connector and that translates the requests to the backend in the specific calls. The connector has to be adapted according to the respective backend system.

Widgets not necessarily depend on the existence of a backend as long as storing or retrieving data from a certain database system is not relevant.

3.2.5 Graphical User Interface

The GUI is the users‘ client on which the widgets will run. This GUI is connected to the server via an adapter providing access to the communication system. As the adapter is also the broker between different programming languages, the widget developer can choose the one he is comfortable with, which simplifies widget development enormously and decreases barriers working with the server. Moreover, it does not matter if the widgets run standalone in a browser or if they are encapsulated in a closed environment where this is responsible for mapping messages to the certain widgets. Nor is it irrelevant if a widget is running as a desktop client as long it is connected to the server. Consequently it is possible to link browser widgets and desktop widgets interacting with each other. It only depends on the adapter and can be accessed via different techniques, e.g. JMS, DWR and others.

3.3 Relation between Modules

In order to provide a clearer picture of the interaction between these modules an use case based example of how the server manages the certain requests is shortly presented.

A calendar widget and a weather widget are loaded by including them into iGoogle, interacting with each other and with the backend system. Using widgets in iGoogle is easy as it only contains a link to the *Storage* in the *Widget Management* package. After the browser has loaded both widgets, the server authenticates them at the *Profile Management*. Then a message is send to the *Message System* for registering the widgets in the *Instance Management*, the pool for all running widgets. Furthermore, they are registered to communication group that allows them to talk to each other.

This is either realised by connecting widgets automatically according to their input and output format or manually by the user. The latter one is only possible as widget developers have specify the input and output format so that the server can map them.

If the user now creates a calendar entry with a certain place, the calendar widget sends a message to the server that distributes it to the common group. The weather widget gets this message, checks if it can interpret the message and updates the display.

By creating the new entry, the calendar widget also sends a save request to the widget server. This is translated into a backend save request to store the calendar dates. The widget server only transfers this message to the backend connector which calls the specific web service or backend system.

4 Conclusions and Outlook

In this paper a widget server has been presented that allows hosting of widgets, intercommunication between widgets and provides an access to a backend system that traces requests to it. It is constructed as a flexible and modular server that allows to connect almost every kind of widgets. Because of the approach of connecting small pieces of functionality the similar based mashup approach of a PLE is fully supported also within the server. This allows for creating more sophisticated and complex functionality by connecting widgets via this server.

With this approach an essential step forward towards personlisable learning environments has been done, especially by providing widget developers a platform for easily creating widgets and using the output data of foreign widgets to provide more sophisticated functionality. Users do not need to care about in which environment, on which client or on which platform they want to run their widgets, the server itself does not have any conceptual limitations concerning this.

But as described above, in order to make the development of widgets as easy as possible but also to be as standard conform as possible, this server concept makes demands on the widgets specification. Hence, the future work concerning this is to provide a specification that allows developers to describe their widgets as exact as possible which is necessary to support linking widgets. Otherwise the mapping of data between widgets would not be possible. Moreover, we have to have a look on the W3C Widget Specification[2] (working draft) and how we can include and validate this standard within our server. In opposition to this W3C standard, we do not limit the server running Web-Widgets. May be a similar specification can be found for desktop widgets and those of other programming languages (e.g. Flex/ActionScript), too.

Acknowledgement

This work was co-funded by the European Commission under the Information and Communication Technologies (ICT) theme of the 7th Framework Programme (FP7) within the Integrating Project MATURE (contract no. 216356).

[2] http://www.w3.org/TR/widgets/

References

1. Wilson, S., Sharples, P., Griffiths, D.: Distributing education services to personal and institutional systems using widgets. In: Proceedings of the First International Workshop on Mashup Personal Learning Environments (MUPPLE 2008) (September 2008)
2. Arfman, J.M., Roden, P.: Project Athena: Supporting distributed computing at MIT. IBM Systems Journal 31(3) (1992)
3. Attwell, G.: Personal Learning Environments for creating, consuming, remixing and sharing. In: Griffiths, D., Koper, R., Liber, O. (eds.) Service Oriented Approaches and Lifelong Competence Development Infrastructures: Proceedings of the 2nd TENCompetence Open Workshop, Institute of Educational Cybernetics, pp. 36–41 (2007)
4. Attwell, G.: The Personal Learning Environments - the future of eLearning? In: eLearning Papers, vol. 2(1) (2007)
5. Attwell, G., Bimrose, J., Brown, A., Barnes, S.-A.: Maturing learning: Mash up Personal Learning Environments. In: Wild, F., Kalz, M., Palmér, M. (eds.) Proceedings of the First International Workshop on Mashup Personal Learning Environments (MUPPLE 2008), Maastricht, The Netherlands, September 17 (2008); In conjunction with the 3rd European Conference on Technology Enhanced Learning (EC-TEL 2008), Maastricht School of Management, Maastricht, The Netherlands, September 18-19, vol. 388. CEUR Workshop Proceedings (2008)
6. Lizcano, D., Soriano, J., Reyes, M., Hierro, J.J.: EzWeb/FAST: Reporting on a Successful Mashup-Based Solution for Developing and Deploying Composite Applications in the Upcoming. In: Ubiquitous SOA, pp. 488–495
7. Hinkelmann, K., Magenheim, J., Reinhardt, W., Nelkner, T., Holzweißig, K., Mlynarski, M.: KnowledgeBus - An Architecture to Support Intelligent and Flexible Knowledge Management. In: Duval, E., Klamma, R., Wolpers, M. (eds.) EC-TEL 2007. LNCS, vol. 4753, pp. 487–492. Springer, Heidelberg (2007)

Communities of Practice as a Support Function for Social Learning in Distance Learning Programs

Ashley Healy

School of Computing, University of the West of Scotland
Paisley Campus, Paisley, PA1 2BE

Abstract. Communities of Practice (CoP) are organic entities, which evolve as a result of the passion for the domain by its members. CoP present opportunities for social learning and supporting distance learning programs.

But, What Is Social Learning?
Simply, learning is the process of moving from not knowing to knowing. The 'social' aspect of the learning relies upon human interaction. Therefore, social learning is constructed from conversations and participation within a CoP. It is about 'how' members learn, rather than the content or 'what' they learn. This is in contrast to the Cartesian vision of learning, which proposes, "I think, therefore I am". Social Learning puts forward the alternative "We participate, therefore we are" (Seely Brown and Alder, 2008).

How Do CoP Support Social Learning?
CoP support social learning through an apprenticeship model. Members take on simple tasks under the guidance of the community, they then progress to more demanding tasks as confidence, skills and knowledge improve. The CoP supports the apprentice by interacting regularly, sharing tales and experiences. Social Learning within a community is 'learning about a subject' as well as 'learning to be a full participant'. The shared passion and the interaction engages the learner and enables them to acquire deep knowledge about a subject and the ability to participate in the community through productive enquiry and social interaction.

Keywords: communities of practice, CoP, learning, social learning, distance learning.

1 Introduction

This paper reports on the potential of Communities of Practice (CoP) to support social learning in distance learning programs. It has long been recognized that distance learning fails to achieve academic respect [1]. There are a number of student and faculty barriers in distance learning but one of the most commonly documented is the "loneliness of the distance learner" [2]. This paper proposes using CoP to overcome such barriers by integrating CoP in distance education programs. Therefore, enabling distance learning to become a success in twenty-first century education, particularly

M.D. Lytras et al. (Eds.): WSKS 2009, CCIS 49, pp. 49–56, 2009.

as technology continues to shape the current generation of learning. The 2007 Sloan Consortium Survey of Online Learning reported that almost 2.5 million students enrolled in online educational programs in 2006 and that over the past 5 years this figure has grown on average by 21.5% per annum [3]. This figure is likely to continue growing with the emergence of new technologies. Therefore, it is critical that teachers and faculty recognize the potential of CoP to transform teaching distance learning programs by creating a learning culture committed to problem solving, inquiry, interaction and innovation.

2 Distance Education and Learning

The literature reveals that there are many definitions available to describe distance education and learning. Despite being almost fourteen years old, the definition by Filipczak is useful as it informs the description of CoP as a support function for social learning. Filipczak defines distance learning as *"getting people…into the same electronic space so that they can help one another learn"* [4]. This is not dissimilar to the definition of CoP provided by Wenger (1998) *"group of people….learn to do it better through interaction"* [5]. Both these definitions exploit the idea of social interaction and people, the very foundations of social learning.

As suggested in the introduction, technology is playing an ever-growing role in education and whilst distance learning is not a new phenomenon, technology is shaping the next generation of learning. The idea of independent study introduced by Moore (1973) is an important foundation of distance learning. This approach implies that learning can take place despite the teacher and student being physically separated. Teachers face the challenge of meeting learner expectations at a distance as highlighted by Cantelon; *"most of higher education will take place off-campus through technological methods of delivery"* [6]. Distance learning is already a part of everyday life for most educational institutions, recognizing the problems and addressing them will be critical to the success of distance learning in the future, particularly as demand increases.

3 Barriers to Distance Learning

There are student and faculty barriers associated with distance learning. This section of the paper summarizes the barriers before moving on to explain why CoP could help overcome and/or minimize many of the barriers.

3.1 Student Barriers

The problems and barriers reported in the literature from the students perspective fall into five main categories; costs and motivators, perceived lack of contact and feedback, lack of support and services, isolation and alienation and technological expectations.

- Costs and Motivations

As far back as 1988, Knapper noted that distance learners will generally have more insecurities about learning than traditional learners because they have to face anxieties about the financial costs of study, trying to integrate studying into home-life and career [7]. As a result of such insecurities and the associated pressure the dropout rates in university level distance learning are definitely higher than those in conventional universities according to Kotsiantis et al [8]. They suggest that dropouts are caused by professional, academic, health, family and personal reasons. They go onto to emphasize the importance of tutors continuously supporting students regardless of the distance between them.

- Perceived lack of contact or feedback from teachers

With little or no regular contact between students and teachers, students often find it difficult to self-evaluate their progress. This suggests that a vital communication link is missing in the current approach to distance education. Tinto [9] and Keegan [10] have hypothesized that distance learners who lack this communication often fail to experience the complete academic and social integration of learning, resulting in drop-outs [9]

- Lack of support or services

Students reported a lack of support and services, for example, access to tutors, academic planners or technical assistance. This lack of support often leaves students feeling isolated which has a negative and complex impact on their learning experience and the learning process.

- Alienation and isolation

All students want to feel part of a larger community, but more so distance students as they long for the feeling of 'belongingness'. For example, if you consider the traditional student, the community surrounding their course tends to form an important part of their social lives and the learning experience. The 'distance' element of distance learning removes many of the opportunities for social interaction that would be present in traditional environments. Additionally, there are practical problems such as contacting academic and support staff because perhaps the course is held out of hours, but there are also reported problems of obtaining study materials and borrowing and accessing library resources. All these aspects leave the student feeling alienated and isolated from the 'scholarly community' associated with traditional study. Again, having a negative impact on their confidence and overall learning experience.

- Technological expectations

Distance learning is generally delivered using technological methods which tacitly assumes that the students are technologically literate. The demographics of distance learning suggest that many are first time distance students or are mature students [11] and many of which have little understanding of the process of studying at a distance or an understanding of the technology. With this in mind, it is essential that adequate introductory training is provided for students to ensure they are familiar with the basics of the operating system and how to use the electronic information.

3.2 Faculty Barriers

Similarly, the literature exposes many barriers from a faculty standpoint. Faculty barriers can be categorized as lack of support, technological experience and teacher acceptance of non-traditional approaches.

- Lack of support by the faculty

With the introduction of distance programs comes a change in faculty roles. This can often be difficult for some teachers, as they must change their teaching style to become a mentor, tutor and facilitator rather than the traditional lecturer role. Additionally, the content may need to be adapted to accommodate the needs of the operating system and the distance learner. Unfortunately, as long as educational institutions see such problems as a burden, distance education will not be taken seriously or allowed to successfully develop.

- Technological experience

Like the student technological barrier, teachers may lack the necessary skills to fully participate and successfully deliver distance education to students. The consequence of the teacher being unfamiliar and uncomfortable with the technology will have a negative affect on the quality of the learning experience for the student.

- Teacher acceptance of non-traditional programs

Some teachers still have some problems understanding and respecting the academics of distance education. Teachers that are enthusiastic about non-traditional approaches are better suited to deliver distance programs. Although motivation and enthusiasm are not success factors in themselves, there is evidence to show "how positive attitudes on the part of the faculty regarding distance learning....correlates to positive student learning outcomes" [12, 13, 14].

4 CoP Offers a Solution

Wenger (1998) provides the definition of CoP as being a *"group of people....learn to do it better through interaction"* [5]. CoP offers a theory of learning that begins by assuming that social engagement is a paramount element of the learning process. Thus, it presents a framework for thinking about learning as a process of social interaction and participation. So how does this resolve the barriers connected with distance learning?

The definition provided by Filipczak (1995) of distance learning: *"getting people...into the same electronic space so that they can help one another learn"* [4], is very much in line with the foundations of CoP; people coming together for a shared cause, sharing, interacting, collaborating, trusting and forming relationships. The definition also implies learner centeredness and control which is essential to education in the twenty-first century. The use of a CoP in distance programs can facilitate the coming together of the people and the social learning which transpires within a community will also supplement the core learning within the program.

One of the main advantages of a CoP aside from the potential social learning, is the 'community' itself. The community will help reduce the problems of isolation and alienation by bringing together the students and enabling them to share and discuss their concerns. Thus, a CoP provides the students with a highly interactive space to develop electronic connectedness. This connectedness allows the students to form relationships, interact socially and generally create the 'scholarly community' that has reportedly been lacking in many of the distance programs delivered to date.

The interaction that takes place within a CoP can also have a positive impact on drop-out rates as the students will have a sounding board and through discussions with peers will come to realize that they are not alone in dealing with issues regarding family, finance and employment. The opportunity to share your woes is invaluable to reduce the dropout rate of distance programs. Fontaine and Millen (2004) cite the quality of advice and trust among community members in their community value research study findings [15]. The feeling that this advice and trust brings to the students generates confidence, motivation, ambition and the 'belongingness' needed to complete a distance program.

The facilitation of communication and social interaction will also help to minimize the perceived lack of contact or feedback as the CoP will restore the communication link. The community will make it possible for the student and teacher to maintain communication regardless of the distance between them. The teacher will also be able to use this interaction as an opportunity to provide feedback, guidance and support to the learners. This will create a sense of belonging and inclusion for the students and will enable the teacher to get a better understanding of student needs and expectations.

The students can use the CoP to develop their own support network to replace that missing from the institution. The students can share their experience and knowledge with one another, as a result creating learning opportunities for each other. Through participation and engagement in the community students can seek support from their peers. As Fontaine and Millen (2004) state in their community value research study findings, the CoP offers the benefits of collaboration, problem solving, empowerment and learning and development [15]. These benefits will enhance the learning experience and learning outcomes for the students.

The CoP can be used to develop a knowledge repository where the students and the teacher can deposit ideas and knowledge, which the members can access, re-use and develop. Wasko and Faraj [16] suggest that the CoP can also be useful for students to solve problems. They postulate that although the process of problem solving may be challenging, the practice of working through and participating will help to strengthen learner's skills and understanding. Additionally, regarding the barriers to technology, the community could be used as a training platform to demonstrate the operating system and to discuss the issues faced by learners, without disrupting academic time. This would be beneficial because it will help the learners to "enhance their own learning and self-efficacy" according to Wasko and Faraj [16].

The CoP will also encourage learners to reflect upon their learning experience and personal development. Rohfield and Hiemstra [17] suggest reflection activities as a method for overcoming the challenges of distributed learning, in that such activities help the learner to "refine their thinking and contribute to the development of new insights" [16]. The empowerment that self-reflection generates will also encourage the learners to help one another and further enhance the learning outcomes.

The CoP can also help the faculty to overcome or minimize some of the barriers that the literature reports. In the same way as the student community, the teacher can use a CoP to support one another, share materials and resources, socially interact and collaborate.

Of course, these advantages are based on the premise that students and teachers would be willing to actively participate in the community. CoP are organic entities, which evolve as a result of the passion for the domain by its members. The benefits presented add considerable value to distance programs and can help combat many of the negative perceptions of distance education programs.

5 But, What Is Social Learning?

Simply, learning is the process of moving from not knowing to knowing. The 'social' aspect of the learning relies upon human interaction. Therefore, social learning is constructed from conversations and participation within a CoP. It is about 'how' members learn, rather than the content or 'what' they learn. This is in contrast to the Cartesian vision of learning, which proposes, "I think, therefore I am". Social Learning puts forward the alternative "We participate, therefore we are" [18].

Social Learning Theory assumes that engagement in social practice is the essential process through which we learn and become who we are. Thus, resulting in a broad framework for thinking about learning as a process of social participation. This is by no means a new concept; Bandura derived the concept of social learning as far back as the 1960's [19]. Bandura was the leading proponent of Social Learning Theory and his belief was that learning occurs within a 'social context' and that people learn from one another [20].

CoP facilitate social learning as a result of the active participation and community members can learn from interacting and observing other people. As a student, one is likely to pick up valuable knowledge from discussing the subject with friends as well as they could from sitting through hours of lectures. Participation is the process of being active in the practice of the community and constructing identities in relation to these communities.

6 How Do CoP Support Social Learning?

CoP support social learning through an apprenticeship model. Lave and Wenger [21] coined the term CoP while studying apprenticeships as a learning model. They suggest that people usually think of an apprenticeship as a relationship between a student and a master, but their research tells of a more complex set of social relationships. They used the term CoP to refer to the community that *"acts as a living curriculum for the apprentice"* [21]. The use of the apprenticeship model means that learning in the community is not limited to novices. The practice of a community is dynamic and involves learning on the part of everyone. Members take on simple tasks under the guidance of the community, they then progress to more demanding tasks as confidence, skills and knowledge improve. The CoP supports the apprentice by interacting regularly, sharing tales and experiences. Social learning within a community is 'learning about a subject' as well as 'learning to be a full participant'. The shared passion and the interaction engages the learner and enables them to acquire deep knowledge about a subject and the ability to participate in the community through productive enquiry and social interaction.

7 Summary

The student and faculty barriers although problematic can be overcome by adopting the principles of CoP by bringing the students together to focus on the domain of the community, in this instance the distance program. The community provides students and the teacher with the opportunity to build rapport, relationships and trust. These elements also create a sense of belongingness and eliminate much of the loneliness experienced by distance learners. As well as supporting distance programs, CoP also initiate social learning. The opportunity to learn socially adds to the overall learning experience and supplements the learning that takes place within the program.

References

1. Galusha, J.M.: Barriers to Distance Education. Interpersonal Computing and Technology 7, 7 (1997)
2. Eastmond, D.V.: Alone but Together: Adult Distance Study Through Computer Conferencing, p. 46. Hampton Press, Cresskill (1995)
3. Sloan Consortium Survey of Online Learning (2007), http://www.sloan-c.org/publications/survey/index.asp (cited 14/07/2009)
4. Filipczak, B.: Putting the Learning into Distance Learning. Training Journal 32(01), 111–118 (1995)
5. Wenger, E.: Communities of Practice: Learning, Meaning and Identity. Cambridge University Press, Cambridge (1998)
6. Cantelon, J.E.: The Evolution and Advantage of Distance Education Facilitating Distance Education, p. 5. Jossey-Bass Inc., San Francisco (1995)
7. Knapper, C.: Lifelong Learning and Distance Education. American Journal of Distance Education 2(1), 63–72 (1998)
8. Kotsiantis, S., Pierrakeas, C.J., Pintelas, K.E.: Preventing Student Dropout in Distance Learning Using Machine Learning Technologies. In: Knowledge-Based Intelligent Information and Engineering Systems. Springer, Berlin (2003)
9. Sweet, R.: Student Drop-out in Distance Education: An Application of Tinto's Model. Distance Education Journal 7, 201–213 (1986)
10. Keegan, D.: The Foundations of Distance Education. Croom, London (1986)
11. Galusha, J.M.: Barriers to Distance Education. Interpersonal Computing and Technology 7, 3 (1997)
12. Webster, J., Hackley, P.: Teaching Effectiveness in Technology Mediated Distance Learning. Academy of Management Journal 40, 1282–1309 (1997)
13. Berger, N.S.: Pioneering Experiences in Distance Learning Lessons Learned. Journal of Management Education 6, 23–68 (1999)
14. Brower, H.: On Emulating Classroom Discussion in a Distance Delivered Course: Creating an Online Learning Community. Academy of Management Learning and Education Journal 2, 1–10 (2003)
15. Fontaine, M.A., Millen, D.R.: Understanding the Benefits and Impact of CoP. In: Hildreth, P., Kimble, C. (eds.) Chapter to be in Knowledge Networks: Innovation through Communities of Practice. Idea Group Publishing, USA (2004)

16. Wasko, M.M., Faraj, S.: It Is What One Does: Why People Participate and Help Others in Electronic Communities of Practice. Journal of Strategic Information Systems 9, 155–173 (2000)
17. Rohfeld, R.W., Hiemstra, R.: Moderating Discussions in the Electronic Classroom. In: Computer Mediated Communication And The Online Classroom, vol. 3, pp. 91–104. Hampton Press, Cresskill (1995)
18. Seely Brown, J., Alder, R.P.: Minds on Fire: Open Education, the Long-tail and Learning 2.0 (2008), http://www.educause.edu/EDUCAUSE+Review/EDUCAUSEReviewMagazineVolume43/MindsonFireOpenEducationtheLon/162420 (cited 13/07/2009)
19. Bandra, A., Walters, R.: Social Learning and Personality Development. Holt, Rinehart and Winston (1963)
20. Bandura, A.: Social Learning Theory. General Learning Press (1977)
21. Lave, J., Wenger, E.: Situated Learning: legitimate peripheral participation. Cambridge University Press, New York (1991)

Do Open Source LMSs Support Personalization? A Comparative Evaluation

Tania Kerkiri[1] and Angela-Maria Paleologou[2]

[1] Phd, Dept. of Applied Informatics University of Macedonia, Hellas
[2] Assist.Prof., Clinical Psychology & Psychotherapy, Ioannina University Psychology Sector, Hellas
kerkiri.tania@gmail.com, angel@ioa.forthnet.gr

Abstract. A number of parameters that support the LMSs capabilities towards content personalization are presented and substantiated. These parameters constitute critical criteria for an exhaustive investigation of the personalization capabilities of the most popular open source LMSs. Results are comparatively shown and commented upon, thus highlighting a course of conduct for the implementation of new personalization methodologies for these LMSs, aligned at their existing infrastructure, to maintain support of the numerous educational institutions entrusting major part of their curricula to them. Meanwhile, new capabilities arise as drawn from a more efficient description of the existing resources –especially when organized into widely available repositories– that lead to qualitatively advanced learner-oriented courses which would ideally meet the challenge of combining personification of demand and personalization of thematic content at once.

Keywords: open source LMSs evaluation, personalization, adaptation.

1 Are Open Source LMSs Properly Designed towards Personalization?

The plethora of current Learning Management Systems (LMSs) and Learning Resources[1] (LRs *per se)* constitutes the very evidence that establishing life long learning is the necessary step in improving the trainees' skills and ease, their trainers' appeal and straightforwardness, and both parties' professional competitiveness in synchronous working environments. Thus the LMSs are rapidly evolving to comprehensively encompass a number of features and capabilities in order to further fulfill these needs. For example, they provide a variety of LR types (docs, html, ppt, etc); they offer the means to change the learner's interface; they supply with a number of helpful utilities such as calendars, to-do lists, forums, etc. In most instances, such features are the direct product of observations-on-demand, and it may be that their being scattered – to say the least in terms of the form they are usually presented – may be due to this *ad hoc* nature of their production and distribution. Despite the existing functionalities, it

[1] **LR** is every digital entity, usually called "learning resource" that can be used for learning, education, or training.

M.D. Lytras et al. (Eds.): WSKS 2009, CCIS 49, pp. 57–66, 2009.

appears that e-users in the realm of educational exchanges, i.e., e-tutors and e-Learners, are still keen in search for organized patterns for accessing knowledge. Judging from certain results on monitoring e-Learners', e.g., academic progress, their search is not just attributable to mere technological 'greed': it seems there is still a lot to cover until LMSs noteworthy growth satisfy both, e-learners' and their tutors' demands on novelties in LRs creation and the formers' objective academic advancement. In this respect, it could be noticed that one of the weaknesses of the existing resources might be their lack of connection to pedagogical theories [1], especially constructivism that would integrate electronically proper features into pedagogically correct interoperability. On the other hand, an increasing body of the literature [2] tends to lately acknowledge that e-Learners do differ in skills, aptitudes, preferences, discernment of information and perhaps particular capacities and needs due to special (dis)abilities they may have [3]. In these respects, the qualitative content of an LR does not automatically or necessarily lead to a satisfactory educational result, unless notions from adaptation theory [4] are seriously taken into consideration. This specifically indicates the need to incorporate sophisticated psychological methods in the quest of adequately improving personalization of knowledge and at the same time personification of demand. Thus, any new practices that have to be built in the LMSs and which must also be supported by their storage mechanisms should ideally satisfy such requisites as well. New provisions presuppose that the system offers adaptation/personalization[2] methods, which, despite the fact that their theories have already been widely applied in LMSs, are still to be refined. The latter are by definition web-based adaptive systems focused on learning: generally, an adaptive system –in contrast to static systems– is a physical system capable of self-adjusting in response to changing environments. An adaptive system is to be equipped in such a way that it can modify itself into different system-states in order to navigate, function and succeed while shifting its performance among different users. As seen in [4], these systems should have a user model, and also be able to adjust their environment and content using this model, so as for the same system to behave differently in direct correspondence to different individual e-users' demands. Adaptation has so far been achieved through i) adaptive navigation support, and ii) adaptive presentation of the required educational material. Pertinent parameters are mainly: user's goal or user's task, user's background, and user's preferences. To these, one more category ought to be added, namely, the provision of personalized content, user-specified and interoperable with the aforementioned criteria for fulfillment. To implement these methods the characteristics of the learner and the features of the learning object should be known. In addition, there should be a function that combines them. Taken together, such requirements compel to re-think issues about systems adaptation. Apparently, these methods greatly depend on suitable descriptions that reflect the LRs/learners properties/abilities. Surely, some of the LR descriptions such as format, language, author, etc, may be visible at once. Some others, though, such as semantic density, different approach of the learning process, e.g. holistic-vs-serial, etc. [3, 6, 7], which are

---

[2] Here, **adaptation** refers to available navigational support and interface modification. The term **personalization** refers to the capability of the LMS to provide different content suitable to different individual learners. **Customization** is the simplest form of adaptation, and it refers to the ability of the users to modify the page layout or specify which links in a page should be displayed as well as how the content should be displayed.

implied by the LR content have to be aptly focused on, and aptly extracted. The extra work needed for the proper LR annotation is added to the dearly costly LR creation: Despite flexibility and expanded capabilities submitted by various works authoring multimedia tools, the LRs creation still remains a hard process: knowledge engineers of weighty expertise in informatics should be involved in their creation and many person-working hours are foreseeable. Nonetheless, there are some additional reasons thought to justify such an extra effort: i) *the LR creation becomes economically lucrative when it can provide a highly qualitative learning result* – with this result being well anticipated due to the carefully designated properties pertaining to its pedagogical capabilities; and ii) *its good value for its worth becomes increasingly apparent to a greater extent as accumulative existent repositories contain progressively larger and superior amount of LRs which are sharable and reusable.* This means that the LMSs have to compatibly export their LRs to other LMSs. The significance of this compatibility magnified when the exported procedure sweep along many other descriptions about the LR, concerning its educational result, its outcomes, the learning style theories it supports, etc. When the properties of the LRs are immediately visible: *i) the tutors are endowed greater competence for planning their course:* with the LRs designed on the basis of inherent capability promotion, the tutor is able to select the most appropriate one to the personality/individuality of the e-learner. Knowing of the LRs pedagogical properties s/he may efficiently create a more appealing course; and knowing of the personality profile of e-learners s/he shall certainly serve resolutely as learner-oriented. *ii) numerous add-ons can be built in the LMSs, with the most critical being the* [7, 8] *suggested intelligent algorithms for adaptation/personalization:* as the LRs properties are traceable, intelligent algorithms implemented in the LMSs, aid in the automated creation of personalized courses and the most apposite material selection from the variety of the resources in the repositories.

2 Which Factors Support Personalization?

Consequently, the most popular open source LMSs were here stepwise evaluated to find out which of them support adaptation/personalization features. This study only focused on open source LMSs, since several universities and educational institutes rely on them to support their curricula. Moreover, the consistent annotation[3] of the enormous amount of available resources/courses may potentially feed repositories of sharable resources. Here, the open sources LMSs were examined by the http://www.edutools.info. Outcomes of [5, 9] also guided the effort. Thus, ATutor [10], Claroline [11], Docebo [12], Dokeos [13], Ilias [14], Moodle [15] were assessed. The adopted methodology dictated not to include java-based LMSs, such as dotLRN and sakai as the complexity for their installation is not affordable –mainly because of: a) degrade of the java/tomcat versions needed for, and b) insufficient guidance to their valid installation. According to these posits, a content-personalized LMS should support:

i) learner's description: the description of the learner should as accurately as possible portray most factors affecting his/her ability in perceiving knowledge, e.g. age, interests, background, abilities, learning style. This can be easily achieved using

[3] **Annotation:** the process of creating metadata about the LRs.

metadata to illustrate the properties of his/her personality. Educational standards (e.g. PAPI [16]) used for these properties' description facilitate interoperability among different LMSs. An "open" standard, such as XML, to depict the metadata can be used towards the same direction. **ii) LR description:** similarly, the LR has to be well-annotated and its properties relating to its educational effects have to be visible. An educational standard (e.g. LOM [17]) is preferable for the description of the LR content, in order to foster reuse and share-ability. In both cases (i, ii) the educational standards also provide structured descriptions using a predefined and commonly accepted vocabulary. **iii) availability/access to a variety of alternatively selectable resources:** a variety of LRs facilitates the selection of the most suitable one according to the personality of the individual and the education scope. Furthermore, the use of a variety of LRs and activities, either built internally in the LMS or selected from an external repository, activates different sides of the human brain during the learning process –hence teaching effect are strengthened. As an example, different kinds of context, e.g. explanation, example, movie, exercise, etc should be able to be used alternatively and/or sequentially. **iv) implementation of a personalization algorithm:** A personalization algorithm should provide a different version of the same course attuned to the individual or target group traits. Obviously, an algorithm based on psycho-pedagogic and educational theories [1], as well as on learning style outcomes [3, 6] is the preferable path to follow. The implementation of this ability is mainly based on the existence of a resource repository. Extra factors assessed here for their contribution to an adaptation/personalization capability of an LMS, included: **v) the ability to modify the LMS interface:** as concluded from the HCI [18] and adaptation theory [4] a system has to *i) provide an immediate understanding of its interface and flexibility during the navigation*, and ii) *allow the user to change the interface according to his preferences and conveniences*. Several additional factors attracting the interest and concentration of the learner have also been studied: the colors being selected as potent to calm/induce/inhibit the learner, the auto-explained icons instantly indicating what they can do (e.g. quiz, LR, survey etc) and directly facilitating understanding of their functionality (e.g. delete, select, modify) as well as assisting navigation into the LMS (e.g. back/forward buttons). **vi) a collaborative space:** cooperative activities that enable each e-learner's participation enhance his/her eagerness to work with the LMS. So, the LMSs that offer a collaborative space, wherein participants may express opinions, clarify issues, propose solutions, and speculate from problems faced by and experience gained from other participants, triggers learning and incites knowledge acquisition. LMS that support tutor-to-learners and/or learner-to-learners interaction stimulates the learners and solidifies the learning outcomes. **vii) the LMS ability to sustain statistics:** statistics can be provided explicitly or implicitly (e.g., by the user's navigation). They may concern the time one was engaged with an LR, one's goal in a test, etc. and may well serve as a recommendation tool for LRs. Collaborative recommendation systems may be used to the same end. Typically, a recommender system compares the user's profile to some reference characteristics, and seeks to predict the 'rating' that a user would give to an item not as yet considered. Recommender systems form a specific type of information filtering technique that attempts to present information items likely to be of interest to the user. These characteristics may come from the item (the content-based approach) or the user's social environment (the collaborative filtering approach) [8]. **viii) import/export of its content using a**

standard: this feature makes the LRs sharable and reusable. A key facet of this criterion is the export in a commonly accepted standard. **ix) the LMSs capability to install new modules** which provide new functionalities: a more-or-less self-evident requirement, inasmuch as heightened educational standards command for inclusion of such LMSs characteristics.

More specifically: For criteria (**i**) and (**ii**) the capability of the LMS to create metadata about the learners/resources/courses that describe their (educational) properties was thoroughly examined. The investigation concentrated on the capability to create metadata about: a) the learners, including: learning style, user role, previous background, level of experience etc and b) LRs, including: semantic density, interoperability, difficulty, learning outcomes etc [3, 6]. These are factors that have been used to personalize courses [7, 8]. Of equal interest was to examine if the LMSs have adopted learning standards to describe the properties of LRs and learners. Likewise, existence of suitable storage-structures in the repositories of the LMSs to amass these properties was an issue that concerned the study. When these structures are not available attention was shifted to the easiness to create a super-structure built above the current ones that may sustain information. For criterion **(ii)** the availability of different properties for the same LRs was checked. In this category, notice was taken of the different formats that may be supported, such as html, doc, pdf, of any different kinds of questionnaires (quizzes, multiple choice, selection etc), and of the capability to use already made LRs, created by different tools such as power point, flash, hot-potatoes etc. For criterion (**iii**) point of reference was the support of repositories and different management capabilities on them. Repository management, containing course-independent resources, is the main tool to personalize content. To satisfy this criterion the capabilities were tested for features as course tree-structure, directories that categorize resources, etc. For (**iv**), the existence of a personalization algorithm that may introduce LRs to learners was the core theme. This, in its simplest form, may well be connoting a test of capacity to create different learning paths or alternative versions of the same e-course available. For criterion (**v**) weight was put to the adaptability of the interface and the ease in changing it: theme-changing; tool-bar existence to organize several capabilities and to facilitate navigation; existence of several student tools, e.g. calendar, various tools for navigation support, e.g. site map, etc; were all under inspection. For this criterion metrics were used as derived form HCI-theory [18], such as flexibility, obvious/consistent interfaces etc. For criterion (**vi**) checking was done in the availability of tools and usual collaborative activities such as: chat, forums, on-line meetings, massive e-mail, posts on issues of common interest, LRs annotation, announcements, to-do lists, assessments, assignments, etc. In these, the assessments of the tutor to the learners, news and announcements, communication via e-mail, etc were also included. For criterion (**vii**) the ability of the LMS to keep statistics was tried, as for keeping the time spent per LR, results from tests, entrance/exit time etc. Whether polls and surveys were involved in this dimension was also one of the study's interests for this criterion. For criterion (**viii**) the import/export capability of the system was inspected along with its potential to support different learning standard formats. The usefulness of this capability is that such an export could extract all the information for the LR including its properties, so that they are "known" to other LMSs, as well. Finally, for criterion (**ix**) the capability of the LMS to install new modules was orderly scrutinized. The new modules change the interface of the LMS and furthermore they modify its functionality. Here, a noteworthy parameter is also the availability of new modules. Lastly, any innovative idea included in the LMSs, as support of distributed virtual classes, different time zones, course scheduling, on-line payment provisions, etc. was checked.

Table 1 presents the evaluation results.

Table 1. The results of the evaluation

		Criteria	ATutor 1.6.2	Moodle 1.9.4	Docebo 3.6.0.3	Claroline 1.8.11	Ilias	Dokeos 1.8.5
Easy installation			5	5	3	3	2	5
i	Learner	Metadata	-	-	-	-	+	-
	Learner	User roles	+	+	+	+	+	+
	Learner	Data structure supporting properties	-	-	-	+	+	+
ii	LR	Autonomous resource-structure	-	-	+	-	-	-
	LR	Metadata	-	-	-	-	+ (LOM standard)	- (some info to tutor)
	LR	Structure to support properties	-	+	Rules for assessment	-	+ (xml compliant)	-
	Course	On Course Authentication	+	-	+	+	+	-
	Course	On Course Management	-	-	+	+	+	+
	Course	General		Outcomes		Properties		
	Resources	html/docs	+	+/+	+/+	+/+	+	+
	Resources	assignments	+	+	+	+	+	+
	Resources	tests	+	+	+	+	+	+
	Resources	questionnaires	+	+	+	+	+	+
	Repository management	LR repository	+	-	+ (even from web)	+	+	+
iii	Repository management	Reuse of LRs	+	+	+	+	+	+
iv	Repository management	Repository search	Keywords / forum/ post/web	List of LRs in a dropdown box	-	Search for users / lessons	LRs, Files, Forums, Tests, media-pools etc	Users/ lessons
	Personalization	Matching algorithm	-	-	-	-	-	-
	Personalization	Course Versioning	-	-	+	-	-	-
	Personalization	Learning path	(Sub)Pages	-	Course Structure, pre-assessment	+	Learning path/Chapters/Preconditions	+ (previous/next)
	Adaptability	Student tools	Parameterized side bars	+ (Available on both side bars)	Certificates, user-portfolio	-	+	- (for administrators only)
	Adaptability	Theme	+	+	+	-	+	+
v	Adaptability	Language support	+	+	+	+	+	+
	Adaptability	Other utilities	Enhanced user administration	-	Feed Reader, Automade reports	-	Enhanced user administration, private notes	Organization of the resources/ Directory support

Table 1. *(continued)*

	HCI criteria	Self-defined icons	4	4	3	4	3	4
		Uniform navigation	5	4	4	4	4	5
		Massive actions	5	2	4	2	4	4
		Easy to understand interface	3	5	3	4	3	4
		Navigation support	Sitemap/ Tree Structure	-	-	Directories	TOC	Tree structure
vi	Collabora-tive Tools	Chat/Forum/Wiki	+/+/+	+	+/+/+	+	+/+/+	+/+/+
		Content Posts	+	-	-	-	-	+
		E-mail	+	+	+	+	+	+
		Announcements	+	+	+	+	+	+
		To-Do list	Reading list	Calendar, Agenda, Assignment	Agenda	Calendar, Agenda, Assessment	Calendar	Agenda, Assessment
		Workgroup support	+	-	+	+(classes of users)	+	+
		Video-conference	-	-	+	-	-	-
vii	Recom-mendation Tools	Reputation	-	-	-	-	-	-
		Statistics	+	+	+	+	+	+
		Polls	+	+	+	-	-	-
		Surveys	+	+	+	+	+	+
viii	Import Export	SCORM	+	+	+	+	+	+
		IMS/ AICC	+	+	-	-	-	+
		XML	-	-	+	-	+	+
		csv	+	-	-	-	-	+
		HTML	+	+	+	-	+	+
ix	New modules, RSS feed		+/+	+/+	-/+	-/-	-/+	-/+

3 Results and Discussion

As seen in Table 1, the methodological steps taken here dictated examination of existence-vs-absence of capabilities in several cases, wherein the +/– symbol was accordingly used. In all other cases, where degree of, e.g., ease, was sleeked for, a Likert-type scale from 0 (the minimum) to 5 (the maximum) was used. The main evaluation standard to measure the capability of an LMS in personalization was the extent to which LR was an autonomous entity; that is, if the LR could be enhanced with metadata which described its properties. Consequently, metadata can be used either into personalization algorithms or manually from tutors to create courses aligned to the learners' characteristics, typical and idiosyncratic [1, 3]. Based on this condition repositories can be created from relevant resources in such a manner that they can be sharable by simultaneously encompassing their psycho-pedagogical properties. Hence, just in this light assessment can result in any meaningful prioritization of the strongest LMSs asset –that is, if personalization is to be taken seriously into planning considerations. Notably, some of the LMSs may be taken as CMSs[4], with a clear example being ATutor, and with Docebo well toggling between the two options. This decision chiefly stems from their interface, and from the fact that their educational

[4] **Content Management Systems (CMSs):** computer applications used to create, edit, manage, search and publish various kinds of digital media and electronic text.

orientation is rather unclear –an impression, mainly formed due to the absence of several capabilities in them, otherwise essential to an LMS, such as the capability to create working-teams to carry out an assignment, devise an experiment, process assessments, etc. To epitomize results of this evaluation, certain focal points should be kept in mind: Moodle is a champion in customization. It offers a great variety of different kinds of LRs and much help for using its platform, as it is lesson-oriented. Notwithstanding its reputation, though, it provides no more personalization facilities. Worst, it uses a dedicated storage structure for each one of its LRs. The same holds for Claroline, with its significantly less capabilities in customization. Dokeos is a pure LMS, it supports learning paths, pre-requisite knowledge, outcomes, many tools for managing students and courses, and is also capable of creating resources as autonomous entities, thus of guiding towards creation of course-independent resources. Ilias design was initiated from personalization theories. Although its installation is a bit hard (missing tables, many error messages) and has not enough error trapping during usage, it grants an impressive set of features: it provides an amazing description for the LRs, using metadata adopted from LOM [17]. It uses LR repositories and provides rich capabilities for search in it. It treats the courses as an autonomous unit to which LRs of several types can be attached. During the course it also uses a hierarchy structure (learning scenario) and organizes resources in prerequisites, chapters per course etc., while exporting its data in XML-format. Still, Ilias and Claroline support only different time zones. ATutor is also a personalization-oriented LMS, but it lacks interface and flexibility. Icons on its left somehow function as a learning path, but its general facade is the one of a CMS. ATutor, like Moodle, uses the course like a holistic unit, thus creating it and providing the environment and proper tools for its manipulation. In its strong points is the creation of repositories of resources. ATutor, like most CMSs, has seminal capabilities in installing new modules and offers a number of additional ones, as, indicatively, the installation for synchronous learning tools (e.g. eluminate!) and learning repositories (e.g. Equtella). Lastly, Docebo, an unconventional LMS of several novelties, albeit being of fewer hubs for academic concerns, it creatively supports the fine distinction of courses in: e-Learning, Blended, and Classroom processes, plus the notion of distributed virtual classes, thus appearing fairly promising for complementing new trends.

4 Conclusions

This paper claims originality in that it sought a systematic inspection of open source LMSs' key-aspects in a comparison to highlight areas of importance, possible weaknesses and improvement orientations on lifelong e-learning highly user-personalised and context-personified experiences. It attempted to evaluate the most popular LMSs using criteria on any of their advanced features with a view to illustrating any of their personalization/adaptation capabilities. Prominent characteristic towards personalization surfaced to be the proper description of the LRs and the learners, provided it is based on palpable and consistent exemplification built on psycho-educational standards of properties in both LRs and learners. Bearing this capacity a number of capabilities may be contained in the LMS, e.g. repositories of reusable and sharable LRs, auto-filling course-hierarchies supporting learning paths, manual or automated personalization algorithms [7, 8]. Also, the LMSs storage structure was explored to trace their actual ability to support these outcomes. Ultimately, this design pursued: a) the most efficient usage of the massive LMSs-relevant resources existing, b) better identification of the learner's personality for implementing intelligent personalization algorithms and, finally, c) means for tutor guidance to create e-courses offering superior learning experiences to e-users. The results prompt for advances in the architecture of the open source LMSs and incorporation of sophisticated capabilities towards personalization. This system could be a tool in the knowledge society. More specifically, certain visible and feasible applications would readily include: a) creation of a structured course that would inherently use this advantageous system as a tool to become even mainstream practice highly appealing to learners; b) independent or subsidiary use of it to facilitate knowledge accessibility by tutors themselves towards in-depth usage and exploitation of sources for their students and trainees; as well as c) a supplementary and/or additional utility to support innovative activities of tele-mentoring and tele-counselling in offering services on self-awareness-, self-control-, commitment-, stress-management-, and, generally, self-regulation- and self- and time-management- boosting sessions – all being potential applications that would serve the student community in an immediate and straightforward manner. To sum up, in an ideal LMS, one would surely choose the highly personalization-oriented Ilias infrastructure due to its advanced features and repository storage-structure, by structurally combining it with the Moodle's customization utilities and interoperability qualities, which, in turn, one would merge with the Docebo's pioneering attributes in order to address pressing demands (courses' allocation, classes' distribution, online payment of fees, etc), which are continuously presented and challenge the contemporary educational system –all, en route for achieving both, personification of quest and personalization of material in chorus.

References

1. Dalgarno, B.: Constructivist computer assisted learning: Theory and techniques, `http://www.ascilite.org.au/conferences/adelaide96/papers/21` (L.V.14/09/08)
2. Lytras, M., Carroll, J., Damiani, E., Tennyson, R.: Emerging Technologies and Information Systems for the Knowledge Society. In: 1st WSKS 2008, Greece, September 24-26 (2008)
3. Jonassen, D., Grabowski, H., Barbara, L.: Handbook of Individual Difference, Learning, and Instruction. Lawrence Erlbaum Associates, Publishers, Hillsdale (1993)
4. Brusilovsky, P.: Methods and techniques of adaptive hypermedia. User Modeling and User-Adapted Interaction 6(2-3), 87–129 (1996)
5. Hauger, D., Köck, M.: State of the Art of Adaptivity in E-Learning Platforms. In: 15th Workshop on Adaptivity and User Modeling in Interactive Systems (2007)
6. Felder, R., Silverman, L.: Learning and teaching styles in engineering education. Engineering Education 78(7), 674–681 (1988)
7. Kerkiri, T., Konetas, D., Paleologou, A., Mavridis, I.: Semantic Web Technologies Anchored in Learning Styles as Catalysts towards Personalizing the Learning Process, IJLIC (in press)

8. Kerkiri, T., Manitsaris, A., Mavridis, I.: How e-Learning systems may benefit from ontologies and recommendation methods to efficiently personalize resources. In: IJKL, vol. 5-2 (2009) ISSN:1741-1009
9. Graf, S., List, B.: An Evaluation of Open Source E-Learning Platforms Stressing Adaptation Issues. In: 5th IEEE Intern. Conf. Advanced Learning Technologies, pp. 163–165 (2005)
10. ATutor, `http://www.atutor.ca` (L.V.: 01/01/09)
11. Claroline, `http://www.claroline.net/` (L.V.: 01/01/09)
12. Docebo, `http://www.docebolms.org/doceboCms/` (L.V.: 01/01/09)
13. Dokeos, `http://www.dokeos.com` (L.V. 01/01/2009)
14. ILIAS, `http://www.ilias.de` (L.V.: 01/01/09)
15. Moodle, `http://www.moodle.org` (L.V.: 01/01/2009)
16. IEEE P1484.2.1/D8,2001-11-25, `http://ltsc.ieee.org/wg12/files/LOM_1484_12_1_v1`
17. LOM, IEEE Standard for Learning Object Metadata. IEEE, Los Alamitos (2002)
18. Dix, A., Finlay, J., Abowd, G.D., Beale, R.: Human-computer interaction, 3rd edn. Pearson Education, London (2004)

The Impact of Organizational Learning on Innovativeness in Spanish Companies

Jesús David Sánchez de Pablo González del Campo[1] and Miha Škerlavaj[2]

[1] University of Castilla-La Mancha, Avda. Camilo José Cela 10, 13071 Ciudad Real (Spain)
jesusdavid.sanchez@uclm.es
[2] University of Ljubljana (Slovenia)

Abstract. Innovativeness is a key factor regarding the survival and progress of a company in modern business environments. The question is how to facilitate innovativeness in organizations. This article studies the impact of the organizational learning process on innovativeness. We understand innovativeness as a combination of (1) innovative culture and (2) technological and process innovation. Organizational learning is a consecutive process of (1) information acquisition, (2) information distribution, (3) information interpretation, and (4) behavioral and cognitive changes. New knowledge obtained through organizational learning improves innovativeness. As a methodological framework, we use the partial least square (PLS) approach to structural equation modeling on data from 107 Spanish companies. The results show that organizational learning has a strong positive direct impact on process, product, and service innovations. In addition, the impact of organizational learning on innovation is also indirect, via innovative culture.

Keywords: Organizational learning; Innovativeness; Behavioral and Cognitive Changes; Spanish Companies.

1 Introduction

Organizational learning is one of the most important sources of a sustainable competitive advantage that companies have [1]. Hence, it is crucial to manage organizational learning processes within these organizations in order to successfully compete [2].

An interesting consequence of organizational learning is the generated knowledge which can facilitate the development of innovations [3]. An innovation can be a new product or service, a new production technology, a new operation procedure, or an enterprise's new management strategy [4], [5]. Innovations have always been essential for an organization's long-term survival and growth. Innovations currently play an even more crucial role in the future of a company following the rapid pace of market evolution [6]. Moreover, a truly innovative firm must be embedded in a strong culture that stimulates engagement in innovative behavior.

There is a vast body of literature that has studied the relation between organizational learning and innovation. The main conclusion is that organizational learning enhances the innovative capacity of an organization and that firms can only innovate

M.D. Lytras et al. (Eds.): WSKS 2009, CCIS 49, pp. 67–76, 2009.

if they develop effective knowledge of their resources, competencies, and capabilities [3], [7], [8], [9], [10]. So, the knowledge is identified as a crucial factor in innovation management.

In this contribution, we aim to introduce the influence of the organizational learning process both directly and indirectly on innovation. We consider that innovation needs an appropriate culture (values) that creates a facilitating environment for innovations to occur. For this reason, we have examined the direct effect of organizational learning on innovative culture and its total effect (direct and indirect) on innovation.

This paper is organized as follows. We begin by analyzing the concepts of organizational learning and innovativeness. Following this, we present our research hypotheses and the methodological framework. This is followed by a presentation of the results of the partial least square modeling. We then discuss the results and their implications, and suggest future research opportunities.

2 Theoretical Framework

Organizational learning is a process that concerns the transformation of information into knowledge and knowledge into action [8], [11], [12], [13], and the reflection of this in accompanying behavioral and cognitive changes [11].

Despite and perhaps because of its importance, organizational learning has numerous definitions and there are many perspectives in the field. One could call the tradition that understands organizational learning as the process of information acquisition, information distribution, information interpretation, and behavioral and cognitive changes, Huber's tradition of information processing upgraded with an action perspective. This tradition is based on information processing and behavioral and cognitive changes as operationalized in, e.g., the OLIMP (the Organizational Learning and Information Management Processes) questionnaire [2], [14], [15].

A key characteristic of learning is that should be continuous, that is, a permanent phenomenon. This requires the creation of a corporate culture that stimulates the process of continuous learning.

The current circumstances (e.g. uncertainty, high risk and volatility) entail that firms need to develop innovations in order to maintain or increase their competitiveness. Innovativeness or the capacity to innovate is among the most important factors that impact business performance [16]. Innovation is a process of turning opportunities into practical use [17] and is such only when it is really adopted in practice [18]. The degree of innovation reflects the extent that new knowledge is embedded in an innovation [19].

The conversion of technical ideas into new business, products, or services can be based on an understanding of the synergies and interactions between the different knowledge possessed by the firm, its technologies, its organizational learning process, and its internal organization [20].

In the literature there are several typologies of innovations. The traditional classification distinguishes between product and process innovation. Innovation does not refer only to a new product or service. Moreover it is possible to modify the way the enterprise obtains the final product (process innovation). Daft [21] analyzes the "extent" of innovation and develops a "dual core" model: (1) administrative innovations

– these occur in the administrative process and affect the social system of an organization; and (2) technical innovations – these pertain to products, services, and the organization's production process or service operations [4]. We consider that a technical innovation needs prior innovation with regard to any business process different from the production, which is considered to be necessary support for the technical innovation that takes place in products, services, or the processes of producing them [4]. In this way, we consider that product innovations are similar to technical ones and process innovations are similar to administrative ones. We try to capture the firms' innovation considering the full set of organizational operations [21]: products (technical innovations) and processes (administrative innovations) innovation.

On the other hand, if firms want to develop effective innovation, it is necessary to improve the innovative culture so that all enterprise employees continuously analyze customer needs and the market evolution in order to introduce novelties and become cutting-edge enterprises. Simpson et al. [22] point out that firms need the ability to innovate continuously, and for this they must have an appropriate set of organization-wide shared beliefs and understanding. Then, firms must modify their culture in order to improve their innovative culture. Due to the importance of cultural factors, we consider it relevant to analyze the influence of having an innovative culture on the relation between the organizational learning process and innovation. This is an important contribution of our paper.

3 Research Hypotheses and Model

3.1 Organizational Learning Process

There are several ways to operationalize the organizational learning concept. For this study, we follow Huber's tradition of information processing [13] and upgrade it with the contribution of Dimovski [14], which claim that learning needs to reflect in behavioral and cognitive changes. In addition, the purpose of this contribution is to add to the generalizability of these research findings by developing, operationalizing, and testing a partial least square model of organizational learning.

Hence, organizations stressing organizational learning processes must first acquire information, distribute it to their stakeholders, interpret it to fully understand its meaning, and transform it into knowledge. At the same time, they must not forget the most important part – to implement behavioral and cognitive changes – in order to convert words into action. We hypothesize that the information processing part of organizational learning will a have strong and positive impact on behavioral and cognitive changes, and will involve four distinct constructs (all forming the organizational learning process).

The first stage in the organizational learning process is information acquisition, which is a firm's ability to identify relevant external information from the total amount of information that surrounds the firm [10]. In order to develop efficient organizational learning, it is necessary to ascribe great importance to the acquisition of information and its subsequent distribution. In the next phase this information needs to be transformed into meaning through the information interpretation phase. For learning to happen, information needs to be acquired, understood, and above all transformed into action [13].

If we want to evaluate the success of an organizational learning process, we should analyze the behavioral and cognitive changes of the company. If these changes have not occurred, we might think that the organizational learning is not fully developed or absent because of the lack of changes. All four phases of the organizational learning process need to be assigned a high level of importance in order to improve the efficiency of the process. On this basis, we propose the first three hypotheses:

H_1: Ascribing greater importance to the acquisition of information leads to the better distribution of information.

H_2: Ascribing greater importance to the distribution of information leads to the better interpretation of information.

H_3: Assigning greater importance to interpreting information leads to more action in terms of behavioral and cognitive changes.

3.2 The Effect of the Organizational Learning Process on Innovation

The impact of organizational learning on innovation has been an important factor in the empirical and theoretical research of recent decades [7], [8], [9]. Stata [7] regards innovation as a result of individual and organizational learning. Snyder and Cummings [23] consider organizational learning to be transformed into an improvement in performance through the generation of innovations. So, firms need continuous innovation, which requires generative and adaptive learning. In addition, a high degree of effective learning capability is required in order for innovation capability to come to the fore in firms [24].

When a firm contains heterogeneous resources, capabilities, competences, and skills, the learning process more quickly and easily delivers new opportunities (internal and external). In this way, learning can contribute to the creativity and innovation of the firm [25] and to fostering an innovative culture through the developed knowledge.

There is an important uncertainty regarding the innovation process. This is the reason a firm must possess an important variety of knowledge in order to evaluate more alternatives in order to develop new products, processes, or services [25]. Due to the importance of knowledge, firms should seek ways to improve the research and development of knowledge, to manage it efficiently and to utilize it effectively [5]. So, the basis of innovation is organizational learning, because this is the way to increase the firm's knowledge. The more knowledge is shared between the employees in a firm, the greater the innovation capacity will be.

Moreover, detecting, confronting, learning from, and overcoming errors and setbacks regarding innovation will generate a significant tradition of learning in the firm, creating professionals in the management of innovation [26]. However, considerable work is required to create a context that encourages and motivates innovation, a context that legitimates innovative behavior, dedicates resources to innovation, and assumes a structure and culture that nurture and nourish the development and implementation of innovation [27]. This aim can be reached by means of rewards for workers who acquire new ideas and methods and share accumulated experience and skills with other employees. So, it is crucial to promote an innovative culture.

Changing the actions and cognitive maps of members of an organization should lead to the understanding that innovative proposals are welcome in organizations, that people are encouraged to experiment in order to be creative, and a higher level of managerial support for innovative ideas and creative processes. In turn, a culture that values creativity, experimentation, and innovation should result in more technical as well as administrative innovations. Thus, we propose the following hypothesis:

H_4: Behavioral and cognitive changes lead to an improved innovative culture.

H_5: Behavioral and cognitive changes lead to improved product and process innovation.

H_6: An improved innovative culture will have a positive impact on product and process innovation.

4 Method

The questionnaire used (OLIMP) has been undergoing constant development and validation for more than 15 years [2], [14], [15], [28]. The measurement instrument used in this study has 46 items for the Organizational Learning Process construct (13 items for Information Acquisition – InfAcq-, 6 items for Information Distribution – InfDis-, 11 items for Information Interpretation –InfInt-, and 14 for Behavioral and Cognitive Changes –BCC-), 13 for Innovation –Innov- (9 items for product and service innovation –PSI-, 4 items for process innovation –ProcI-) and 4 items for Innovative Culture –IC-. Pre-testing procedures were conducted in the form of interviews and pilot studies with managers and focus groups with academic colleagues.

Several researchers agree that organizational learning is difficult to measure due to the complexity and dynamics of situations and problems arising due to the difficulty and costs of data collection [29]. Each item used was measured on a 1 to 10 point Likert scale. When measuring InfAcq,InfDist, Innov and IC we asked the respondents about their degree of dis/agreement with a certain statement. When measuring InfInt we measured perceived importance and, with BCC items, the respondents were asked about changes in the last three years in terms of their increase or decrease.

With respect data collection, during March and April 2008 questionnaires were distributed to Spanish companies with more than 50 employees. We decided to send out electronic questionnaires to 1,000 Spanish companies and therefore to use a systematic sampling procedure. The questionnaire was addressed to the CEOs or chairpersons of the companies, who were instructed to fill out the questionnaire themselves or forward it to a competent person within their organization. We obtained 107 full responses.

More than two-thirds of the selected Spanish companies had between 50-250 employees. The main industry in our sample is manufacturing at 28%, followed by construction at 10.3%. The questionnaire was completed mostly by people from the middle-management level (directors of functional departments), and top management members were represented more than lower management.

5 Results

Having gathered the data, the next step in the process relates to the development of a valid and reliable measurement model. We consider that content validity was done via pre-testing procedures by the inclusion of several academics knowledgeable in the field. Moreover, construct validity was reached in our sample because it have convergent and discriminant validity. In order to evaluate convergent validity, we used AVE, which is a measure of the average variance shared between a construct and its measures. Bagozzi and Yi [30] and Hair et al. [31] considered that the cut-off value most often used for this tool is 0.50. In our case, the AVE values are above 0.65. Furthermore, the AVEs of the latent variables are greater that the square or the correlations among the latent variables. So, as Fornell and Larcker [32] suggested our model has discriminant validity. We have evaluated the reliability of the measures of the constructs through standardized loading. On the other hand, construct reliability was analyzed by means of composite reliability. All tools obtained higher scores than recommended by Hair et al. [31] and Forner and Larcker [32]. To summarize, in assessing the measurement model it was established that the measures and constructs used are valid and reliable enough to continue with further analysis.

In order to analyze the quality of the structural equations we use cv-communality and cv-redundancy indices. We can state that our model shows predictive validity and presents a good fit because all latent variables have positive cv-redundancy and communality indices.

Figure 1 represents models of the impact of the organizational learning process through behavioral and cognitive changes on innovative culture as well as on innovation. In this figure, standardized solutions of the path coefficients (with the corresponding t-values in brackets), standardized loadings of the measurement variables, and structural multiple correlation coefficients are presented, which will allow us to test the hypotheses. The results of the analysis offer support for all hypotheses because all relations are significant.

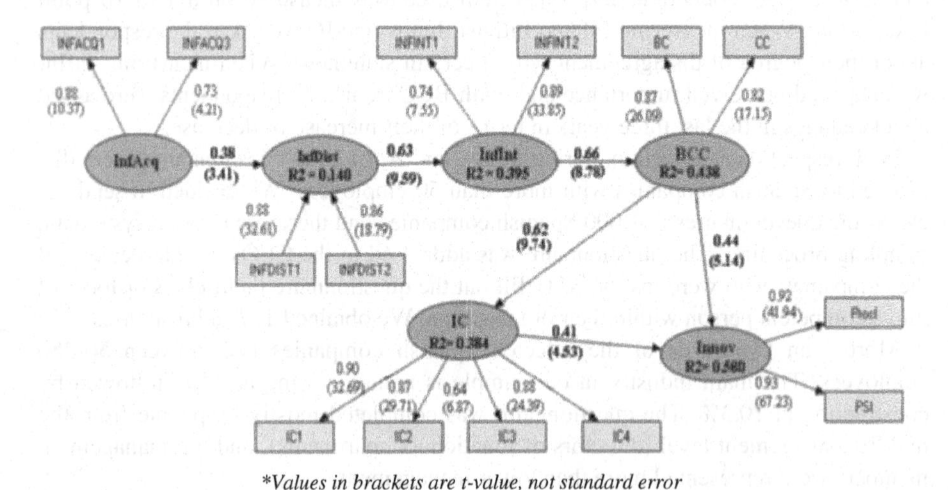

**Values in brackets are t-value, not standard error*

Fig. 1. Research model (standardized solution)

H_1 assumes a statistically significant impact of InfAcq on InfDist, which was also the case in our analysis. So, InfAcq demonstrated a statistically significant (t = 3.414), positive, and strong impact (standardized value = 0.375) on InfDist. So, we obtain empirical support for H_1. Furthermore, valuing the distribution of different information types leads to a better understanding and interpretation of the distributed information. The effect is very strong (standardized value = 0.628), statistically significant (t=9.587), and positive. Thus, H_2 is supported. Placing a high level of importance on various channels of information interpretation leads to greater action in terms of behavioral and cognitive changes (standardized value =0.662, t=8.783), meaning that more learning has actually occurred. So, H_3 is supported, as well.

The BCC construct has a statistically significant, positive, and strong effect on IC and Innov. This effect is a bit stronger directly on Innov (0.69, t= 5.140), while it is also strong on IC (0.620, t= 9.739). However, the direct effect of BCC is stronger on IC than Innov, but the indirect effect in the second relation does increase this relation. Moreover, IC has a positive, strong effect and has a statistically significant effect on Innov (0.411; t=4.543). So, there is empirical support for H_4, H_5, and H_6.

Our model explains variance in observed endogenous constructs very well because all coefficients of determination (R^2) are significant (R^2 appears in Figure 1).

6 Discussion and Implications

Organizational learning leads to innovation. All six hypotheses were supported by data from Spanish companies. The results show that better organizational learning in terms of acting upon information acquired, distributed, and understood, leads to a more developed innovative culture and consequently to more innovation. The direct link from behavioral and cognitive changes to innovation is even stronger.

Specifically, the results in relation to H_1 suggest that information acquisition has a strong impact on information distribution within companies. Hence, better management of information acquisition methods and techniques, providing opportunities for information distribution, collecting information from internal and external sources, etc. – all these lead to changes in the way companies operate. For its part, H_2 was empirically supported. Therefore, the more information is exchanged within firms, the more meetings are held in order to inform employees, the more best practices are transferred between various areas o work, and the more employees work in several teams or projects group, etc., the better will be the interpretation of information which was acquired internally or externally. Moreover, there is a significant and indirect effect between information acquisition and information interpretation.

However, in order to complete the organizational learning process, it is necessary that information interpretation leads to changes in the firm. H_3 confirms this relationship. Therefore, the way that information acquired is well distributed and interpreted will effects the way employees perceive and understand the new situation, and will lead to cognitive changes as well as changes in the way organizational members behave. Having information and understanding this information leads to behavioral and cognitive changes in order to adapt to and even to create changes in the business environment, which completes the organizational learning process. Good information interpretation involves generating new knowledge which can improve the ability to recognize entrepreneurial opportunities. Behavioral and cognitive changes entail transforming words into actions and seizing these opportunities, which completes the organizational learning cycle [2]. In this way, the organizational learning process is completed, as mentioned by [8], [12], [14]. Hence, we consider that our work has relevant implications in the current knowledge society. It reveals that new knowledge, generated through organizational learning process, in turn generates innovations. In order to reach the innovativeness goals it is necessary to foster the development of new knowledge which causes changes in firms. Therefore, organizational learning is a crucial facilitator and enabler of innovations than organizations in knowledge society strives for.

After analyzing the organizational learning process, we tested what the link is between the organizational learning process and innovativeness. Our results confirm that behavioral and cognitive changes lead to an improved innovative culture. Therefore, the greater the adaptability to environmental pressures (changes in the quality or number of products/services), efficiency in team meetings, or employees' level of understanding of the company's strategic orientation or of the problems in the company, or the number of novel products and services, the more innovative proposals, new ideas, and rewards will be fostered in the company. Therefore, we have shown that fostering an innovative culture needs an efficient organizational learning process.

Higher-level behavioral and cognitive changes improve innovation results from the perspectives of product/service innovation and process innovation. This link between constructs is very strong and has direct and indirect effects. As mentioned above, these changes foster an innovative culture and this culture causes improvements in innovation. Therefore, we obtained empirical support for H_5 and H_6.

The managerial implications of our research show organizational learning to be a crucial antecedent and tool for augmenting innovativeness within organizations. Investing time, effort, and resources in various forms of information acquisition (both external as well as internal), developing channels for information distribution (personal and electronic), creating opportunities for people to interact and interpret the information at-hand, and, above all, causing organization to change in accordance with the business environment, all lead to more innovation. In addition, this also stimulates the values that are necessary in order to create an innovative culture, and in turn additionally reinforces product, service, and process innovation.

7 Conclusion

The results confirmed the proposed relations in our conceptual model. If we assume that effective organizational innovation is the key to maintaining competitive advantage in a constantly changing environment, companies need to know how to improve their innovativeness. Some recent literature presents organizational learning as the key factor needed in order for innovation to occur [9], [24].

Specifically this paper defines and measures organizational learning as a process with four phases: (1) information acquisition, (2) information distribution, (3) information interpretation, and (4) behavioral and cognitive changes. After confirming the positive and significant relations between organizational learning and innovation, we consider that it is necessary for companies to promote and develop effective organizational learning in order to reach their goals with regard to innovation. The results of our work in Spanish companies are consistent with previous research which analyzed organizational learning process-innovation links [7], [8], [9], [28].

Therefore, our results show that new knowledge obtained through organizational learning process is fundamental to develop innovations. In this way, this work increases the relevance of knowledge in the current competitive environment.

While this study only considered that organizational learning is a facilitator of innovation within organizations [1], there are most certainly others. Future work will need to test other antecedents to organizational innovativeness. It will also need to test the organizational learning-innovation link in other contexts (other countries), and at other levels. Longitudinal observations may add new understanding to the field by observing the time lag between antecedents and innovativeness.

References

1. De Geus, A.P.: Planning as learning. Harvard Business Review 88, 70–74 (1988)
2. Škerlavaj, M., Indihar Štemberger, M., Škrinjar, R., Dimovski, V.: Organizational learning culture—the missing link between business process change and organizational performance. International Journal of Production Economics 106, 346–367 (2007)
3. Caballé, S., Juan, A., Xhafa, F.: Supporting effective monitoring and knowledge building in online collaborative learning systems. In: Lytras, M.D., Carroll, J.M., Damiani, E., Tennyson, R.D. (eds.) WSKS 2008. LNCS (LNAI), vol. 5288, pp. 205-214. Springer, Heidelberg (2008)
4. Damanpour, F.: Organizational innovation: a meta-analysis of the effects of determinants and moderators. Academy of Management Journal 34(3), 555–590 (1991)
5. Liao, S., Fei, W.C., Liu, C.T.: The relationship between knowledge inertia, organizational learning and organizational innovation. Technovation 28, 183–195 (2008)
6. Santos, M.L., Álvarez, L.I.: Innovativeness and organizational innovation in total quality oriented firms: the moderating role of market turbulence. Technovation 27, 514–532 (2007)
7. Stata, R.: Organizational learning: The key to management innovation. Sloan Management Review 30, 63–74 (1989)
8. Argyris, C., Schön, D.A.: Organizational learning: a theory of action perspective. Addison-Wesley, Reading (1978)
9. Akgün, A.E., Keskin, H., Byme, J.C., Aren, S.: Emotional and learning capability and their impact on product innovativeness and firm performance. Technovation 27, 501–513 (2007)
10. Fosfuri, A., Tribó, J.A.: Exploring the antecedents of potential absorptive capacity and its impact on innovation performance. Omega. The International Journal of Management Science 36, 173–187 (2008)
11. Crossan, M.M., Lane, H., White, R.E., Djurfeldt, L.: Organizational learning: dimensions for a theory. The International Journal of Organizational Analysis 3, 337–360 (1995)
12. Fiol, C.M., Lyles, M.A.: Organizational learning. Academy of Management Review 10, 803–813 (1985)
13. Huber, G.P.: Organizational learning: The contributing processes and the literature. Organization Science 2, 88–115 (1991)

14. Dimovski, V.: Organizational learning and competitive advantage. Ph.D. Thesis, Cleveland State University (1994)
15. Dimovski, V., Škerlavaj, M., Kimman, M., Hernaus, T.: Comparative analysis of the organizational learning process in Slovenia, Croatia, and Malaysia. Expert Systems with Applications 34, 3063–3070 (2008)
16. Hurley, R., Hult, G.T.M.: Innovation, market orientation, and organizational learning: an integration and empirical examination. Journal of Marketing 62, 42–54 (1998)
17. Tidd, J., Bessant, J., Pavitt, K.: Managing innovation. Wiley, Chichester (1997)
18. Schumpeter, J.A.: Theory of economic development: an enquiry into profits, capital, interest and the business cycle. Harvard University Press, Cambridge (1934)
19. Dewar, R.D., Dutton, J.E.: The adoption of radical and incremental innovations: an empirical analysis. Management Science 32, 1422–1433 (1986)
20. Guadamillas, F., Donate, M.J., Sánchez de Pablo, J.D.: Knowledge management for corporate entrepreneurship and growth: a case study. Knowledge and Process Management 15, 32–44 (2008)
21. Daft, R.L.: A dual-core model of organizational innovation. Academy of Management Journal 21(2), 193–210 (1978)
22. Simpson, P.M., Siguaw, J.A., Enz, C.A.: Innovation orientation outcomes: The good and the bad. Journal of Business Research 59, 1133–1141 (2006)
23. Snyder, W.M., Cummings, T.C.: Organization learning disorders: a conceptual model and intervention hypotheses. Human Relations 51, 873–895 (1998)
24. Park, Y., Kim, S.: Knowledge management system for fourth generation R&D: knowvation. Technovation 26, 595–602 (2006)
25. Rodan, S., Galunic, C.: More than networks structure: how knowledge heterogeneity influences managerial performance and innovativeness. Strategic Management Journal 25, 541–562 (2004)
26. Drucker, P.E.: Innovation and entrepreneurship: practice and principles. Heinemann, London (1985)
27. Senge, P.M., Roberts, C., Ross, R.B., Smith, B.J., Kleiner, A.: The fifth discipline fieldbook. Doubleday, New York (1994)
28. Dimovski, V., Škerlavaj, M.: Performance effects of organizational learning in a transitional economy. Problems and Perspectives in Management 3, 56–67 (2005)
29. Spector, J.M., Davidsen, P.I.: How can organizational learning be modeled and measured? Evaluation and Program Planning 29, 63–69 (2006)
30. Bagozzi, R.P., Yi, Y.: On the evaluation of structural equation models. Academy of Marketing Science 16(1), 74–94 (1988)
31. Hair, J.F., Anderson, R.E., Tatham, R.L., Black, W.C.: Multivariate data analysis, 5th edn. Prentice-Hall, London (1998)
32. Fornell, C., Larcker, D.F.: Evaluating structural equation models with unobservable variables and measurement error. Journal of Marketing Research 25, 186–192 (1981)

Educatronics, New Technologies and Knowledge Society: A Critical Approach

Olmo A. Moreno-Franco[1], Prudenciano Moreno-Moreno[2], and L.A. Muñoz-Ubando[3]

[1] Universidad Modelo, Ingeniería Mecatrónica
Carretera a Cholul, 200 mts. después del Periférico, Mérida, Yucatán, 97305, México
kingalo@ciesd.net
[2] Universidad Pedagógica Nacional, CCAA: Políticas Públicas y Educación. Área I
Carretera al Ajusco No. 24, Col. Héroes de Padierna, México, D. F., 14200, México
pmoreno@ajusco.upn.mx
[3] The Robotics Institute of Yucatán, http://www.triy.org/ (CITI), Calle 60 Norte # 301
Colonia Revoluciones, Mérida, Yucatán, 97118, México
alberto.munoz@triy.org

Abstract. This paper provides an explanation of the development of NCITs in their impact on the education sector under the name "Educatronics", as an essential step for the world today, but concludes with a critic to the purely "technical" approach to propose that *educatronics* become a paradigm to integrate educational training on a larger scale, to provide a highly educational, but not dehumanized as claimed by the educational competency model, but with "human face".

1 Introduction

The constant change and growth in information technologies make it almost impossible to keep a continuous line of work to cover all technical aspects encompassing the New Communication and Information Technologies (NCITs). In the past, scientific studies in this area were exclusive to large research and businesses institutions that have a rich budget for the experiments, NCITs now have invaded most of the classrooms in public and private schools and are found in the academic grid title as "Automation and Control", "Mechatronics", "Development of Artificial Intelligence", among others. These proposals are attractive to arouse public interest in general, to come to know the development technologies with which they operate, and encourage new generations of young people to explore and develop scientific material in a matter of technology.

However it is necessary to insert this scientific-technological progress as an important issue within a broader socio-anthropological context in order to not repeat the mistakes that led into a modern techno-economic-financial-industrial automated nightmare without human vision.

2 Explanation of the Techno-pedagogical Perspective

One interesting proposal that advocates the linking of educational development and new technology is *Educatronics*.

M.D. Lytras et al. (Eds.): WSKS 2009, CCIS 49, pp. 77–86, 2009.

The term *Educatronics* was developed by Ruiz-Velasco [1], he presented a pedagogical perspective of robotics, which seeks to introduce a new technological and global culture. The result applies in the integration of different areas of knowledge, a structural enriched thinking, logical and formal. The *Educatronics* could be defined as an object of study, focusing on those areas in New Communication and Information Technologies, and these machines and devices that process information are characterized by their immateriality, interactivity, instantaneous, high image quality and sound, automation and networking [1].

The computer systems include the use of NCIT in the classroom, home and office, now is the end-user who programs the computer to generate their own applications and adjust their software to the need required by him. *Educatronics* in principle is that students develop their own scientific activities through guided exploration for knowledge achievement; this concept has been the work of a group of experienced scientists in education who have tried to demonstrate that the use of NCIT in the cognitive process optimizes student covered education under certain parameters.

Papert [2] [3], one of the creators of programming language for education called Logo [4], widely used as a tool to teach programming, takes Jean Piaget [5] concept of children as "builders of their own mental structures." Papert belongs to the group that supports constructionism thesis, which holds that the child creates his knowledge in an active shape and that education should provide tools for activities that promote this activity [2]. The Epistemology and Learning Group at MIT, led by Mitchel Resnick [6], which in turn was pupil of Papert, he was deeply influenced by Piaget's constructivism, extended by the own Papert under the name of constructionism [7].

According to this perspective, instead of instructing the student by providing recipes and techniques (instructions) [8], it is better to enhance learning by creating an environment where students can perform activities of engineers and inventors as a way to access fundamental principles of science and technique, because this is how it develops the own way of scientists thinking, students are really interested in their own work and try to learn to solve problems that are found. Constructivist and instructionist theories plus NCITs open the paradigmatic model of *Educatronics*, by the use of pedagogical robotics as a teaching and learning mechanism of the NCITs that lies primarily in embedded systems, kernels and hearts of innovation technologies, as well as autonomous agents.

3 The Paradigm of Educatronics

For Murphy [9], a paradigm is a philosophy or a set of assumptions and / or techniques that characterize an approach to a class of problems. It is as the same time a way of seeing the world and involves a number of tools for troubleshooting. With this definition is analyzed the *Educatronics* based tools. Four major disciplines are the backbone of the *Educatronics* concept: Electronics, Mechanical, Computer Science and Epistemology, who participated in an education debate on the role of NCITs in education, this education debate on the use of NCIT is conformed by main ideas as the "Learn by Making" part of the New-School (Dewey) [10], discovery learning (Piaget) [11], problem solving, (Beynon) [12], concrete experience (Whitehead) [13], participation of students (Makka cited by Beynon) [14], computational knowledge (Papert) [3] with his construccionist theory that participated in the history of Computer Education and Informatics Training.

Figure 1 shows the synergy of these areas given rise to a series of scientific subsets as applied sciences.

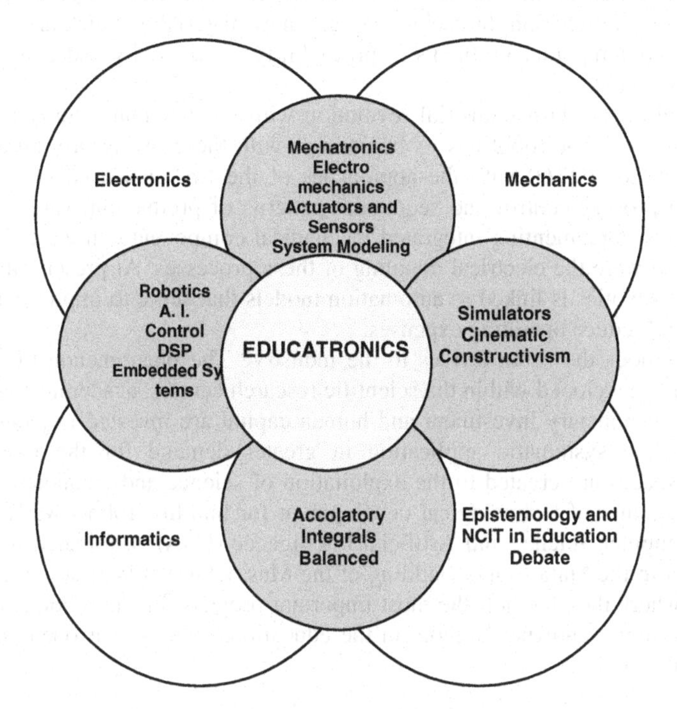

Fig. 1. The *Educatronics* Paradigm

One of the applied sciences with more weight in this paradigm is the Mechatronics. The definition of mechatronics has evolved, to Bishop [14] the concepts presented by Yasakawa [15], Harashama, Tomizuka, and Fukada (cited by Harashama in 1996) [16], Ausländer and Kempf [17] suggest that all the definitions and statements about mechatronics are precise and informative, but each in itself can not capture all of mechatronics totality. This lack of consensus is a healthy sign. It says the field is alive; it is an issue of youth. Even without a doubt a definitive description of mechatronics is to understand the definitions beyond their own personal experiences of the philosophical essence of mechatronics. Exposed the concept of mechatronics is easy to see the tension that is synergistic with other areas of knowledge, and thus become a science applied to education.

The semantic roots of the *Educatronics* conceptual chain defined as follows: the study generates knowledge, application of the study generates more knowledge, and new knowledge focused on education decreases the technology gap and generates levels of academic competitiveness. By emphasizing the concept earlier discussed is analyzed the technological evolution of NCIT for reach the *Educatronics* main idea as the globalization of science in generating NCITs in Pedagogical Robotics as well, the historical profile of the latter.

4 Evolving Technology: Educatronics, a Long Preparation Way

The increase in the incorporation into the production of technological development is included with the concept of the Third Industrial Scientific and Technologic Revolution. This is based on the use of new technologies (computers, telecommunications, programmable automation, biotechnology and new materials), which are broadcast to respond to contemporary problems of profitability, productivity and competitiveness of enterprises [18].

A branch of the third industrial revolution with greater momentum in the last 60 years is robotics. The robot has its beginnings with the growing pro-industry model that dates back to 1760 with the approaches of the first industrial revolution. The automation process control and sequential patterns of production were significantly environments implementing integrated mechanical components, in the era of electric power, those were the electrical coupling of these processes. At present, the industrialization of robotics is linked to automation models that serve to improve the production time and reduce operating expenses.

But how does the robot arrives to the industry? The phenomenon of technology production is developed within the scientific research centers, academic and corporate ones. Strong monetary investment and human capital are invested in technology development and systematic application in greater demand for the industry, thus dedicated spaces are created to the exploitation of science and technology. The first recognized center of technological development for building robots was the Laboratory of Computer Science and Artificial Intelligence (CSAIL), founded in 1959 [19] and located in the Stata Center building of the Massachusetts Institute of Technology complex, where they had left the most important projects for the robotics industry as well as the most renowned authors in the educational theory that opens the road to educational robotics.

5 Pedagogic Robotics

Pedagogic Robotics basic proposal sets the creation and use of teaching prototype tools which can be playful tools, technology or both, where the use of engines, interfaces, and the computer control programs, enhance the integral environment for education. Pedagogic Robotics is an alternative proposal in this regard tends to provide explanations of reality, building bridges between technology and the accelerated environments in which students act as the answers given in education, referring to a humanistic paradigm in which education is a process where there is enjoyment of the senses [1].

For Sanchez [20], the presence of technology in the classroom, aims to provide interdisciplinary learning environments where students acquire skills for structuring research and solve specific problems, develop people with the ability to develop new skills, new concepts and give efficient response to changing the world today. A learning environment with educational robotics, is an experience that helps to develop creativity and thinking of students.

6 The NCIT in Educatronics

With the miniaturization of components and research into new materials, a growing number of innovative technologies could be found in the laboratory, classroom, and market. For the developing branches of *Educatronics* there are three major aspects: hardware, software, and a mixture of both called embedded systems [21].

Technology innovation is the reason for coverage through the medium of teaching robotics, attacking the technology gap in educational institutions because the rejection by the use of these technologies. The peculiarity of these technologies is the composition of the operation, being based on a hardware and a software component, which means that the hardware architecture is described through a programming language, designed after the architecture this can be downloaded to the chip that works as a raw material, and is armed with the design set, giving way to systems and embedded Programmable Logic Devices and introducing students to the standard of current technology at the international level.

In this way the analysis, design and implementation of embedded systems in *educatronics*, becomes the most convincing mechanism to teach new technologies in almost all educational levels. Since the manipulation of the physical medium (real environment), the student understands the processes of interaction between man-machine and machine-world, becoming the embedded system into a tool of assessment and study approach on real-world signals.

7 Stages of Implementation of the Educatronics

Increasingly, more educational institutions that bind into the academic grid subjects on technological development in the field of robotics, artificial intelligence and aerospace engineering at the various educational levels. At the international level, one event that brings together the largest community of developers in robotics is the RoboCup [22], which is an international initiative to promote the participation of talents in the field of AI and robotics, and whose mission objective is to address mechatronic equipment to play football in different categories, depending on the complexity of hardware and software requirements, the interesting thing about this initiative is the variety of ages of the contestants, as there are classes geared for the very young and for the experienced researchers. The U.S. FIRST Robotics Program (For Inspiration and Recognition of Science and Technology) [23], which is a unique sport of the mind designed to help young high school students to discover how interesting and rewarding it can be the life of engineers and researchers. FIRST Robotics Competition brings together teams of youth and their mentors to solve a common problem within a duration period of six weeks using a standard "kit of parts" and a common set of rules. The FIRST initiative was designed by Dean Kamen, Dr. Woodie Flowers, and a committee of engineers and other professionals. FIRST redefines winning for these students because they are rewarded for excellence in design, demonstrated team spirit, professionalism, maturity elegant, and the ability to overcome obstacles. At the University of Massachusetts Lowell (UMASS Lowell) [24], Dr. Fred G. Martin [25] has devoted part of their research models to improve functional-interactional robotics for a wide range of users, their group is known for providing high level scientific knowledge to students and / or professionals who do not necessarily belong the engineering profile. After PhD. graduation from the MIT in 1994 and under the supervision of Papert, Fred G. Martin created the Enganging Computing Group [26], which defines it as the computational processes that are committed to the world, thinking about embedded systems. Computational environments has created assets, and a physical-multimedia design for a unusual set of users, including children, artists and other non-engineers. The personal interest of Dr. Martin is to allow the public to play fast development of prototypes, exploring their ideas to implement them immediately, get feedback to see what they do and performing iterations, and repeat all the process.

In Mexico there are few educational institutions that adopt the area of NCIT and *educatronics* in elementary, middle and high school levels, but it is increasing the incorporation of academic activities around these theme, so in 2008 Dr. Luís Alberto Muñoz Ubando [27] creates The Robotics Institute of Yucatan (TRIY) in southeastern Mexico [28], which is a unique school in its class, far away from teaching a classic LEGO NXT workshop , the TRIY aims to provide a philosophy of design and interaction with robots through targets, aimed at solving local and national issues with the support of technology tools and design methodologies. The initiative of "The Robotics Institute of Yucatan" is to contribute to the formation of skills for the early development of scientists and specialists in the field of robotics. Information Technology, Communication and Control, Mechatronics and Cibernetics join together to undertake the understanding of the machinery and tools of the future. The tutoring of Robotics demand the use of strategies particularly suited to the age and abilities of each student, without forgetting the importance of the teamwork. Among the plans are forming a Museum of Robotics, a Documentation Center and a space where university students of all levels can come and share a passion of study of the multi-disciplinary most exciting engineering and accurate science. In the medium term, TRIY offers the first Master's degree in Robotics in Mexico with the support of the most renowned specialists worldwide. Within the academic activities that take place at TRIY it is noteworthy that students are aged between 11 and 15 years, those who exploit their cognitive abilities in solving problems proposed by them, making use of scientific equipment such as the Handy Board, Arduino and The Tower, among others. In addition there is team of students and advisors who started the project in the RoboCup at the Small Size League and 3D simulation mode, other purpose at TRIY is break into the design and development of Pico-satellite CubeSats.

Another initiative to break into the area of *educatronics* has been proposed by the Government of Yucatán, through the Ministry of Youth [29], who since the mid-2008 and 2009 has given the workshop called Basic Robotics as free and open to the general public wishing to explore into the mechatronics area. The result has been obtained interesting data on the academic track of the participants through a survey applied to the end of each course, 51% of them are aged between 18 and 21 years, 46% of students who have not entered into a university wishes to study Mechatronics, Electronics 31% and 23% Robotics; for the subjects proposed that students would like to take into another course were 32% Programming, 28% Digital Systems, 19% Robotics and lower percentages English, Mathematics and Chemistry. When the students were questioned about to pay for the course 87% is defined that it would be willing to pay if it is possible to provide the academic quality of the course. The above results show the growing interest of the public school by NICT and *educatronics*, to solve this need for knowledge which can be easily implemented at an early age to develop human resources who will be individuals able to solve problems set to work in team and participate in different stages of a research project.

Finally a scenario with successful projects *educatronics* is at the Institute of Progressive Education The Workshop located in Merida City[30], this educational institution is focused on primary level students to become bilingual and provides a series of workshops in various verses of which is the Robotics Workshop, designed for activities that involve knowledge of autonomous agents by category, by the operation of robots, design of structural models, programming of agents across 3D platforms and principles of electronics. This model helps children to meet and reaffirm concerns in matters of technology, since it is common that new generations of students in medium and large cities are familiar with the use of technologies, by this way young students have fun and a complement academic training in the future may help to define its role.

8 The Knowledge Society and NCITs

The premise that education, science, technology and linkage with the productive sector of goods and services in each country will form a national innovation system, the central hypothesis is handled in the ideology of the economy and society of knowledge.

"The approach is not only more widespread, but its main tenets are accepted both by those decision makers in the public sector, like a few actors from the business sector" [31] this is part of the following premises:

1. "It is said that the knowledge society is a paradigm in which the economy identifies factors associated with education, technology and innovation as key elements associated with economic growth and development." (Ibid).
2. It guides the development of economic and public policies toward labor in technical training, research and development, creation of new manage areas structures and organizational work, applying new information technologies, and adapting *educatronics* to the educational scheme based on technical skills - functional - instrumentals.
3. In the 90s, the technological application in the production of the OECD has more than doubled (reaching a 20-25%). Over 50% of GDP in these countries is based on the knowledge economy (Ibid).
4. Believers in the knowledge society paradigm that "... almost nobody refuses that one of the most impetuous of the country, if not most, is economic growth ... everyone sees economic growth as something desirable" [32].
5. Adjacent concepts are indicative of the paradigm of its anti-humanism: to compete successfully, competitive advantages, fierce global struggle, for its overwhelming speed, struggle to attract such investment flows... etc. (Ibid).

With different nuances, this thesis underpinning the knowledge society is shared by Latin American authors of outstanding today: Cimoli [33], Villavicencio [34], Bazdresch Parada [35], Dutrenit [36], Puchet Anyul [37] Fernández Zayas [38], Villa Rivera [39] and González Brambila [40].

9 From Modernity to the Hypermodern

In a work of the sixteenth century, "New Atlantis", Francis Bacon, pioneer of advanced thought of the birth of modernity, made for the first time the utopia of a perfect human society through science and technology. Despite hopes that these techniques led to the dehumanization of hypermodern society, this thesis still stands up today. "Even I would say that today we witness a strong return of that aspiration of the first moderns" [41].

But the reality is that the situation has become uncertain and the saturation of messages according to which the knowledge society (understood as a simple society informatics, telematics, microelectronics, biotechnology, robotics and aerospace, with NTIC and *educatronics*) is the only road to a full education in particular and humankind as a whole, is revealed as an outright error, because is conceived in the human as a consumer and an input to production, being human capital as psycho-physical, intellectual or digital, but as a simple capital, not as a "human being" but as a "know-how". This is the main weakness of the ideology known as hypermodern knowledge society, an extremely simplistic gray view of what is knowledge and society.

Thus the educational model "mainstream" was flooded with the same vision of technical training, ignoring the formation to understand deeper human needs and causing such a vacuum of competence model of education, new technologies (*educatronics*) and standardized evaluation, are met with Ritalin and Prozac, as substitutes for genuine undesirable ethical and transcendental meaning of "knowing how to be."

This paper is inclined to unlink the competence educational model and *educatronics* proposed by the wrong vision of the knowledge society, to become it into a broader one, proposing an educational model based on *educatronics* in one half, but the other half an educational model focused on the development of the "being".

10 Conclusions

The technical aspect and the social-cultural anthropology must be integrated into an educational model that allows to organize, manage and develop an education of quality human, as proposed by the competency model and use of NCITs has only quality technique.

The imposition of the ideological concept of the information society or knowledge society, the educational model for competencies, the NICTs and standard evaluation of school knowledge, have acquired the condition of validity for themselves, leaving the actors of the education system their right to reply, making dark the creative and thinking capacity of citizens and obstructing the exercise their right to a debate that is necessary to ventilate the "oxidation" of vision with clear signs of being blocked solutions to the crisis of identity the actors and educational imbalances, psychosomatic and personality of social subjects, to which the hypermoderns prefer to close their eyes and wait for a miracle to solve such problems without the intervention of the human and education for "learning to be." Only interdiscipline and transdicipline may account for the educational research of the twenty-first century.

References

1. Ruiz-Velasco, E.: Educatrónica: Innovación en el Aprendizaje de las Ciencias y la Tecnología. In: de Santos, D. (ed.), México (2007)
2. Papert, S.: Mindstorms: Children, Computers, and Powerful Ideas. Basic Books/Harper Collins Publishers, New York (1980)
3. Papert, S.: Constructionism: A New Oportunity for Elementary Science Education. A MIT Proposal to the National Science Foundation, Boston, Massachusetts, USA (1986)
4. Resnick, M.: MultiLogo: A Study of Children and Current Programming. Master Thesis. Massachussets Institute of Technology, Boston, USA (1988)
5. Piaget, J., Inhelder, B.: La Psychologie de L'enfant, Paris, France (1966)
6. Resnick, M.: LEGO, Logo, and Life. In: Langton, C. (ed.) Artificial Life. Addison-Wesley, Boston (1989)
7. Camon: Lego Mindstorms Introducción al sistema de robótica Lego Mindstorm dando unas breves pinceladas de historia y programación de estos. Publicado por Comisariado de Nuevas Tecnologías 21 de agosto de (2008), `http://www.tucamon.es/contenido/lego-mindstorms/`
8. Ackermann, E.: Piaget's Constructivism, Papert's Constructionism: What's the difference? MIT Media Laboratory, Boston, Massachusetts (2001), `http://learning.media.mit.edu/content/publications/EA.Piaget%20_%20Papert.pdf`
9. Murphy, R.: Introduction to AI Robotics. Massachusetts Institute of Technology, Boston, Massachusetts, USA (2000)
10. Dewey, J.: Educación y Escuela. En el libro de Moacir Gadotti, Historia de la Ideas Pedagógicas, Ed. Siglo XXI, México (2003)
11. Piaget, J.: Teoría de la Inteligencia, Ed. Amorrortu, Argentina (1990)
12. Beynon, J.: Learning to Real Technology. Under standing Technologies in Education. Falmor Press, UK (1991)
13. Whithead, J.: La Técnica.En el libro de William Barrett, La Ilusión de la Técnica. Ed. Cuatro Vientos, Argentina (2001)
14. Bishop, R.: The Mechatronics Handbook. University of Texas at Austin. CRC Press, USA (2002)
15. Kyura, N., Oho, H.: Mechatronics—an industrial perspective. IEEE/ASME Transactions on Mechatronics 1(1), 10–15 (1996)
16. Harshama, F., Tomizuka, M., Fukuda, T.: Mechatronics—What is it, why, and how?—an editorial. IEEE/ASME Transactions on Mechatronics 1(1), 1–4 (1996)
17. Auslander, D.M., Kempf, C.J.: Mechatronics: Mechanical System Interfacing. Prentice-Hall, Upper Saddle River (1996)
18. Corona, L.: Teorías Económicas de la Tecnología. Ed. Jus, México (1999)
19. Chiou, S., Music, C., Sprague, K., Wahba, R.: A Marriage of Convenience: The Founding of The MIT Artificial Intelligence Laboratory. MIT Lab, Boston, Massachusetts, USA (2003), `http://mit.edu/6.933/www/Fall2001/AILab.pdf`
20. Sánchez, M.: Implementación de Estrategias de Robótica Pedagógica en Las Instituciones Educativas, Universidad de los Andes, Colombia (2004), `http://www.eduteka.org/RoboticaPedagogica.php`
21. Broekman, B., Notenboom, E.: Testing Embedded Software. Addison-Wesley, Great Britain (2003)
22. RoboCup Offical Site (2009), `http://www.robocup.org/`
23. For Inspiration and Recognition of Science and Technology (2009), `http://www.usfirst.org/`

24. UMass Lowell: University of Massachusetts Lowell (2009), http://www.uml.edu/
25. Fred, G.: Martin Personal Web (2009), http://www.cs.uml.edu/~fredm/
26. Engaging Computing Group Official Web (2009), http://www.cs.uml.edu/ecg
27. Luís Alberto Muñoz Ubando Personal Web (2009), http://www.triy.org/ENG/Alberto.htm
28. The Robotics Institute of Yucatán (2009), http://www.triy.org/
29. Secretaria de la Juventud del Estado de Yucatan (2009), http://www.sejuve.gob.mx/
30. Instituto de Educación Progresiva The WorkShop (2009), http://www.workshop.edu.mx/
31. Valenti, G.: Ciencia, Tecnología e innovación: Situando la agenda de los sistemas nacionales de innovación, Hacia una agenda de política publica. Serie: Dilemas de las políticas públicas en Latinoamérica. Ed. FLACSO, México (2008)
32. Chapela, G.: Las prioridades en educación, ciencia, tecnología e innovación, México (2008)
33. Cimoli, M.: Las políticas tecnológicas en América Latina, México (2008)
34. Villavicencio, D.: Cambios institucionales y espacios para la investigación científica y la innovación en México, México (2008)
35. Bazdresch, C.: El financiamiento del sistema nacional de ciencia y tecnología, México (2008)
36. Dutrénit, G.: Políticas de financiamiento e investigación y desarrollo para endogeneizar la innovación en el sector productivo y empresarial, México (2008)
37. Puchet, M.: Incentivos, mecanismo e instituciones económicas presupuestales en el ordenamiento legal mexicano vigente para la ciencia y la tecnología, México (2008)
38. Fernández, J.: Política científica para el siglo XXI, México (2008)
39. Villa-Rivera, J.: Instituciones educativas y exigencias de formación de recursos humanos de alto nivel, México (2008)
40. González, C.: Un análisis de productividad de la comunidad científica mexicana, México (2008)
41. Jacques, D.: La revolución técnica. Ensayo sobre el deber de la humanidad. Ed. Jarale, México (2003)

Effectiveness of Game-Based Learning: Influence of Cognitive Style

Miloš Milovanović[1], Miroslav Minović[1], Ivana Kovačević[1], Jelena Minović[2], and Dušan Starčević[1]

[1] Faculty of Organizational Sciences, Laboratory for Multimedia Communications, University of Belgrade
{milovanovicm,mminovic,ivanak,starcev}@fon.rs
[2] Belgrade Banking Academy, Faculty for Banking, Insurance and Finance, Union University, Serbia
jminovic@ien.bg.ac.yu

Abstract. Today students have grown up using devices like computers, mobile phones, and video consoles for almost any activity; from studies and work to entertainment or communication. Motivating them with traditional teaching methods such as lectures and written materials becomes more difficult daily. That is why digital games are becoming more and more considered to have a promising role in education process. We decided to conduct a study among university students. Purpose of that study was to try to find some empirical evidence to support the claim that educational games can be used as an effective form of teaching. We also invested an effort to measure effects of different teaching approaches with the respect of individual differences in cognitive styles. Initial results provide a good argument for use of educational games in teaching. In addition, we reported some influence of cognitive style on effectiveness of using educational games.

Keywords: Learning, Educational game, Cognitive style.

1 Introduction

Today students have grown up using devices like computers, mobile phones, and video consoles for almost any activity; from studies and work to entertainment or communication. This has probably altered the way in which they perceive and interact with the environment, both physically and socially [1]. Nevertheless, most teaching strategies ignore these social changes and remain anchored in traditional text based instructional formats, provoking problems like arising lack of students' motivation [2]. Organizing a course to respect these notions is by no means a simple task. There are usually many limitations in conducting a course by use of games, either video or standard games. These limitations include group size, adequate interaction, sharing responsibilities between students. Our interest in this matter triggered when realizing the drop of motivation and interest in the subject among students at the course of Computer Networks. We discussed an idea of modernizing the course through the use of educational games. The need for better student involvement occurred. In order to

M.D. Lytras et al. (Eds.): WSKS 2009, CCIS 49, pp. 87–96, 2009.

increase the motivation of students, better understanding of the subject matter as well as improving collaboration, new form of teaching was required. Best possible way of animating new generation of "digital natives" [1] is by approaching them in their own language, the language of video games.

Research conducted and described in this paper had a task of comparing the effect of video games integrated in to curriculum with the effects of traditional teaching. In addition, we took in to the consideration a specific cognitive style of each student and the effect this way of conducting course had on their results. Initially we have two presumptions. First is that the use of educational game will increase the overall results of the students. Second is that the specific cognitive style of a student will influence the results based on whether the student had the opportunity of using educational games as a part of the learning process or not.

2 Literature Review

Computer games are widely accepted form of entertainment and their popularity increased over the past three decades. There were a lot of different studies concerning computer games and their influence on cognitive performance. It has been shown that computer games have great potential as a learning tool for the following reasons [3]:

- They can affect much more users than 'normal' lesson
- They can be played anytime, anywhere
- They are designed according to effective learning paradigms
- They stimulate chemical changes in the brain that promote learning

Studies comparing video game teaching effectiveness to the classic lecture show positive improvements (for the example, Supercharged! [4], the game that was used to teach students the principles of electromagnetism)

In the same time, games were associated with the concept of fun, while, with learning, it seemed that it was not always the case [1]. Games used several different techniques that kept the player motivated. Also, it has been noted that play improves ability to reason and understand the world [5]. Our opinion is that 'classic' learning can benefit from positive aspects of the computer games and, most notably, the concept of fun.

It was clear that different types of games have effect on different sets of skills. Platform and action games developed motor coordination and reflexes. Some games helped players to relax. Finally, complex games, like strategies and simulations, had influence on the development of intellectual skills. Authors came to the conclusion that games like Sims can be used as an example of social interaction, while games like Civilization could be used as strategic and historical simulation.

All researches stated that games had positive effect on concentration, the decision – making process, problem solving skills, logical thinking creativity, team work and, of course, computer skills [6]. Estallo claimed in his work [7] that people who play games have more developed intellectual skills than those who don't. Considering all this, there is a reason to believe that there is a strong connection between different types of games and improvement of a particular skill. Even more interesting are those games that affect intellectual skills. There are several reasons why computer games can be an effective learning platform [8]: Scope; Anytime, anywhere; Interesting;

Brain stimulation and the level of subject understanding is almost 30 percent higher if games are used.

Cognitive style is generally defined as "characteristic manner in which cognitive tasks are approached or handled" [9] more specifically, it is "a habitual and distinctive way of attending and processing perceptual and cognitive information" [10]. By applying cognitive style concept we can cover not only the situations of "adoption of particular ways of learning" but also its "impact upon problem solving of work place activities" [11], which is very similar to the given educational situation where students learn through simulation.

When cognitive style is seen in context of human-computer interaction, other measures of intellectual functioning are used more frequently. We consulted studies that consider meaning of cognitive style aspects in computer –based education, using: Sternberg's Thinking Style Inventory [12], Kirton Adaptation-Innovation Inventory – KAI [13], Pask`s holism/serialism dimension [14], Riding`s Imagers/Verbalizers and Wholist/Analytic dimensions [15] and most often Witkin`s field dependence/independence dimension [14, 15]. We decided to use MBTI as an instrument for measuring cognitive style, although it is more often used as a tool for assessment of one`s method of decision making.

3 Research Methodology

The success of the study depends on choosing the right parameters. Creating a quality environment for the study is most essential. We decided to conduct our study among students that attend the course of Computer networks and telecommunications. That course conducts during the first semester of the third year of study at the Faculty of organizational sciences. Agenda of the course should enable students to understand basic principles of computer networks. Our course attendants major in informational systems and their profile is mainly an engineering type. Our group of subjects consisted out of 125 students. There were 66 male subjects and 59 female subjects. Slight supremacy in number of male subjects is noticeable, but the main reason for that is the fact that the course belongs to an engineering department, which usually shows greater number in male attendants.

We divided students randomly in to two separate groups: control and experimental. Control group consisted out of 34 male and 33 female subjects. Experimental group was made of 32 male and 26 female subjects.

Experimental group got the assignment of designing a educational game that covers the area of computer networks while control group was involved in doing a programming assignment from the area of computer networks (protocols, distributed systems, services etc.). Students in experimental group were divided into smaller groups (design teams) consisting of two or three members. Every two weeks, teams were given a set of questions from computer networks area. Their task was to choose some of those questions and use them in their game design. Questions were to be modeled like problems in the game. Students were free to modify the problems any way they see fit. Main problem was writing a specification/scenario for an educational game. The specification/scenario had to be very detailed: it included dialogs, scene descriptions, character descriptions, etc. Students were free to choose any type of game. Upon ending a phase in their game developement they used a framework developed by research team [16,17] to materialize their game. Digital materials they collected or created as well as their scenarios were included in a educational video game that was the result of their work. Graphical editor was used to create the flow of the game while knowledge repository was created by the use of Knowledge manager application. All the design teams had a menthor assigned whose assigment was to overlook the progress every week and give advice. Control group was also divided in to smaller groups (programming teams) consisted of two to three person. At the begining of the semester they were given an assignment to develop an application that uses the benefits of a computer network. That application had to posses its own network protocol. There were a variety of different application types(P2P File sharing, P2P messaging, Voice communication, Video communication, Remote control etc.). Programming teams also had a menthor assigned.

In this research we use self-report MBTI questionnaire adapted and translated on Serbian language. The MBTI F form has 95 forced-choice items that forms four bipolar scales: Extraversion-Introversion (EI), Sensing-Intuition (SN), Thinking-Feeling (TF) and Judging-Perception (JP). A combination of these dimensions builds 16 different types of cognitive functioning. As MBTI is well theoretically conceptualized [18] and metrically evaluated instrument [19, 20], we thought that it might be useful to apply it on problem of learning by computer games. Actually, metric characteristics of scale are mostly adequate. Carlson [19] examined great body of reliability tests for this scale and found that coefficients for split-half reliability goes from .66 to .92, and test-retest reliability shows that results are relatively stable (coefficients in different studies are ranging from .69 to .89).

a. Aims

Educational games are a target of much debate. There are many opposing opinions on that matter. Some researchers claim that traditional methods of teaching are most effective and produce the best results, while others state strong advantages of educational games. In addition, the specific psychological profile of learner was subject of much consideration regarding educational games effect. On the other side, there is very little empirical evidence supporting any of these claims.

Aim of this study is to provide some information about effect of educational games on improving general knowledge and results. In addition, we aim to find out whether the cognitive style of the learner has any effect on the usability of educational games in education. Specific cognitive style of the learner could have an effect on whether educational games are applicable and in what extent.

b. Hypothesis

Use of educational games in teaching is by no means a new topic. Educational games aim to pass knowledge to learners while playing. Fun during learning should improve motivation and enrich learning process. This brings us to our first general hypothesis: H1 “Learning through game design is more effective and improves final results more than learning by traditional methods.”

When choosing a teaching method specific profile of the learner must be taken in to account. Every person is different, has different needs, personality, different motivational factors etc. It is reasonable to presume that some people will benefit more from learning through educational game design while others might benefit less. This raises our second general hypothesis: H2 "Effectiveness of learning through game design depends on specific cognitive style of learner."

Determining the effect of cognitive style on the benefits educational games have on learners presents a strong challenge. Proving or even disproving our second hypothesis requires decomposition. We will decompose that hypothesis on four sub-hypothesis. By proving any of those four, we will realize that we cannot disprove our general hypothesis. The division on different dimensions of cognitive functioning mentioned earlier places persons personality in four different categories. Each person belongs to one of the two opposing categories in every group. Person can be either extrovert or introvert. One can be either sensor or intuitive etc.

Since extroverts like variety and action, working and learning through interaction with others and they are often impatient to see the results of their activities, we presume that their results will benefit from learning through educational game design. In that light, we come to our first sub-hypothesis: H2a "Extroverts, if learning through educational game design, reach better results than introverts do."

Analyzing sensers, we realize they prefer to learn details, nurturing and establishing order, sticking to the routines and avoiding ambiguous situations. This led us to believe that designing an elaborate system such as educational game should improve their results. This forms our second sub-hypothesis: H2b "Sensers, if learning through educational game design, reach better results than intuitive do."

If we compare next two decision-making styles, we can claim that feeling style has more to gain from learning through games than thinking style, since fun is involved. Fun should have less positive effect on thinking style since it is more prone to awards of overcoming logical hurdles and performing deductive analysis. Next sub-hypothesis: H2c "People that poses feeling style, if learning through educational game design, reach better results than those with thinking style do."

Finally, comparing judgers and perceivers, we presume that general openness to different options and ability to adapt to complex situations gives perceivers the upper edge when it comes to educational game designing. Thus, we develop our last sub-hypothesis: H2d "Perceivers, if learning through educational game design, reach better results than judgers do."

c. Procedures

We performed the study during three-month period that constitutes one school semester. At the beginning of our study, we captured the cognitive style of all our subjects by use of a questionnaire. Earlier, we described methodology for determining cognitive styles. All subjects attended traditional classes held by the professor, and they usually contained the theory in area of computer networks. In addition, our control group performed programming assignment, while experimental group had the task of designing an educational game that covers topic of computer networks. We performed evaluation of students in several steps. The final product of their project, depending if they belonged to control or experimental group, was educational game or adequate network application. Projects marks were on a scale of up to 50 points. Teaching team evaluated the projects based on student's involvement, creativity and quality of solution. At the end of the semester, they were required to take an electronic test that covered the theoretical side in the area of computer networks. By performing this test, they could earn maximum amount of 40 points.

Finally, in order to measure how deep they understood the topic of computer networks, we gave them a practical test, which had a form of a case study. We introduced students to a concrete problem from the area of computer networks, and they needed to propose a solution based on their acquired knowledge. Maximum mark for that practical test was 10 points. Final mark is determined as a sum of all previous marks, and that mark we used in our statistical analysis.

After the participants completed the course, they were asked to fill out questionnaire. Questionnaire included questions about subjective satisfaction the applied form of teaching. Questions required them to rate the effectivenes of the taken form of teaching. Questions about subjective satisfaction were presented using five points semantic differential rating scale from positive impression to negative impression (for example 1 = completely disagree 5= completely agree).

In addition, we formed a focus group consisted out of 20 participants that were a part of experimental group. The purpose of that focus group was to record the informal impressions of our subjects about the described teaching method.

Due to the limited size of the paper we will concentrate on the results acquired with our statistical analysis, while the result of the questionnaire and recorded notions in our focus group will be shortly addressed to.

4 Results

The study means and standard deviations, as well as correlations between all factors are given in Table 1. Group1 to Group4 represent cognitive style factors that carry values 1 or 2, depending on cognitive style inside group. TotalPoints represent final score on course, and its values scale from 0 to 100. DesignGame factor defines whether student belongs to experimental or control group. Experimental group designed a game while control group performed programming assignment. Gender is a demographic factor.

For verifying correlation significance, we performed a two-tailed significance test, and presented P-values in parentheses below correlation coefficient value. Significant correlation is noticeable between Group1 and Group2 at 10% confidence level, Group2 and DesignGame at 1% confidence level and DesignGame and TotalPoints at 10% confidence level. Thus, we can anticipate significant interaction between aforementioned factors.

We did not find significant correlation among other factors, thus we can reject hypothesis H2a, H2c and H2d. Further analysis will be focused on interaction between DesignGame and Group2 regarding to TotalPoints.

Table 1. Descriptive statistics and Correlation matrix

	Mean	S.D.	1	2	3	4	5	6	7
1. Group1 (1) introverts (2) extroverts	1.54	.500	1						
2. Group2 (1) sensers (2) intuitive	1.29	.455	.157 (.081)	1					
3. Group3 (1) thinking (2) feeling	1.03	.177	.075 (.405)	.085 (.345)	1				
4. Group4 (1) judgers (2) perceivers	1.27	.447	-.018 (.843)	.008 (.927)	.093 (.301)	1			
5. TotalPoints between 0 and 100	79.4200	9.79084	-.016 (.862)	-.127 (.158)	-.068 (.448)	-.018 (.842)	1		
6. DesignGame (1) yes (2) no	1.54	.501	.018 (.844)	.308** (.000)	-.104 (.247)	.028 (.757)	-.166 (.064)	1	
7. Gender (1) male (2) female	1.47	.501	-.003 (.973)	-.070 (.435)	-.081 (.370)	-.146 (.105)	.078 (.385)	.044 (.624)	1

** Correlation is significant at the 0.01 level (2-tailed).

Since Group1 to Group4 factors represent cognitive style group, correlation between groups one and two is not of any interest for our research and will not be further analyzed. In order to examine possible interaction between DesignGame –TotalPoints, and Group2 – DesignGame, we conducted one-way ANOVA (Analysis of Variance) between given factors. An inspection of the ANOVA table (See Table 2) indicates that the interaction was significant between DesignGame and TotalPoints at 10% confidence level, as well as between DesignGame and Group2 at 0.1 % confidence level.

Table 2. One-Way ANOVA between correlated factors

Factors	DesignGame TotalPoints	DesignGame Group2
F	3.502 (.064)	12.923 (.000)

Hypothesis H1 suggested that Learning through game design is more effective and improves final results more than learning by traditional methods. Mean value of TotalPoints for students that did design game is 81.1638, which is greater than mean value for students that did not design game where N=77.9104. Since F-value is 3.502, with P-value 0.064 (See Table 2), this hypothesis was not rejected.

Also, we notice significant interaction between DesignGame – Group2 at 0.1% level, which refer to H2 hypothesis. In order to analyze this phenomenon in more details, we conduct Univariate Analysis of Variance between DesignGame and Group2.

Table 3. Descriptive Statistics for DesignGame and Group2. Dependent Variable:TotalPoints.

DesignGame	Group2	Mean	Std. Deviation	N
Yes	Sensers	82.5200	7.92050	50
	Intuitive	72.6875	8.36206	8
	Total	81.1638	8.61475	58
No	Sensers	77.2436	10.17600	39
	Intuitive	78.8393	11.14056	28
	Total	77.9104	10.53671	67

Table 4. Tests of Between-Subjects Effects. Dependent Variable:TotalPoints.

Source	Type III Sum of Squares	df	Mean Square	F
Corrected Model	1037.289[a]	3	345.763	3.856 (.011)
Intercept	469586.643	1	469586.643	5237.149 (.000)
DesignGame	3.713	1	3.713	.041 (.839)
Group2	328.777	1	328.777	3.667 (.058)
DesignGame * Group2	632.907	1	632.907	7.059 (.009)

a. R Squared = .087 (Adjusted R Squared = .065).

Performed analysis of variance confirmed strong interaction between factors DesignGame (1-Yes, 2-No) and Group2 (1-Sensers, 2-Intuitive) with F-value 7.059 at 1% significance level (.009). Difference between sensers and intuitive students inside experimental and control group was statistically significant, too, with F=3.667 at 10% level (0.058), thus we cannot reject hypothesis H2b which claim that sensers, if learning through educational game design, reach better results than intuitive do. This implies that we cannot reject our general H2 hypothesis too.

On the other side, we did not find any significant difference between DesignGame groups regarding to Group2 groups. Thus, we cannot claim that sensers who designed a game performed better on final score than sensers who did not design a game. Same is true for intuitive students. Finally, we provided a plot, which graphically presents estimated marginal means for Group2, depending on DesignGame factor.

5 Discussion

Idea that educational games can be used as an effective teaching tool is what inspires our work in this field. Purpose of this study was to try to find some empirical evidence to support this claim. As seen in our results section we found that our general hypothesis H1 cannot be rejected. That provides a strong reason to continue our research. On the other side, since the study we conducted is novel there is a lot of room for question and improvement. That is why it is essential to take in to account subjective thoughts of our subjects about this method of teaching. According to the questionnaire we've administered at the end of our study subjects were generally satisfied with this new teaching style. Majority of subjects found this method interesting and fun. Also, questionnaire gave us some perspective in what can be improved in this method "Requirements of this project lack a bit of structure so it is quite hard to realize what is expected" and "Lack of time presented a strong issue in realizing this project, as well as lack of personal consultations with the teaching staff".

In this specific research it is important to emphasis that cognitive style is found to be in correlation with learning style. So, we could comment our results showing that sensers are better performer using computer games as a learning tool in context of Felder-Silverman`s learning style model that implies that sensing learners are more prone to details and more practical than intuitive ones who prefer to learn general principles rather than to involve into concrete learning action. Learning trough computer games seems to be more "field-dependent" activity, with computer game plot giving actual context (background) and it was proved that field-dependent learners like those sensing learners, prefer concrete material, compare to field-independent and intuitive category of learners which both prefer and are more capable to use abstract material differentiating it from the given background [21]. We could also compare our findings with the results of Graf et al. study showing that so called field-dependent learners "have difficulties in learning text-only material and benefit more from material that contains text as well as graphics". Further, students with so called high global level cognitive styles are more effective when learning materials in less structured manner. So, we could conclude that sensers achieve better results due to the fact that they are cognitively predisposed for learning tasks on this subject and that computer games are suitable medium for learning the material in matter. Nevertheless, it has to be emphasized that cognitive style is especially important due to the fact that, unlike the expertise, it is a relatively stable quality and that it creates a "bridge" between cognition and personality.

References

1. Prensky, M.: Digital Game-Based learning. Paragon House Publishers, NY (2007)
2. Sancho, P., Fuentes-Fernandez, R., Fernandez-Manjon, B.: NUCLEO: Adaptive Computer Supported Collaborative Learning in a Role Game Based Scenario. In: Eighth IEEE International Conference on Advanced Learning Technologies, ICALT, pp. 671–675 (2008)
3. Mayo, M.J.: Games for science and engineering education. Communications of the ACM 50(7), 31–35 (2007)
4. Squire, K., Barnett, M., Grant, J., Higginbotham, T.: Electromagnetism supercharged! Learning physics with digital simulation games. In: Proceedings of the 2004 International Conference of the Learning Sciences. UCLA Press, Santa Monica (2004)

5. Gilkey, R., Kilts, C.: Cognitive fitness. Harvard Business Review, Boston (2007)
6. De Aguilera, M., Mendiz, A.: Video games and Education. Computers in Entertainment 1(1) (2003)
7. Estallo, J.A.: Los videojugos. Juicios e prejuicios. Planeta, Barcelona (1995)
8. Mayo, M.J.: Games for science and engineering education. Communications of the ACM 50(7), 31–35 (2007)
9. Reber, A.S., Reber, E.: The Penguin Dictionary of Psychology. Penguin Books, London (2001)
10. Entwistle, N., Peterson, E.: Learning styles and Approaches to Studying. In: Spielberger, C. (ed.) (2004)
11. Roberts, A.: Cognitive styles and student progression in architectural design education. Design Studies 27, 167–181 (2006)
12. Workman, M.: Performance and Perceived Effectiveness in Computer-based and Computer-aided Education: Do cognitive style makes a difference? Computers in Human Behavior 20, 517–534 (2004)
13. Cheng, M.M., Luckett, P.F., Schultz, A.K.: The Effects of Cognitive Style Diversity on Decision-Making Dyads: An Empirical Analysis in the Context of a Complex Task. Behavioral Research in Accounting 15, 39–62 (2003)
14. Cegarra, J., Hoc, J.-M.: Cognitive Styles as an Explanation of Experts Individual Differences: A Case Study in Computer-assisted Troubleshooting Diagnosis. International Journal of Human Computer Studies 64, 123–136 (2006)
15. Frias-Martinez, E., Chen, S.Y., Liu, X.: Investigation of Behavior and Perception of Digital Library Users: A Cognitive Style Perspective. International Journal of Information Management 28, 355–365 (2008)
16. Minović, M., Milovanović, M., Lazović, M., Starčević, D.: XML Application For Educative Games. In: Proceedings of European Conference on Games Based Learning, ECGBL 2008, Barcelona, Spain, pp. 307–315 (2008)
17. Jovanovic, M., Starcevic, D., Stavljanin, V., Minovic, M.: Educational Games Design Issues: Motivation and Multimodal Interaction. In: Lytras, M.D., Carroll, J.M., Damiani, E., Tennyson, R.D. (eds.) WSKS 2008. LNCS (LNAI), vol. 5288, pp. 215–224. Springer, Heidelberg (2008)
18. Briggs-Mayers, I., McCaulliey, M.H., Quenk, N.L., Hammer, A.L.: MBTI Manual. A guide to the Development and Use of Myers-Briggs Type Indicator. Consulting Psychologists Press, Palo Alto (1998)
19. Saggino, A., Kline, P.: Item Factor Analysis of the Italian Version of the Myers-Briggs Type Indicator. Journal of Personality and Individual Differences 19(2), 243–249 (1995)
20. Murray, S.W.: Testing the Bipolarity of Jungian Functions. Journal of Personality Assessment 67(2), 285–293 (1996)
21. Ford, N., Chen, S.Y.: Individual differences, hypermedia navigation and learning: an empirical study. Journal of Educational Multimedia and Hypermedia 9(4), 281–312 (2000)

Goneis.gr: Training Greek Parents on ICT and Safer Internet

Nikos Manouselis[1], Katerina Riviou[2,*] Nikos Palavitsinis[1], Vasiliki Giannikopoulou[1], and Panayotis Tsanakas[1]

[1] Greek Research & Technology Network (GRNET S.A.), Athens, Greece
{palavitsinis,nikosm}@grnet.gr
[2] Doukas School S.A., Athens, Greece
kriviou@doukas.gr

Abstract. Children's use of the Internet has significantly risen in the last decade. Nevertheless, children spend a lot of time online which makes them susceptible to various threats (such as inappropriate material, offensive language, etc). Parents are the last frontier to this menace but they also need to be educated and trained in order to protect their children. Goneis.gr is an initiative launched by the Greek government that aims to educate parents on safer Internet and the use of parental control software. Parents are also entitled to distance learning courses covering basic computer skills. This paper presents the results of two separate surveys that took place in the last few months (December 2008-January 2009). The first survey targeted the parents that have completed the programme and the second one the educational providers that participate in the programme and offer the training to the beneficiaries.

Keywords: training, parents, ICT, internet.

1 Introduction

Nowadays, more and more children use the Internet as a source of communication, as well as in the context of their homework. More specifically, in United States, American teens are more wired now than ever before [1]. 93% of all Americans between 12 and 17 years old use the internet (up from 87% in 2004 and 73% in 2000), while in the European Union this figure reaches 85%. Additionally, the parents are hardly aware of their children activities on the Internet. The number of children using the Internet varied considerably across Europe. The proportion of parents who thought that their child used the Internet was the lowest in Italy (45%), Greece and Cyprus (both 50%). In all other Member States, at least two-thirds of the parents answered that, as far as they knew, their child used the Internet: from 68% in Portugal to 94% in Finland [2].

The phenomenon of offensive and inappropriate content reaching the children when using the Internet becomes more and more obvious as statistics indicate. Namely:

* Funded by GRNET during the implementation of this study.

M.D. Lytras et al. (Eds.): WSKS 2009, CCIS 49, pp. 97–106, 2009.

- 28% Dutch and 24% UK parents said that, when their child asked for their help, this was because they had been contacted by a stranger, were bullied or harassed online or saw violently or sexually explicit images online [2].
- 32% of all teens and 43% of teens active in social networking have been contacted online by a complete stranger [3].
- 69% of teens regularly receive personal messages online from people they don't know and most of them don't tell a trusted adult about it [3].

In this context, parents are the last line of defense against this threat. Statistics show that parents are more or less concerned about what their children do online, but at the same time they are unaware of what they can do to protect them (e.g. install and configure a parental control program). As stated in a recent survey [4]:

- 66% of mothers of teens are just as or more concerned about their teenagers' online safety as they are about drunk driving (62%) and experimenting with drugs (65%).
- 43% of the children have closed or minimized the browser at the sound of a parental step.
- Only 15% of parents use a software program to monitor what their kids do online.

Parents who did not use the Internet themselves (but their child did), answered that more and better teaching and guidance about Internet use in school and training sessions for parents organised by NGOs would contribute to a safer and more effective use of the Internet by their child [2].

In Greece there have been some major initiatives dealing with these issues and educating the parents regarding the need for a safer Internet such as Safer Internet [5], and Digital Awareness & Response to Threats task force (D.A.R.T.) [6]. Similar initiatives take place in European level such as Safer Internet Plus (SCAD) [7] and Safer Internet [8] where countries from all around Europe participate (e.g. Italy, Germany, France, United Kingdom, Spain, etc). Interest on these issues has risen in the United States such as i-Safe Inc [9]. The current situation showcases that the issue of safer Internet is here to stay and should be addressed.

To this direction, a new major initiative called "Goneis.gr: Training the Parents of High school Pupils on ICT and Safer Internet" educates and trains the parents of high school children on the issue of safer Internet and how they can protect their children from online threats. This initiative is co-funded by the Greek Ministry of Economy and Finance and the European Union. It concerns the training of the parents of students that enrolled in the obligatory second grade education of Greece (middle school – from 11 to 14 years old) for the school year 2008-2009. It is being implemented by the Greek Research & Technology Network (GRNET), in cooperation with the Ministry of Economy & Finance and the Ministry of Education. The initiative has a budget of 21MEuros and has provided training to over 28.300 parents so far, through 826 Educational Service Providers. It started in the summer of 2008 and is ongoing until the end of spring 2009, aiming to train about 50.000 parents all over Greece.

Beneficiaries of the programme are the parents of pupils that are enrolled in one of the three years of any type of high school (daily, experimental, musical, etc.) private or public. Beneficiaries of the programme are also the parents of students that study in schools for challenged children. This initiative aims at familiarising the parents with Information and Communication Technologies (ICT), focusing on the Internet, its safer use and its educational applications.

In this paper we present initial results from the ongoing evaluation of the "Goneis.gr" project. More specifically, we present results from the application of two separate surveys. The first one aiming at the parents that completed the programme and the second aiming at the Educational Providers that offer the training.

2 Goneis.gr Initiative

The initiative's beneficiaries are provided with free-of-charge home training from specialized instructors, as well as with access to educational packages for autonomous learning through the Internet (e-learning courses). The duration of the home based training is at least five (5) hours and it can take place in more than one visits to the Beneficiary's house. The content of the e-learning courses has a duration of at least forty (40) hours. In addition, the Beneficiaries have the option to apply for a pre-paid high-speed internet connection (ADSL) for at least two (2) months. In order to participate in the programme, the Beneficiary must have a computer (either a laptop or a desktop) with an Internet connection of any type.

The training of the parents is carried out through the cooperation with education providers that are responsible for the entire training process of each parent as far as the educational aspects are concerned. The Educational Providers are participating in the program according to some pre-defined eligibility criteria. For any Educational Provider to participate in the programme they should be able to provide to each Beneficiary home-training carried out by specialised trainers, as well as the access to educational packages of autonomous learning through the Internet (e-learning courses). The duration of the home training should cover at least five (5) teaching hours and it can be completed in one or more visits (depending on the Beneficiary's availability). The content of the e-learning courses should cover at least forty (40) teaching hours and to offer the possibility to check its completion by the Beneficiary.

The home training that the Educational Providers are entitled to offer covers the following topics:

- Basic concepts on the use of the Internet, communication and information search.
- Safe use of the Internet and child protection from malicious and inappropriate online content.
- Educational applications of the Internet and services of the Panhellenic School Network (PSN) [10]. For a presentation of the services in the PSN see [11].

Additionally, the e-learning courses that the Educational Providers undertake:

- Cover at least forty (40) teaching hours
- Provide the Beneficiaries with all the necessary knowledge so as to be able to participate in "Basic ICT Knowledge and PC skills" certification exams.

After the completion of the e-learning courses, the Beneficiaries are entitled to participate (free of charge) in certification exams so as to acquire a certificate on "ICT knowledge and PC skills" in at least three of the units of basic knowledge (i.e. Basic concepts of ICT, Use of PC and File Management, Word Processing, Spreadsheets, Databases, Presentations and Internet Services).

3 Goneis.gr Web Portal

A crucial element that will define the success and dissemination of the Goneis.gr initiative is the deployment of a web portal that contains all the necessary information for all the involved parties (Beneficiaries and Educational Providers). The Goneis.gr web portal addresses an audience that is not well familiarised with the Internet and ICT. For this reason, the structure of the web portal is fairly simple and easy to comprehend and allows even the most inexperienced users to navigate through the various pages and get all the necessary information regarding their participation in the initiative.

The overall structure of the Goneis.gr web portal is the following:

- The Homepage that presents the logos of the initiative and the participating organisations. It also gives a brief but comprehensive outline of the initiative. If the user chooses the hyperlink "Description" they can get more details on the training offered by the initiative.
- The Navigation Menu contains hyperlinks to all the main pages of Goneis.gr web portal and it is included in all the web portal pages. Its options include:
  - The Action, with information regarding the initiative, its goals and its Beneficiaries.
  - The Educational Packages, with information regarding the available educational packages for each area/region of Greece and the capability of searching through them by providing the area/region of residence for each Beneficiary.
  - Beneficiaries, with information regarding the categories of Beneficiaries that can participate in the action, the way in which they can participate as well as the ways they can get informed about the educational packages offered.
  - Training, with specific information on the training that the Beneficiaries will receive and more specifically: the analytical description of the home training and the description of the e-learning courses. Additionally, information is provided regarding their participation in certification exams, the procedure of acquiring a broadband connection and finally the categories of grants that are offered to the Educational Providers by the initiative in order to train the Beneficiaries.
  - Educational Providers, with information regarding their participation in the initiative and the specifications of the educational packages, the e-learning system and the home training. This page also provides details on the Educational Providers' obligations regarding the certification exams, the details on the support of the beneficiaries through the establishment of a helpdesk and the categories of the

grants offered to the Educational Providers for the training of the Beneficiaries. The Quality Control Mechanism is also presented in this section.

- o Frequently Asked Questions (FAQ), with answers in frequently asked questions by the Beneficiaries and the Educational Providers.
- o Communication, with information regarding the ways that the Beneficiaries and Educational Providers can communicate with the coordinating organisation of the initiative. It also includes the advertising material of the initiative, available for everyone to download.

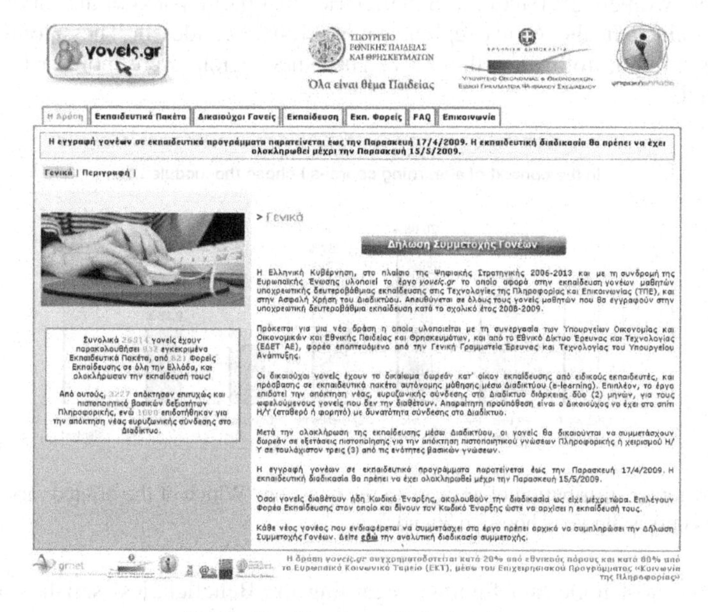

Fig. 1. The Goneis.gr Web Portal

4 Evaluation of Users' Satisfaction

This section aims to analyze the answers provided by the Beneficiaries and the Educational Providers in the survey that evaluated the initiative through relevant questionnaires deployed. The feedback was gathered through telephone communication in the case of the Beneficiaries and through an online procedure in the case of the Educational Providers.

4.1 Beneficiaries' Survey

From the overall population of about 28.300 parents that have completed their training, approximately 1.500 have been contacted through telephone and 667 responses (completed questionnaires) were collected. The collection of this data took place during December 2008, thus corresponding to the intermediate evaluation of the programme.

In the next pages the most important facts and figures from the Beneficiaries' survey are presented. The evaluation and commentary on each set of graphs and tables, takes place just after the end of each set.

From a total of 667 questionnaires, 30 of them were not completed in full (0,04% of the total) so they were not taken under consideration whatsoever. When using the term "Beneficiary" in this part of the paper, we refer to the Beneficiaries that filled out the questionnaire.

From the total of 667 that participated in the survey, 184 (29%) were men and 453 (71%) were women. 28,1% of the Beneficiaries that participated in the survey, resides and was trained in the Attica region, while 11,46% resides in Thessaloniki region. The Larissa region follows with 4,55%. The other regions were more or less equally represented.

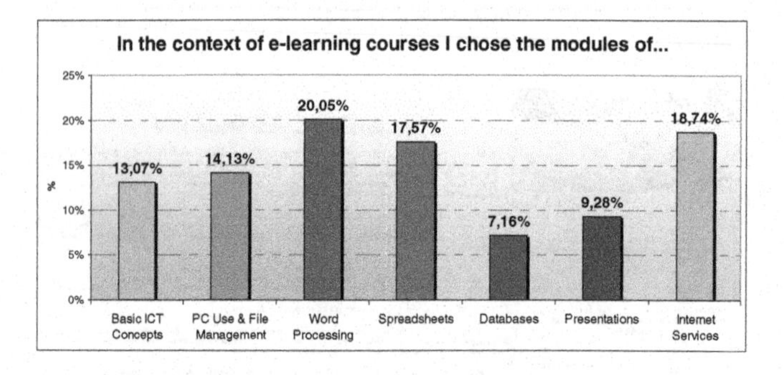

Fig. 2. Percentage distribution of answers to the question "Which of the offered modules in the e-learning courses did you choose to attend?"

Among the most important findings regarding the Beneficiaries' satisfaction on the training procedure:

- 52% of the Beneficiaries believe that the new knowledge acquired from the home training is very useful, whereas 44% state it is useful, while a mere 3,50% thought of the acquired knowledge to be indifferent.
- The most popular module of the e-learning courses is the Word Processing with 20%, while Internet Services follow with 19% and Spreadsheets with 17% (Fig. 2).
- 26% of the Beneficiaries are fully satisfied with the duration of the e-learning courses (rated with 5 out of 5), while a similar percentage declared that they are very satisfied (rated with 4 out of 5).
- 53% of the Beneficiaries are fully satisfied with the content/material provided with the e-learning courses (rated with 5 out of 5), whereas 36% rated it with 4 out of 5.

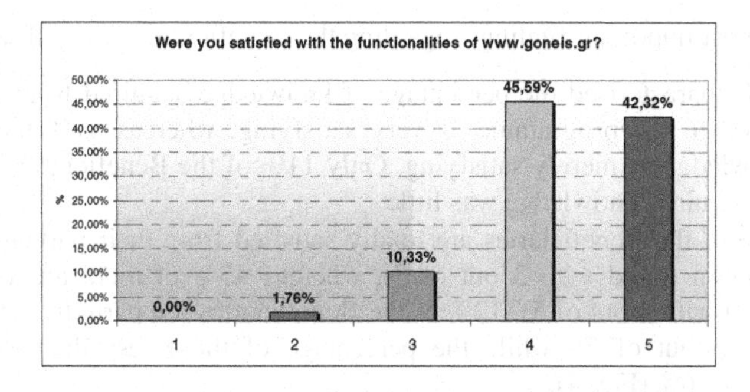

Fig. 3. Percentage distribution of answers to the question "Please indicate the level of your satisfaction on the operation of the Goneis.gr Web Portal"

Among the most important findings regarding the Beneficiaries' satisfaction from the supportive infrastructure and the certification process:

- Percentages of 34% and 30% of the Beneficiaries think that their training prepared them fully and fully enough respectively, so that they could handle the certification process.
- 70% of the Beneficiaries visited the Goneis.gr Web Portal, whereas 23% did not visit the portal as they did not need to do so. Only 7% of the Beneficiaries were not informed about the portal's existence.
- 42% of the Beneficiaries graded with "excellent" their degree of satisfaction on the operation of the Goneis.gr Web Portal (rated with 5 out of 5), whereas 46% are very satisfied (Fig. 3).
- 48% of the Beneficiaries did not need to contact the helpdesk, whereas 38% of them did contact the helpdesk to address their questions regarding the initiative. 14% declared that they did not know about the existence of the heldesk.
- The majority of the Beneficiaries are satisfied with the operation of the helpdesk as 78,7% of those that contacted the helpdesk characterized their degree of satisfaction as "excellent" (rated with 5 out of 5), whereas 17% of them are very satisfied (rated with 4 out of 5).

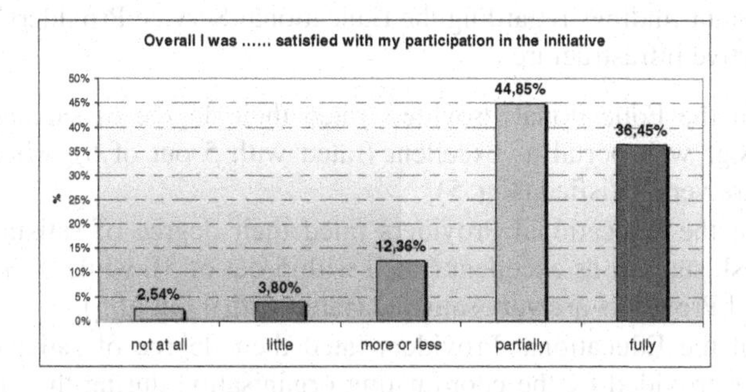

Fig. 4. Percentage distribution of answers to the question "Please rate your overall level of satisfaction on your participation in the initiative"

Among the most important findings regarding the Beneficiaries' overall satisfaction:

- 38% characterised the percentage of knowledge acquired from their participation to the programme as very satisfying, whereas 51% thought of the knowledge as merely satisfying. Only 11% of the Beneficiaries thought that the acquired knowledge was little.
- 37% of the Beneficiaries are totally satisfied from their participation in the program (rated with 5 out of 5), whereas 45% of them are very satisfied (rated with 4 out of 5). 12% of the Beneficiaries are partially satisfied (rated with 3 out of 5), while the percentage of those less than satisfied only reaches 6% (Fig. 4).

4.2 Educational Service Providers' Survey

The Educational Service Providers that participated in the initiative were 826. After contacting all of them via e-mail, 335 provided their feedback in an online version of the evaluation questionnaire up until late January 2009. In the next pages the most important facts and figures from the Educational Providers' survey are presented.

From a total of 335 questionnaires that were collected, 72 of them were partially filled out (0,22%) and were not taken under consideration.

Among the most important findings regarding the Educational Service Providers' satisfaction on the training procedure:

- 43% of the Educational Providers rated their degree of satisfaction with the content of the training as excellent (rated with 5 out of 5), whereas 45% of them are very satisfied with the training content (rated with 4 out of 5).
- 27% of the Educational Providers rated their degree of satisfaction on the methodology of the training process (home-based and e-learning courses) as excellent.
- 22% rated the degree of their satisfaction from the duration of the home-based training as excellent, whereas a smaller percentage (13%) has the same opinion regarding the duration of the e-learning training.
- 12,5% of the Educational Providers thought that it is very easy (rated with 5 out of 5) to gather and organise the necessary paperwork to apply for the grant given by the initiative, whereas 19% found the process fairly easy.

The most important findings regarding the Educational Service Providers' satisfaction from the supportive infrastructure:

- 44% of the Educational Providers rated their degree of satisfaction by the Goneis.gr web portal as excellent (rated with 5 out of 5), whereas 36% of them are very satisfied (Fig. 5).
- 47% of the Educational Providers rated their degree of satisfaction by the helpdesk support as excellent (rated with 5 out of 5), while 34% of the Educational Providers are very satisfied (rated with 4 out of 5).
- 43% of the Educational Providers rated their degree of satisfaction by the support provided by the coordinating organisation during the preparation of the necessary paperwork for the grants' claim as excellent (rated with 5 out of 5), 31% are very satisfied (rated with 4 out of 5).

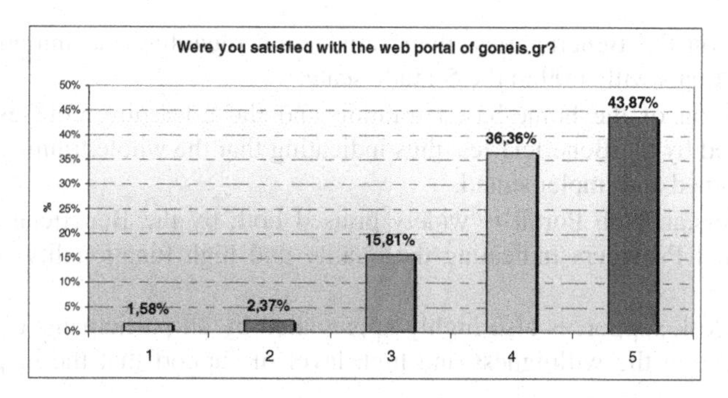

Fig. 5. Percentage distribution of answers to the question "Please rate the degree of your satisfaction on the Goneis.gr Web Portal"

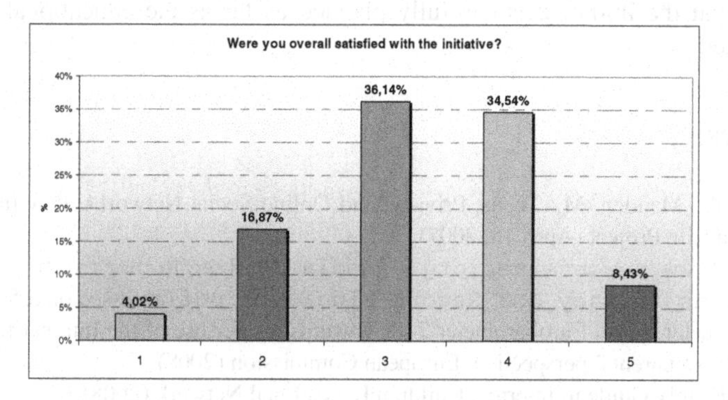

Fig. 6. Percentage distribution of answers to the question "Were you overall satisfied with the iniative?"

The most important findings regarding the Educational Service Providers' overall satisfaction:

- 8% of the Educational Providers are fully satisfied with the initiative as a whole (rated with 5 out of 5), whereas 35% of them are very satisfied (rated with 4 out of 5). 36% of the Educational Providers are partially satisfied, while 21% of them are less than satisfied (Fig. 6).
- 12,5% is fully satisfied with the rules that governed the operation of the initiative (rated with 5 out of 5), whereas 31% of the Educational Providers are very satisfied (rated with 4 out of 5) and 27% satisfied (rated with 3 out of 5).

5 Conclusions

Summarising the main points that concern the evaluation of the Goneis.gr initiative:

- The respective percentages of Beneficiaries and Educational Providers that are more than satisfied with the initiative are over 80%, which indicates that the initiative is successfully deployed in all of its aspects (promotion, implementation, administrative, etc.)

- Focusing on the Beneficiaries, their satisfaction regarding the initiative and its services equals with 4,09 in the 5-grade scale.
- The duration of the home-based training and the e-learning courses is highly appreciated by the Beneficiaries, thus indicating that the whole training process is well-designed and implemented.
- The Goneis.gr Web Portal is widely praised both by the Beneficiaries and the Educational Providers indicating the quality and high functionality of the Web Portal.
- The helpdesk support is also highly appreciated by all participants of the initiative indicating the willingness and high level of support that the helpdesk staff provides.
- The Educational Providers are also satisfied with the content and duration of the training, as well as from the implementation of the training program. This shows that the initiative is carefully planned as far as the educational aspect is concerned.

References

1. Lenhart, A., Madden, M.: Teens, Privacy, and Online Social Networks. Pew Internet and American Life Project (April 18, 2007), http://www.pewinternet.org/~/media//Files/Reports/2007/PIP_Teens_Privacy_SNS_Report_Final.pdf.pdf (accessed on 5/5/2009)
2. Eurobarometer. Flash Eurobarometer 248: Towards a safer use of the Internet for children in the EU – a parents' perspective, European Commission (2008)
3. CEN. A Family Guide to Internet. Children Educational Network (2008), http://kidsafe.com/wp-content/uploads/2009/01/cen-online-safety-guide.pdf (accessed on 5/5/2009)
4. Harris Interactive-McAfee. Survey on the Usage of the Internet by Teens (2008), http://www.mcafee.com/us/about/press/corporate/2008/20081022_095000_x.html (accessed on 9/2/2009)
5. Safer Internet, http://www.saferinternet.gr
6. Digital Awareness & Response to Threats task force (D.A.R.T.), http://www.dart.gov.gr
7. Safer Internet Plus (SCAD), http://ec.europa.eu/information_society/activities/sip/index_en.htm
8. Safer Internet, http://www.saferinternet.org
9. i-Safe Inc., http://www.isafe.org
10. Panhellenic School Network, http://www.sch.gr
11. Kalochristianakis, M.N., Paraskevas, M., Varvarigos, E.: Asynchronous Tele-education and Computer-Enhanced Learning Services in the Greek School Network. In: Lytras, M.D., Carroll, J.M., Damiani, E., Tennyson, R.D. (eds.) WSKS 2008. LNCS (LNAI), vol. 5288, pp. 234–242. Springer, Heidelberg (2008)

Educational Accountability and the Global Knowledge Society – What Can We Learn from the EU Experience?

Kornelia Kozovska[1], Rossana Rosati[1], and Daniele Vidoni[2]

[1] European Commission - JRC, Centre for Research on Lifelong Learning (CRELL), Ispra, Italy
kornelia.kozovska@jrc.ec.europa.eu, rossana.rosati@ec.europa.eu
[2] Italian National Institute for Educational Evaluation (INVALSI), Rome Italy
daniele.vidoni@invalsi.it

Abstract. The national education systems of the European Union Member States operate in a rather unique framework which sees EU level monitoring and creation of benchmarks in the framework of the Lisbon Agenda interact with national priorities and social characteristics. This paper examines the use of benchmarking in education and training and the presence of different performance levels of groups of countries, using Sapir's European social models classification. There is, in fact, evidence for consistent trends of similar performance within groups and clear distinction of performance levels between groups of countries. This gives indication for the existence of strong structural differences and implies that one-size-fits-all policies do not necessarily respond to the specific characteristics and needs of different countries.

Keywords: Knowledge Society, European Union, Open Method of Coordination, Lisbon objectives in Education and Training, Benchmarking.

1 Introduction

The impact of globalization and new technologies on economic competition within and among countries has been of great importance in shaping the new perceptions of how to achieve economic competitiveness around the world. A number of recent large-scale education reforms have found their rationale in the pursuit for an increased competitiveness of the labor force, underlining the general assumption that higher levels of economic competitiveness are closely linked with the way citizens acquire knowledge, skills, and attitudes for civic success and the knowledge-based economy. [1] Knowledge and education feed off each other and education is considered the central force which supports the culture of knowledge production. Any attempt to establish a precise definition of the knowledge society would result difficult, limiting and indeed, not necessary. The discussion raised by the concept of knowledge society has had impacts on all levels. Starting from the micro-level, knowledge has entered the production function and become an essential factor for the competitiveness of firms and locations. At the macro-level, it has entered the policymaking discourse as a crucial element for the prosperity of nations underlining the importance of knowledge, knowledge production and knowledge sharing. Easier access to knowledge has

M.D. Lytras et al. (Eds.): WSKS 2009, CCIS 49, pp. 107–118, 2009.

intensified knowledge production while at the same time the emphasis on knowledge, both in terms of potential and production, as essential for competitiveness and prosperity has introduced higher propensity for comparison of performance levels and benchmarking.

This paper will focus on one of the crucial elements and the driver of the knowledge society - education, and discuss it in the context of the Lisbon Agenda. The national educational systems of the European Union (EU) Member States operate in a rather unique framework which sees EU level monitoring and creation of benchmarks as part of the Lisbon process interact with national priorities as well as other informal constraints (welfare models, culture, social values, etc.). Taking a principle-agent theory outlook, it will look into the ways in which the European Union establishes benchmarks in the fields of education and training. It will consider the performance of Member States in light of regional differences and will offer some insight as to the ways in which the use of benchmarking could be used as well as the intrinsic limits it possesses.

2 Benchmarking in a Principal-Agent Perspective

The principal-agent theory has been developed in the context of analysis of private firms and the relationship between shareholders and managers, where shareholders are 'principals' as they own the firm and managers are 'agents' appointed by them to run the business with asymmetric information and power distribution governs their relationship. In the principal-agent theory framework, the overarching argument is that targets facilitate the accountability of agents to principals. Transferring this framework to the public sector, principals could define policies and priorities for the agents' activities and monitor adherence and achievement. In principal-agent relationships, benchmarking can take advantage of this capacity of performance measurement to influence behavior. [2] The principal can communicate the policies and priorities it has set for the agent(s) in a variety of ways such as policy guidelines, setting up of priorities, etc. On the other hand, in the nature of the principal-agent relationship is the fact that the agent has a certain freedom in the interpretation of such communication tools. Benchmarking exercises have, in fact, been analyzed often times as reinforcing methods that could be used by the principal to assure the success in the attainment of the desired objectives.

Vink (2000) points out that many countries have used benchmarking as a tool for setting targets in job creation, percentage of unemployed, etc. in the relationship between ministries and national employment service organizations. [3] Another common use in the education field is the creation of performance league tables for schools, as in the case of the United Kingdom. Often time it is not so much the level or strength of the principal's authority that could influence an agent's behavior, but the availability to the wider public of performance results in relation to a given benchmark.

In the context of the European Union, where we have the European Commission often times having a surveillance role in relation to Member States (MS), a practical tool for cooperation among MS with a view to convergence of national policies towards given common objectives is the open method of coordination (OMC). Under this intergovernmental method, Member States are evaluated by one another (peer pressure), with the European Commission's having a supervisory role. The open method of coordination takes place in areas which fall within the competence of the Member States, such as employment, social protection, social inclusion, education, youth and training. It has been used as an instrument of the Lisbon Strategy (2000) and monitoring of the attainment of the Lisbon objectives. [4] The European Commission's White Paper on European Governance defines it in the following manner [5]:

> "*The open method of co-ordination is used on a case by case basis. It is a way of encouraging co-operation, the exchange of best practice and agreeing common targets and guidelines for Member States, sometimes backed up by national action plans as in the case of employment and social exclusion. It relies on regular monitoring of progress to meet those targets, allowing Member States to compare their efforts and learn from the experience of others.*
>
> *In some areas, such as employment and social policy or immigration policy, it sits alongside the programme-based and legislative approach; in others, it adds value at a European level where there is little scope for legislative solutions. This is the case, for example, with work at a European level defining future objectives for national education systems.*" (p.21)

OMC is based essentially on identifying and defining objectives to be achieved, jointly establishing measuring instruments (statistics, indicators, guidelines), benchmarking as a comparison of the Member States' performance and exchange of best practices, as monitored by the Commission. As such it can be perceived easily in a principle-agent framework where the agent's behavior is influenced both by the level or strength of the principal's authority as well as the availability to the wider public of performance results in relation to a given benchmark.

3 The Role of the European Commission in Education and Training Policies

Education and training are critical factors for the development of the knowledge economy and their role for the development of the knowledge-based society been clearly recognized by EU Member States. The Lisbon Agenda, a unique project, given the geographical dimension and level of ambition, sees EU Member States agree to put their common energies and resources to become "*the most dynamic and competitive knowledge-based economy in the world capable of sustainable economic growth with more and better jobs and greater social cohesion, and respect for the environment by 2010.*" It was adopted by the European Council in 2000 and re-launched in 2005. As part of achieving the objectives of the Lisbon Agenda, education and training have been recognized as crucial for the realization of the long-term potential for competitiveness and social cohesion. In its new conceptualization, the Strategy entails a stronger mobilization of all appropriate national and Community resources and a strengthening of monitoring procedures to give a clearer picture of national implementation of the strategy. The re-launched Lisbon strategy focuses on competitiveness, growth and productivity and strengthening social cohesion. Even more than in its first phase, the revised Lisbon strategy places strong emphasis on knowledge, innovation and the optimization of human capital [6].

Such emphasis on education and training as critical factors for the development of EU's long-term potential for competitiveness as well as for social cohesion requires the joint effort of all Member States. Still, similar to the US, EU Member States have not delegated to the Union the responsibility over education. Thus, the EU can only exercise a coordination role by means of art.126 of the Treaty of Maastricht, which indicates that "*the Community shall contribute to the development of quality education by encouraging cooperation between Member States and, if necessary, by supporting and supplementing their action, while fully respecting the responsibility of the Member States for the content of teaching and the organization of education systems and their cultural and linguistic diversity.*" Moreover, art. 127 points out that the European Union shall also "*...implement a vocational training policy which shall support and supplement the action of the Member States, while fully respecting the responsibility of the Member States for the content and organization of vocational training.*"

This puts the national education systems of the EU Member Sates in a rather unique framework. On the one hand, national priorities interact with other informal constrains such as economic priorities, welfare models, culture, social values, etc. On the other hand, EU membership brings along EU-wide objectives and the monitoring of single countries' performance against them.

4 Differences as a Source of Improvement: The Coordination Role of the European Commission

The implementation of the Open Method of Coordination in education has led to the identification of 5 benchmarks for education and training. By adopting these five European benchmarks in May 2003, the European Council has undertaken a political commitment by setting up measurable objectives, thereby, indicating in which policy areas, in particular, it expects to see clear progress. The five EU reference levels of average performance to be achieved by 2010 are the following:

- ➢ No more than 10% of young people (aged 18-24) should be early school leavers;
- ➢ Decrease of at least 20% of the share of low-achieving pupils in reading literacy;
- ➢ At least 85% of young people should have completed at least upper secondary education;
- ➢ Increase of at least 15% in the number of graduates in mathematics, science and technology, with a simultaneous decrease in gender imbalance;
- ➢ At least 12.5% of the adult population (aged 25-64) should participate in lifelong learning.

Together with eleven further core indicators, these constitute the Coherent Framework on Indicators and Benchmarks in the field of Education and Training, adopted by the European Council in 2007. The eleven additional indicators are: participation in preschool education, special needs education, literacy in mathematics and science, ICT skills, higher education graduates, cross-national mobility of students in higher education, educational attainment of the population, investment in education and training, civic skills, professional development of teachers and trainers, adults' skills, language competencies, learning to learn skills.

The annual report on "Progress towards the Lisbon Objectives in Education and Training," a yearly publication of the European Commission jointly produced by the Directorate General for Education and Culture, the Centre for Research on Lifelong Learning – CRELL, and EUROSTAT, monitors the progress of European educational systems towards reaching these objectives.

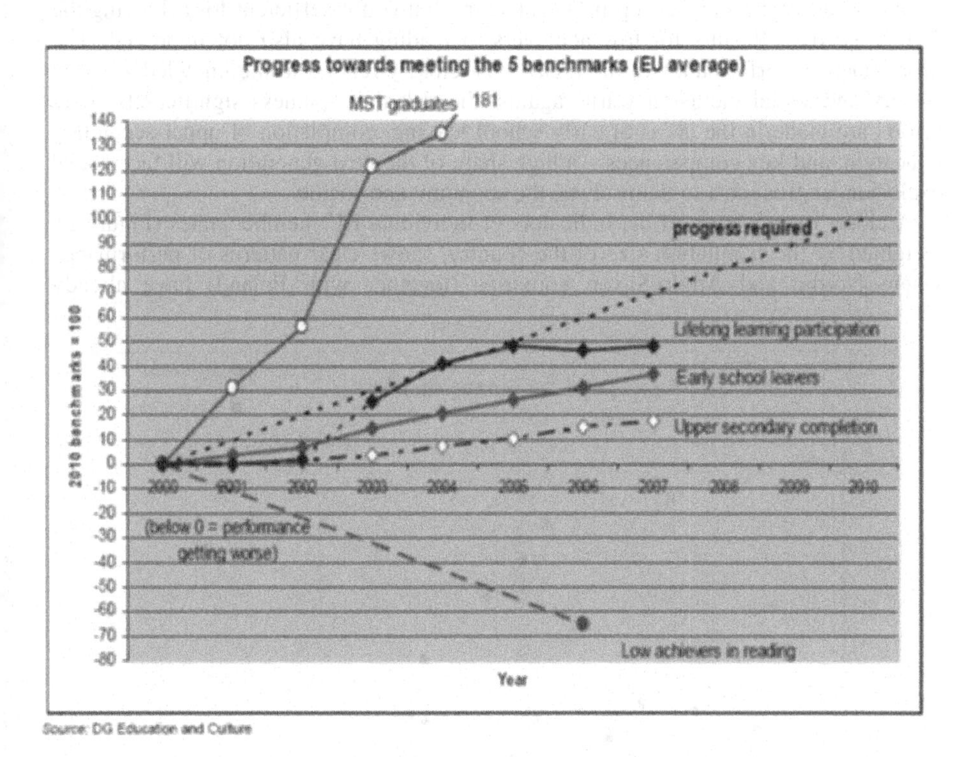

Fig. 1. Overview of EU average performance levels and progress in the five European benchmarks for 2010 [12][1]

As we can see from Figure 1, the only benchmark already achieved is the increase of at least 15% in the number of graduates in mathematics, science and technology. However, progress in reducing the gender imbalance is still limited. Lifelong learning participation has been broadly on track until 2005, but only as a result of breaks in the

[1] The starting point in the year 2000 is set in the graph as zero and the 2010 benchmark as 100.The results achieved in each year are thus measured against the 2010 benchmark (=100). A diagonal line shows the progress required, i.e. each year an additional 1/10 (10%of total) of progress towards the benchmark has to be achieved to reach the benchmark. If a line stays below this diagonal line, progress is not sufficient, if it is above this line progress is stronger than needed to achieve the benchmark. As regards lifelong learning it should be considered that there have been many breaks in time series, which overstate progress, especially in 2003, therefore the line 2002-2003 on LLL participation is dotted. For low achievers in reading (data from PISA survey) there are only results for 16 EU countries and for two years (new data will become available in December 2007).

data-series for several countries, which led to identifying higher (but more accurate) participation rates and overstate overall progress. In 2006 there was even a slight drop in lifelong learning participation in EU 27. There is constant improvement as regards early school leavers, but faster progress is needed in order to achieve the benchmark. As regards upper secondary attainment there has been slow but constant progress. Progress has slightly picked up in recent years, but is not sufficient for achieving the 2010 objective. Results for low achievers in reading have also not improved. The unsatisfactory performance in the areas most closely related to the knowledge-based society and social inclusion warns against the risk that – unless significantly more efforts are made in the areas of early school leaving, completion of upper-secondary education, and key competences – a high share of the next generation will face social exclusion, at great cost to themselves, the economy and society.

A closer examination of the indicators of individual EU member states (Figure 2), weighted by the population size of the country, shows clear patterns of performance where Nordic and Anglo-Saxon countries (together with Poland) have already

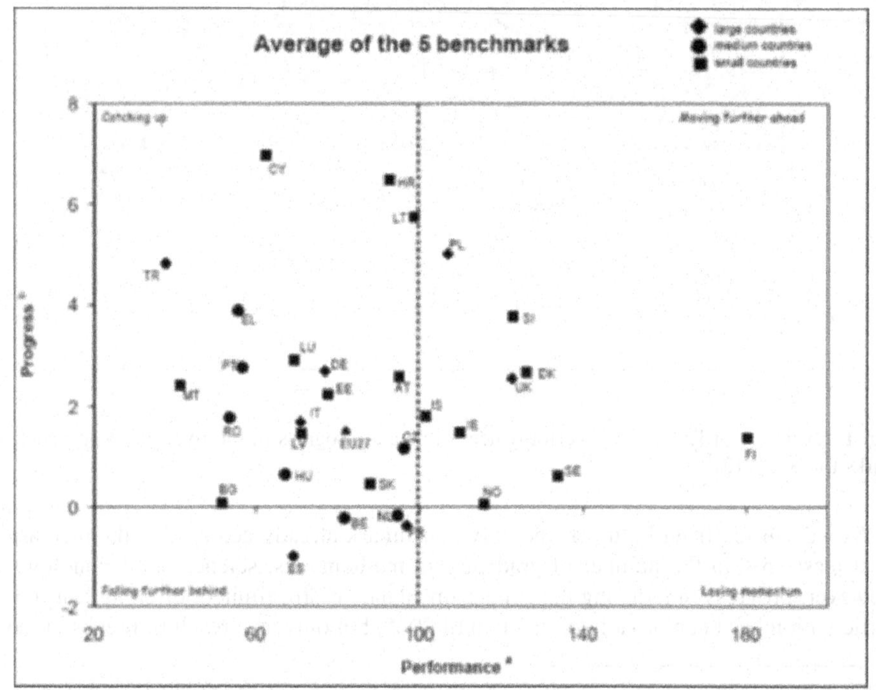

Source: CRELL/Joint Research Centre, 2008.

Fig. 2. Average levels of country performance (2006) and progress (2000-2006) [12][2]

[2] Benchmark for 2010= 100 (Performance); a - average performance (2006); b – average annual growth (2000-06) % (average yearly growth across the five benchmarks). In the case of the indicators on Low achievers in reading literacy and Early school leavers, the average growth rate is multiplied by (-1) to take into account that a negative growth rate is a plus for the country.

reached the benchmarks and are moving further ahead while all Continental, Mediterranean and Eastern European countries are lagging behind. While in any case most countries are catching up, two large countries (Spain and France) and three medium countries (Belgium, the Netherlands, and Romania) are loosing momentum with respect to the starting levels of year 2000.

A useful framework for looking at these patterns of performance is relating them to the European models proposed by André Sapir (Figure 3). Sapir claims that the notion of a single European Social Model is largely misleading as, even though common values are shared, there are still many differences in single countries' social models. He divides European economies into four main models – Nordic, Anglo-Saxon, Continental and Mediterranean, based on the trade-off between efficiency and equity. A model is considered efficient if it provides sufficient incentives to work, and to consequently generate relatively high employment rates. It is deemed equitable if the poverty risk is kept low.[7] Using similar lenses for analysis can challenge to an extent the use of EU-wide benchmarks to promote better performance as they fail to capture the different social characteristics which describe certain groups of countries.

As emerging from our previous considerations, certain group-specific patterns of performance are visible even when looking simply at the education and training benchmarks. In order to further examine this aspect, we have investigated performance levels of groups of countries, as defined by Sapir, taking into consideration the single benchmarks. Nordic countries are Denmark, Finland, Sweden, and the Netherlands; Anglo-Saxons are Ireland and the United Kingdom; Continentals are Austria, Belgium, France, Germany and Luxembourg; Mediterraneans are Greece, Italy, Portugal, Spain, Malta and Cyprus. The accession of the Eastern European countries requires the introduction of a fifth model, Eastern Europeans, with characteristics similar to the Mediterraneans, but taking stock of the specific history and tradition and justifying a separate category. Similar exercise shows clearly that the performance of EU Member States on the education and training benchmarks follows a similar logic and the corresponding groups are a valid categorization.

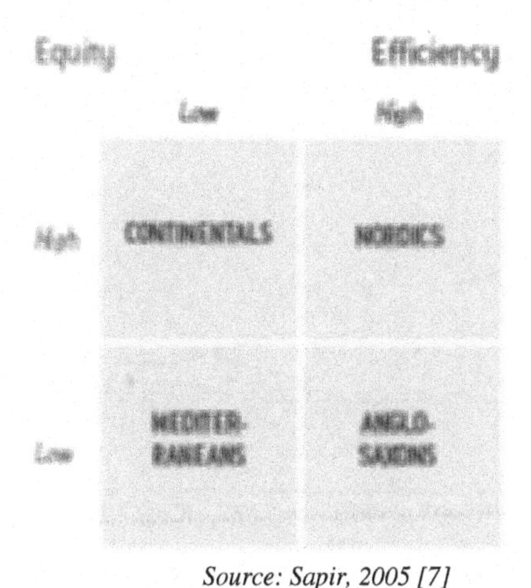

Source: Sapir, 2005 [7]

Fig. 3. A typology of European social models

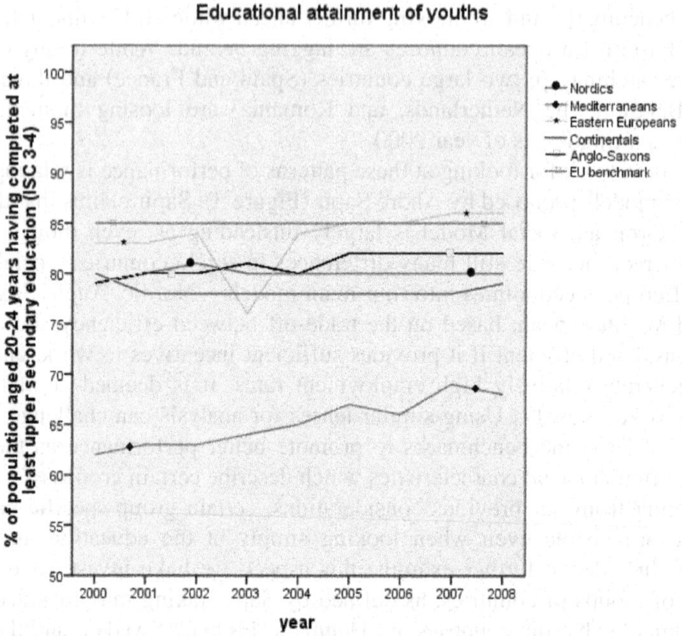

Source: Authors' elaboration on data from the Progress Reports.

Fig. 4. Educational attainment of youths benchmark (average among countries in different groups)

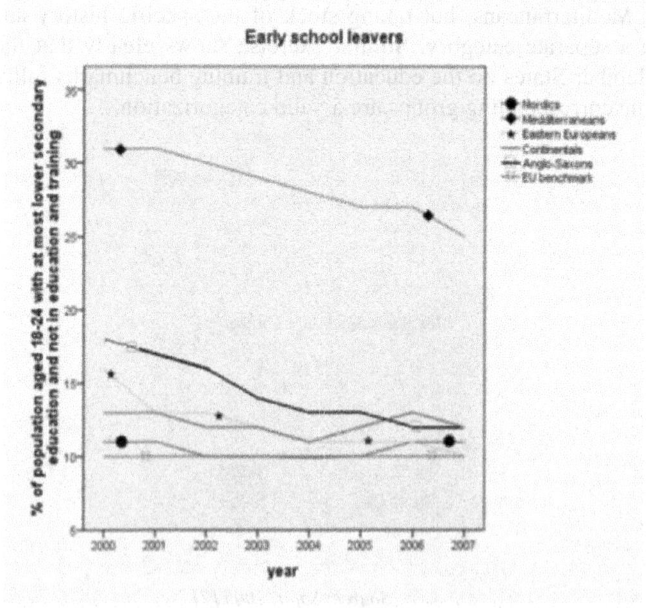

Source: Authors' elaboration on data from the Progress Reports.

Fig. 5. Early school leavers benchmark (average among countries in different groups)

As we can see from the following Figure 4 on the performance of MS on the educational attainment of youths benchmark, there is a clear pattern which sees Mediterraneans demonstrating much lower results. Eastern European countries and Anglo-Saxons (UK and Ireland) have the best results. Interestingly, we see that new Member States are much further ahead than others, are the only group of countries which has already achieved the performance level suggested by the benchmark.

Figure 5 shows the performance of Member States on the early school leavers benchmark, where again we have the group of the Mediterranean countries lagging significantly behind while all other groups have more or less similar levels of performance, with Nordic countries having the best results.

Looking at the lifelong learning benchmark (Figure 6), Nordic and Anglo-Saxon countries are the two groups which have achieved the benchmarked results, even though with a slight decrease in the last years. Mediterranean and Eastern European countries are, on the other hand, lagging behind significantly but have demonstrated a slight progress in the last couple of years.

Similarly, even though the MST benchmark has been already achieved, we can still see a significantly higher performance of Anglo-Saxon countries (Figure 7).

On the reading literacy benchmark, most countries have not had a significantly better performance than in the beginning of the monitored period (2000). The only exceptions are the Anglo-Saxon and Mediterranean countries. With regards to the actual performance level, we can not also here the much better performance of Nordic and Anglo-Saxon countries, and the rather low results of Mediterranean and Eastern European countries.

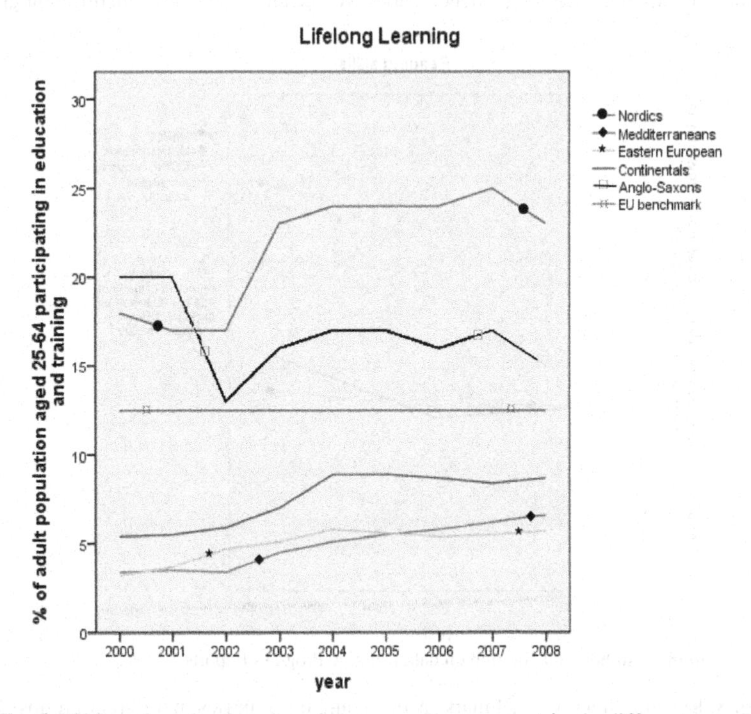

Fig. 6. Lifelong learning benchmark (average among countries in different groups)

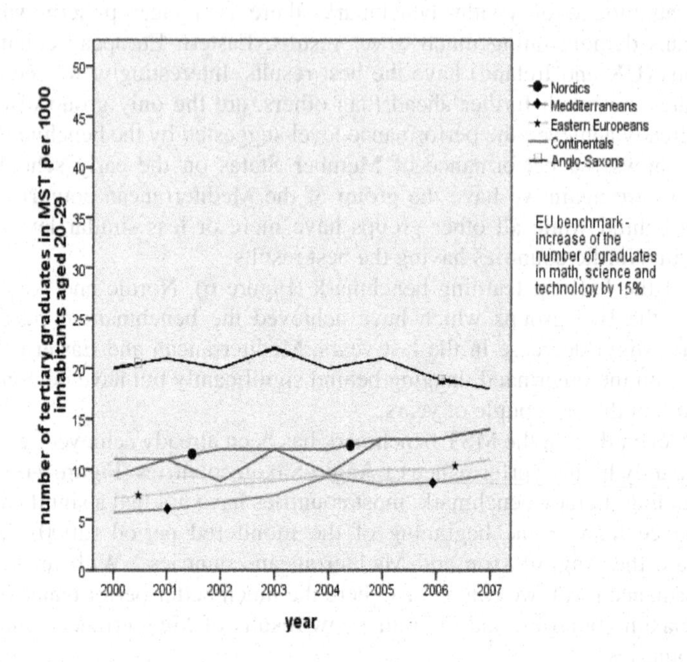

Fig. 7. Math, science and technology benchmark (average among countries in different groups)

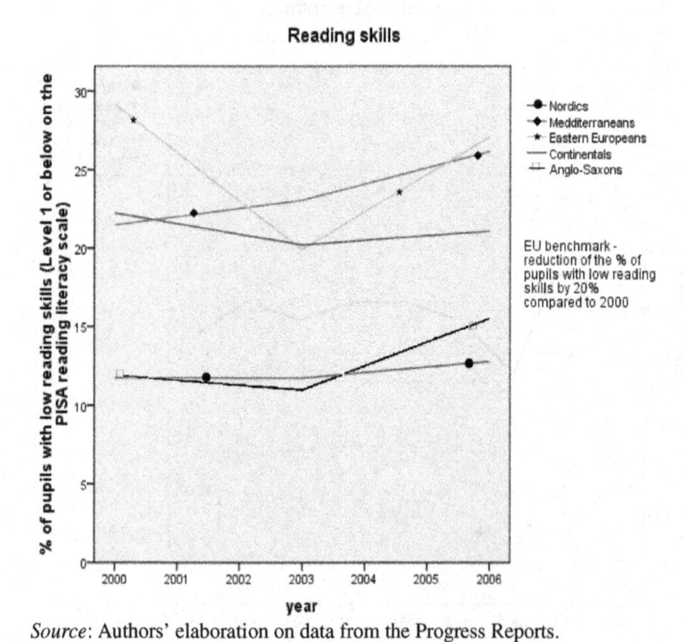

Source: Authors' elaboration on data from the Progress Reports.

Fig. 8. Reading literacy benchmark (average among countries in different groups)

5 Conclusions

Globalization poses significant pressure on countries to create conditions in which citizens could acquire the knowledge and skills for successfully participating in the knowledge-economy. In this context, the example of the European Union and the use of the open method of coordination as an instrument for achieving the Lisbon Agenda, in particular in the field of education and training, offer an interesting insight on possible ways of using benchmarks and stimulating better performance among a number of countries. On the other hand, monitoring performance purely on benchmarks could create expectations which do not necessarily take into consideration social and welfare models, which inevitably have an impact on the possibilities and limits of national education systems. In fact, a simple examination of the performance of groups of countries on the five education and training benchmarks shows consistent trends of similar performance within groups and clear distinction of performance levels between groups. In this sense, the existence of strong structural differences impedes the application of one-size-fits-all policies and suggests the need of cooking ad hoc recipes closely agreed upon with the stakeholders, who take on the independent and legitimate role of interpreting social needs and of launching cooperation strategies at territorial level.

In part, the European Union has acknowledged this need in the Council Conclusions on a strategic framework for European cooperation in education and training ("ET 2020") which re-affirm the use of European benchmarks for the period 2010-2020, but point out that these benchmarks should not be considered as concrete targets for individual countries. [8] Member Sates are invited to consider the ways in which they can contribute to the collective achievement of the European benchmarks through national actions, taking into account their national priorities and economic circumstances. The risk that the current instruments do not allow for identifying correctly the needs of the different countries and do not acknowledge the structural differences existing in Europe still exists. Hence, the steps ahead are related to the development of benchmarking instruments which are better tailored to the specific characteristics of countries or groups of countries and are, thus, more effective in stimulating improvement and facilitating the advancement of the knowledge economy.

References

1. Sahlberg, P.: Rethinking accountability in a knowledge society. J. Educ. Change (2008)
2. Andersen, B., Henriksen, B., Spjelkavik, I.: Benchmarking applications in public sector principal-agent relationships. Benchmarking: An International Journal 15(6), 723–741 (2008)
3. Vink, D.: The development of benchmarking in Dutch social services. In: European Social Services Conference, Madrid, Spain (2000)
4. Lisbon European Council Conclusions (2000), http://www.consilium.europa.eu/uedocs/cms_data/docs/pressdata/en/ec/00100-r1.en0.htm

5. European Commission. White Paper on European Governance (2001)
6. European Commission. Progress towards the Lisbon Objectives in Education and Training: Indicators and Benchmarks (2008)
7. Sapir, A.: Globalization and the Reform of European Social Models, Bruegel, Brussels (2005)
8. Council Conclusions on a strategic framework for European cooperation in education and training ("ET 2020") (May 12, 2009)

Participatory Design of the Participatory Culture: Students' Projections of e-Learning 2.0

George Palaigeorgiou[1], George Triantafyllakos[2], and Avgoustos Tsinakos[1]

[1] Department of Industrial Informatics, Technological Educational Institute of Kavala, Kavala, 65404, Greece
[2] Computer Science Department, Aristotle University of Thessaloniki, Thessaloniki, 54124 Greece
gpalegeo@gmail.com, gtrianta@csd.auth.gr, tsinakos@teikav.edu.gr

Abstract. The participatory culture of Web 2.0 and the implicit empowerment of the learners have not been yet associated with participatory design projects that involve learners in the design and development of the new mediating tools. In this paper, we examine students' projections of Web 2.0 in higher education. Ninety seven undergraduate students participated in 20 design sessions exploiting two needs' elicitation techniques with the aim of envisioning of a course website that meets their learning particularities, that incorporates and exploits their new technological habits and which can be harmoniously situated in the daily routine of a modern, active student. 583 needs were produced and their abstract categorization is presented. Students proved that they had refined views about the elements that can render successful the next wave of e-learning applications and provided directions that can help designers and researchers in developing more informed designs. Students are the main agents of educational change and, hence, they deserve a more active and contributive role in the knowledge society.

Keywords: Web 2.0, elearning 2.0, participatory design, participatory culture, learning management systems.

1 Introduction

Only recently, formal research on Web 2.0 in higher education has started to get published and there is no evidence that the promised revolution has been unlocked. Relevant studies concentrate on specific technologies such as podcasts, tagging, blogs (e.g. [1],[2],[3]) while holistic approaches that embrace the pedagogy transformation, are either non-existent or rare. Simplistic hypotheses about the discrepancies between a slow and cumbersome university and the spontaneous and ever-evolving Web 2.0 [4], and between the stereotypical labeling of students as "digital natives" and the ascertained lack of the required computer literacy, call for a more thorough and systematic investigation of the emerging opportunities and alternatives, along with their requirements and effects.

The introduction of Web 2.0 technologies requires a significant shift in students' and learners' practices and, we should not forget that change is not an instantaneous

M.D. Lytras et al. (Eds.): WSKS 2009, CCIS 49, pp. 119–128, 2009.

incident but a complex and subjective learning/unlearning process for all concerned [5]. A "culture shock or skills crisis" may take place when "old world" educators are forced to introduce novel approaches such as networking, blogging, podcasting, etc. with which they may lack experience and confidence [6]. Hence, a pragmatic e-learning 2.0 should be concerned not solely with affordances but also with the inter-crosses between stakeholders' expectations, motivations and skills.

Interestingly, the participatory culture of Web 2.0 and the implicit empowerment of the learners have not been associated with participatory design projects that involve learners in the design and development of the new mediating tools. Students' input, until recently, has been constrained to the evaluation of prototypes or priorities, and the assessment of satisfaction and attitudes and there are no studies focused in exposing their thoughts and desires from a design perspective. However, students do have high expectations of how learning should take place and which technologies and learning environments best meet their needs [7]. Given the dominance of constructivist and participatory approaches to learning, keeping students out of the design of the new educational platforms creates a paradox: while we seek active involvement in learning, we decrease students' freedom to make decisions about their own learning [8].

Several participatory design methods involve students directly in the software development, based on the assumption that "as a result of their extensive experience with common educational tasks, [students] (1) are able to easily recall, state and elaborate on their prior problems and needs, (2) have unconsciously or deliberately thought of and formed solutions and proposals concerning those educational processes, (3) are willing to collaborate with their colleagues on engineering joint solutions to their problems and, consequently, (4) may produce numerous diverse ideas for the construction of prototypes in a short amount of time" [9]. Students' willingness to lead a process of re-conceptualizing existing pedagogies in the light of new opportunities and engaging in the co-formulation of their future has been documented in several case studies [8, 9].

By adopting the view that educational transformation must involve the concerned stakeholders and not be imposed upon them, we aimed at eliciting students' proposals for the design of an ideal course web-site. Our main objectives were to

a) portray students' notion of the "ideal course website",
b) reveal their projections of Web 2.0 in their learning,
c) examine their suggestions and priorities concerning the required learning affordances,
d) provide a more holistic student-initiated perspective of priorities and power relationships before bringing in questions about current pedagogies and institutions.

We believe that the resulting knowledge can help designers and researchers in their efforts to develop more informed designs, which will positively influence student engagement with learning and its outcomes.

2 Methodology

We followed the participatory design framework of the We!Design methodology [9] which designates that educational requirements can be extracted by conducting iterations of concise and highly-structured collaborative design sessions with different students. The iterations ensure the representativeness of the students' needs while their short duration renders them attractive to many students without significantly disturbing their primary educational obligations. We conducted 20 sessions with the participation of 97 undergraduate students (41 male and 56 female) studying in a Greek Technological Educational Institution. Students were on the third or fourth year of study (with a mean age of 22.8 years old), fulfilled the methodology's requirements for intense computer experience, were engaged in social networking, and Web 2.0 technologies, and had extensive educational experience in tertiary courses and corresponding LMS. Hence, it was expected that they had refined predispositions towards the weaknesses and the opportunities of institutional e-learning services. Each design session lasted for approximately 2 hours and 30 minutes and was comprised of four to six students and two coordinators who guided the students throughout the design process and provided support when needed. The design sessions included two phases: the introductory phase and the needs' elicitation phase. Two different approaches were employed for the elicitation of students' needs. The first technique was applied in twelve sessions while the second was applied in eight sessions.

A video camera captured the design sessions' setting in order to provide a detailed documentation of the whole process. After each session, students were asked to evaluate the final list of needs in terms of the perceived significance for the learning process and their innovativeness in a 5-point Likert scale (1-5) and assess the design process and the coordinators influence. Students, in another context, also completed a computer experience questionnaire.

2.1 Introductory Phase

At the beginning, a detailed description of the design problem was presented, namely, the envisioning of a course website that meets students' learning particularities, that incorporates and exploits their new technological habits (e.g. social networking) and which can be harmoniously situated in the daily routine of a modern, active student. Students were then asked to play the role of a scriptwriter and develop their own fictional character – we use the term "design alter egos" – for whom they would be asked to create scenarios during the rest of the design process. The process started with the selection of a photograph from a set of photographs carefully collected from cultural magazines. The participants were then asked to shape their design alter ego' s physiological, sociological and psychological traits through a process of introspection, recollection and organization of personal experiences, and, at the same time, reflection on other user's attitudes and characteristics, and document them using a detailed predesigned form. The characters should (a) refer to a student, preferably close to their age, with explicit learning interests and (b) be someone with whom they can either relate to or simply be able to talk about. When the students completed the creation of their design alter egos, they presented them to the rest of the group. The design alter ego concept was introduced in order to liberate students from the fear of straightforwardly exposing themselves and to offer a mediating artifact to think for and with.

2.2 Needs' Elicitation Technique I

Five sequential activities were conducted: (a) Students were provided with a set of textual and visual stimuli, in the form of nouns, verbs, questions and photographs regarding the diverse contexts where their design alter egos could spend time during the course of a day (e.g. "in the morning, in the afternoon, at night" or "alone, with my roommate" etc.), together with exemplary imagery that included photographs of a lecture room, a student's room etc. They were then asked to create short scenarios in respect to the specified contexts and recall existing problems and needs. (b) In the next activity, students focused on their fictional character's personality traits and behaviors, and searched for well-situated needs, directly linked back to their psychological characteristics. Relevant textual stimuli was provided (e.g. "When would she use her favorite motto while using the course website?" etc.). (c) Next, students were asked to situate their design alter ego in an advanced technology context. Textual stimuli in the form of questions (e.g. "What kind of opportunities do Web2.0 technologies offer in the design of the course website for your design alter ego?") was provided. (d) Afterwards, students were asked to look for features or services that might be helpful in open-source or commercial LMS. A set of printouts depicting existing course websites were given to them. Students evaluated known solutions through their design alter ego's perspective and incorporated them (or not) into the needs pool. (f) During the last activity, students were shown a 5-minute video comprised of segments from well-known Sci-Fi movies and were asked to envision the ways in which the educational system or the social scenery might change and how the new status quo could affect the requirements of the corresponding course website.

2.3 Needs' Elicitation Technique II

The second approach was structured as a board game. The main elements of the game were a round board, pawns and dices, one for each participating student. The board was divided in 20 slices, each one accompanied by a description card. Each slice referred to different exploration activity designated to assist students in creating scenarios. Students were given points whenever they managed to fulfill what was asked of them and the number of points related to the number of needs they recorded. The slices were organized in seven categories: (1) *The learning category* (4 positions) which provided different sets of verbs (e.g. learn, understand, explain etc.) and nouns (e.g. lecture, assessment etc.) extracted from learning theory books and aiming at reinforcing the essence of learning. (2) *The time category* (3 positions), which presented to the students different periods of time in design alter ego's daily routine: morning, afternoon and night. (3) *The context category* (3 positions), which provided images of three distinct contexts where their agent might live: school/university, home-students' rooms, recreational spaces (coffee shops, clubs, etc.). (4) *The technology category* (3 positions), which asked students to envision the ways in which upcoming technological innovations could affect their design-alter-ego's needs. The cards presented either verbs (e.g. change, foresee, imagine etc.) or questions regarding the use of novel technology products in the classroom (e.g. Tablet PCs, interactive whiteboards etc.). (5) *The divergent category* (2 positions) consisted of two creativity techniques. The first one asked students to imagine that the course website was replaced by a human agent and then try to think of ways in which this agent would act to fulfill their design alter ego's needs and requirements. The second technique was similar but students were asked to imagine that the course website was transferred to another medium such as cinema, theater, radio etc. (6) *The existing solutions category* (2 positions) in which existing LMS were presented. (7) *The extras category* (3 positions) which allowed students to use cards from whichever category they wanted while being rewarded double points for each scenario proposed.

In both techniques, students, after being presented with the tasks, thought alone at first and presented and discussed their scenarios afterwards elaborating their view on their significance.

3 Results

3.1 Students and the Process

The participants could be considered as a representative sample of computer literate students since they used computers for an average of 5 hours per day, shared equally this time between academic work and personal interests and have been using computers for 6,52 years. Their most frequent computer tasks were listening to music, e-mailing, social networking, synchronous communication with their friends (text-voice), seeking for entertainment material (YouTube, etc.), watching movies, reading news on the internet, searching for consumer goods and playing games.

Students were excited with their participation and evaluated very positively both the products of the design sessions and the process. They characterized the resulting needs of their session as "innovative", "interesting", and "complete" and the design process as "satisfying", "unexpectedly enjoyable", and "efficient". They also underlined the friendly, collaborative and creative atmosphere that prevailed throughout the sessions. Without any significant statistical differences in the perceived satisfaction and effectiveness, the two different techniques produced similar needs that converged to the categories analyzed later in the paper.

3.2 Needs

Students produced 583 distinct needs (duplicates in each session were removed). The suggested needs were initially organized based on their content, and similar needs were grouped and rephrased in order to form a set of discrete categories. These categories are presented in Table 1. The columns correspond to the total number of proposed needs, the percentage of needs in each category that were considered to be innovative (their mean assessment was greater than 4), and the mean perceived significance of the needs in each category.

***Course Syllabus*:** Students underlined the need for a better contextualization of the domain into their world, their professional and academic ambitions. They asked for the integration of the course description with further educational prospects (post-graduate studies, opportunities for diploma theses, etc.), artistic work (films, literature, etc.) which could stimulate the pursuit of the philosophical roots and pragmatic consequences of the subject matter, and with a detailed portrayal of the skills they would acquire (even showing videos of workers practicing in industry or other institutions). Two non-typical needs were identified: an estimated study load chart, namely, an informal calendar of the anticipated load of students during the semester in order to be able to better organize their priorities; and a frequent strategic learning mistakes list which could help them avoid common misunderstandings. Students seemed to look for informal views and empirical hints from their professors that could inform their decision making processes.

Table 1. List of proposed needs

Needs Categories	*#*	*I*	*S*
Course Description	44	23%	3,85
News and Updates	59	19%	4,02
Communication	59	41%	3,99
Content Delivery	173	52%	3,89
Participation	28	43%	3,75
Networking	46	49%	3,51
Projects	58	33%	3,96
Exams	24	37%	4,38
Course assessment	14	36%	4,00
Usability	65	42%	3,69
Entertainment	6	17%	2,21
Secretarial integration	7	29%	4,03
Total	*583*	*40%*	*3,9*

***News and Updates*:** Students suggested typical news and updates services including information feeds in the form of emails, SMS and RSS. However, the most unanticipated requirement was the pursue of a wider variety of information services; students considered as self-evident that the course website should function as an informational portal for the subject matter by providing industry and research news (extra-curricular info about developments, discoveries, new software, press releases, etc.), political and environmental news (new laws, policies, research roadmaps) and related activities that take place in the academic environment or the city they live (seminars, conferences, lectures, competitions). Students essentially asked for the opportunity to scent the idiosyncratic characteristics of the domain through information services that could give them more chances for participation and learning.

***Communication*:** Students proposed typical services of synchronous and asynchronous communication with their peers and their professors. The acknowledgement of others e-presence was requested in all sessions and for all stakeholders. They sought for a way of submitting public questions to the instructor in order to force him answer with immediacy and developing an open knowledge base. They wanted to exploit the transparency of the medium in order to render it as a regulative channel that binds teachers and learners.

***Contents and presentation*:** This category gathered the majority of students' needs (31% of the total needs), providing validation of the high value students attribute to online learning resources. More specifically:

A) They proposed live broadcasting of lectures and the provision of the respective recorded versions. They also asked for podcasts which could be played offline when walking or exercising. In two sessions, the format of documentaries was suggested as an intriguing form of presenting the learning material.

B) They asked for two types of summaries, one which should describe the issues that would be discussed in upcoming lectures, and one which should address the key elements of each prior lecture. Multiple formats were proposed for these summaries as well (e.g. textual, podcasts, vodcasts etc.).

C) They requested extensive supportive material for each lecture. They mostly referred to video-presentations from the workplace (e.g. for an instructional theory course, students asked for video recordings of classroom lessons that convey the studied concepts). They even asked for live connections and teleconferencing with corporations, industries, schools, etc.

D) They pinpointed the value of practicing the acquired knowledge through simulation and modeling software. They envisioned electrical circuit simulators or instructional planners and seemed puzzled about their absence in existing websites. Game-like simulations were also discussed in combination with rewards system for the students with the best performances.

E) They requested historic/background information for the subject domain, its evolution, and its projections in the future, along with information about the most important figures that determined its development. A similar proposal concerned the presentation of videos in the form of "as today".

F) They underlined the significance of study material beyond the scope of the course that could support those who might want to tackle more with the topic (literature, instructions, links, articles, advanced projects and cases).

G) Students acknowledged that most subject domains demand the use of one or more software applications. Hence, they asked for a space dedicated to alternative software, guidelines and updates. They acknowledged that the process of learning about software is one of the most complicate problems they confront every day.

H) They asked for psychometric and learning style questionnaires which could help them detect by themselves their unique characteristics and develop their study strategies.

I) Finally, they proposed the creation of a dictionary in the form of a wiki that could prove to be a valuable source for peer-to-peer learning.

***E) Course co-formulation*:** Students were willing to create shareable resources such as class notes, bookmarks, and links to relative articles. Additionally, they asserted a role in shaping the course by selecting and voting their preferred way of teaching, taking the responsibility to provide a topic of interest as the theme of a lecture that would be elaborated by the instructor, organizing mentoring sessions with the teacher or ex-colleagues, initiating extracurricular student projects relevant to the domain with no control from the instructor, and finally, by voting on student initiated requests, such as the change of the exams dates. Students did not want to control or direct the core of the learning process but asked for further opportunities for involvement and participation in the decision making processes.

***F) Course networking*:** Students anticipated multiple networking opportunities. They wanted the web course site to belong in a network of similar courses of different universities, a network with explicit possibilities for sharing resources and which could exceed the existing organizational and computational infrastructures. They expected a variety of video presentations, notes and cases studies coming from different instructors, search services that index all relevant resources, and plentiful and innovative communication/cooperation opportunities. They claimed that the opportunity to communicate with other professors seemed especially attractive.

They also asked for extensive networking opportunities with professionals as sources of authentic information. Students exhibited a genuine interest about the respective labor market, and proposed the offering of employment ads relative to the course domain, as a way of gaining a deeper understanding of the labor market orientations and assessing the contents of the course from a different perspective.

Students asked for the integration of their identity in existing social networking sites, such as Facebook©, with their course profile as a mechanism for familiarizing with their fellow students. Students alternatively suggested the creation of local social networks in course level, with the presentation of personal information for each student and feedback about his behavior on the website. Students wanted to socialize more with their colleagues in a way similar to that exploited in existing social networks. They asked for similar networking opportunities with ex-students who had successfully attended the course or even graduated from the department, so as to discuss and share their views about the value of the course knowledge.

They were interested to learn more about their instructors asking for their biography, their close associates, their publications, and so on. In a way, they were asking to become their friends and exceed the prescribed walls of formality inherent in their relationship. Many of them had already requested to become friends with those professors who had active accounts in social networking sites.

***G) Projects*:** Students extracted needs related to the typical tasks of managing projects, and put emphasis in the creation of a project pool. Some students asked for personal video presentation of the projects so as to make them more personal and attractive. Many needs concerned group work which, according to students, should be better supported by integrated project workspaces. They wanted to have the choice to manage their groups' synthesis and illustrated a tool which could enable them to select teammates from an available "market". They hinted that their social relationships do not allow them to participate in teams as flexibly as they wished.

***H) Exams*:** The students stressed the importance of a question bank with exemplary answers. They expected to view electronically their grades and to study teacher's feedback on their answers in order to learn from their mistakes. They also wanted to be able to start an e-argumentation in case of dispute.

***I) Course assessment*:** Students wanted to be able to evaluate both the instructor and the course. They differentiated their approach by proposing informal evaluations during the semester which would have a direct effect on the course's progression. Other forms of evaluation proposed included a regular column of complaints, a form for improvements suggestions, and an anonymous blog devoted to assessment. Students also asked for the establishment of an online session dedicated to discussions about the course and proposals for its improvement. They wanted to be able to co-formulate the course and adapt it to their needs and interests. They also said that they were willing to video-record informal assessments which could be useful for other students before selecting the course.

***K) Usability and HCI*:** Students asked for notifications of changes on the website content, for a customizable user interface, and accessibility tools for students with special needs. They emphasized the need for simplicity in the interface and criticized several usability issues of existing LMS. They indicated the need for a client application that could download automatically the website's learning material, allowing them to work offline while retaining the same web interface. Students also asked for mashup flexibility either in the form of incorporating a concise version of the course site to other portals or by providing the possibility of including mashups of other applications into the course website (such as email, news feeds eth.). The students pursued the unification of their information channels with the course's updates as a motivation for closer attendance of the corresponding activity. They also asked for a personal space which would host their profile pages, and proposed the implementation of a personal calendar and a notes folder.

4 Discussion

Students did not seem eager to challenge the dominant paradigm of Learning Management Systems, an observation which is in agreement with research results concerning Web 2.0 tools that appear to extent, rather to challenge, current pedagogies. Their proposals were in a close relation to the participatory culture [10] and pedagogy 2.0 [6] where there are greater opportunities to: (a) initiate and influence curriculum (they wanted to propose lecture subjects, organize mentoring sessions, negotiate procedures and learning directions, etc.), (b) produce and share learning material as authors (they were willing to initiate extracurricular projects, produce video-assessments for the course, share their projects, links, etc.), (c) connect to the world as a whole (they asked for networking opportunities with instructors, fellow students, ex-students, other courses, students of the same course in different departments, professionals, labor market, existing social networking sites), and (d) create a community of practice among teachers, learners and professionals in order to familiarize with professional practices and exchange ideas, products and interests. They blamed the isolating experience of much text-based traditional education by asking for multiple forms of presenting the course contents [6] and focused on a better contextualization of all domain knowledge in real life.

However, overall students tried to avoid the initiative of learning. They did not emphasize self-direction and focused mainly on improving existing practices of self-studying pre-organized material. Despite the opposing references, they did not concentrate on learning experiences that are short and opportunistic. They kept their learning and personal spaces apart and did not challenge the role of the academic institution, nor its fundamental organization. Students pinpointed inter-university constellations as means towards improving dramatically the quality and the quantity of the courses' contents and the opportunities of networking. They attributed to the instructors the role of authoritative sources of information, a secure intermediate layer between them and the available web resources.

Our study demonstrated that students had refined views and ideas about the elements that can render successful the next wave of e-learning applications. Thus, it would be safe to claim that the participatory culture of e-learning 2.0 must be combined with participatory efforts for co-designing its characteristics with the students, in an open, democratic, empowering atmosphere that can also address change management prerequisites. Students' visions of education either concerning an LMS, a note-taking tool or an assessment application are fundamental for the harmonious integration and exploitation of educational technology in everyday learners' life. Students deserve a more active role in co-formulating their future in a truly participative knowledge society which promotes contribution and involvement, highlights change management and not abrupt interventions and comes in opposition to homogenization and passivity of the learners and learning environments [8].

References

1. Lazzari, M.: Creative use of podcasting in higher education and its effect on competitive agency. Computers & Education 52(1), 27–34 (2009)
2. Yew, J., Gibson, F.P., Teasley, S.D.: Learning by Tagging: The Role of Social Tagging in Group Knowledge Formation. MERLOT Journal of Online Learning and Teaching 2(4) (1), 275–285 (2006)
3. Kim, H.N.: The phenomenon of blogs and theoretical model of blog use in educational contexts. Computers & Education 51, 1342–1352 (2008)
4. Jones, C.: Infrastructures, institutions and networked learning. In: Proceedings of the 6th International conference on Networked Learning, pp. 666–674 (2008)
5. Scott, G.: Effective change management in higher education. Educause Review 38(6), 64–80 (2003)
6. McLoughlin, C., Lee, M.J.W.: The 3 P's of Pedagogy for the Networked Society: Personalization, Participation, and Productivity. International Journal of Teaching and Learning in Higher Education 20(1), 10–27 (2008)
7. Conole, G., Creanor, L.: In their own words: Exploring the learner's perspective on e-learning. JISC (2007), http://www.jisc.ac.uk/media/documents/programmes/elearningpedagogy/iowfinal.pdf
8. Siozos, P., Palaigeorgiou, G., Triantafyllakos, G., Despotakis, T.: Computer based testing using "digital ink": Participatory design of a Tablet PC based assessment application for secondary education. Computers & Education 52, 811–819 (2009)
9. Triantafyllakos, G., Palaigeorgiou, G., Tsoukalas, I.: We!Design: A student-centred participatory methodology for the design of educational applications. British Journal of Educational Technology 39(1), 125–139 (2008)
10. Jenkins, H.: Convergence Culture: Where Old and New Media Collide. New York University Press, New York (2006)

Towards an "Intelligent" Tagging Tool for Blogs

Juraj Frank[1,2], Renate Motschnig[1], and Martin Homola[2,3]

[1] University of Vienna, Research Lab Educational Technologies, Austria
[2] Comenius University, Faculty of Mathematics, Physics and Informatics, Bratislava, Slovakia
[3] Fondazione Bruno Kessler, Trento, Italy

Abstract. Tagging allows people to effectively organize web resources such as images, bookmarks or blog articles. Things are found easier by browsing tag clouds relying on the tags that have been assigned before. The success is by large determined by the quality and relevance of tags assigned to content – and so it is dependent on people who do the tagging. We investigate mental processes that underlie tagging. In order to improve quality of tagging, we provide guidelines for users of tagging systems and in addition we suggest features that an "intelligent" tagging tool should bear in order to facilitate the tagging process.

Keywords: tagging, categorization, blog, usability.

1 Introduction

Tag is a label associated with something for the purpose of identification. Tagging is a process of assigning such labels or keywords to objects for sake of future identification. It is possible to tag anything what makes sense, however, these days tagging is a popular categorization method for photos, bookmarks or blogs articles. But why is tagging so popular? During the history, first major successful catalogs of web content used simple hierarchical categorization schemes, where one object could usually be a member of one category only. The catalog was exclusively edited by its editors. Tagging is different in two ways. First, tag is much more a label than a folder: apart from classical categories, many labels are possibly associated with each object. Second, Everyone is allowed to tag things. These two little tweaks made tagging more accessible and attractive to everyday people. Tagging is useful to them: it enables to find things by fishing them out of tag clouds and it allows to effectively organize things with little effort. Thus, tagging has become a novel decentralized way of organizing, sharing, and structuring information in the knowledge society [1].

However, several practical problems are associated with tagging. Different people usually use different tags to describe the same topic or type of object (e.g., where one would assign tag *fruit*, someone else assigns the tag *food*). Also, people sometimes forget to assign some particular tag that is of some importance (e.g. *fruit* is missed out for an article about oranges, apples and pears). In order to improve quality of tagging we try to answer the following three research questions:

M.D. Lytras et al. (Eds.): WSKS 2009, CCIS 49, pp. 129–136, 2009.

– What are the basic mental processes that underline tagging?
– How should people be educated and guided during the process of tagging?
– How should tagging systems be improved in order to support the users during the process of tagging?

Literature study shows that both prototype theory [2] and the concept of the basic level categorization [3] are relevant approaches to explain tagging. We have conducted interviews with blog users. Based on this research, we have put together guidelines for tagging and we have identified features that intelligent tagging tools should have in order to improve the quality of tagging.

2 Small Empirical Study

Interviews were conducted with users of the `blog.matfyz.sk` portal.[1] First, a key user was interviewed in a face-to-face session to provide encompassing and detailed feedback on the initial version of the questionnaire. Eight users have completed the questionnaire. The conclusions drawn in this section are based on these interviews, and they provide insight into human motivation and habits concerning blogging and tagging. However, due to the smaller number of participants and special purpose of our portal these conclusions should not be taken as universally applicable and exhaustive. Let us quote some of the responses to the three research issues we consider the most central.

2.1 Why Do People Write Blogs?

The questionnaire opened with the question: "Why do you write blogs?" Typical answers were:

– Because I have to and sometimes because I want to.
– Sometimes I confess and sometimes I write something useful for students, eventually useful for the world.
– I wrote blogs as a part of my course on web-design.

Besides for the obvious motivation connected to the educational purpose of our portal [4], the users have suggested the following: writing an article on topic of interest for oneself or for others; writing something useful. This clearly points to the use of blogs to organize and share knowledge. Interestingly, one user pointed out she was writing blogs in order to confess, therefore blog can be seen as a psychological tool that may help to deal with aspects of one's personality.

2.2 What Strategy Do People Choose for Tagging?

The survey included several questions about tagging strategy: "What kind of words you usually use as tags?" "What do these words mean?" "Do you use

[1] A community portal used at the Comenius University, Faculty of Mathematics, Physics and Informatics.

names of people as tags?" "Do you use abbreviations as tags?" "Do you use more tags to name the same thing?" "Are the tags that you assign to one blog posting related, or are they unrelated?" Typical answers were:

- I use more tags to name the same thing, but I try to avoid such use.
- I do not use more tags to name the same thing, however I use hyponyms.
- I use the name of a person as a tag, in case it's a famous person.
- I use abbreviations, definitely. Tag "FMFI" is a much better tag than "Faculty of Mathematics, Physics and Informatics".

Some of our users try to avoid using more tags to name the same thing and *hyponyms* can be used to avoid this. If the article is about a person, it can be topically tagged by the name of the person. The interesting point suggested by one of our users is that it is reasonable especially if this person is known to others. Clearly this answer makes sense, it is not very likely that others would search using this tag if they do not know the person. People also prefer to use abbreviations if they are commonly used in speech or in writing.

2.3 What Are the Properties of a Good Tag?

We are also concerned with how to determine which tags are good. The following questions were relevant in this respect: "What kind of words do you use as tags?" "What do these words mean?" "What are the properties of a good tag?" "How would you define a well tagged blog article?" "When you choose which words to select as tags, do you consider that good selection may later facilitate search for the article?" We have learned the following answers:

- I am using nouns, sometimes proper nouns and names.
- A good tag should be a word, that comes first to mind to the majority of the users.
- A good tag should be precise, appropriate, intuitive, grammatically correct.
- A well-tagged article should have tags that are appropriate and brief, but not too specific neither general. It shouldn't include too many synonyms.
- A well-tagged article should have tags, that describe each important part or topic of the article with the most general terms and at the same time with the most appropriate terms.
- Tags are supposed to facilitate later search for the article.

According to the responses of our users, a tag should be a word, usually a noun, which is appropriate in regards to an article. It should be also intuitive and the word that comes first to mind (activated immediately). A well-chosen tag should be grammatically correct and it should enable future retrieval of tagged content, many users agree, that this is one of the goals of tagging.

In this section we have presented the results of a survey we have conducted with users of our blog portal. We have learned some interesting observation and users' views on blog publishing and tagging. However, we have also learned that some of the users are unaware of how to properly tag the content in order to assure good navigation through articles in the future. In order to build "intelligent" tagging tools, not only users' but also experts' opinion needs to be considered.

3 Psychology of Tagging

Categorization is one of the most fundamental and pervasive cognitive activities [5]. Categories are studied since the works of Plato and Aristotle. The classical view on categorization assumes, that each object can be a member of one category. This corresponds with a standard hierarchical categorization scheme, where each category can contain objects and one or more categories, however each object can be associated with only one category. This scheme is usually used to categorize books in a library or in early online catalogs of web content (e.g., Yahoo.com). Although tagging is closely related to this "classical" view on categorization, in current tagging systems we usually assign more than one tag to particular object, hence it belongs to more than one category.

Theories based on the work of Rosch [2], Lakoff [6] and Tversky [3] change the "classical" view on categorization by introducing the prototype theory. Prototype theory assumes that some members of a category can be more central than the others, e.g., *a raven* is more central to a category *bird* than *a penguin.* Systems of categories can be different for each individual, they are based on the experience, so different cultures and individuals can have different categorization systems, which also coresponds with Piaget's views on constructivism [7].

Considering prototype theory and different levels of categorization, there is one important level of categorization - *basic level.* To illustrate basic level categorization, consider the following question: *What are you sitting on?* People usually say that they're sitting on a *chair* rather than on a piece of *furniture* (more general category) or on a *kitchen chair* (more specific category). The basic level of categorization is also learned first by children [8]. Categories on the basic level are very informative and easy to imagine (e.g., car, dog). The categories are based on semantic features that are related to our perception of the real world [3]. Understanding human categorization through basic level theory is essential to user-centered design of taxonomies, ontologies and tagging systems [9].

In order to get a better insight on what mental processes are active during tagging, we should first consider some common motivations to tagging. There are basically two categories of motivations: organizational and social [10]. Tagging helps people to manage their information and knowledge, it enables people to express themselves and to have fun. Tagging also gives information management a social dimension by enabling people to share information and perform these activities collaboratively [11].

Inspired by current opinions in the blogging community [12], we hypothesize that the actual process of tagging an object consists of two stages. First, semantic concepts related to the object are activated immediately. Semantic relationships of activated concepts were discovered during cognitive experiments where subjects were asked to write semantic associations with an object. However, not all of activated concepts are suitable as a tag. Therefore a second stage is needed, during which one decides to use a particular concept as a tag or not, considering relations of the concept (tag) to the tagged object.

The process of tagging a blog article can also be divided into multiple subtasks, that user needs to perform, so assigned tags are relevant and informative.

The user of the tagging system should be able to indicate the main topic and subtopics of the blog article and express it by using the correct words. In addition, a brief search for synonyms of these words should be performed. Since added tags will help other users to find the article, thinking about the usual searching behavior of other users may be also useful. Therefore, tags that others may use to find the article should be added as well. Finally, one should spell-check and evaluate the relevancy of each assigned tag to the article.

4 Guidelines for Improving Tagging Systems

Based on our research presented so far, we suggest that tagging systems should follow specific recommendations. This serves to the main objective of our effort, that is, to improve the quality of tag-based navigation, in other words, to facilitate the process of finding the desired information when tags are later used to access content. The recommendations we present here aim to support this goal by increasing the accuracy of tagging carried out by the user. They concentrate on two essential areas: educating the users so that they know how to tag better and enhancing the system in order to support users during the tagging task.

4.1 Educating the Users

Usually a lot of redundant and low-quality tags are assigned as tags, because users are confused and they do not have clear ideas regarding tagging. Users need to understand what exactly tagging is, how it works and what purposes it serves. This information should be provided to then in a suitable way (e.g., a help link near the tagging tool). Apart from that, we can further increase the quality of tagging by simply providing recommendations, best practices or examples of tagging. This is an important usability feature of any system that employs tagging. We suggest the following guidelines to be presented to the users:

Think topic-wise, do not miss any topic. One of the most common mistakes in tagging occurs when people only concentrate on the main/intended topic and fail to realize that the article also covers other topic.

Think specific and general at the same time. An article is only described properly by specific and general words together. This has to do with the importance of base-line categorization with respect to tagging.

Do not be afraid of synonyms. Different people may use different tags than you referring to the same concepts when searching for your article.

Think of what tags others would use to find the article. Certain words are more often used then others, consider how likely is that you would use the word when searching (e.g., more likely one would use *car* than *automobile*). Other commonly used words are names of people and places, etc.

Some tags are better in plural. Many times it is natural to name tags in plural. This is not a universal rule however (e.g., *cars* is a good tag but *home* is better than *homes*). This is more related to the use of these words as categories than to the actual number of things in the picture/article.

Enter multi-word tags correctly. Multi-word tags are also useful (e.g., *modern art, flight attendant, Philip K. Dick*). Users should not be afraid to use them, on the other hand they should be cautioned how to enter them correctly since different syntax is employed in different tagging systems.

Use established forms consistently. Consider what form of the desired tag is used by others. Which of the synonyms, whether singular of plural form is used, and consider also case in names and acronyms. It is a mistake to miss a tag which is frequently employed as such tags are given prominent place on web sites and hence assure that the content is found by other users.

Follow the community conventions. While these guidelines may be useful in general, it is more important to follow established patterns and community conventions in tagging. Only so the tags assigned by all users of the system can be used to navigate content consistently and efficiently.

Other recommendations are possibly found, however each community may prefer a different strategy and so it is important to follow community conventions above all. These recommendations should be presented to the users in an intuitive way so that they are not overlooked. This is no easy task for the designers. The following is suggested. Help pages that explain tagging conventions ought to be accessible directly from the tagging interface. Assisting tools should be implemented within the system that analyze user's inputs and display warnings/hints when needed. Example should be given by distinguished users which are prominent in the community. Every community has such users. They should be spotted and educated so they would serve as example to the others.

4.2 Enhancing Tagging Systems

Even if we educate users and recommend the best ways to tag their content, if we in addition provide supporting tools that facilitate tagging the outcome shall be greatly improved. In fact it is possible to support computationally a great number of mental tasks that are associated with tagging. The interface of an "intelligent" tagging tool should include features for this sake. These features interact with users in two ways: either they mark up suspicious tags so that the user can easily spot them and correct them if needed or they suggest additional tags to be considered. Below we list the most essential of such features:

- spell check – misspelled tags should be marked up;
- unrelated tags – tags that assumed unrelated to the text of article should be marked up as well;
- missed tags – additional related tags extracted from the text of article missed by the user should be suggested;
- plural form, and possibly other established forms (e.g., special capitalization) may be suggested for certain tags;
- synonyms should be suggested;
- more general concepts derived from tags already used should be suggested, thus increasing the chance of hitting the basic level (see Sect. 3).

Spell check is easily supported by building in one of the open source spell checkers that are available (e.g., Aspell). By applying standard machine learning based term extraction and concept extraction techniques we are able to extract characteristic concepts from the text of the article being tagged. By consequent application of term similarity metrics and comparison with the tags provided by the user we are able to estimate whether some concepts are missing or whether there are some possibly unrelated tags suggested by the user. In addition we will keep track on history of tags that have been already used in the past by different users. If for some tags there are preferred forms, e.g., plurals, capitalization, we will suggest these forms to the user. A ready to use tool to handle synonyms, more general and more specific concepts is WordNet. In addition tagging systems that track the history of used tags may employ ontology learning techniques [13] and may try to incrementally learn such relation between tags that have been previously used. One such method has been proposed in [14].

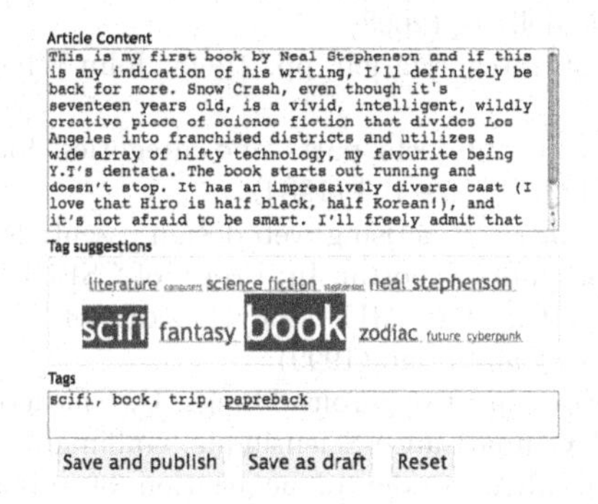

Fig. 1. Prototype of an enhanced tagging tool

In Fig. 1 a prototype of a tagging interface is depicted. This interface includes tag suggestions. Note that suggestions are displayed in enlarged type if they are considered more relevant with respect to the article text by the system. If the user selects some of the suggested tags, these are immediately copied below into the "Tags" input. Since multiple tags can be entered into the "Tags" input, users should be aware of how to separate them – in our case, users can separate multiple tags by using commas. However some systems use other separators, such as space, which may cause difficulties with entering multi-word tags (e.g., *modern art*). In that case, multiple words in one tag can be joined by underscores or hyphens.

5 Conclusion

We present a study in questions such as why humans write blogs, how they tag blog articles, what are the psychological processes that underline tagging, etc. It appears that both prototype theory [2] and basic level categorization [3] are relevant to explain tagging. Based on this research and based on interviews that we have conducted with blog users, we present guidelines for tagging and we identify essential features for intelligent tagging tools in order to increase quality of the tagging process.

The next step will be to design and implement such an intelligent tagging tool to be used on our portal. We plan to employ ontology learning techniques [13] to discover semantic relations between tags and to record these relations in an ontology. With an ontology in the background, we hope to be able to develop intelligent tag suggestion based on semantic relations between tags.

References

1. Derntl, M., Hampel, T., Motschnig, R., Pitner, T.: Inclusive social tagging: A paradigm for tagging-services in the knowledge society. In: Lytras, M.D., Carroll, J.M., Damiani, E., Tennyson, R.D. (eds.) WSKS 2008. LNCS (LNAI), vol. 5288, pp. 1–10. Springer, Heidelberg (2008)
2. Rosch, E.: Principles of Categorization, pp. 27–48. John Wiley & Sons Inc., Chichester (1978)
3. Tversky, B., Hemenway, K.: Objects, parts, and categories. Journal of Experimental Psychology: General 113(2), 169–193 (1984)
4. Homola, M., Kubincová, Z.: Practising web design essentials by iterative blog development within a community portal. In: Procs. of CSEDU 2009 (2009)
5. Wilson, R.A., Keil, F.C.: The MIT Encyclopedia of the Cognitive Sciences (MITECS). MIT Press, Cambridge (1999)
6. Lakoff, G.: Women, Fire, and Dangerous Things. U. Chicago Press (1990)
7. Piaget, J.: Logique et Connaissance Scientifique. Gallimard, Paris (1967)
8. Mervis, C.B., Crisafi, M.A.: Order of acquisition of subordinate-, basic-, and superordinate-level categories. Child Development 53(1) (1982)
9. Rorissa, A., Iyer, H.: Theories of cognition and image categorization: What category labels reveal about basic level theory. J. Am. Soc. Inf. Sci. Technol. 59(9), 1383–1392 (2008)
10. Marlow, C., Naaman, M., Boyd, D., Davis, M.: Position Paper, Tagging, Taxonomy, Flickr, Article, ToRead. In: WWW 2006, Collaborative Web Tagging Workshop (2006)
11. Smith, G.: Tagging: people-powered metadata for the social web. New Riders Publishing, Thousand Oaks (2007)
12. Sinha, R.: A cognitive analysis of tagging, blog article (2005), `http://rashmisinha.com/2005/09/27/a-cognitive-analysis-of-tagging/`
13. Buitelaar, P., Cimiano, P., Magnini, B.: Ontology learning from text: methods, evaluation and applications. IOS Press, Amsterdam (2005)
14. Frank, J., Homola, M.: Ontology-driven categorization of blog postings: A scenario. In: Kognice a umělý život VIII (2008)

ICT in the Classroom Microworld - Some Reservations

Ioannis Papadopoulos[1] and Vassilios Dagdilelis[2]

[1] Hellenic Ministry of Education, Primary Education
ypapadop@otenet.gr
[2] University of Macedonia, Thessaloniki, Greece
dagdil@uom.gr

Abstract. Despite the promising perspective of the usage of ICT in education, contrasting opinions state that education has been barely influenced. In this paper we present a critical confrontation of aspects relevant to the usage of ICT in the classroom and categorize main difficulties that arise when ICT is applied in education.

Keywords: ICT in Education, Microworlds.

1 Introduction

New Technologies have changed our daily life as individuals and *citizens*: they support new ways for communication and expression and finally contribute to the reshape of both, our identity and the political landscape. In this new digital era citizens are able to participate in public affairs in a more active way contributing thus to an essential social progress. However the deep multimodal multiliteracy of the citizen is a sine qua non condition for his/her participation. The integration of ICT in education and their teaching effectiveness constitute a main component for this multiliteracy. Our question (in accordance with several researchers) is: Are ICT adequate from the educational point of view? Do they support this multiliteracy? Information and Communication Technology (ICT) constitute the main aim of most educational systems worldwide, at least in the countries which can afford the cost necessary for the usage of digital technologies. Even though ICT has been adopted as an essential tool and educational object, there exists a lot of criticism concerning among others its effectiveness, cost, or its abuse in the everyday teaching practice in the classroom. Neil Postman (1985) [13] very early criticized this pedagogical point of view that presented education simply as an entertaining activity, an aspect that was strengthened in the era of multimedia (especially of multimedia educational applications). In the 90's, when the enthusiasm about the new possibilities of New Technologies and Internet was at its peak, Clifford Stoll (2008) [15] stated that the situation in education (as well as in governmental services) would not change because of ICT. A similar point of view is included in Larry Cuban's (2002) [3] and Tedd Oppenheimer's (2003) [10] work who additionally stated the question: "What will be then?" Many of these critiques focused on the general characteristics of the new media and their consequences in the way people learn and attribute meaning to the new knowledge. They adopt a critical attitude against ICT considering them as an important cultural component that

M.D. Lytras et al. (Eds.): WSKS 2009, CCIS 49, pp. 137–145, 2009.

influences the educational system [1]. However in a series of less known relevant research projects, a question of the effectiveness of ICT in teaching practice is also raised. They specify the problems caused by the usage of ICT in the micro-world of the classroom and at the same time in the macro-world of the educational system. And what is important in these studies, is the perspective that ICT constitutes mainly a kind of didactical and learning tool rather than that ICT is not so effective. Thus, its usefulness is essential only under certain circumstances – otherwise not only it does not support teaching and learning but on the contrary, sometimes it can play a negative role. Examining these studies would help to clarify these circumstances. We present some of the results concerning the usage of ICT in the classroom and the teachers' training. We focus on aspects that demand a critical confrontation. Some findings are in accordance with ones of previous research and studies, while others could be regarded as original. We consider these findings neither idiosyncratic nor occasional. On the contrary, they can be met in various didactic and cultural environments and consequently they represent somehow a more general value. In the next section we present briefly some of the critiques that have been expressed from time to time about the usage of ICT for educational purposes. After that we present in categories the main (didactical) difficulties that arise when ICT is applied in education. In the last section we summarize the most important of our findings and we extract some general conclusions.

2 Some Critical Aspects about ICT

ICT are considered as a main factor of progress. However contrasting opinions claim that ICT create a new kind of gap among the privileged and non-privileged people, the so-called 'digital divide'. Moreover some researchers express a more intense criticism about the uncontrolled spreading of ICT, claiming that these technologies contribute to the destruction of the local cultural elements – particularly those that are not compatible with these technologies [1]. The landscape in education is almost similar. Very often the usage of ICT (or the announcement of this usage made by the politicians and those who are responsible about educational reform) is accompanied by triumphal statements that promise the radical upgrade of education (which in the last decades is in permanent crisis). However, as Larry Cuban [3] notices, it is about an almost permanent phenomenon that is reiterated periodically during the last century: radio, television and video are typical instances indicating that the society reposes its hope for educational revival in technology up to its disproof by the next technological discovery. Up till now none of the existing technologies managed to change radically the educational system and the way of teaching (in the same extent as it happened in commerce, transaction business, press or scientific research). On the other hand, the continuous thoughtless usage of new technologies results in a mode of operating (for both teachers and students) which gives an emphasis on the usage of the technology itself rather than its effectiveness. Oppenheimer's aspect (ibid) is more critical since he claims that this continuous and thoughtless usage means that the students progressively loose the capability to emphasize, to pay attention or to think creatively. The students become information consumers as the tv-viewers become show consumers instead of critical thinking users. This criticism concerns even newer technologies such as the so-called Web 2.0. Andrew Keen (2007) [9] claims that this "cult of instant" in information and knowledge, the massive participation of everybody to everything, the cultivation of the absolute amateurism, finally undermine the western civilization itself, despite its long and hard process up to the establishment of procedures and prescriptions for the production, update, validation and spread of information and knowledge via specific kinds of press (i.e., newspapers, scientific magazines, encyclopedias). Returning to education, 25 years after the massive usage of ICT in teaching and learning, the results are not the expected ones. Obviously it is unthinkable to talk today about education without technology. However, it seems that finally it is required a very careful design of the way ICT will be used in the teaching process so as to have substantial learning outcomes. There are many reasons to do that. First of all, educational software must be well designed and adapted to the students' real needs. However, current software, in most cases, constitutes a commercial product more adapted to the market needs rather than to the pedagogical and didactical requirements. Very often the level of didactical aim is low (for example memorization) and the adopted methods rather ineffective and old-fashioned (consequence of the behaviorism model). Besides, educational software cannot frequently face the complexity and variety of the teaching necessities and are ineffective under the usual classroom conditions. The amount of money needed for buying, maintaining and upgrading the software is extremely high and it is probable that most of the national economies cannot afford this. The same applies to the amount of money for maintaining and upgrading hardware, networks and the whole infrastructure necessary for using ICT in education. Open Software and Open Code Software could be a solution to the problem, as well as the new generation of inexpensive PCs, but for the time being this is just a potentiality. However, in any case, the usage of ICT in education, presupposes an important amount of training time for learning the software which is not negligible or costless. Initially the students themselves must devote time to learn all these different environments. But the teachers face the most important problem. They eventually have to learn the way these software function and additionally to use them effectively in their teaching practice. This aspect becomes even more important since the new generation of educational software is considerably different from the previous ones. These new environments give emphasis to searching and composing information from the students themselves; they favor the communication among the students and the usage of multiple ways of elaborating, organizing and presenting information. Thus new skills are required for both, the students and the teachers who will guide them. According to Burbules et als (2000) [2], ICT is not a passive and neutral way to achieve our teaching aims. On the contrary, through its special features, it redefines all the processes that are afforded to the teacher. Simultaneously it re-assigns a meaning to the information and to the new knowledge which is constructed by the student. So the teacher has to operate in a new environment, to renegotiate a new kind of knowledge and meta-knowledge and the truth is that he rarely is prepared for that. Maybe this is the most important didactical problem that is posed by ICT. The problems presented in this section are of somewhat general character. However these problems acquire a very specific and practical character when ICT is used in the classroom. The teaching practice in the classroom is depended among others on a set of interrelated parameters that determine the nature of teaching as well as the quality of learning. When ICT is added in this list of parameters, then the system is "perturbed" and choices must be taken so as the equilibrium is obtained.

These choices are profoundly related to the above mentioned parameters that constitute the teaching system. They sometimes cause a series of difficulties that have a rather negative impact on the teaching process. We categorized these difficulties into four categories. In the first one some reservations are presented concerning how technology is used in the classroom environment. In the second we refer to the time management inside the classroom or to the training time. Next we focus on the necessity of the close relation between curriculum and integration of technology in the teaching practice as well as on the official support that has to be offered to teachers so as to incorporate technology in their classroom. What follows is issues related to the interface, menu commands and compatibility. Finally we present some evidence indicating the direct link that exists between technology and level of knowledge. It is worth mentioning that what we present here are examples of an ongoing phenomenon and not its complete description.

3 The Necessity to Use Technology as a Tool

The idea of the computer as a tool is related to two specific aspects. The necessity to use technology a) only after the development of certain skills relevant to the concept that is going to be taught and b) as a tool only if it can solve a specific problem. It is common thesis that the concepts taught to the students have a "double life". Initially a concept is itself the object of the learning process. Later the same concept can be used as a tool in order for another concept to be acquired. For example, in the 5th grade students in their first lessons in geometry are taught the concept of 'height' (i.e., what it is, how to draw the height in a triangle or a parallelogram, how it can be measured). Later, during the same year and in the topic of area of known shapes the height constitutes a tool in order to calculate the area of these shapes (it is a necessary component of the formula $E=(b*h)/2$ or $E=b*h$). In Dynamic Geometry Environments students have the possibility to draw automatically the height of the selected shape. However it is very important to use these environments if only the teacher has confirmed that the students have acquired the concept of the height and they have developed the relevant skill of drawing it in any shape. Otherwise this premature introduction to the dynamic geometry environment could result to misunderstandings and misconceptions about the nature and the function of the concept. Besides we must also pay attention to the possibility that the early introduction of technology could lead to a complete loss of some skills, such as the usage of geometric tools. Papadopoulos (2004) [12] refers to a case where students in a computational environment used to draw the height of a triangle as a simple segment connecting a vertex of the triangle with the opposite side. The segment seemed to be perpendicular to the side. However the students did not use the suitable tools ('Perpendicular line') of the software (Cabri Geometer). They also did not check whether the angle between their 'perpendicular' line and the side of the triangle was a right one (90°). Thus, it is obvious that there was not an established knowledge of the concept of height. Technology can be used for almost any purpose. So it is not rare to see students in a passive role in a computational environment or to see technology serving memorization of facts or rules. It is not a rare phenomenon also to see teachers using presentation software (Power Point for instance) for presenting solely the outline of their teaching or just reading what is written on the slides, wasting thus time and usage of technology.

4 Issues Related to Time

In that we could mention three kinds of reference: time management (inflation), time consuming and time investment. Any kind of teaching innovation usually causes an inflation of time. In the case of the introduction of technology, this inflation could result from the distance that exists between the expected duration of time needed to complete a pre-designed activity (or a series of activities) and the "real" time, necessary for its application in the classroom. Moreover technology usually demands more time for its integration compared to the time needed for the teaching of the same concept in a traditional way. Schneider (1999) [14] reported on a teaching based on the use of TI-92 about logarithms and exponentials which took 40 hours of teaching instead of the usual 9 hours. From the students' point of view sometimes is noticed that they cannot accomplish the task posed to them, because of their inability to perform specific computational operations necessary for its completion. Despite that the students have perceived the correct procedure, they cannot apply it. This results to a time consuming effort in order for the student to response. In a geometry project concerning area of irregular shapes [12], some students realized that it was necessary to apply cut-and-paste in order to transform the shape so as to estimate its area. Despite that they had chosen the correct strategy it was impossible for them to apply the required cut-and-paste. They wasted the available time and finally missed the task. This raises the issue of the students' training for an adequate period before they are asked to use technology. This time investment for training concerns teachers and also students. It constitutes an important factor that could facilitate both the usage of ICT in teaching practice and to a great extent to guarantee the success of the endeavour. However this perspective has as a consequence a cost of learning how to use technology and it should be taken seriously in mind before making any decision. Given that one software is not enough to cover a lesson-usually more than one software are combined for the same lesson- the dimension of the cost (time and money) for learning either multiple software or a super-software that incorporates the existing ones, is worth to be taken into account.

5 Technology vs. Curricula and Official System Support

James Kaput (1994) [8] took a strong position that technology should be thought of as an infrastructural, not as particular applications and tools. In this spirit, he said that 'Technology without curricula is worth the silicon it is written on'. This statement represents actually the reality in many cases, since there is an effort to integrate technology in the educational system, but this effort is not accompanied with an analogous reform of the curricula. Therefore very often technology is accepted uncritically by the teachers, leading to sometimes awkward marriages between learning environment and technological innovation. Sometimes this uncritical acceptance has its origin to the fact that there are not any teachers updated about the possibilities offered by technology. Teachers do not always know how to choose among the available options the one that supports their didactical aims in the best way. It should be clear that it is a challenge for the teachers to provide their students with the suitable software for each case, in order to help them to cope with the task they face [6]. When teaching in a computational environment, there is an intermediary system (technology, interface) between the user and the so-called *educational milieu.* So the teacher must be prepared to make the connection between the typical knowledge as it is stated by the scientific community and the knowledge that is constructed by the students through their interaction with the computational environment. These two kinds of knowledge are not always identical [11]. The meaning constructed by the students possibly could be different from the one that was the teacher's aim. So the teacher's interventions are of critical importance and this lack of updated teachers must be addressed by the official system.

6 The Role of Menu and Interface: The Technological Dimension

It is not always certain that what the student 'do', or try to do is interpreted by the interface in a way that preserves the same meaning. Very often students find in the menu toolbar words that are familiar to them from their every day life – just to give an example. So they use to attribute them this common meaning and not the one that is adopted by the system. Papadopoulos and Dagdilelis (2006) [11] report instances of this behaviour of students during computer aided teaching. For example they present how students tried to apply the '*Rotation*' tool of the Cabri software based on the meaning of their daily life rather than its mathematical one as it is adopted by the software. In the same spirit, we could mention the incompatibility that exists between what the students usually do and how technology is responded. For example, students are accustomed to work with the two-dimensional formulas in geometry (such as $E = \frac{B+b}{2} * h$). However the keyboard usually functions in a linear manner. So the students in the latter case have to express the formula using parentheses something that is more complicated and makes the needed arithmetic operations less obvious. The notion of incompatibility could be met also when the teacher is working on different versions of the same software. The teacher moreover has now to face the extra amount of work that is needed for designing tasks for the classroom. Up till now the teacher was responsible for the content of the didactic material. But now and due to the introduction of technology, the teacher is also responsible for the management of the layout, the appearance of the material (fonts, colours, borders, sounds) etc. Another aspect related to the interface is the *teaching noise* [5]. Teaching noise refers to undesirable side effects that can overshadow the real objective of a lesson. When the student is working in a computational environment and the task demands the usage of certain menu commands or buttons, it is possible for the student to waste significant amount of time to detect them. Even more impressive is the fact that the students sometimes search for tools that do not exist. In our work with 5th and 6th graders (11-12 years old), a task asked the students to transform an irregular shape to a known one. The accomplishment of the task needed a series of cut and paste actions. However there were students who insisted to search through the various options in the program menu the tool 'Transform' in order to cope with the task. There is a final question, whether some activities in a technological environment could lead to a weakening of the learning. The question is based on the fact that the computer screen functions to 'bully' the student in front of it. The screen constantly demands for actions either with the mouse or the keyboard. We had some cases where students were forced to make illogical actions. They were simply playing with the software but no strategy was hidden behind their actions. Assuming that some concepts demand mental effort we conclude that the students in the above mentioned cases wasted valuable time which could be better dedicated to mental effort in a non-technological environment.

7 Impact on Students' Knowledge and Conceptualization

In this part we deal with three facets. First, the fact that in a computational environment the concept that is to be taught is usually mediated. Second, the theme of learning control, and third (that is closely related to the previous one) the 'monism' of the applications. In the case of mathematics in different computational environments the students have in front of them shapes that seem to have exactly the same appearance and consequently it is expected by the students to have also the same behaviour. However in the modern educational software the interface allows the management of microworlds which in essence are representations or simulations of a system. So in different environments we have different representations of the same shapes. This means that in each environment some aspects of the shape's behaviour are present and others are hidden [7]. What the students usually do is to try to transfer and apply knowledge from one environment to other(s). This causes a difficulty in the level of conceptualisation since the new environment cannot response. This leads to a difficulty in the level of knowledge. The repetitive use of the same software progressively forms an intuitive knowledge that covers partially the concept (procedural knowledge) and this knowledge rarely coincides with the knowledge described through the official curricula (declarative knowledge). Equally important could be considered the issue of the learning control i.e., to what extend the student has control over the content of learning. Obviously the level of control is different in simulation environments compared to drill and practice ones. But in general none of the existing computer programs gives full learning control to its users. All computer programs currently available on the market satisfy partly the concept of control of learning (LC) from the user incorporating simultaneously the control from the program (PC) itself. The existing literature about LC does not confirm either its beneficial effects on students or the improvement of instructional effectiveness when a higher degree of LC implied in a computer program. The research findings range from the strong positive effect of LC to lack of any effect or even to a negative effect on learning outcomes, students' academic achievement and motivation. This previous issue of Learning Control is related directly to the issue of 'monism'. The monism presents a difficulty with regard to the applications or more precisely with regard to the effort for solving mathematical problems based only on one computational application. As Dagdilelis and Papadopoulos (2004) [4] report in their work concerning the usage of educational software for teaching area and its measurement, at least in such elementary level, one software is not enough or, more precisely, no software is enough for the teaching of mathematical concepts – probably any complex concept. From the teacher's point of view there were always restrictions in the software's capabilities which prohibited the realization of certain desired ways of software usage. This is maybe unavoidable since behind each software exists its designer who did not have in mind the particular teacher or the particular situation in which a teacher decides to use it.

8 Synthesis and Conclusions

ICT changes dramatically many of the aspects of the modern societies as well as our daily life. However, our findings indicate that ICT and their application in education do not produce (at least for the moment) the expected teaching results that would strengthen the digital multiliteracy of the citizens. It seems somehow paradoxical that the domain that needed 'desperately' this change (i.e., teaching, learning, education in general) actually was barely influenced. We tried to show that in our days this seems to be less paradoxical since we have comprehended the great complexity that is associated with the (didactically) effective usage of new technologies. This usage does not simply mean the addition of another tool in the classroom (among others such as the logarithmic rule, dictionaries, and geometrical instruments). The new digital means are not "neutral"; can not be approached with simple terms such as 'technophilia' and 'technophobia'; can not be used effectively by merely following 'simple' instructions. It seems that their systematic usage influences so deeply the education action, that it demands an essential educational reform. Consequently this demands a detailed, careful, and time consuming design of their incorporation in the educational system, which presupposes both a holistic approach and a lot of resources (financial, human, time). Another factor that seems to be essential for the successful usage of ICT in education is the in-depth comprehension of the terms and consequences of its usage in the daily teaching practice. According to the relevant research literature (even in the micro-level of the usage of educational software for problem solving purposes) its usage could facilitate or hamper the students from constructing new knowledge. It could also restrict or expand the time required for the solution of the problem and even more to become a learning obstacle. It could broaden the gap between high and low achievers (a hypothesis that has not been object of a research study yet, but it seems to be reasonable). Moreover the usage of these new educational environments and their exploitation for teaching purposes demand not only the learning of these environments but also the re- design of the course. However the most complicated problem is not stemming from the modern digital technologies but from the future ones: In what way could the teachers be prepared to face this future? (since we are ignorant of this future or more precisely we are ignorant of its technological evolution). The limited effect of ICT on learning, despite the considerable amount of money and resources that have been spared, is due to the fact that it is required further systematic and broader research of the above mentioned issues (always according to our findings and our line of thought). Otherwise there exists the danger to continue to observe in the education domain, that the tremendous possibilities offered by ICT remain unexploited

References

1. Bowers, C.A.: Let Them Eat Data: How Computers Affect Education, Cultural Diversity and the Prospects of Ecological Sustainability. University of Georgia Press (2000)
2. Burbules, C.N., Callister, T.A.: Watch IT! The Risks and Promises of Information Technologies for Education. Westview Press (2000)
3. Cuban, L.: Oversold & Underused: Computers in the Classroom. Harvard Univ. Press (2002)

4. Dagdilelis, V., Papadopoulos, I.: An open problem in the use of software for educational purposes. In: McKay, E. (ed.) Proceedingf ICCE 2004, Australia, pp. 919–924 (2004)
5. Dagdilelis, V.: Principles of Educational Software Design. In: Mishra, S., et al. (eds.) Multimedia in Education and Training, pp. 113–134. Idea Group Publishing (2005)
6. Gawlick, T.: On Dynamic Geometry Software in the Regular Classroom. ZDM 34, 85–92 (2002)
7. Hollebrands, K., Strasser, R., Laborde, C.: Technology and the learning of geometry at the secondary level. In: Heid, K., et al. (eds.) Research on technology and the teaching and learning of mathematics, pp. 155–206. Information Age Publishing (2006)
8. Kaput, J., Thompson, P.: Technology in mathematics education research: The first 25 years in JRME. Journal for Research in Mathematics Education 25(6), 676–684 (1994)
9. Keen, A.: The Cult of the Amateur: How Today's Internet is Killing Our Culture. Nicolas Brealey Publishing, London (2007)
10. Oppenheimer, T.: The Flickering Mind. Saving Education from the False Promise of Technology. Random House (2003)
11. Papadopoulos, I., Dagdilelis, V.: The Theory of Transactional Distance as a framework for the analysis of computer aided teaching of geometry. The International Journal for technology in Mathematics Education 13(4), 175–182 (2006)
12. Papadopoulos, I.: Geometry problem solving in a computational environment: Advantages and reservations. Paper presented at ICME10, Copenhagen, Denmark (2004), `http://www.icme-organisers.dk/tsg18/S22IOANNIS_PAPADOPOULOS.pdf`
13. Postman, N.: Amusing Ourselves to the Death, Public Discourse in the Age of Show Business. Penguin Books, USA (1985)
14. Schneider, E.: La TI-92 dans l'enseignement des mathematiques des enseignant(e)s decouvrent la didactique des mathematiques. In: Guin, D. (ed.) Actes du congres Claculatrices symboliques et geometriques dans l' enseignement des mathematiques, pp. 49–60 (1999)
15. Stoll, C.: The Internet? Bah! Hype alert: Why cyberspace isn't, and will never be, nirvana. Newsweek (2008), `http://www.newsweek.com/id/106554` (last visited: August 1, 2008)

Quality Development in Education

Spiros Panetsos[1], Anastasios Zogopoulos[1], Odysseas Tigas[1], and Albina Gubaidullina[2]

[1] Department of Electronics Engineering Educators
School of Pedagogical and Technological Education ASPETE
Athens, Greece
s.panetsos@aegean.gr, tzogx@hotmail.com, ody_cute@hotmail.com
[2] Elabuga State Pedagogical University
Russian Federation
alban-gi@rambler.ru

Abstract. The article suggests that people involved in quality development need a specific competence, called quality literacy, in order to successfully improve learning processes. Quality literacy is viewed as a set of competencies that are needed for professional quality development. Quality literacy emphasizes the importance of professionalism as a necessary component for quality development, in addition to structural quality management models. Quality development is a co-production between learners and their learning environment. This means that the educational process can only be influenced and optimized through participation and not steered externally. Quality strategies cannot, therefore, guarantee a high quality of learning processes but rather aim at professionalisation of the educational process. This article suggests participation and negotiation between educational participants (clients and providers) as a main condition for quality development.

Keywords: Quality literacy, Participatory quality development, Educational quality.

1 Introduction

Quality in education has become a main aspect in educational policies, an imperative for practitioners, and a huge demand for learners. Achieving high quality is a much debated and sought-after goal in all segments of education. It is, however, not so much characterized by its precise definition but rather by its positive connotation.

The search for quality in education is often addressed in the way of finding a suitable approach for controlling or steering the pedagogical process. Yet, this view ignores the fact that the relation between cause and effect in the field of pedagogical practice is rather open and insecure [1],[2],[3]. It is one of the few secured results of educational research so far that pedagogical practice is much more characterized by insecurities and situational interpretations than through systematic cause-effect relations. In particular, psychologically oriented educational research tried for some time to determine the exact cause-effect relation between media attributes (screen colors, length of dynamic learning objects, etc.) and learners' learning progress in order to

M.D. Lytras et al. (Eds.): WSKS 2009, CCIS 49, pp. 146–153, 2009.

derive consequences for the design of learning environments. However, such research designs proved to be too complex, and we can conclude that, not the media characteristics alone, but rather the underlying learning methodology and instructional arrangement facilitate learning success [4]. Today it is clear that knowledge, information, and learning media do not have an inherent learning quality but rather carry a quality potential, which has to be released in co-construction processes during the learning phase. In this article, we discuss a new understanding of quality development in education.

Quality development should not rely solely on structural models and strategies but take into consideration the professionalization of quality development — especially in light of its technological deficit. The main assumption of this article is that there are certain competencies for professional-quality development, and that these apply to both the learner/client side and the teacher/provider side. Quality development in education is viewed as the result of quality competence of the participants. This competence is termed quality literacy. It is viewed as a critical factor for success of every quality development activity in education. The concept builds on earlier work and develops a theoretical foundation based upon educational theories and terminology for the concept of quality literacy [3]. The scope, the validity of described concepts, and the reach of this concept have to be understood within this theoretical framework. Quality development is defined, from an educational point of view, as a co-production and a participative concept. Evidently, a theoretical contribution with this focus has restrictions in scope: Economic and/or technological models are not integrated into the argument.

Although e-learning is the general context in which the concept of quality literacy has been developed, wedo not distinguish between education and e-learning in this article. We believe that e-learning is an educational innovation and has a number of specific challenges to it. When introduced to educational scenarios, it often functions like a magnifying glass and reveals immediately deficits in pedagogical planning or teaching/learning organization. However, the concept is of a generic nature and addresses quality development issues from their very core — and thus does not make a distinction between "e"-learning as the field of quality development and "non-e"-learning. Although there are a number of specific challenges which differ between e-learning and non-e-learning, it is argued that the concept of quality literacy addresses issues that are the same in both fields. In this sense, the concept is a generic concept and is equally applicable to the field of e-learning vs. education as well as to the different educational sectors.

2 Quality in Education

Quality development is a constant negotiation process in which the educational community should participate in a common effort to define and implement quality in a continuous, improved way. In order to empower the participants in the educational process — be it as teachers or learners — and to orient every educational interaction towards improvement, the participants have to be *quality literate*.

In this section, two characteristics of educational quality development are described: the multidimensional nature of quality in education, and the need for rethinking quality as a participatory process that must be facilitated as a co-production between educational stakeholders. Both aspects emphasizes that continuous improvement processes in education are of an unforeseeable and dynamic nature, which demands a certain ability of the involved participants to respond to these challenges. This ability is described as a competence rather than as a reproducible knowledge.

2.1 Quality as a Multidimensional Concept

Quality in education is a multidimensional concept. Therefore, different approaches to define quality are available. Berkel [5] suggests a three dimensional scheme, originally for service quality, which has been adapted to the field of educational processes in the following description. It locates quality within three dimensions:

- objective vs. subjective: Berkel addresses the question of who is defining quality criteria and values. If the quality value is defined only through the performance indicators of a product, Berkel terms it objective quality. The quality characteristics then have to be a part of the respective good, which is only partially true for the field of education. For education, the quality characteristics are usually defined through individual persons or committees in a subjective way. The definition of quality requirements through clients or learners is a subjective quality definition.
- inherent vs. instrumental: This dimension relates to the questions of where quality can be observed and when it becomes explicitly measurable. Inherent quality relates to the quality of a product that can be observed as lasting and innate. If quality reveals itself only through a service process, and thus the participation of clients, we refer to it as an instrumental quality. Often objects with inherent quality characteristics (e.g., Learning Management Systems, learning materials, etc.) are used in an instrumental way.
- endogenous vs. exogenous: If organizational processes and structures are taken into account when evaluating and/or assuring educational quality, we say they are of endogenous quality. If the educational institutions or organizations are not part of a quality evaluation, we say they are of exogenous quality. The quality evaluation of education requires an active process. Endogenous and exogenous can be used to distinguish between quality assessments that are either directed to the surface structure (exogenous) or to the deep process structure (endogenous) of an educational service.

The quality of education is therefore constituted only through mutual interaction of learners with their learning environment [6], and the evaluation of quality is influenced by organizational processes within which the educational process takes place (endogenous).

2.2 Participation and Co-production in Educational Quality Development

Classical service theory conceptualizes the interactive relationships between the roles of people-oriented services and the categories "production" and "consumption". It is argued that education is a symbolically mediated, productive active interaction as well as a production process. This process involves learners together with other participants (learners, teachers, etc.). It therefore has to be conceptualized in the form of a pro-sumption rather than a production-consumption relationship. The addressees of educational services are therefore conceptualized as active "co-producers" and not as passive receptors.

For the design of high-quality learning environments, this view bears some consequences: Learning environments — a term that is used here in the broad sense, referring to the sum of all processes that constitute the learning opportunity and including all resources and persons that are part of it — have to be designed in a way that makes it possible for learners to express their demands and preferences as part of the construction process. Only then can learners bring forth their experience, backgrounds, and demands, thus enabling providers to design learning environments in a way that allows active learning, problem solving, and competence development oriented towards the learners' individual needs. The assurance of quality exclusively reached through predefined, static frameworks (e.g., standard evaluation questionnaires) often does not sufficiently address this particular necessity of co-production in educational settings. From this perspective, it is important that the development of quality strategies takes into account an active negotiation process as a specific condition of quality development and supports it proactively. Quality management concepts therefore have to include a negotiation component. This requires an extended understanding of process-oriented quality-development models, and asks for competence development and staff professionalization components within quality strategies.

From a socio-structural point of view we can moreover observe that clients' identity structures change and standard biographies become more and more heterogeneous, and therefore lose their prognostic value for planning educational processes. Quality concepts that are still based on concepts of traditional biographies are losing their analytic powers over educational processes. If the described necessity of individualization of educational processes is taken seriously, then it is difficult to formulate fixed and prescriptive quality standards for progressively heterogeneous situations. They have to be compared to flexible negotiation frameworks that allow consideration of the learners' situation and perspective in a co-productive process. To use a participatory quality strategy means to support or hinder negotiation processes but not to substitute them through management processes any longer.

3 Quality Literacy — Competencies for Quality Development

The concept of quality literacy is based on the assumption that quality in education is the result of competent behavior of participants involved in an attempt to develop quality. In this section we describe the theoretical background of the concept and the methodology that has been used to construct the concept of quality literacy. We define a set of skills that are necessary to perform quality development processes. The concept is embedded in the view that quality has to be defined in a participatory way.

3.1 Theoretical Background and Approach of a New Concept

Quality literacy is a concept that is much related to the philosophy of total quality management. Within this approach, quality is seen as a continuous improvement process, involving all participants in the process of a permanent assessment and quality improvement. One element is of key importance — the introduction and development of a quality culture into an organization. This has two dimensions. First, a managerial dimension that is of a rather technocratic nature and deals with implementing tools and instruments to measure, evaluate, enhance, and assure quality. This is usually facilitated though a top-down process. Second, a dimension of quality commitment focuses on an individual level. It relates to the individual commitment to strive for quality, using tools and instruments for quality development. First and foremost, however, it focuses on changing attitudes and values, and developing new skills and competencies in order to make a permanent improvement of quality possible. Individual abilities, attitudes, and values add up to a collective level, which in turn leads to a quality competent organization. This dimension relates to a bottom-up process.

The ability for an individual to competently use, modify, and further develop existing tools, instruments, and strategies, or to introduce them or develop them new in order to pursue a permanent quality orientation in an educational setting shall be called quality literacy. Quality literacy is not a freefloating concept, but can be rooted in and connected to many already long-existing theories and approaches.

In an organizational context, quality literacy is a set of skills that enables individuals to take part in the development of a quality culture. For individual learners, the same set of skills enables them to pursue permanent improvement processes of their own learning and development processes, using quality instruments and concepts. Quality literacy thus applies to both sides' — actors on the providers' side of educational processes (teachers, tutors, media designers, or administrative staff) and actors on the clients' side of educational processes (learners). It is a set of generic skills that applies in both contexts and has to be adapted to the specific situation.

This concept is comprehensively introduced in this article for the first time. It is based on the belief that quality improvement is the result of the (quality) competent action of individuals. It is of complementary nature to external organizational quality strategies that are seen as an important but not sufficient component for achieving high quality in education. Quality literacy manifests itself in the ability of actors of an organization or of an individual learner to use quality strategies and tools, and incorporate the changed and new beliefs and values they inherently carry into their everyday professional behavior and procedures. Only then will educational quality development be successful.

Quality Literacy in this sense is seen as a basic prerequisite to acting professionally in quality development contexts. On the first step, information about quality and quality development or related fields is interconnected and linked to knowledge. On the second step, they are applied and result in abilities. This is the step where individuals have practical experiences with applying or using quality strategies, tools, or instruments. These abilities are transformed in activities through motivation and will. Competence, however, demands an additional evaluation about whether the performed activity is suitable in a given context. For this, an individual usually needs standards against which he or she can assess whether something is suitable in a specific context. For quality development, these can be societal norms, legal rules, criteria that are agreed on in the specific organizational context, or set of standards for individual behavior.

In general, it has to be noted that quality literacy applies to all forms of knowledge, information, and learning of technology-related educational concepts, such as e-learning, blended learning, and presence courses. There are commonalities and differences between "traditional" educational scenarios and e-learning. Concerning quality development, however, we have to note that it is a process of negotiation with the goal of providing successful education in both educational fields. Of course, additional areas of knowledge apply here. In principle, however, quality development requires the same competencies.

In conclusion, we can state that the concept of quality literacy builds upon existing concepts and aims to describe skills that enable individuals to perform quality development competently. Sometimes these situations are very complex. Sometimes, though, there is little complexity when only one specific quality instrument is applied to perform quality assurance (e.g., a questionnaire at the end of a program or course). Quality literacy, moreover, is a concept that cannot exclusively be learned by means of books or training, but requires experience and practice. It is a concept that is subject to constant change, as the means and forms of technology-enhanced education change as well.

3.2 Dimensions of Quality Literacy

Quality literacy can be seen as a set of four central competencies that contribute to carrying out successful quality development in education. They do not constitute distinct factors of quality literacy, but rather differentiate the inner structure of the concept of quality literacy. A more precise description of the inner structure and coverage of the concept are presented and elaborate upon the four dimensions the concept contains.

Dimension 1: Quality knowledge
This dimension addresses the "pure" knowledge about the possibilities of today's quality development and up-to-date quality strategies in elearning and education. The term "quality strategies" refers to all guidelines, structures, rules, tools, checklists, or other measures that have the goal of enhancing the quality of an educational scenario. There are two sub-dimensions to quality knowledge:

a) **informative:** The informational dimension refers to information and knowledge about quality systems, tools, and procedures. It is about having access to information resources, primary as well as secondary, and understanding the system of quality development.

b) **instrumental:** The instrumental dimension refers to the knowledge of how to use and apply a specific tool, such as an evaluation questionnaire, or how to use a list of criteria or guidelines for a specific context. The instrumental dimension does not, however, relate to the competence of implementing a quality system with a certain intention, such as reducing a course's drop-out rate. That is covered through the dimension of quality experience.

Dimension 2: Quality experience
This dimension describes the ability to use quality strategies with a certain intention. It is based on the experiences that actors have with quality development and with applying quality measures and strategies to educational scenarios. It can be differentiated from the instrumental knowledge dimension because it refers not only to the pure application of quality strategies or tools but also covers the processes of feedback analysis and initiating improvement. That means that, in addition to the instrumental knowledge of quality strategies, this dimension also carries with it an intention and a

goal. Quality experience refers to the ability to use (existing) quality strategies (e.g., guidance and consulting concepts) to generate data about educational processes in order to improve them. It answers questions such as: How can I use quality strategies in a certain way to improve the educational process?

Dimension 3: Quality innovation

This dimension relates to the ability that goes beyond the simple use of existing instruments and strategies. It refers to the modification, creation, and development of quality strategies and/or instruments for one's own purpose. An innovative and creative aspect is important for this dimension. Within this dimension, "adaptation" and "creativity" mean further development and reorganization of existing quality strategies within a given context. "Innovation" means thinking up and developing new strategies for quality development.

a) **Adaptation:** This sub-dimension refers to the ability to adapt an existing quality strategy or tool to one's own context. It goes beyond the pure usage of an existing tool, requires deeper understanding of it within the given methodological framework, and demands creativity.

b) **Creation/innovation:** The creation/innovation dimension describes the ability to think beyond existing strategies and go further than just modifying them. It also describes the ability to invent a complete new quality system. Such self-developed systems are often used for an organization's internal purposes when existing approaches do not cover the specific goals and requirements. An example would be the development of a new evaluation questionnaire for the assessment of a course when existing tools fail to analyze the desired aspects. Also, it could be the development of a new method of consultation with learners before a course starts in order to assess their needs and goals.

Dimension 4: Quality analysis

Quality Analysis relates to the ability to critically analyze the processes of quality development in light of one's own experiences and to reflect upon one's own situation and context. It enables actors to evaluate different objectives of quality development and negotiate between different perspectives of participants. To critically analyze means to differentiate between and reflect upon existing knowledge and experiences in light of quality development challenges. For learners, this means being aware of their responsibility for quality in education as a co-producer of learning success. For providers, this means enabling flexible negotiation processes in educational offerings and respecting individual objectives and preferences as well as societal contexts and organizational structures in their definition of quality objectives for education. Two sub-dimensions can be differentiated: analytic and reflexive.

a) Analytic Quality Analysis: The analytic dimension covers the process of analytically examining the meaning and the debate of quality in education in general. It is the ability to move within the framework of quality discourse, to contribute analysis, and to understand the different influences, starting from the market perspective and business models, taking into account technical aspects, and not forgetting the pedagogical aspects.

b) Reflexive Quality Analysis: The reflexive dimension is directed towards the analysis of one's own situation. It is the ability to set quality goals for one's own individual or organizational context, and to position oneself in the quality debate. The reflexive dimension emphasizes the ability to understand future challenges in educational quality development, rethinking one's current quality situation, and developing a strategy to meet future challenges. A typical field of the reflexive quality analysis competence is the development of future goals and strategies either for oneself as the individual learner or for an organization.

4 Conclusions

Quality development in education aims to improve educational processes. These are the result of a co-production between learners and their learning environments, and in principle cannot be defined prescriptively. This means that in the end, the result of an educational process cannot directly be influenced and optimized like a production process. Quality strategies therefore cannot guarantee high-quality learning processes but should rather aim to professionalize the quality development process, both on the client's side and on the provider's side.

The quality model relates to theoretical work that has been done in the field of service quality and combines it with concepts of negotiation, participation, and co-production. However, a comprehensive empirical validation of the described concepts has so far not been undertaken.

In an educational setting, quality literacy is a prerequisite for quality development for both the client and the provider. The described competencies allow clients and providers to act in a competent way in the field of quality development and to enter into a process of stimulating a quality culture with the aim of continuous improvement. Enhancing competence is a move toward professionalization of the quality debate.

References

1. Fink, L.D.: Creating Significant Learning Experiences. Jossey-Bass, San Francisco (2003)
2. Moslehien, S.M.: A glance at postmodern pedagogy of mathematics: Philosophy of mathematics education (2003), `http://www.ex.ac.uk/~PErnest/pome17/contents.htm` (retrieved March 10, 2009)
3. Taylor, M.: Generation NeXt Comes to College. A Collection of Papers on Student and InstitutionalImprovement 2, 19–23 (2004)
4. Russel, T.L.: The No Significant Difference Phenomenon (1999), `http://teleeducation.nb.ca/nosignificantdifference` (retrieved March 10, 2009)
5. Van Berkel, H., Schmidt, H.G.: On the additional value of lectures in a problem –based curriculum. Education for Health 18, 45–61 (2005)
6. Brindley, J.E., Walti, C., Zawaki-Richter, O.: Learner Support in Open, Distance and Online Learning Environments. BIS, Oldenburg (2004)
7. Ehlers, U.-D.: A Participatory Approach to E-Learning-Quality: A new Perspective on the Quality Debate. Journal for Lifelong Learning in Europe XI (2005)

Capitalizing Knowledge in Communities of Practice of e-Learning: Ontology-Based Community Memory

Akila Sarirete[1], Azeddine Chikh[2], and Elizabeth Noble[1]

[1] Effat University, PO Box 34689, Jeddah 21478, Saudi Arabia
{asarirete,enoble}@effatuniversity.edu.sa
[2] Information Systems Department, College of Computer & Information Sciences, King Saud University, Riyadh, Saudi Arabia
az_chikh@ccis.ksu.edu.sa

Abstract. Together members of a community of practice (CoP) share their expertise and mutual understanding about a common domain to develop greater knowledge and build best practices. Due to the informal character of learning within a CoP, most of the knowledge is mainly tacit and needs to be reified and capitalized. The aim of this paper is to model the CoP to include the community memory in order to manage its knowledge. A CoP is considered as a learning organization where members are interacting, producing, exchanging and adapting resources and knowledge to meet their needs. The CoP memory is based on a framework with several ontologies including knowledge, domain and task ontology. The application of the proposed framework is done through a CoP of e-learning.

Keywords: CoP, organizational learning, ontology, memory, knowledge management, e-learning.

1 Introduction

Organizational learning is the focus of considerable attention and is considered a powerful tool to improve the performance of an organization. The knowledge base of an organization is recognized as a vital source of competitive advantage; however, there is little understanding of how organizations actually create and manage knowledge dynamically. The aim of knowledge management (KM) is to facilitate the preservation of knowledge, its reuse and its actualization; and more importantly the organizational learning which involves the creation and integration of new knowledge at the organizational level. Historically, there have been a number of tools facilitating KM practices [1]. Nevertheless, most of the KM efforts were based on a typical top-down approach where knowledge was seen as a separate entity and the focus was associated with the creation of central knowledge repositories. Researchers are investigating how KM data/objects can be fused with e-learning practices to produce and distribute meaningful, organized, effective, and performance-enhancing solutions through the online delivery of information, education, and training. Recent research on knowledge management (KM) clearly recognizes the importance of

M.D. Lytras et al. (Eds.): WSKS 2009, CCIS 49, pp. 154–163, 2009.

communities of practice in the creation and maintenance of knowledge within organizations [1-3] to name but a few. Indeed, since the nineties, Communities of Practice (CoPs) have attracted an increased number of academics and professionals from both the private and public sectors. A CoP is defined as a unique combination of three fundamental elements [4]: (i) a domain of knowledge, which defines a set of issues around which the community engages; (ii) a coherent group of people who care about this domain, and each other's learning needs and aspirations; and ultimately (iii) a shared practice that is being constantly developed and adapted in their domain. The aim of this paper is to model the CoP to include the community memory and apply it to the e-learning domain – specifically instructional engineering – in order to manage the CoPEs (CoPs of E-learning) resources.

The remainder of this paper is organized as follows: Section 2 introduces the research issues and positions this work within the field. Section 3 reviews organizational memories and especially ontology-based memories. Section 4 proposes a community-based memory framework and section 5 shows an application of the framework to the CoPEs.

2 Research Issues

CoPs are dependent on technology to survive and maintain contact with their members. The world has become closer together and communication has become easier through the Internet and several other means of communication provided by the new information technologies (IT). The virtual nature of such groups that bond together somewhere in cyberspace is only related to its means of communication and accessibility. The activities of such groups result in the collaborative construction of knowledge, practices and improvement of their skills as professionals as well as individuals. In this paper, the focus is on CoPs that rely on IT and cyberspace and rely less on face-to-face meetings. Together, the members of the community share their expertise and mutual understanding about the domain to develop greater knowledge and build best practices. Due to the informal character of learning within a CoP, most of the knowledge is mainly tacit and needs to be reified and capitalized. Knowledge creation is seen as constructivist learning process where knowledge is constructed collaboratively amongst CoP members.

In the framework of the PALETTE[1] project [3] several knowledge management services were proposed to support CoPs. These services rely on a semantic web-based approach using ontologies [5]. An ontology provides common semantics that can be used to improve communication between either humans or computers. The present work goes in the same direction proposing a general framework for the CoP memory using different forms of ontologies as a backbone for the memory and showing how they are integrated. A CoP-dedicated ontology named O'CoP was proposed within PALETTE This ontology consists of CoP relevant concepts and relations and is used to annotate CoP resources. The O'CoP ontology is divided into several models such

[1] PALETTE project aims at facilitating exchanges and learning in CoPs (http://palette.ercim.org). It acknowledges the importance of CoPs in knowledge management (KM) and therefore looks at different aspects of KM in the CoPs and tries to give solutions and a foundation for further research in the field of communities of practice.

as a learner profile model, a competency model, a lessons-learnt model, an activity model, etc. These models will be used by the knowledge management tools and services within PALETTE such as a semantic wiki called “SweetWiki”. SweetWiki relies on the semantic web technologies using “OWL Lite”[2] ontology and “social tagging” approach which consists of allowing any CoP member to tag his content [6]. SweetWiki’s ontology defines the structure of the wiki and describes its concepts and relationships such as “Page”, “Web”, “Link”, “Author”, etc. To reify practices Daele [7] proposed to use a SweetWiki. He argues that besides the tools of communication and collaboration, a CoP needs some tools for the reification of its practices. Reification means giving concrete form to something that is abstract. Daele discusses a case study about a CoP of teachers training: Learn-Nett[3] (LEARNing NEtwork for Teachers and Trainers). The study showed that the tool helped the CoP members to reify their practices by reflecting on them and having an established process to write about their practices. In the same track, Arnold et al. [8] suggest using a public wiki as a community memory where they invite researchers and practitioners in their field of interest to participate and collaborate. They insist on the support of Web 2.0 technologies for CoPs. To support a CoP of tutors in online delivery of courses, Garrot et al. [9] propose a web-based platform. This platform combines several tools such as forums, email, and a database to store the data. The tutors’ good practices (threads of forum or produced documents by members) are saved in a predefined database. A user interface represented by a matrix of parameterized activities related to the tutor function is used to store or retrieve tutors’ experiences. This work insists more on the socialization and the participation process within the community than the reification or capitalization of knowledge.

In the present paper, we argue that an online CoP needs a fully fledged framework or platform that will help manage and create knowledge within the community. Indeed we consider a CoP as an organization where members are interacting, producing, exchanging and adapting resources and knowledge to meet their needs.

3 Organizational Learning Approach

Four elements are considered when talking about Organizational Learning (OL) [10]: knowledge acquisition, information distribution, information interpretation, and organizational memory. Most often the organizational learning is considered as a process and the learning organization as a structure where this process is integrated as the organization culture. A Learning Organization as defined by Senge [11] is "a group of people continually enhancing their capacity to create what they want to create". Senge adds that a Learning Organization embed processes in its culture such as to encourage learning at individual, group and organizational level. According to Malhotra [10] a Learning Organization is an "Organization with an ingrained philosophy for anticipating, reacting and responding to change, complexity and uncertainty." For Dodgson [12], a Learning Organization builds structures and strategies to enhance and maximize OL. OL participates to the sustainability of the organization since it involves the creation and integration of knowledge at the

[2] http://www.w3.org/TR/owl-features/

[3] http://learn-nett.org

organizational level. OL occurs when individuals' memories become part of the organization memory and knowledge is integrated and shared.

3.1 Organizational Memory Concept

The organizational memory is a place where the organization's knowledge assets are found. The word 'memory' is often used as a metaphor in organizations to refer to the organization's information and knowledge assets and the processes of capturing, retaining, accessing and using such assets as stated by Guerrero and Pino [13]. Such memory can be named in different ways such as "corporate memory" and "organizational memory" and is present in different forms (project memory, individual memory, group memory). Dieng-Kuntz et al. [14] define an organizational memory as:

> *"...an explicit, disembodied, persistent representation of knowledge and information in an organization, in order to facilitate its access and reuse by adequate members of the organization for their tasks".*

The main goal in building such memory is to preserve and reuse knowledge, experiences and behaviors providing the user with relevant information and leaving him the decision to make appropriate interpretation depending on the context he is in [15]. The purpose of an organizational memory is to better exploit assets especially "knowledge". These assets are better exploited if the memory is used within a learning organization. A learning organization is an organization where learning is encouraged at the individual, group, and organizational level ([11], [16]). The process of knowledge capitalization is the main factor in building the memory, maintaining it and reusing its contents.

When an organization tries to make its knowledge explicit, it begins to see what it is missing and what it doesn't know. To minimize the risk of turn-over and loss of knowledge, an organization should start capitalizing its important knowledge. This can be done through the use of an organizational memory. Most often organizational memory is based on ontologies. The use of ontologies helps the organization to become a "semantic learning organization" as stated Sicilia [17]. These ontologies become the technological backbone of the learning organization and help performing intelligent activities and producing artifacts that are semantically interpreted.

MEMORAe project (Organizational Memory Applied to the e-learning) showed the importance of using ontologies to represent an organizational learning memory. Interest was focused on the capitalization of knowledge and competencies in the context of an e-learning training [18] and for a community of learners [16]. The learning content is indexed by knowledge organized by means of ontologies. MEMORAe distinguishes two levels of ontologies: an application ontology and a domain ontology. The application ontology concerns a specific course containing notions to learn while the domain ontology is about the domain of training in general.

4 Ontology-Based CoP Memory: General Framework

We consider a CoP as a "semantic learning organization" where members are the main actors and knowledge assets and community resources should be capitalized based on the organizational memory concept. Figure 1 shows the general architecture of such memory showing the different layers. The CoP memory is based on several ontologies:

- a generic ontology describing the CoP concepts such as members, their profiles, the community, the common activities within the CoP, etc.
- a domain ontology, describing the concepts related to the CoP domain of interest (its common practice); it is used also to annotate semantically different resources of the CoP;
- a knowledge ontology describing the different sorts of knowledge acquired by the community members such as their skills, competencies, their lessons learned, their experiences, and their routine knowledge [19]; and
- a task ontology specifying problem solving processes related to identified tasks by the community members [20]. The tasks are divided into sub-tasks and the ontology specifies the concepts and relations appearing in the task of interest.

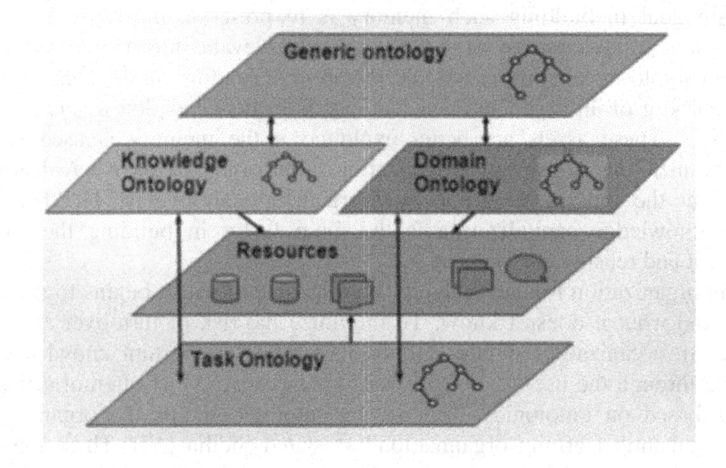

Fig. 1. CoP Memory Ontologies

This paper argues that at any time of the life of a CoP, there is a need to solve a problem or to execute a task. To solve these issues, there is a need to provide the right knowledge at the right time as stated by Dieng-Kuntz et al. [14] (a corporate memory should provide "the good knowledge or information to the good person at the right time and to the good level").

The proposition here is that the use value[4] of knowledge should be done through a semantic search or help through a task ontology that is used to model the task of the member and relates it to some specific domain knowledge through the knowledge and domain ontologies or connecting it to the right member in the community by checking his competencies.

[4] Learning something in a context where it is immediately useful.

5 Building the CoP Memory: Saga Approach

OL can utilize the process of single-loop learning or the utilization of a double-loop learning processes [21]. The single loop is a linear approach and would appear in information systems and decision systems which are narrowly constructed within one field of expertise; i.e. finance. The double loop system would encompass socio-economic constructs with multidimensional analyses which are both qualitative and quantitative. The saga is the contextual element of any Organizational Learning model. This context enables sorting, and labeling of issues to accommodate the Organization. The contextual factor is critical to understanding, to interpretation and to utilization of learning.

Each CoP member has its own individual view of the CoP. When solving a task within the CoP, the member brings his knowledge and uses part of the CoP memory (the artifacts he already knows). The good news is that in a CoP, members are not bond to strict organizational rules instead they have a vision they share and are having similar learning goals. This makes it easier to adapt the double-loop learning and help their CoP to sustain itself.

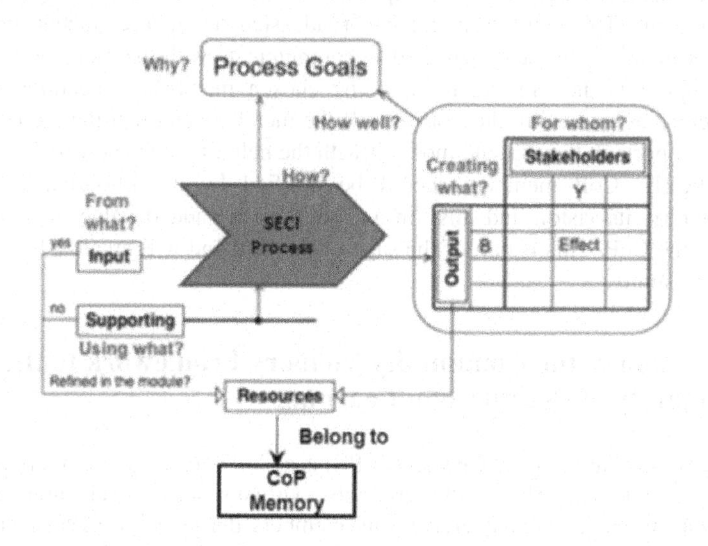

Fig. 2. The SECI Process within the CoP with its Inputs, Outputs, Goals, Supporting CoP memory and its resources (Adapted from [23])

In earlier work, the authors adapted the SECI[5] model of Nonaka [22] for knowledge creation to CoPs as a functional model where CoP members participate to build knowledge [20]. This paper reuses the process model approach from Naeve et al. [23] and applies it to the CoP context based on inputs, support resources, goals and outputs. The SECI process is considered as a subclass of "Process Module" in the

[5] SECI model involves four processes of knowledge transformation between tacit and explicit knowledge: socialization, externalization, combination and internalization

process model. Figure 2 shows the interaction between the inputs, outputs, the CoP resources such as its memory in the SECI process of creation of knowledge within the CoP. This process is done fulfilling certain goals and for certain stakeholders (CoP members). The outputs will have some effect on the CoP members. The SECI process is supported by some technology tools such as community building tools such as forums, wikis, email, concept mapping tools, integration tools etc. This process can be done at the individual level as well as the CoP level where members exchange knowledge and interactions among them. Each member can access his individual memory or his group memory at anytime in the process of communication. Several scenarios can be possible:

- CoP members decide to reify some of their practices and share them with others. The community memory should provide them a way to define their task related to the given task ontology. They will be asked to relate their task to some of the existing knowledge in the memory. This is done through collaboration among the members. This is done in a design for reuse methodology – being able to design for later use.
- A CoP member might ask some question to other members through the CoP environment. The request should be broadcasted to other members and some specific members might be selected to communicate with the member requestor. The request is then multi-casted to the chosen members depending on their competencies (related to their profile). If the members feel that the request or the task is important, they should model it with the help of the framework.
- Periodically, CoP members should be notified by the importance of their community, its vision and mission and what is the added value to them to be among the CoP. This is part of the saga of the CoP and will create a bond within the community.

6 Application of the Community Memory Framework to the Community of Practice of e-Learning

A Community of Practice of e-learning (CoPE) is a CoP which gathers actors involved in the e-learning domain. This CoPE represents a virtual space for exchanging, sharing, and resolving problems encountered by its members during all phases of an online learning system (OLS) life cycle [24]. An OLS is a set of linked elements (methods, tools, services, procedures, and routines) having the objective to produce individual and collective competencies, and assist in learning. An OLS goes through several instructional engineering phases. The CoPE members will consider each phase as a part of a learning process where they share their knowledge and make use of it. Figure 3 shows the different phases with support of different resources and models from the CoPE memory such as the supporting ontologies, pedagogical models.

During the different phases of the OLS, CoPE members negotiate questions related to the whole process. For example in the analysis phase they plan relevant tasks such as the curriculum review, lectures, syllabi preparation, teachers teaching styles, students learning styles. These tasks can be governed by some conceptual mapping where the instructional engineers conceptualize their ideas in concept maps and share

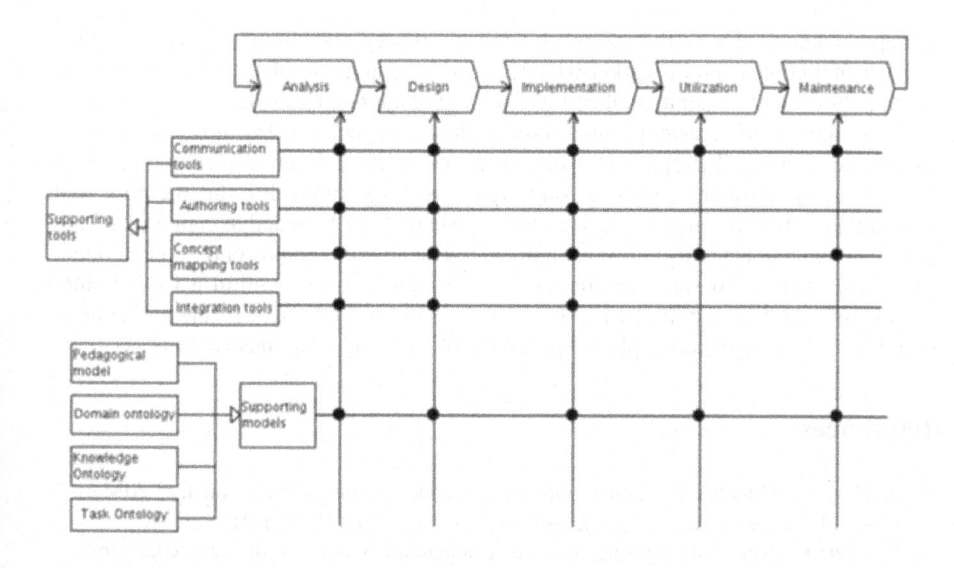

Fig. 3. Online Learning System Phases associated with Supporting Tools and Models

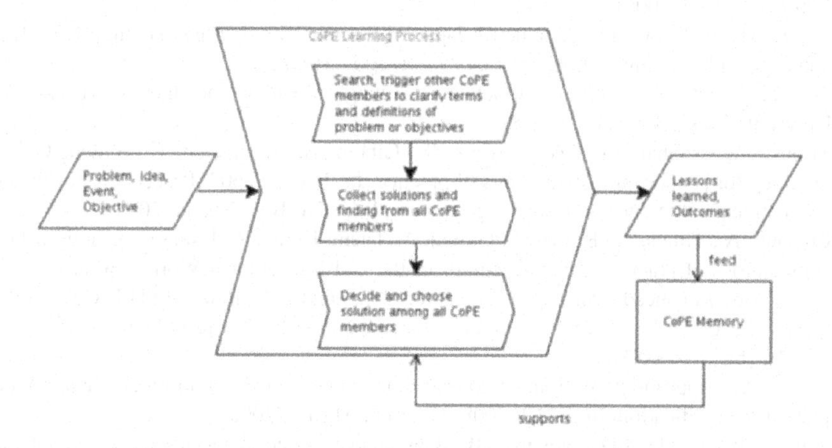

Fig. 4. Generic Scenario for a CoPE Learning Process

them with others. When negotiations are done, they may decide to represent them in the CoPE memory. Figure 4 shows a generic learning process within a CoPE. The CoPE members are always in negotiation to choose a solution based on a given problem or needs.

7 Conclusion and Future Work

The Knowledge Society is committed to advance the relationship of technology to the expansion of knowledge, and to information utilization. This paper addresses the contextualized element which differentiates knowledge from information. The concept of knowledge is that body of research, theory and discussion which added together in a context produces knowledge. The accumulation of facts is information, which is sterile unless contextualized into a body or cannon of work.

CoPs and social learning may have a huge impact on learning as well on knowledge sharing. This present work proposed a general framework for the CoP memory using different forms of ontologies as a backbone for the memory and showing how they are integrated. It also suggests that the knowledge capitalization is important as well as the use of Saga approach. The authors are presently conducting a field force analysis to show the importance of having a saga within a CoP. Future studies are addressing completing the domain ontology of a CoPE and integrate it with the task ontology and apply the proposed framework using these ontologies.

References

1. Kimble, C., Hildreth, P.: Communities of practice: Going one step too far? AIM, 1–7 (2004), http://www.cs.york.ac.uk/mis/docs/AIM14.pdf
2. Wenger, E.: Knowledge management as a doughnut: Shaping your knowledge strategy through communities of practice; Reprint # 9B04TA03; IVEY MANAGEMENT SERVICES (January/February 2004), http://www.iveybusinessjournal.com (retrieved March 2006)
3. PALETTE: Pedagogically sustained Adaptive LEarning Through the exploitation of Tacit and Explicit knowledge (2006), http://www.palette.org
4. Wenger, E.: Communities of Practice: Learning, Meaning and Identity. Cambridge University Press, New York (1998)
5. Tifous, A., El Ghali, A., Dieng-Kuntz, R., Giboin, A., Evangelou, C., Vidou, G.: An ontology for supporting communities of practice. In: K-CAP 2007: Proceedings of the 4th international conference on Knowledge capture, Whistler, BC, Canada (2007)
6. El Ghali, A., Tifous, A., Buffa, M., Giboin, A., Dieng-Kuntz, R.: Using a semantic wiki in communities of practice. In: Proceedings of the 2nd International Workshop on Building Technology Enhanced Learning solutions for Communities of Practice (TEL-CoPs 2007), Crete, Greece (2007), http://palette.ercim.org/images/Publications/using_telcops07.pdf
7. Daele, A.: Appropriation d'un Wiki par une communauté de pratique: analyse des pratiques de reification. In: EACP 2008, Tlemcen, Algeria (2008)
8. Arnold, P., Smith, J.D., Trayner, B.: Conference paper about narrative, community memory and technologies – or from piles of books around a sofa to an ongoing collaborative literature review in a wiki. In: Stillman, L., Johanson, G. (eds.) Constructing and Sharing Memory: Community Informatics, Identity and Empowerment, September 2007, pp. 1–12 (2007)
9. Garrot, E., George, S., Prévôt, P.: A Platform to Support a Virtual Community of Tutors in Experience Capitalizing. In: Web Based Communities Conference (WBC 2007), Salamanca, Spain, pp. 103–110 (2007)
10. Malhotra, Y.: Organizational Learning and Learning Organizations: An Overview (1996), http://www.brint.com/papers/orglrng.htm
11. Senge, P.: The Firth discipline. The Art and Practice of the Learning Organization. Doubleday, New-York (1990)
12. Dodgson, M.: Organizational Learning: A review of some literatures. Organization Studies 14, 175–194 (1993)

13. Guerrero, L.A., Pino, J.A.: Understanding Organizational Memory. In: XXI International Conference of the Chilean Computer Science Society (SCCC 2001), Punta Arenas, Chile, November 7-9, pp. 124–132. IEEE Computer Society Press, Los Alamitos (2001)
14. Dieng-Kuntz, R., Corby, O., Gandon, F., Giboin, A., Golebiowska, J., Matta, N., Ribière, M.: Méthodes et outils pour la gestion des connaissances: une approche pluridisciplinaire du knowledge ma- nagement, 2nd edn., Dunod (2001)
15. Benayache, A.: Construction d'une mémoire organisationnelle de formation et valuation dans un contexte E-learning: le projet MEMORAe. PhD Thesis. Université de Technologie, Compiègne (2005)
16. Leblanc, A., Abel, M.H.: E-MEMORAe2.0: an e-learning environment as learners communities support. International Journal of Computer Science & Applications 5(1), 108–123 (2008); Technomathematics Research Foundation
17. Sicilia, M.A.: Ontology-based competency management: Infrastructures for the Knowledge Intensive Learning Organization. In: Lytras, M.D., Naeve, A. (eds.) Intelligent Learning Infrastructure for Knowledge Intensive Organizations: A semantic web perspective, pp. 302–324. Information Science Publishing, Idea Group Inc. (2005) ISBN 1-59140-503-3
18. Abel, M.-H., Barry, C., Benayache, A., Chaput, B., Lenne, D., Moulin, C.: Ontology-based Organizational Memory for e-learning. Educational Technology & Society 7(3) (2004)
19. Sarirete, A., Chikh, A.: Knowledge Capitalization in the Communities of Practice of Engineering. In: Proceedings of EACP 2008 workshop: Learning Environments for Communities of Practice held in Tlemcen Algeria, November 18-20, pp. 9–22 (2008)
20. Mizoguchi, R.: Tutorial on Ontological Engineering: Part 1: Introduction to Ontological Engineering. New Generation Comput. 21(4) (2003)
21. Smith, M.K.: Chris Argyris: theories of action, double-loop learning and organizational learning, the encyclopedia of informal education (2001), `http://www.infed.org/thinkers/argyris.htm` (last update: February 5, 2009)
22. Nonaka, I., Toyama, R., Konno, N.: SECI, Ba and Leadership: a Unified Model Knowledge Creation. Long Range Planning 33 (2000)
23. Naeve, A., Sicilia, M.-A., Lytras, M.D.: Learning processes and processing learning: From Organizational Needs to Learning Designs. Journal of Knowledge Management, Special Issue on Competencies Management 12(6), 5–14 (2008)
24. Chikh, A., Berkani, L., Sarirete, A.: Modeling the Communities of Practice of E-learning "CoPEs". In: Proceedings of the Fourth Annual Conference of Learning International Networks Consortium (LINC), Jordan, October 27- 30 (2007)

An Empirical Study on the Use of Web 2.0 by Greek Adult Instructors in Educational Procedures

John Vrettaros, Alexis Tagoulis, Niki Giannopoulou, and Athanasios Drigas

NCSR DEMOKRITOS,
Institute of Informatics and Telecommunications
Net Media Lab
Ag. Paraskevi, 15310, Athens, Greece
{dr,jvr}@iit.demokritos.gr, alextagoulis@hotmail.com,
ngianloy@gmail.com

Abstract. In this paper is presented an empirical study and its results. The empirical study was designed through a pilot training program which was conducted in order to learn if Greek educators can learn to use and even adopt the use of web 2.0 tools and services in the educational process and in which extend, where the type of learning is either distant learning, blended learning or the learning takes place in the traditional classroom.

Keywords: empirical study, Web 2.0, e-learning, social software, wikis, blogs, youtube, facebook.

1 Introduction

The web in the 90s helped to reduce the barriers regarding time, place and cultural boundaries, and contributed in the effective user communication and access to information. The philosophy behind web 2.0, aims not only in the effective communication but also in the development of collaboration and networking of the users, which led for example to the creation of new terms like social networking. This evolution led to the transformation of learning, from the use e-learning platforms to the use of tools of the social web such as blogs, wikis, and forums, to a collaborative and cooperative learning framework. In this framework can be formed groups of learning in various levels of collaboration which can be small group, an organization, a society and finally the global community.

Since adult education is based in learning in groups, the new technologies of social networking can enhance learning effectiveness through the use of tools and services, with the appropriate learning scenarios, learning strategies, and adult education learning theories.

The technologies that support collaboration support the learning goal of changing stance related to learning based on learning groups. These technologies provide a reach, shared virtual learning space, where interactions are not taking place only between learns and technology but among learners, who share a mutual aim. These interactions can be facilitated and coordinated by an instructor.

M.D. Lytras et al. (Eds.): WSKS 2009, CCIS 49, pp. 164–170, 2009.

There have been many research and development efforts for the creation systems that support collaborative learning, but there is a research gap in how can we enhance learning effectiveness in adult education through the use of collaborative learning theories and techniques and collaborative tools and services provided by Web 2.0.

To be more specific, several web 2.0 tools have been widely used for collaborative learning purposes and the results have proved to be very fruitful. A wiki – based project was incorporated in a media technologies class where the learners developed the M/Cyclopedia (Media/Culture), a wiki – based encyclopedia, using MediaWiki of Wikipedia [1]. The educational use of wikis in distance learning classes has been analyzed in the context of a symbolic logic course [2] and wikis have also been used as an environment empowering cooperation among the distant learners [3]. Blogs have also been widely used in education. The creation of learning blogs has been attempted with the use of a group of 31 students that participated in an Information Systems class in Hong Kong University and this study finally proved, that blogs are a powerful tool when it comes to predicting final student performance [4]. Additionally, two groups (one consisted of professors and one consisted of students) were used in order for a research to be conducted aiming at determining the value of use of weblogs in education. The results made it more than clear that blogs were approved as a very useful tool in supporting physical presence in class [5]. Finally, Harvard University, Stanford University and Texas State University have set up virtual classes using Second Life via which learners participate in online courses. It is widely believed that Second Life can provide numerous possibilities when it comes to educational applications [6].

2 Empirical Study

For the wider and appropriate application of the web 2.0 tools and services in the learning process it is necessary the preparation of the adult trainers and also their opinion about these tools and services.

An approved way of investigating their stance and eliciting their opinions is testing these tools and services themselves, and take part in a learning process which comprises social networking tools and services. For the purpose described above an empirical study was designed which is described below

The aims is studying if the adult trainers have a positive stance against the web 2.0 tools and services, if they have used it, and if they can use the collaborative tools and services in an educational process designed by themselves.

<u>Methodological Framework</u>
Design of a learning community of practice
<u>Formal characteristics</u>
Year of realization of the training program: 2008 -2009
Place of realization: Greece
Training program title:
Technology used: LMS Moodle
Tools and services: social networking tools e.g. Blogs, wikis, social networking tools and services

The degree of self-governance and the type of the training program, if it's a open one (learner participation in defining the learning process) or a closed one (there is little or no participation in defining the learning process)

The training program comprises features from open and closed type training programs, where the learners will be able to interfere and shape a new path for the learning process if they expect that it will be more effective in learning, not only in the training program core but also in the activities.

Training program content/curriculum

Aims of the model implementation:

- The learners will get to know others ways of learning
- The learners will explore e-commerce implementation
- The learners will be introduced to e-commerce and they will learn how can they develop a business activity
- The learners will explore the capabilities that e-commerce provides
- The learners will adopt a positive stance against the use of the new technologies in business activities.

Principles of the model implementation

The main principle that governs the community is the respect of the opinion of the peers. Every peer can freely think and express themselves. Through dialog the creative argument, the analysis and the synthesis of oppositions and positions is promoted. Through the interaction among peers and the collaboration of all stakeholders, proceed to aggregative actions.

2.1 Process/ Methodological Approach

2.1.1 Actions/Methods of Instruction and Learning Process

The actions considering the methods of instruction and learning will be based on the Project method. In the first phase the previous experiences and tacit knowledge of the learners will be explored, and the core of the training program will be configured.

In the second phase based on resources the educational material will be obtained. The learning groups will be defined and the activities or roles will be assigned to learning groups.

2.1.2 Assessment (Assessment Criteria)

The assessment criteria will be based on how useful the learners find the web 2.0 collaborative tools and services, and on how they can integrate e-commerce in their working life.

These questionnaires will be handed first, as a pilot to a smaller sample, for the respondents to co-design the questionnaire, in order to take their final form, which in this form the questionnaire will be handed the total sample.

Based on the statistical factor analysis that will follow, it is expected that important conclusions will be extracted, relevant to the factors significance considering the facilitation of the web 2.0 tools and services on learning, and in how these factors affect the learning effectiveness. Finally the correlation between these factors will be examined.

2.1.3 Process of Data Collection

The trainees that participated in the survey got an invitation by email, which explicitly describes the form and the cause of the survey.

The questionnaire filling by the respondents took place electronically, and with the help of blogs there query elaboration took place in order to record every respondent question and remark.

Due to the fact that the training of the educators took place two months earlier than the survey, there are shots of the educational process accompanying the questionnaires, in order to help them recall easily the correspondent parts that are relevant to the survey. The respondents that took part in the survey are familiarized with the ICTs.

2.1.4 Sample Description

The sample consists of 30 learners who attended the realized training programs.

The demographic data of the correspondents are in our possession.

The sample consists of correspondents, aged between 30-55, who are adult trainers, certified that they have attended the National Train the Trainers Program. Their competences and skills in computer and internet use vary.

2.2 Data Collection Tool

The questionnaires consists of closed type questions, where in each question is asked the comment of the correspondent for each answer, in order to extract information that cannot be expressed in the answer itself.

The questionnaire is designed in decreasing Likert scale (4 = very much, 3 = little, 2 =average , 1 = no)

The 4rth Likert degree was chosen in order to avoid having answers by undecided correspondents that usually express a medial position.

2.2.1 Questionnaires

Sex: Male, Female

Studies: Theoretical, Practical

Experience of work in Adult Education: Less than 5 years, 5 to 10 years, More than 10 years

<u>Before the training program</u>

Knowledge:

- Do you use e-learning platforms in teaching?
- Are you familiar with the tools below and do you know if they have implementations in teaching?

Competences:

- Are you familiarized with the use of tools and services of web 2.0?

Stances:

- What's your opinion about the implementation of the web 2.0 tools and services in education?

After the completion of the training program

- Was it difficult handling the web 2.0 tools and services, regarding the registration and the participation in the learning process?
- Did you need the help of an expert user to handle the web 2.0 tools and services?
- Based on the experienced acquired by your participation in the training program, what's your opinion about the educational value of the web 2.0 tools and services (blogs, wikis, youtube, facebook)?
- Do these tools satisfy the principles and the needs of adult education (blogs, wikis, youtube, facebook)?
- Do these tools lead to the elicitation of the tacit knowledge (blogs, wikis, youtube, facebook)?
- Do you think these tools can support learning by doing (blogs, wikis, youtube, facebook)?
- Do you think these tools can enhance collaborative learning (blogs, wikis, youtube, facebook)?
- Do you believe you can design and develop activities for adult education with the use of these tools (blogs, wikis, youtube, facebook)?
- (BLOG, WIKI, YOUTUBE)
- What's your opinion in the usefulness of facebook as a mean for the learners' first introduction?
- Do you think facebook can foster the creation of group atmosphere in a virtual class?

Knowledge:

- Do you think you acquired knowledge in using these tools and services (blogs, wikis, youtube, facebook)?
- Competences:
- Did you get familiarized with the use of these tools?

Stances:

- Would you use these tools in class (virtual or traditional)?

3 Questionnaires Analysis

In the first phase we analyzed the frequency of every question and some general remarks and in the second phase we will study the crossing of various questions (Crosstabs), and with the statistical check X^2 possible causal relations.

For the first demographic data we didn't any further analysis, as the composition of the two training classes was such, that couldn't be found an equal allocation for the sexes, but the numbers statistically allow us to cross variables than weren't considered significant.

The second element, studies is artificial so there wasn't a question, but because the researcher knew the learners, divided them in these categories, and categorized as practical those who had previous experience in computers.

Although the number of the correspondents is small and precarious to make crosstabs, it's very important because we study the effectiveness of tools and services of web 2.0 to non specialists in informatics educators.

In the research 75% of the educators has many years experience in adult education, as its shown by the answers in the 3rd demographical question.

In the next question and after explaining to the correspondents that it is asked only the use of e-learning as an educator, a positive answer came from the practical educators, but they had all used e-learning as learners during their participation in the Training the Trainers Program.

The next question considers mainly their familiarization about the web 2.0 tools and services and their implementations in education. We observe that almost half of the educators know few about these tools, even the practical educators that have computer's experience knew few. Because the practical educators sample is small, we can have a secure estimation on that subject, and we must remark that there wasn't a negative answer in the familiarization with these tools.

From the four web 2.0 tools that they are asked for the most popular tool, the question concerns their opinion as simple users and not experts. In what concerns Facebook the answer a little means that they have browsed Facebook but they haven't created their own profile.

The last question is about their stances, which is quite determining for the survey, because its positive change during the training course and the survey indicates that the educators are ready to use the web 2.0 tools and services for educational purposes, and it also indicates the success of the pilot training program, not as a statistical but as an educational value.

Finally from a statistical point of view, the question is appropriate for 2χ checks and variable independence, because in every cross cell, we must have over five observations. On the specific question its distribution for all the categories is over five observations (8.16,6), so we select this variable which is appropriate for crossing, for statistical and pedagogical reasons.

<u>After the completion of the pilot training program</u>

From this question we observe that most educators comprehend the function of the web 2.0 tools and services and its educational implementation without particular difficulties.

In this question we observe that a large percentage needed the help of an expert to use the web 2.0 tools and services, 63%, which from a competences point of view leads as to the conclusion that the educators weren't ready to use the tools, from a competences point of view. Taking also under consideration the fact that 15% of the educators are specialized in informatics, we come to the conclusion that the educators' participation in other training programs in web 2.0 tools and services in education would help, for the successful implementation by the educators, in order to acquire the competences needed for the use of the tools and adoption of the use of the tools.

As we can see from the question concerning the educational value of the tools, only one respondent answered positively, where he has used the web 2.0 tools and services in class.

Generally the respondents, 55%-90% have a more positive stance, after their participation in the training program, against the matters that affects the adult training.

One of the most important answers is the one that indicates that, through a short training program they gained the self efficacy, and the competences to develop their own learning activities through the use of the web2.0 tools and services (67%).

Facebook became very popular among educators. Apparently, its use for the educators' acquaintance as trainees, which created a positive atmosphere in class, had an important role.

The educators' answers also show that their where greatly familiarized with the tools and services during the pilot training program, and recognized their educational value.

4 Conclusions

Through the empirical study presented we can understand the stances of Greek adult educators against the web 2.0 tools and services, but also the competences they can acquire through the use of the tools and services, but also the need for greater dissemination of the capabilities that the web tools and services in education. There is also the need for more training programs concerning web tools and services, and education in order, to be adopted by a large audience of educators.

An essential conclusion of this empirical study is that the trained educators, is that the use of the web 2.0 tools and services as its possible adoption in class by a large percentage of educators is generally easy and approachable even from educators that are not trained in ICTs.

References

1. Bruns, A., Humphreys, S.: Wikis in teaching and assessment: The M/Cyclopedia project. In: Proceedings of the 2005 International Symposium on Wikis, San Diego, CA, USA, October 16-18, pp. 25–32 (2005)
2. Byron, M.: Teaching with Tiki. Teaching Philosophy 28(2), 108–113 (2005)
3. Tsinakos, A.A.: Collaborative student modelling- A new perspective using wiki. WSEAS Transactions on Advances in Engineering Education 3(6), 475–481 (2006)
4. Du, H.S., Wagner, C.: Learning with weblogs: an empirical investigation. In: Proceedings of the 38th Hawaii International Conference on Systems Sciences (2005)
5. Barbosa, C.A.P., Serrano, C.A.: A blog as a construction tool for the cooperative learning and knowledge (2005)
6. Wongtangswad, J.: Uses of Second Life in Higher Education: Three Successful Cases. In: Richards, G. (ed.) Proceedings of World Conference on E-Learning in Corporate, Government, Healthcare, and Higher Education 2008, pp. 1389–1391. AACE, Chesapeake (2008)

Estimation for Up/Down Fluctuation of Stock Prices by Using Neural Network

Toyohide Watanabe and Kenji Iwata

Department of Systems and Social Informatics,
Graduate School of Information Science, Nagoya University
Furo-cho, Chikusa-ku, Nagoya 464-8603, Japan
watanabe@is.nagoya-u.ac.jp

Abstract. In general, it is not always easy to estimate stock prices exactly and get profits. Until today, many researchers have attacked to this subject, but could not report the successful estimation methods even if various approaches or many heuristics were applied in our knowledge-oriented society. This is because the fluctuation of stock prices is inherently characterized as random walk. In this paper, we address a short-term-specific up/down fluctuation estimation method of stock prices. Our approach is first to select 16 brand companies in Japan Stock Market as the fundamental stock features, and then to define analytically 8 stock attributes as input parameters for our 3-level neural network. We used 32,000 samples of 2,000 days from 16 brands: the first 1,000 days samples were used as learning data for our neural network; and the last 1,000 days samples were as test data. Our experiments showed that the up/down fluctuation estimation method in the short-term from the end value of today to the start value of tomorrow functions effectively.

1 Introduction

E.F.Fama published the paper "Efficient Capital Market" in 1970 [1]. The proposition in his paper insists that the stock prices always keep the reasonable situation: namely, the fluctuation of stock prices is characterized as random walk. Even if the fluctuation was not random walk, it is difficult to formulate explicitly the cause-effect relationships for the fluctuation of stock prices because the up/down fluctuation is directly dependent on the buy-sell actions of investors. On the contrary, until today the challengeable estimation methods of future stock prices have also been investigated with respect to various utilizations of engineering knowledge/technologies. For example, the computational methods of neural network, genetic algorithm, genetic programming, etc. which have been usefully developed in research fields of soft-computing, are applied; unfortunately, they did not always make its effectual results clear yet. However, we think that these methods might possibly apply to some restricted situations for the fluctuation of stock prices if the adaptable parameters were appropriately selected by experts. At least, the subject on the estimation of stock prices is one of most wonderful and exciting topics on e-Business, like auction, and is typical trend in knowledge society.

M.D. Lytras et al. (Eds.): WSKS 2009, CCIS 49, pp. 171–178, 2009.

In this paper, we address the estimation method of future stock prices from a viewpoint of the restricted usages of some parameters or the rigid interpretation of fluctuation features, so that even un-experts can challengingly invest to some stocks with the minimum risk. Our experimental method is to estimate the up/down trend of future stock prices by applying a neural network to the accumulated time-series data, because it is difficult to formulate the fluctuation of stock prices. Also, we do not construct a long-term estimation method, but develop a model which is adaptable to the short-term fluctuation. Generally, the efficient capital market proposed by E.F.Fama means that it is difficult to certainly estimate the fluctuation of stock prices in the future on the basis of only the current situation. However, we develop an estimation method for up/down trend of stock prices to be possibly observed in the next day, based on the accumulated time-series data. Therefore, the expected return may be generally small in comparison with our long-term fluctuation, even if the expectation were just suitable. Our approach is different from many currently investigated ones, which are actually based on the effectiveness theory of market. Of course, we agree with this theory from a viewpoint of the features, observed ordinarily from the global fluctuation. We focus on the most short-term-specific phenomenon as the analytic feature: our research interest makes up the estimation from the difference between the end values of stock prices in yesterday and the start values of stock prices in today.

2 Approach

Traditionally, the technical analysis method and fundamentals analysis method have been typically used as the prediction means of stock prices. The technical analysis method estimates analytically the fluctuation of future stock prices, using the past stock prices and past profits on the basis of technical index. The main feature in this method is to make the market phenomenon clear and also to predict the future trend. Additionally, this analysis viewpoint is short-term-oriented. In this case, it is assumed that the fluctuation patterns in the future should be similarly derived from those in the past. However, this analysis method has the following disadvantages:

- The effectiveness based on the technical index is not sure.
- The necessity to define the technical index flexibly runs short.

On the other hand, the fundamentals analysis method is an analysis means of stock brands, attended with company potentiality, on the basis of fundamentals which specify most primitive factors such as business administration process, economic situation of nations or companies, etc., though the technical analysis method concentrates on the fluctuation of stock prices from a viewpoint of the surface worth. The fundamentals analysis method is furthermore divided into two methods:

- top-down method, based on the global information derived from macro-economics;
- bottom-up method, based on individual information about each company.

These basic methods concerning to the search range, reference scope, focus view, factorized origin, etc. have been practically investigated through the integration with computing means. For example, as for the application of neural network the improvement from traditional technical analysis method [2], computer support for buying/selling stocks [2], estimation of stock prices [3-5] and so on have been reported.

While, the estimation of stock prices by genetic algorithm [6], the estimation of stock prices by genetic programming [6], the estimation of stock prices by applied planning means [2], the approximation of fluctuation of stock prices by support vector machine [7], the classification of time-series patterns of stock prices by Fuzzy theory [8], and so on were investigated on various paradigms. Until now these theoretical approaches did not sufficiently make the adaptabilities valid in practice.

The estimation methods predict exactly the stock prices in the future, but in our investing activities it is necessarily better to grasp the fluctuation situation for brand stocks, which investors buy or sell. Even if the exact prediction of stock prices were not absolutely possible in the future, we can discuss the up/down trend for short-term period, but do not address directly a long-term-based estimation. Our approach is to apply the neural network as a computation model because in general it is not easy to formulate the fluctuation of stock prices. The computation in the neural network is closed from the external observation process as a black-box, but this computation mechanism keeps input-output patterns, which are fit to given problems by suitable machine learning. If we make use of neural network with a view to estimating the fluctuation of stock prices, it is not difficult to keep the relationship between the past time-series data of stock prices and the up/down trend, which is sometimes observed between before-processed data and after-processing data.

3 Attributes for Stock Price

The raw time-series data which we manipulate in this research is composed as a sequence of 3 items: "date", quadruple of ("start value", "highest value", "lowest value", "end value") and "turnover". These data can be easily accessed through Web contents such as Yahoo Finance [11], Pan Active Market Database [12], etc., and also are plotted as a chart with a view to making the understandability for fluctuation of stock prices clear. This chart is called a candle-leg chart, as illustrated in Figure 1. The vertical axis indicates the time sequence and also the horizontal axis represents the stock price. Individual candle-legs represent the meanings, as shown in Figure 2. The candle-leg, which has respectively the highest value, lowest value, start value and end value of stock prices, observed in one day, is categorized into explicit and implicit forms. The explicit form points out that the end value became higher than the start value during the day; and the implicit form represents that the end value became lower than the start value.

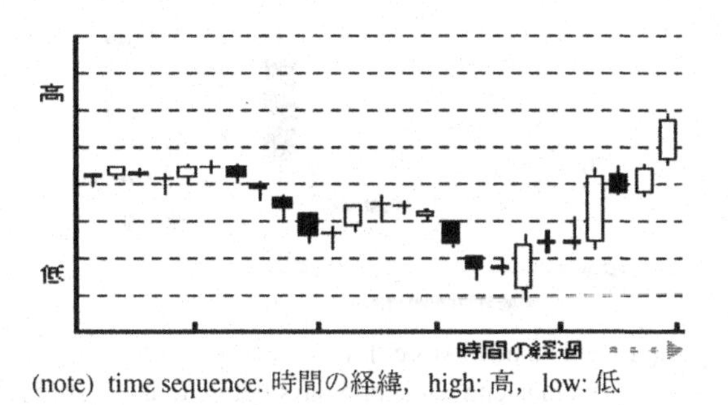

(note) time sequence: 時間の経緯, high: 高, low: 低

Fig. 1. Candle-leg chart

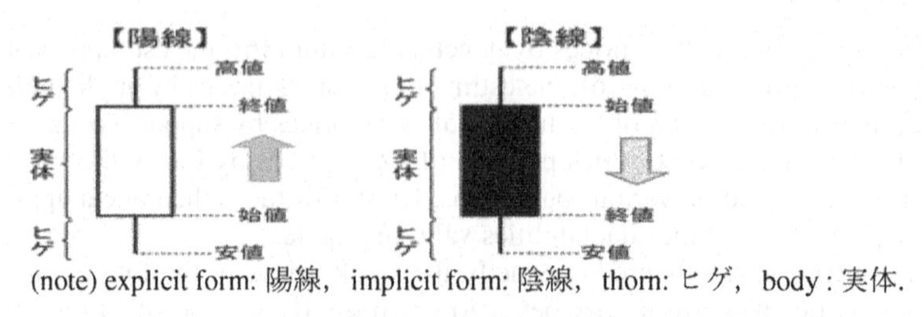

(note) explicit form: 陽線, implicit form: 陰線, thorn: ヒゲ, body : 実体.

Fig. 2. Meanings of candle-leg

Next, we extract the characteristics of fluctuation phenomenon from the time- series data of stock prices in order to estimate the up/down trend for before/after- observed values from the past stock prices. In this case, we assume that the stock prices of individual brands are independent. Moreover, to propose the short-term estimation, we must find out the short-term-specific characteristics from each brand in each day. Thus, we chose the following 8 attributes in 3 categories:

(1) Value movement in today

- Body = (end value of today – start value of today) / end value of yesterday
- Upper = (highest value of today – ***MAX***(start value of today, end value of today)) / end value of yesterday
- Lower = (***MIN***(start value of today, end value of today) – lowest value of today) / end value of yesterday

(2) Value movement from yesterday

- Open = (start value of today – end value of yesterday) / end value of yesterday
- Close = (end value of today – end value of yesterday) / end value of yesterday
- Volume = (turnover of today – turnover of yesterday) / turnover of yesterday

(3) Trend of stock price

- divMA(5) = (end value of today – average end value of 5 days before today) / average end value of 5 days before today
- gradMA(5) = (average end value of 5 days before today – average end value of 5 days before yesterday) / average end value of 5 days before yesterday

Figure 3 illustrates such 8 fluctuation attributes with respect to the candle-leg, except the attribute Volume. Volume is not the fluctuation of stock prices, but the up/down turnovers from yesterday to today. Finally, these 8 attribute data are individually normalized and then are put, unit-by unit, in our neural network, as one unit [9].

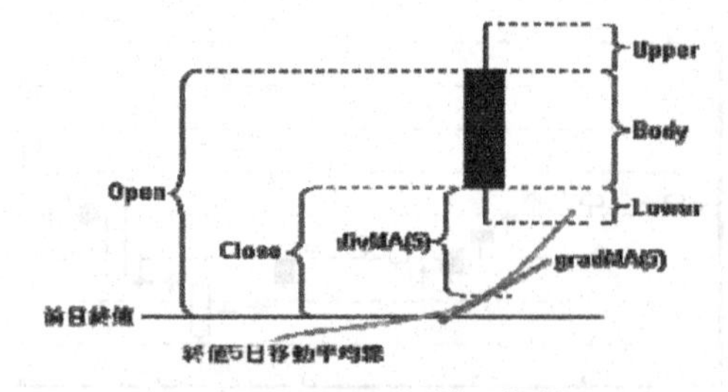

(note) end value of yesterday: 前日終値,
average end value of 5 days before today: 終値5日移動平均線

Fig. 3. Meanings of 8 attributes

4 Up/Down Trend Estimation of Stock Prices

Our neural network is 3-level hierarchical structure. This neural network makes the back-propagation learning possible [10]. The number of neurons in the input and middle layers is 8, and that in the output layer is 1. Our propagation mechanism among neurons was implemented by the sigmoid function. Also, in the back-propagation mechanism it is necessary to set appropriately the initially combined weight values among neurons, as values of learning parameters. As the first step, the initially combined weight values were at random preset independently in each edge between neurons, and then many learning processes were respectively performed after the learning parameters have been set to smallest value. Thus, the computation could be successfully controlled so that the long-term effect does not depend on the initially combined weight values and also the vibration in the near range or in the minimum value of errors should be distributed.

5 Experiment

We experimented to estimate the up/down trend of tomorrow stock prices on the basis of time-series stock data, collected analytically until today. The number of used data is accumulatively 32,000 samples (= 16 brand companies × 2,000 days; from September 16, 1999 to October 31, 2007), which were retrieved from Pan Action Market Database. These 16 brands were selected from 30 brands, nominated by Mainichi-Newspaper in Japan. Additionally, those data were moreover divided into 2 classes: one is September 16, 1999 to October 7, 2003 and another is October 8, 2003 to October 31, 2007. The former was used to construct our neural network as the learning data, and also the latter was to evaluate the performance as the test data. Figure 4 and Figure 5 show the transitions of teacher signals and output values in the learning period. Finally, we show respectively the transitions of output values in the test period in Figure 6 and Figure 7.

In Figure 5, the output values are centered near to 0.5 with respect to 3,000 learning steps because the improvement process of combined weight values attains to a stable state in about 3,000 steps. The reason is as follows: first, many of teacher signals are distributed from 0.4 to 0.7, though those of 0 or 1 are too little, as shown in

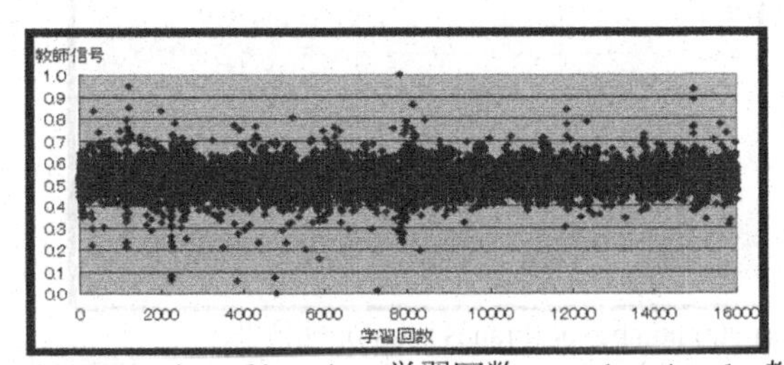

(note) number of learning: 学習回数, teacher signals: 教師信号

Fig. 4. Transition of teacher signals in learning phase

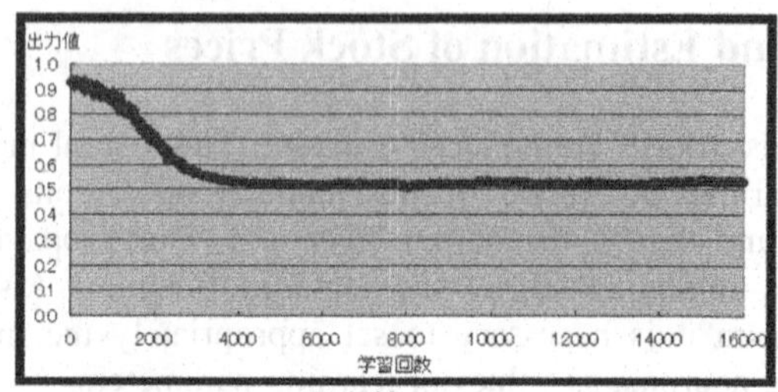

(note) output value: 出力値

Fig. 5. Transition of output values in learning phase

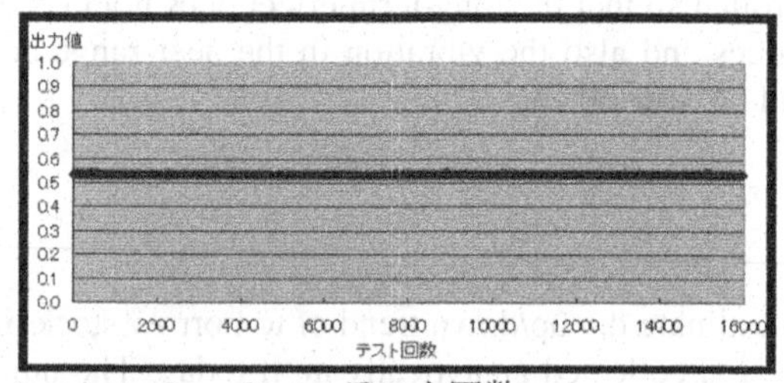

(note) number of tests: テスト回数

Fig. 6. Transition of output values in test phase

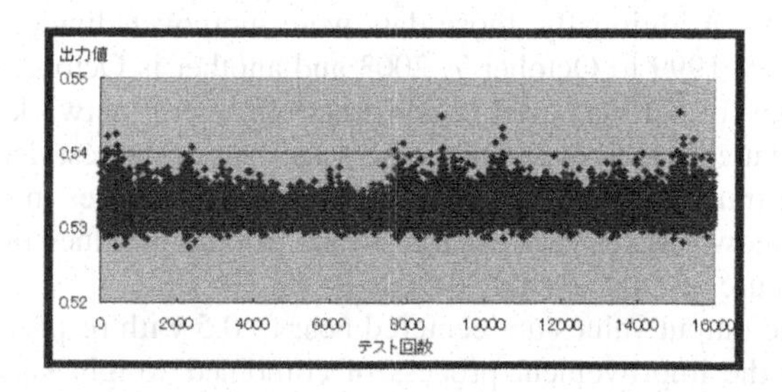

Fig. 7. Transition of output values in test phase (expended version)

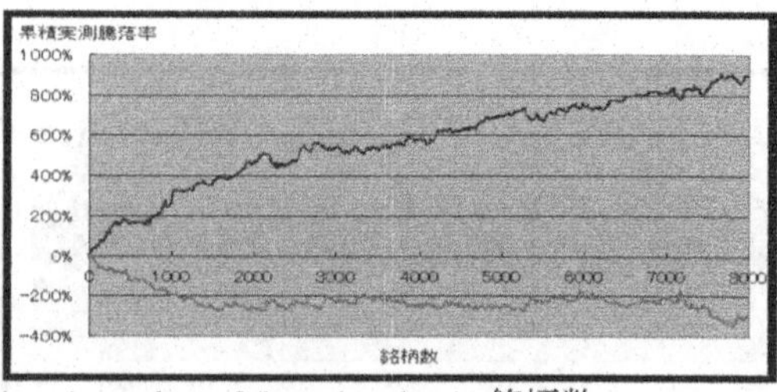

(note) number of (brands ×days): 銘柄数,

really accumulated up/down ratios×100: 累積実測騰落率

Fig. 8. Accumulated up/down ratios for higher/lower classes

Figure 4. Though we scattered the teacher signals uniformly over the range [0, 1] by means of normalizing teacher signals, the output value is yet remained near to 0.5, as shown in Figure 5. Second, the different teacher signals were given to the same or similar input patterns in every learning phase because the time-series data of stock prices cannot be formulated explicitly on the basis of the cause-effect relationship. Also, the output value in the test phase is almost 0.5, as shown in Figure 6. In this expectation, our neural network should estimate the up/down trend of stock prices by the fact that the output values are to be distributed uniformly from 0 to 1.

However, though the originally designed neural network could not be constructed, we can indirectly interpret our output values from a viewpoint of more direct inspection. Figure 7 shows the expended situation for the output value 0.5. The output values are approximately distributed from 0.53 to 0.54: the difference is very little as 0.01. Our idea is to focus on this very little difference value, and estimate the up/down fluctuation of stock prices. For the test data, 16,000 samples (= 16 brands ×1,000 days) were used, and 16 output values are generated in one day with respect to the fluctuation difference among 16 brands. These 16 output values are divided into 2 classes: the group A of 8 higher output values and the group B of 8 lower output values. As for 2 groups we computed the up/down ratios of stock prices. Figure 8 shows these accumulated up/down ratios for 2 classes. The group A is completely separated from the group B, and the difference becomes large in accordance with the number of evaluated samples. This means that our neural network could provide the powerful discrimination ability explicitly, and that we can get +0.08% profits per one day in average if we buy 8 brands of group A by half investing money and also sell 8 brands of group B by half investing money.

Additionally, we experimented to estimate the up/down trend of stock prices for tomorrow end value and the stock prices of 2 days after today, by using the time-series data of today after. All results are almost 0.5 as well as the estimation of tomorrow start value. This result makes it clear that our method is applicable to the short-term estimation of stock prices, but not effective to the long-term estimation.

7 Conclusion

In this paper, we addressed an experimental estimation method for the up/down fluctuation of stock prices, which is effective to the short-term prediction of stock prices. Our basic estimation means is to use the neural network and also to make use of time-series data of stock values for 16 brands, as the input data. In this proposal, our method was not successful in the long-term estimation, but was effectual only to the short-term estimation because the very small difference in the final output values makes the gap between the group A of 8 higher output brands and the group B of 8 lower output brands clear. As a result, the effect is clearly distinguished if we focused on the difference carefully. However, we have some remained issues to make the estimation ability and the availability successful:

1) to develop the long-term estimation method,
2) to optimize input values,
3) to structure optimal neural network.

References

1. Fama, E.F.: Efficient Capital Markets: A Review of Theory and Empirical Work. Journal of Finance, 383–417 (May 1970)
2. Baba, N., Nishida, M., Kai, Y.: A Trial for Improving the Traditional Technical Analysis which Utilizes Neural Networks. IEEJ. Trans. on EIS 126(11), 1324–1331 (2006)
3. Roman, J.: Backpropagation and Recurrent Neural Networks in Financial Analysis of Multiple Stock Market Returns. In: Proc. of 29th Annual Hawaii International Conference on System Science, January 1996, vol. 2, pp. 454–460 (1996)
4. Kimoto, T.: Stock Market Prediction System with Modular Neural Network. In: IEEE International Conference on Neural Networks, vol. 1, pp. 1–6 (1990)
5. Kutsurelis, J.E.: Forecasting Financial Markets Using Neural Networks: An Analysis of Methods and Accuracy, Thesis of Naval Postgraduate School (September 1998)
6. Dempster, M.A.H.: Computational Learning Techniques for Intraday FX Trading Using Popular Technical Indicators. IEEE Trans. on Neural Networks 12(4), 744–754 (2001)
7. Yang, H.: Support Vector Machine Regression for Volatile Stock Market Prediction. In: Yin, H., Allinson, N.M., Freeman, R., Keane, J.A., Hubbard, S. (eds.) IDEAL 2002. LNCS, vol. 2412, pp. 391–396. Springer, Heidelberg (2002)
8. Lee, C.-H.L., Liu, A., Chen, W.-S.: Pattern Discovery of Fuzzy Time Series for Financial Prediction. IEEE Trans. on Knowledge and Data Engineering 18(5), 613–625 (2006)
9. Hopfield, J.I.: Neural Networks and Physical Systems with Emergent Collective Computational Abilities. In: Proc. of Natural Academic Science in USA 1979, pp. 2554–2558 (1982)
10. Rumelhart, D.E., McClelland, J.L., The PDP Group: Parallel Distributed Processing: Explorations in the Microstructure of Cognition, vol. 2. MIT Press, Cambridge (1986)
11. Yahoo Finance, http://quote.yahoo.co.jp/
12. Pan Active Market Database, http://www.panrolling.com/pansoft/amarket

Defining Malaysian Knowledge Society: Results from the Delphi Technique

Norsiah Abdul Hamid[1] and Halimah Badioze Zaman[2]

[1] Humanities, College of Arts and Sciences, Universiti Utara Malaysia, 06010 Sintok, Kedah, Malaysia
norsiah.abdulhamid@gmail.com

[2] Department of Information Science, Faculty of Information Science and Technology, Universiti Kebangsaan Malaysia, 43600 Bangi, Selangor, Malaysia
hbz@ftsm.ukm.my

Abstract. This paper outlines the findings of research where the central idea is to define the term Knowledge Society (KS) in Malaysian context. The research focuses on three important dimensions, namely knowledge, ICT and human capital. This study adopts a modified Delphi technique to seek the important dimensions that can contribute to the development of Malaysian's KS. The Delphi technique involved ten experts in a five-round iterative and controlled feedback procedure to obtain consensus on the important dimensions and to verify the proposed definition of KS. The finding shows that all three dimensions proposed initially scored high and moderate consensus. Round One (R1) proposed an initial definition of KS and required comments and inputs from the panel. These inputs were then used to develop items for a R2 questionnaire. In R2, 56 out of 73 items scored high consensus and in R3, 63 out of 90 items scored high. R4 was conducted to re-rate the new items, in which 8 out of 17 items scored high. Other items scored moderate consensus and no item scored low or no consensus in all rounds. The final round (R5) was employed to verify the final definition of KS. Findings and discovery of this study are significant to the definition of KS and the development of a framework in the Malaysian context.

Keywords: Malaysia, knowledge society, definition, Delphi, knowledge, ICT, human capital.

1 Introduction

The concepts of 'knowledge society' (KS) has been widely discussed and debated nationally and internationally since 1960s by researchers [1][2][3][4][5][6][7][8] and well-known organizations [9][10][11]. The term is not only used by the academics, but also popular among politicians, futurists and decision makers. The chronological change in the society, as written by [12], starts from the agricultural age (first wave), followed by the industrial revolution (second wave) and now the information and knowledge revolution (third wave). Societies no longer totally depend on the

M.D. Lytras et al. (Eds.): WSKS 2009, CCIS 49, pp. 179–189, 2009.

resources such as land, labour, capital and natural resources, but has been complementing by the potential usage of knowledge [13] in every sphere of lives. Other scholars and researchers also agree that our society is now living in the age of knowledge, in which the element is the most critical resource. Thus, knowledge becomes the most important commodity for productivity [4][14][15][16][17][18], and major contributor towards the economics and social development [5][19]. Researchers and scholars tend to discuss and explain the transition and change in society by using various buzzwords, such as global information society [20],[21], knowledge society [7],[10],[15],[22],[23], network society [24], post-capitalist [4] and post-industrial society [2].

The emergence of Information and Communications Technologies (ICTs) has changed the way society lives and works. Many scholars and organizations believe that ICT has and will play a tremendous role in the development of a KS [25],[26],[7],[23]. However, ICT alone could not develop a holistic KS. Any technology introduced in the market needs users or human beings to make it successful. Thus, KS needs to integrate between ICT and human capital. In order to function effectively and efficiently, human requires knowledge, which can be nurtured through education and training. [27] suggested that the KS is a program dedicated to action, which involves elements like education, the sciences, culture and communication at one and at the same time, while [28] stressed that KS can be viewed from social and economic perspectives and it involves knowledge and learning infrastructure and strategies plus emerging technologies. [7] stated that "*...knowledge and human capital are essential to all aspects of development* (p.9).

2 Malaysia's Vision 2020

The government of Malaysia aims to becoming a developed and united nation by the year 2020. The Vision 2020 which was launched in 1991 is a 30-year long-term plan and formed as a guideline for Malaysia to achieve the status of a Knowledge Society (KS) in its own mould – be it politics, economics, social, spiritual and cultural. This vision which is also known as National Vision Policy has outlined nine strategic challenges in order to become a developed nation. It is supported by major national policies, namely National Development Policy, five-year Malaysian Plans and the Second and Third Outline Perspective Plans. Currently the country is running its 9th Malaysian Plan, from the year 2006 to 2010. The policies are particularly planned for the development of economics and social of a nation. All recent Malaysian policies have clearly recognized the roles and impacts of ICT in the nation's development, besides taking into account the globalization, liberalization and other issues. Hence, the government plans to improve the ICT infrastructure, content, usage and applications among its society so as to develop a fair and just society, and narrow the gap between and among its citizens. As stated by [29], a crucial component of the nation's knowledge economy envisions the availability of digital content exchange or content infrastructure. Without the right infrastructure, effective and efficient access and delivery of information can be hampered (p.537).

3 Problem Statement

Although numerous researches and documents about KS have long been discussed and debated at international, regional and national levels, there has been little effort in defining what a KS is and there is no agreement about what it means [19]. Thus, it leads to difficulties in understanding or evaluating something that does not have a unified definition [30]. There are some studies or reports which have focused on KS, such as the Arab Human Development Report by UNDP (2003), however they still deficient in discussing the KS concept [31]. In measuring the progress of a nation, policy makers do not define and conceptualize up front the specific concept being studied [32]. Instead, they start their design process with variables and indicator levels, they try to determine which factors are measurable, and they overlook what is truly meaningful in a particular context [32] as the concept is initially unclear and undefined. The tools and methods used by the advanced countries can sometimes be used as models for the emerging countries. However, differences in political systems, stages of economic development, and cultures have to be taken into account [33]. Therefore, it is less significant to rely solely on the existing definitions, tools and measurements developed by the advanced developed countries.

In Malaysian context, the focus is more towards the concept of Knowledge Economy rather than KS. This is proven by a well-defined term of K-based Economy by the Institute of Strategic and International Studies (ISIS) Malaysia, which is proposed in the report of '*Knowledge-Based Economy Master Plan*' [34]. The definition of k-based economy is "*an economy where knowledge, creativity and innovation play an ever-increasing and important role in generating and sustaining growth*" *(p.2)*. The definition was quite simple and brief, however, it is argued that there are other elements that need to be taken into consideration in order to reach the status of KS. In order to become a fully developed country in the year 2020, a framework called 'National IT Agenda' (NITA) was proposed by National IT Council (NITC) of Malaysia [35]. The centre of the framework is an outcome in which to become a knowledge-based society by combining three important elements, namely people, infostructure, and content and applications. To date, however, there is no standard or well-accepted definition of KS being proposed by the government [36].

4 Conceptual Framework

The objective of this study is to discover the perception of experts with regards to the definition of KS especially in Malaysian context and to identify the determinants that contribute towards the definition. In order to develop the conceptual framework, comprehensive review of literature has been done on the definition and the existing models of KS to explain what dimension have been emphasized in discussing the KS. Based on the literature reviewed, the emphasis of KS can be categorized into three main interrelated dimensions, namely knowledge, ICT, and human capital, as shown in Table 1. It shows evidence that the three dimensions are the essential and most crucial determinants to any KS. Failure to address these dimensions appropriately will result in failure of Malaysian KS development by the year 2020.

Table 1. Occurrences of dimensions in the literature reviewed

Dimensions	Literature reviewed
Knowledge	[7],[15],[16],[19],[23],[28],[37],[38],[39]
Technology, particularly ICT	[7],[15],[16],[18],[28],[38],[39],[40],[41]
Human capital	[7],[15],[16],[18],[23],[28],[38],[39],[40]

5 Methodology

This study employed a five-round modified Delphi technique as a method to answer a research question pertaining to the definition of Malaysian knowledge society. Generally, the Delphi technique is a procedure to "obtain the most reliable consensus of opinion of a group of experts…by a series of intensive questionnaires interspersed with controlled opinion feedback [42]. [43] suggested some criteria in which a study is suitable to employ the Delphi technique, such as the problem does not lend itself to precise analytical techniques but can benefit from subjective judgments on a collective basis and the individuals needed to contribute to the examination of a broad or complex problem have no history of adequate communication and may represent diverse backgrounds with respect to experience or expertise. It should be noted that there are also limited resources available to facilitate decision making or generate idea with regards to the definition of Malaysian KS. Since all these criteria exist in the current study, therefore the Delphi technique is suitable to be employed in defining the Malaysian KS. In Delphi, some essential characteristics have been identified, for instance, it involves a group communication process, there is a panel of experts who agreed to participate, and few rounds, multi-stage of questionnaire distribution. Moreover, [44] suggested three criteria as basic characteristics of the Delphi technique, namely anonymity, iteration and controlled feedback, and statistical group response.

The technique used in this study is considered as a modified Delphi due the nature of Round One (R1) questionnaire which provides an initial definition of Malaysian KS and justifications on the definition were also provided and only new items were put in R4 questionnaire. The development of the initial definition is based on the reviewed literature, especially on the existing KS and information society (IS) definitions. In R1, the experts were required to state their level of agreement on the initial definition based on a four-point scale, from 1=Disagree, 2=Partly Disagree, 3=Partly Agree through to 4=Agree. Based on inputs and opinion received from the experts in Round One, the researchers then analyzed and formed list of statements on Malaysian KS in R2, and the panel rated on a five-point Likert scale of relevance, from 1=Extremely Irrelevant, 2=Irrelevant, 3=Undecided, 4=Relevant through to 5=Extremely Relevant. The results of R2 were keyed-in in SPSS v.12 to identify *medians* and *quartile deviations* (QD) for each item. The QDs are used to identify the level of consensus among experts and based on [46]: '$QD \leq 0.5$=high consensus'; '$0.5 < QD \leq 1.0$=moderate'; and '$QD > 1.0$=low and no consensus'. The level of consensus was yielded for each item and all high and moderate consensus items were brought to R3. This round has also added 17 new items derived from R2. Again, the panel rated the items on a five-point Likert scale, and the level of consensus is obtained using the guidelines by [46]. R4 was employed to confirm the ratings of the panel of experts for 17 new items that were first time rated in R3. Lastly, R5 was used to verify the definition of KS which was improvised.

Experts were selected based on field of expertise in review of publications [47],[48], individuals who had direct relationship with the topic studied [49],[50], and using the 'snowball' technique [51]. Since expert opinion is sought, a purposive sampling is employed where people are selected not to represent the general population, rather their expert ability to answer the research questions [52]. There are four requirements for expertise, including knowledge and experience with the issues under investigation; capacity and willingness to participate; sufficient time to participate in the Delphi; and effective communication skills [53]. This study nominated the experts based on their position as policy makers, academicians and practitioners in order to get a balanced combination of the three groups of people. [54] argued that a fairly heterogeneous panel will ensure the full spectrum of views is represented and yield a rich database for in-depth analyses and identifying options to effective decision making. Based on a thorough searching, twenty experts were invited to participate in a panel of experts. After a given date, ten experts replied stating their agreement to participate. [55] noted that "good results can be obtained even with small panels of 10-15 individuals" (p.14). [56] stated that a suitable minimum size is seven with accuracy deteriorating rapidly with smaller sizes and improving more slowly with larger numbers. [57] argued that the optimum size of the panel is 7 to 12 members, while [58] suggested a typical Delphi has about 8 to 12 members.

6 Findings and Discussion

Out of ten experts involved, seven were males and three females. Four experts aged between 41-50 and four between 51-60. One expert aged between 30-40 and the other one is over 61 years old. In terms of academic qualifications, all experts obtained a PhD. With regards to types of employer, one expert involved with the government agency, one is an ex-CEO of a government agency and the other one used to be a consultant with the Malaysian government agency. Seven experts are from institutions of higher learning, in which three are Professors, three Associate Professors, and one a senior lecturer. One of the Professors is also currently a head of department. In terms of working experience, five of them have between 15 to 20 years of experience, three between 21 to 30, and the other two have between 31 to 40 years of working experience. The first round has generated many inputs from the experts based on the initial definition of Malaysian KS proposed. The initial definition was:

> *"Knowledge Society is a society in which information and knowledge are the major inputs in all aspects of people's activities. It is well-supported by three major sectors, namely knowledge sector, ICT sector, and human capital sector which are the drivers and prime inputs, while KS is the targeting outcome. Knowledge sector consists of knowledge creation, knowledge infrastructure, knowledge management and knowledge mediation. ICT sector involves ICT goods and services, while human capital sector consists of activities in education, training and skills."*

All ten experts responded in R1, in which eight were partly agreed with the initial definition, and the other two did not state their level of agreement. However, all experts gave written input and feedback. These inputs were then analysed according to suggestions by [59] and [60]. Based on the main points, key words and phrases identified in the inputs, 73 items which converted into statements, were derived for R2 questionnaire. Four sub-sections were formed to categorize the statements, namely (i) general statements on KS (15 items), (ii) statements on knowledge dimension (19 items), (iii) statements on ICT dimension (26 items), and (iv) statements on human capital dimension (13 items). Nine experts responded in R2. R3 questionnaire consists of three parts; part one relates to the previous 73 old items that were to be re-rated, while part two consists of 17 new items which derived from R2 feedback. The total items in R3 were 90. Part three pertains to the experts' demographic profile. The data were analyzed in the similar way as in R2. Eight experts responded in R3. R4 was simpler as the questionnaire only involved 17 items which need to be re-rated. Eight experts responded in R3. Lastly, R5 was employed which required the panel to verify the improved definition of KS based on all the inputs from R1 to R4. Nine experts responded in R5. The results of the Delphi study are summarised in Table 2.

Table 2. Summary of results from Round Two to Round Four

Part		R2		R3		R4	
		HC	MC	HC	MC	HC	MC
I	General statements on KS	10	5	16	14	6	9
II	Knowledge Dimension	13	6	12	7	-	-
III	ICT Dimension	22	4	22	4	-	-
IV	Human Capital Dimension	11	2	13	2	2	-
Total		56	17	63	27	8	9
Grand Total		73		90		17	

HC=High Consensus; MC=Moderate Consensus.

6.1 The General Statements on KS

This part pertains to the statements which are broad in nature and should generally constitute the KS. It consists of 15 items in R2, excluding one item that was an open-ended statement [Knowledge Society should comprise of other dimension(s). Please state it:]. Ten items scored high consensus, while five items scored moderate consensus. The open-ended item was not analysed using median and QD due to its open structure and brought forward to R3. In R3, out of 30 items including the new items, 16 scored high and 14 moderate consensus. In R4, out of 15 items, six scored high and nine scored moderate consensus. One of the experts commented that "a KS is a community of interacting individuals and groups. Knowledge gets better when it is shared. Thus an ecosystem is needed that promotes knowledge sharing, leading to self-sustaining feedback loops being formed". Results concluded that all the items generated high and moderate consensus among experts.

6.2 Knowledge Dimension

This dimension consists of 19 items related to the knowledge construct with no new items generated in any rounds. In R2, 13 items scored high, while 6 items scored moderate consensus, while in R3, 12 items scored high and 7 scored moderate consensus. The results of R3 showed a very slightly changed in item K07 [Knowledge is a prime input towards KS], in which it changed from high to moderate consensus.

However, the medians were not changed. All panel members agreed on the statements pertaining to 'knowledge access, sharing, utilization and application are parts of knowledge dimension'. For item K01 [Knowledge access is a part of knowledge dimension], there are experts whom agreed on the statement, such as "*Access is an important component, but not the only one*" and "*K-access is a vital component of K dimension*". However, one expert argued that "*I do not really think one can access knowledge, only information. Knowledge is partly tacit and can't be 'accessed' via ICTs*". Other items that have been commented is K07 [Knowledge is a prime input towards KS], in which one panel member said "*Perhaps, but I am not sure it is the prime one as basic needs (health, etc.) are also important*", while the other said "*Seems to imply that only knowledgeable people can participate in KS, which I doubt*". It is agreed that knowledge is not the only input towards KS due to many other human basic needs, but human *do* need knowledge in order to survive and improve their quality of lives. Six of the experts stress that knowledge should be measured both in quantitative and qualitative manners (items K15 and K16). However, one comment seems to disagree with the statement. One panel member stated that: "*...I am suspicious of attempts to measure knowledge because they rely on definitions that are not universally acceptable*". This is true if we analyzed the nature of knowledge, in which it is abstract and usually lies in human mind, thus it makes it more difficult to conceptualize and/or measure [61].

6.3 ICT Dimension

For ICT dimension, there are no new items suggested by experts in R2, hence the number of items is equal in R2 and R3. Results from both rounds indicate that there is a very similar pattern of consensus, where 22 items scored high and 4 items scored moderate consensus, with slightly increased or decreased in values of medians and QD between rounds. In R3, there are two statements which were reworded based on suggestions by the experts. The items are T01 [Original statement "Technology is the driver towards KS" has been changed to "Technology is a driver towards KS"] and T02 [Original statement "ICT is the driver towards KS" has been changed to "ICT is a driver towards KS"]. The rewordings of these two statements have shown an increase in the median scores. Analysis of results of the ICT dimension shows that majority of the items scored high consensus, in which the very strong consensus is item T25 [ICT should be fairly distributed among all Malaysians]. This item scored 5.0 for median in R2 and R3, and QD .0. The panel also believed that ICT is an enabling technology and a sector that helps in achieving the KS. This is similar to [25] whom emphasized that ICTs are a means to advance the knowledge society and serve as a transmission belt to generate, access, disseminate and share knowledge, data, information, communications and best practices at all societal levels. Item T17 [Local content is crucial in ICT application] also get high consensus. As stated by [23], many developing countries are experiencing difficulties in identifying the types of knowledge they possess, in boosting their value and in making their potentials work for their development. Therefore, Malaysia should preserve and highlight its local content especially in ICT application. Similar to knowledge dimension, two of the experts stress that ICT should be measured both in quantitative and qualitative manners. The panel also agreed that item T26 [Digital divide is an obstacle towards KS] is very relevant in the development of Malaysian KS. This is inline with [23] whom indicates that digital divide is an obstacle especially in developing countries. The panel rated high on the ICT measurement which should consist of ICT access, ownership, usage, adoption and the value to society's life.

6.4 Human Capital Dimension

There are two new items suggested by experts in R2. Results from R2 and R3 indicate that there are no changes on the level of consensus among the experts. Out of 13 items in R2, eleven scored high consensus and two scored moderate, while in R3, out of 15 items, 13 scored high and only two scored moderate. In fact, only the value of medians had increased or decreased, but the values of QD are all remained the same. There are two statements which were reworded in R3 based on suggestions by the experts and the rewordings have changed the median scores, in which the median score for item H02 has increased from 4.0 to 4.5, but the median for item H01 decreased from 5.0 to 4.5. Human capital is another crucial dimension in the development of Malaysian KS. In fact, one expert comments the item H02 [Human capital is an input towards KS] as "*Yes, extremely relevant*" and the other stated that "*Human capital is the main component but not the only one*". This is true as KS is also supported by knowledge and ICT dimensions. In terms of basic literacy (item H07), one expert augmented that basic literacy is not necessary in developing the Malaysian KS, but it is very important. He further gave an example of Malaysian-produced PCs named "Makcik PC" where it helps people to access information without literacy. For the new items derived in R2, the panel was highly agreed on both items, H14 and H15 which they scored 4.0 for medians and .5 for QDs.

The study has discovered that all the three dimensions are important to the development of Malaysian KS; hence, crucial to the definition itself. The panel believes that these three dimensions can contribute towards the Malaysian KS. Based on the overall results, it can be claimed that the initial definition of Malaysian KS which is proposed in R1 needs modification and therefore, a new improvised definition which has been verified by the panel of experts is:

> *"Knowledge Society (KS) is a society in which its members appreciate knowledge, in which it is seen as a significant commodity of a nation's growth. KS shows concern in the holistic development of the human capital that encompasses physical, intellectual, and spiritual aspects of growth; while knowledge is seen as a value-added dimension. KS also propagates fairness and equality in all spheres of human lives which must take precedence. This includes the generation, access, use, and sharing of existing and innovation of new knowledge through relevant ICT infrastructure and infostructure in tandem with the needs of that society."*

7 Conclusion

The concept of Knowledge Society has emerged since the realization of the importance of knowledge to the development of nations and the rapid progress in technology especially ICT. KS in Malaysian context needs not only ICT and knowledge, but also the human capital dimension to make it a reality. Thus, this study was conducted to identify the important dimensions of Malaysian KS and proposed three dimensions, namely knowledge, technology especially ICT, and human capital to define the term 'knowledge society'. Results indicated that all the three dimensions scored high and moderate consensus. In conclusion, this study has achieved its objective which is to define the Malaysian KS using the three dedicated dimensions. Future study should yield a more solid definition of the term based on the knowledge, ICT and human capital statements provided in this study, and look at the KS from broader perspectives including economics, social, politics and cultural dimensions.

References

1. Machlup, F.: The Production and Distribution of Knowledge in the United States. Princeton University Press, Princeton (1962)
2. Bell, D.: The Coming of Post-Industrial Society: A Venture in Social Forecasting. Penguin Peregrine Books, Harmondsworth (1973)
3. Masuda, Y.: The Information Society as a Post-Industrial Society. World Future Society, Maryland (1980)
4. Drucker, P.F.: Post-Capitalist Society. HarperCollins, New York (1993)
5. Stehr, N.: Knowledge Societies. SAGE Publications, London (1994)
6. Webster, W.: Theories of the Information Society. Routledge, London (1995)
7. Mansell, R., When, U.: Knowledge Societies: Information Technology for Sustainable Development. United Nations Commission on Science and Technology for Development, New York (1998)
8. Duff, A.S.: Information Society Studies. Routledge, London (2000)
9. United Nations Educational, Scientific and Cultural Organization (UNESCO): Measuring and Monitoring the Information and Knowledge Societies: A Statistical Challenge. UNESCO Publications, Montreal (2003)
10. European Foundation for the Improvement of Living and Working Conditions.: European Knowledge Society Foresight: The Euforia Project Synthesis (2004), `http://www.eurofound.europa.eu/pubdocs/2004/04/en/1/ef0404en.pdf`
11. United Nations on Department of Economic and Social Affairs (UNDESA): Expanding Public Space for the Development of the Knowledge Society: Report of the Ad Hoc Expert Group Meeting on Knowledge Systems for Development (2003), `http://www.brint.org/UnitedNations.pdf`
12. Toffler, A.: The Third Wave. Bantam Books, London (1980)
13. Economic Planning Unit (EPU): The Third Outline Perspective Plan 2001 - 2010. Prime Minister's Department, Kuala Lumpur, Malaysia (2001)
14. Evans, P.B., Wurster, T.S.: Strategies and the New Economics of Information. Harvard Business Review, 71–82 (September 1997)
15. Evers, H.: Towards a Malaysian Knowledge Society. In: The 3rd International Malaysian Studies Conference (MSC3). UKM Bangi, Selangor (2001)
16. Al-Hawamdeh, S., Hart, T.L.: Information and Knowledge Society. McGraw-Hill, Singapore (2002)
17. Clarke, M.: e-Development: Development and the New Economy. UNU World Institute for Development Economics Research, Policy Brief No. 7. United Nations University, Helsinki (2003), `http://www.wider.unu.edu/publications/policybriefs/en_GB/pb7/`
18. Britz, J.J., Lor, P.J., Coetzee, I.E.M., Bester, B.C.: Africa as A Knowledge Society: A Reality Check. The International Information & Library Review 38, 25–40 (2006)

19. Rohrbach, D.: The Development of Knowledge Societies in 19 OECD Countries Between 1970 and 2002. Social Science Information 46(4), 655–689 (2007)
20. Martin, J.W.: The Global Information Society. AslibGower, Hampshire, England (1995)
21. Malek, J.A.: Pembangunan Bandar Pintar dan Identiti Masyarakat Global Bermaklumat di Malaysia: Kajian Kes Putrajaya dan Subang Jaya. Doctoral Thesis, Institute of Graduate Studies, University of Malaya, Kuala Lumpur (2005)
22. Nassehi, A.: Note on Society: What Do We Know About Knowledge? An Essay on the Knowledge Society. Canadian Journal of Sociology 29(3), 439–449 (2004)
23. United Nations Educational, Scientific and Cultural Organization (UNESCO): UNESCO World Report: Towards Knowledge Societies. UNESCO Publishing, Paris (2005)
24. Castells, M.: The Rise of the Network Society: The Information Age: Economy, Society and Culture, 2nd edn., vol. 1. Blackwell Publishers, Oxford (2000)
25. D'Orville, H.: Towards the Global Knowledge and Information Society: The Challenges for Development Cooperation (2000), http://ncsi-net.ncsi.iisc.ernet.in/cyberspace/societal-issues/131/info21.htm
26. Taylor, R.D.: The Malaysia Experience: The Multimedia Super Corridor. In: Jussawalla, M., Taylor, R.D. (eds.) Information technology parks of the Asia Pacific: Lessons for the regional digital divide, pp. 64–118. M.E.Sharpe, Armonk (2003)
27. Portella, E.: Introduction: Indications of the Knowledge Society. Diogenes 50(5), 5–7 (2003)
28. Lytras, M.D., Sicilia, M.A.: The Knowledge Society: A Manifesto for Knowledge and Learning. Int. J. Knowledge and Learning 1(1/2), 1–11 (2005)
29. Manaf, Z.A.: Establishing the National Digital Cultural Heritage Repository in Malaysia. Library Review 57(7), 537–548 (2008)
30. Adams, M.K.: Defining Creative Scholarship and Identifying Criteria for Evaluating Creative Scholarship Using A Modified Delphi Technique. Doctoral thesis, Graduate School, University of Wyoming, Wyoming (2004)
31. Zahlan, A.B.: Arab Societies as Knowledge Societies. Minerva 44, 103–112 (2006)
32. Barzilai-Nahon, K.: Gaps and bits: Conceptualizing Measurements for Digital Divide/s. The Information Society 22(5), 269–278 (2006)
33. Jussawalla, M.: Bridging the "global digital divide". In: Jussawalla, M., Taylor, R.D. (eds.) Information technology parks of the Asia Pacific: Lessons for the regional digital divide, pp. 3–24. M.E.Sharpe, Armonk (2003)
34. Institute of Strategic and International Studies Malaysia.: Knowledge-Based Economy Master Plan. ISIS Malaysia, Kuala Lumpur (2002)
35. Wahab, A.A.: A Complexity Approach to National IT Policy Making: The Case of Malaysia's Multimedia Super Corridor (MSC). Doctoral thesis, School of Information Technology and Electrical Engineering (ITEE). University of Queensland, Queensland (2003)
36. Shariffadeen, T.M.A.: Personal communication via e-mail (December 26, 2008)
37. United Nations Development Programme (UNDP): Arab Human Development Report 2003: Building a knowledge society. Regional Bureau for Arab States, United Nations Development Programme (2003), http://hdr.undp.org/en/reports/regionalreports/arabstates/Arab_States_2003_en.pdf
38. Asha, R.P., Ramachandran, R.: Emerging statistical concepts and definitions in the information era (2001), http://www.stat.go.jp/english/info/meetings/iaos/pdf/asha.pdf
39. Evers, H.-D.: Transition Towards a Knowledge Society: Malaysia and Indonesia in Comparative Perspective. Comparative Sociology 2(2), 355–373 (2003)

40. Lor, P.J., Britz, J.J.: Is A Knowledge Society Possible Without Freedom of Access to Information? Journal of Information Science 33(4), 387–397 (2007)
41. Forfas.: e-Business Monitor Report (2003), http://www.forfas.ie/media/forfas031209_ebusiness_monitor.pdf
42. Dalkey, N., Helmer, D.: An Experimental Application of the Delphi Method to the Use of Experts. Management Science 9, 458–467 (1963)
43. Linstone, H.A., Turoff, M. (eds.): The Delphi Method: Techniques and Applications. Addison-Wesley Publishing, Reading (1975)
44. Rowe, G., Wright, G.: The Delphi Technique as A Forecasting Tool: Issues and Analysis. International Journal of Forecasting 15(4), 353–375 (1999)
45. Dalkey, N.: An Experimental Study of Group Opinion: The Delphi Method. Futures 1(5), 408–426 (1969)
46. Razak, N.A.: Computer Competency of In-Service ESL Teachers in Malaysian Secondary Schools. Doctoral thesis, Universiti Kebangsaan Malaysia, Bangi (2003)
47. Miller, G.: The Development of Indicators for Sustainable Tourism: Results of A Delphi Survey of Tourism Researchers. Tourism Management 22, 351–362 (2001)
48. Hsu, C.C., Sandford, B.A.: Minimizing Non-Response in the Delphi Process: How to Respond to Non-Response. Practical Assessment, Research & Evaluation 12(17), 1–6 (2007)
49. Jones, C.G.: A Delphi Evaluation of Agreement Between Organizations. In: Linstone, H.A., Turoff, M. (eds.) The Delphi method: Techniques and applications. Addison-Wesley Publishing Company, Reading (1975)
50. Anderson, D.H., Schneider, I.E.: Using the Delphi Process to Identify Significant Recreation Research-Based Innovations. Journal of Park and Recreation Administration 11(1), 25–36 (1993)
51. Bogdan, R.C., Biklen, S.K.: Qualitative Research for Education: An Introduction to Theory and Methods, 2nd edn. Allyn and Bacon, Boston (1992)
52. Fink, A., Kosecoff, J.: How To Conduct Surveys: A Step-by-Step Guide. SAGE Publications, London (1985)
53. Adler, M., Ziglio, E. (eds.): Gazing Into the Oracle: The Delphi Method and Its Application to Social Policy and Public Health. Jessica Kingsley, London (1996)
54. Loo, R.: The Delphi Method: A Powerful Tool for Strategic Management. International Journal of Policy Strategies and Management 25(4), 762–769 (2002)
55. Ziglio, E.: The Delphi Method and Its Contribution to Decision-Making. In: Adler, M., Ziglio, E. (eds.) Gazing Into the Oracle: The Delphi Method and Its Application to Social Policy and Public Health. Jessica Kingsley, London (1996)
56. Linstone, H.A.: The Delphi Technique. In: Fowlers, J. (ed.) Handbook of Future Research. Greenwood, Westport (1978)
57. Philips, R.: New Applications for the Delphi Technique, vol. 2, pp.191-196. Annual Pfeiffer and Company (2000)
58. Cavalli-Sforza, V., Ortolano, L.: Delphi Forecasts of Land-Use – Transportation Interactions. Journal of Transportation Engineering 35(4), 324–339 (1984)
59. Glaser, B.G., Strauss, A.L.: The Discovery of Grounded Theory. Aldine Publishing, Chicago (1967)
60. Thomas, J.R.: The Practice of Local Economic Development: Expert Identification of Trends and Issues Affecting the Practice of Economic Development at the County Level in Ohio. Doctoral dissertation. The Ohio State University, Ohio (2001)
61. Huseman, R.C., Goodman, J.P.: Leading with Knowledge: The Nature of Competition in the 21st Century. SAGE Publications, Thousand Oaks (1999)

Cultural Aspects of Secrecy in Global Economy

Enric Serradell-Lopez[1] and Victor Cavaller[2]

[1] Department of Business Sciences
[2] Department of Information and Communication
Open University of Catalonia
eserradell@uoc.edu, vcavaller@uoc.edu

Abstract. The main objective of this paper is to provide greater understanding of the nature of secrecy in firms. It presents an effort to develop some links between management of the secrecy and its relationship with culture. Using measures from Hofstede's work, we have linked some dimensions of national culture with CIS 3 UE survey database. The results show that some attributes of the culture as Masculinity and Uncertainty Avoidance have impact on the tendency of the firms for not to patent and maintain secrecy of their innovations.

Keywords: secrets, patents, management of the secret, innovation.

1 Introduction

In the current society of knowledge speak on the secret in the companies could be considered outdated. In the actuality, the trend is to speak on the diffusion of the information, free knowledge and open software. Exists a true consensus on the virtues of knowledge management under the style to share everything, but in the practice, the companies have to decide some key appearances, as for example, decide which knowledge is important and in which quantity, looking for the balance. A lot of information in the companies is critical for his survival and thus it is necessary to establish of a clear way the people or groups of employees with access to sensitive information.

The firms have to pose which part of his knowledge protect by means of formal methods of protection, as for example the patents, and what splits to maintain secretly to save of the surveillance and scrutiny of other companies competitors, by means of informal or strategic actions, where included the commercial secrets.

The aim of this work direct to review some key concepts related with the management of the secret in the firms.

Several causes exist for which the companies opt for the secret instead of the patents. We can consider two types of causes: economic and cultural. In the first part of the work we will revise some aspects of legal and strategic protection that includes secrecy. In the second part, the objective is describing some dimensions of the national culture studied in the classic work of Hofstede [5] and linking theses dimensions regarding the decision of maintaining results of the innovation secretly.

M.D. Lytras et al. (Eds.): WSKS 2009, CCIS 49, pp. 190–199, 2009.

2 Theoretical Discussion and Hypotheses

2.1 Some Aspects of Intellectual Property Protection: Legal and Informal Methods

A study of the American Society for the Industrial Security and PwC estimated that 1000 companies of the list Fortune lost 59 trillion dollars in 2001 due to the robbery of intellectual property and commercial secrets of property (Financial Times, March 14, 2003: 27).

The companies before the threat that supposes that their most valuable assets pass to other people's hands have a series of protection methods, which can be classified in two big groups: legal or formal, as for example patents, rights of property and commercial marks and strategic or informal, which trust the secret, the complexity of the design and in the proportionate advantage as first mover [9].

The rights of intellectual property can be considered as a group of rights that protect applications of ideas and information that have commercial value. They have three main objectives:

- To provide incentives for the creation of knowledge
- To increase the knowledge inside of a certain company's culture
- To protect a distinctive identity [8]

A patent can be a sword of double edge. On one hand it can be an useful tool to explode and to protect the intellectual property of a company, but it could also be harmful, if for example a company doesn't have the means to defend the patent in the tribunals. In such a way that the process of decision of patenting or it is not complex and it should keep in mind the strategy of administration of the information and the complexity that presents the commercial context of the company.

2.2 Formal Mechanisms of Protection: Patents

The cost of discovering the knowledge generated by a patent depends on the force of that patent. It is very probable that a strong patent is maintained if it provides its proprietor a wide protection environment and guarantees against third.

When the patents are not strong, a company will probably trust the use of commercial secrets or some combination of patents and secret. This is maybe a reason why Cohen et al. [2] found in their study that in most of the sectors the secret percentage was equal or superior to the protection offered by patents, like half of appropriation of the benefits.

Table 1 shows the strategies of protection of a company depending on the characteristics of the invention with regard to the force of the patent and the effectiveness of the secret.

A naked idea invention is one where the critical invention is easily observable in the product or service that embodies that invention. Without a strong patent of protection, these inventions can be imitated easily. The advantage usually resides in the speed of being the first mover in a business. The speed in the introduction of these ideas is critical for the viability and the success in the launching of the product.

Table 1. The effectiveness of the secret concerning the strength of the patent Source: Adapted from Anton et al. [1]

		Patent strength	
		Strong	**Weak**
Secrecy Effectiveness	None *(Naked idea)* Key invention is observable in commercial/product/service and technology readily imitable	Patent all observable inventions *Clasp locker zipper*	Patent but rely on first-mover or complementary asset advantages *Business idea such as a Greek-Chinese fusion restaurant*
	Know-how hidden (Black box inventions) Key invention cannot be reverse engineered but performance observable	Patent most, but concern about protecting future generation inventions *Semi-adhesive post-it note*	Rely heavily on secrecy *Coca-Cola formula*
	Know-how and extent hidden (Unobserved inventions) Use of invention not directly observable	Patent most and use know-how disclosure as a signal of extent of remaining secret know-how *"Reflow" process to aid in manufacture of MOS circuits (Jackson 1997)*	Rely on secrecy, but use know-how disclosure as signal of extent of secret know-how *Cost-reducing process invention*

A black box invention is an invention for which the added performance is obvious when the product or service is observed, but the means by which the performance is achieved cannot be readily discerned or reverse engineered. Software often has some of these characteristics.

Unobserved inventions include process inventions that allow a previously offered product to be manufactured for a greatly-reduced cost. In this case, the competitors could not know what an innovation exists, although they could suspect it if the descent in the price of the product lowered significantly. Such inventions should be kept secret.

Weak patents motivate to the innovating firms to rely more heavily in secrecy, reducing the amount of knowledge publicly disclosed.

2.3 Informal Mechanisms of Protection

In this section we find four mechanisms: trade secrets, lead-time strategy, complexity of the design and defensive publishing. The main mechanism in this topic is trade secret.

a) Trade secrets: They can be described as secrets or proprietary information of commercial value. The trade secrets usually present three characteristics:

- It is not generally known to a relevant portion of the public;
- It has commercial value;
- Reasonable action is taken to maintain its secrecy.

Although trade secrets are not protected by law in the same way as patents or trademarks, the secrets can enjoy some level of legal protection. So, t is necessary to distinguish among legal procedures to obtain a trade secret (e.g., by reverse engineering) and the inappropriate or illegal means (e.g., industrial espionage). Some authors show more reasons to opt for the secret in detriment of the patents [2] :

- Difficulty in demonstrating the novelty of an invention.
- The amount of information disclosed in a patent application.
- The cost of applying
- The legal cost of defending a patent in court
- The ease of legally inventing around a patent

b) Other informal mechanisms

Leader strategy: Also called lead-time strategy. This strategy denotes a behaviour by which the company relies on being consistently more innovative than its competitors. Applying this strategy, the company always has an improved version of its products prepared for its launching before possible actions done by a competitor that has copied the available version of the product protected.

The strategy on relying on the complexity of the design: The composition and design of some products can be so complex that the competitors would incur in disproportionate costs to copy the product. As with the previous strategy, patenting process could show enough hints to other firms to reproduce the innovation.

Defensive publishing: By means of this strategy the firm presents in public its inventions. This strategy can be applied when the firm considers that obtaining a patent can be too expensive, or too risky, and there are reasons to believe that other firms will be able to carry out processes of reversed engineering and patent it themselves.

Concerning the informal mechanisms of protection Cohen et al. [2] consider that the use of the secret allows the appropiability from the innovations to a smaller cost and more quickly than other methods such as time reduction or complementary capacities' exploitation. Afterwards, we will expose some characteristics related with the use of secrets in the administration of the firms.

2.4 Management of Secrecy

The policy of the firms in relating to the secrets can be summarized in three points [3]:

a) Making of secrets
b) The protection of secrets
c) The stealing of secrets

The foundations of all secrets, whether related to government or business is to protect an information asset perceived to be of high value, -whether tactical or strategic - [3]. Especially in those industries where value is embedded in knowledge creation and exploitation activities, especially those activities related with technology. The secret is shown like one of the most reliable organizational mechanisms to prevent the copying and imitation of emergent elements of intellectual property by rivals [6], [3].

2.5 Culture and Its Relation with Secrecy

Often, culture is defined as a shared set of values and beliefs [5] this author considers these beliefs affect all the relations in the companies. We explore the effects of national culture along the dimensions proposed by Hofstede [4]: power distance, individualism, masculinity and uncertainty avoidance.

Power distance: is the extent of power inequality among members of an organizational society.

Culture sets the level of power distance at which the tendency of the powerful to maintain or increase power distance and the tendency of the less powerful to reduce them will find their equilibrium.

The power distance norm can be used as a criterion for characterizing cultures (without excluding other criteria).

High levels of power distance are associated with centralized decisions, large proportion of supervision, the information flows are constrained. The first hypotheses can be stated as follows:

H1: Companies based in higher power distance countries exhibit higher level of secrecy.

Uncertainty avoidance: is the extent to which members of an organizational society feel threatened by and try to avoid future uncertainty or ambiguous situations.

Uncertainty avoiding societies are rule- and routine-oriented and should generally find it more difficult to adapt to novel social and environmental demands and practices. There is evidence that uncertainty-avoiding societies are less innovative because championing roles that overcome organizational inertia to innovations are less likely to be accepted in those societies [10], [11].

The stronger a culture's tendency to avoid uncertainty, the greater its need for rules [5] pp: 147).

The authority of rules is something different from the authority of persons. The first relates conceptually to uncertainty avoidance; the second, to power distance.

Rules are the way in which organizations reduce the internal uncertainty caused by the unpredictability of their members' and stakeholders behaviour [5], pp:147. Good rules can set energies free for other things; they are not necessarily constraining.

Low levels of UAI represent innovators less constrained by rules, and managers directly implied in strategy. Zarkeski [14] compared 256 annual reports of companies in seven countries and found that the (negative) relationship between UAI and disclosure of information.

Thus, we hypothesize:

H2: Companies bases in lower uncertainty avoiding countries exhibit a higher level of secrecy.

Individualism and Collectivism: which describes the relationship between the individual and the collectivity that is reflected in the way people live together.

This is the anchor at one end of two poles, where the other anchor would be collectivism. This is the extent to which individuals are supposed to be self-reliant and look after themselves, versus being more integrated into a group.

Individualism is generally defined as the cultural belief, and corresponding social pattern, that individuals should take responsibility primarily for their own interests and those of their immediate family [4], [13].

In this context we test the hypothesis H3:

H3: Companies bases in higher individualism countries exhibit a higher level of secrecy.

Masculinity and femininity: the extent of roles division between sexes to which people in a society put different emphasis on work goals and assertiveness as opposed to personal goals and nurturance.

Highly masculine societies place low value on caring for others, on inclusion, cooperation, and solidarity. Career advancement, material success and competition are paramount.

Cooperation is considered a sign of weakness. In a study of 1,846 entrepreneurs from seven countries Steensma et al. [12] found that those from more masculine countries have a lower appreciation for cooperative strategies.

The working hypothesis for this dimension is:

H4: Companies based in more masculine countries exhibit a higher level of secrecy.

3 Data and Measures

This study is designed to evaluate how national differences in cultural values impact on firms' secret tendency. To address the question we utilize the Community Innovation Survey (CIS 3) information. The survey was carried out in all 15 EU Member States, Iceland and Norway. Data from culture come from Hofstede [5][1]. Hofstede dimensions of national cultural values have become the standard in cross-cultural research and are being extensively used in broad studies of science articles.

These four dimensions provide a framework not only for analyzing national culture, but also for considering the effects of cultural aspects on management and organization. It can be especially useful for understanding the appropriate mechanisms in coordinating activities inside organizations

Dependent variable

The dependent variable in our analysis is the proportion of companies that use secrecy as a method to protect their innovations. Each item represents the proportion of the firms on each country that use secrecy as a method to protect their innovations.

---

[1] Hofstede identified and additional dimension "long-versus short-term orientation" that was found in the answers of student samples from 23 countries around 1985 to the Chinese Value Survey -CVS- [5]. This study is oriented to European countries and we haven't used it.

Explanatory variables

The explanatory variables of interest to this study are four dimensions of national culture: power distance, uncertainty avoidance, individualism and masculinity, using information from Hofstede [4].

Power distance

This measure ranges between -90 and +210. In the maximum, everyone afraid to disagree with their managers.

Uncertainty avoidance

This measure ranges between -150 and +230: In the maximum all think that rules should not to be broken, everyone wants to stay more than 5 years in the company, everyone always feel nervous.

Individualism vs collectivism

Both concepts can be considered as the two poles of a dimension of national culture.

Individualism stands for a society in which the ties between individuals are loose. Collectivism stands for a society in which people from birth onwards are integrated into strong, cohesive in-groups [5] pp: 225.

Scores were standardized for each country.

Masculinity

This measure ranges from 0 to 100. In the maximum, for example, managers expected to be decisive, firm, assertive, aggressive, competitive and just (see table 5).

Concept of innovation

An *innovation* is the implementation of a new or significantly improved product (good or service), or process, a new marketing method, or a new organisational method in business practices, workplace organisation or external relations. The minimum requirement for an innovation is that the product process, marketing method or organisational method must be new (or significantly improved) to the firm [7].

Innovation activities are all scientific, technological, organisational, financial and commercial steps which actually, or are intended to, lead to the implementation of innovations Innovation activities also include R&D that is not directly related to the development of a specific innovation [7].

Thus, a company is considered innovative if it has introduced an innovation during the considerate period under review.

4 Results

Table 2 presents estimation results of ordinary least squares regression models.

Model 1 is a specification containing all the explanatory variables (Hofstede measures) introduced at the same time into the model for all firms of the survey. In this model, there is no any variable significantly associated with the level of secrecy. The global model is not significant.

In model 2 we introduce Hofstede's measures to explain the values of secrecy for innovative enterprises only, using forward method. The estimation results provide strong support for H2 and H4. Levels of UAI are significantly negatively associated with levels of secrecy. Levels of masculinity are significantly positively associated with levels of secrecy. We haven't find any evidence that levels of power distance and individualism affect secrecy in organization, so we can't accept Hypotheses 1 and 4.

In model 3 we introduce Hofstede's measures to explain the values of secrecy for non innovative enterprises only, using forward method. The estimation results provide support for H2 and H4. National values affected the level of secrecy in organizations for this type of firms. The results are similar to the model 1 with little differences in the coefficients.

Table 2. OLS regression analysis on secrecy on organizations

Variable	Model 1		Model 2		Model 3	
Hofstede cultural dimensions	**Un.Coef**	**St.Coef.:**	**Un.Coef**	**St.Coef.:**	**Un.Coef**	**St.Coef.**
Masculinity	0,001 (0,01)	0,203	0,003**	0,557**	0,001**	0,658**
Power distance	-0,03 (0,03)	-0,710				
Individualism	0,01 (0,02)	0,252				
Uncertainty Avoidance	0,01 (0,03)	0,357	-0,003** (0,001)	-0,622**	0,000**	-0,589**
Constant	0,100 (0,180)		0,356** (0,079)		0,065** (0,024)	
R-squared	0,348		0,403		0,450	
Adjusted R-squared	0,058		0,303		0,359	
F-value	1,201		4,045		4,915	
Signification	0,374		0,045**		0,028**	

Notes: Standard errors in parentheses; * p<0,10; **p<0,05; ***p<0,001.
Un.Coef.: Unstandardized coefficients.
St.Coef.: Standardized coefficients.

5 Conclusions

This paper sets out to develop the theoretical linkages between national culture measured with Hofstede's dimensions and the level of secrecy in European companies. We have found that two dimensions of national culture analyzed, namely masculinity and uncertainty avoidance have significative influence in the level of secrecy of European countries. No significant effect was borne for the other dimensions of culture, collectivism and power distance. The analysis realized show that innovative and non innovative companies are affected with the same dimensions of culture. The direction of the effect is opposite, so masculinity affects in a direct form to the level of secrecy, and uncertainty avoidance has a opposite effect to the secrecy. The adjusted R-squared shows that the cultural factors explain between 30% and 36% of the secrecy level. So, economic factors seem to have more weight than cultural factors.

The consequences of national differences for organizations are summarized in table 3. This table provides a reference guide for managers to analyze cultural influences in companies.

Table 3. Consequences of cultural influences and the use of secrecy in organizations

Consequences of masculinity index (high values of index)
Challenge and recognition in jobs important
High job stress
Belief in individuals decisions
Work very central in a person's life space
Consequences of uncertainty avoidance index (low values of index)
Weak loyalty to employer; short average duration of employment
Preference for smaller organizations but little self-employment
Scepticism toward technological solutions
Innovators feel independent of rules
Top managers involved in strategy
Power of superiors depends on position and relationships

There are, however, some limitations to the exploratory study. First, the sample utilized represents adequately European countries, but we haven't introduced control variables as the size of the companies. It is hoped to conduct in the future a complementary study introducing these variables and also confirming the results with other available cultural index.

References

1. Anton, J., Greene, H., Yao, D.: Policy implications of weak patent rights. In: Jaffe, A., Lerner, J., Stern, S. (eds.) Innovation policy and the economy, vol. 6, pp. 1–26. MIT Press, Cambridge (2006)
2. Cohen, W.M., Nelson, R.R., Walsh, J.P.: Protecting their intellectual assets: Appropriability conditions and why U.S. manufacturing firms patent (or not). n.7552, http://www.nber.org/papers/w7552
3. Dufresne, R.L., Offstein, E.H.: On the virtues of secrecy in organizations. Journal of Management Inquiry 17(2), 102–106 (2008)

4. Hofstede, G.: Culture's consequences: International differences in work-related values. Sage, Newbury Park (1980)
5. Hofstede, G.: Culture's Consequences: Comparing Values, Behaviors, Institutions, and Organizations Across Nations. Sage, Newbury Park (2001)
6. Hannah, D.R.: An examination of the factors that influence whether newcomers protect or share secrets of their former employees. Journal of Management Studies 44, 465–487 (2007)
7. OECD.: Oslo Manual.Guidelines for collecting and interpreting innovation data. Third Edition, http://www.ttgv.org.tr/UserFiles/File/OSLO-EN.pdf
8. Radauer, A., Streicher, J., Ohler, F.: Benchmarking national and regional support services for EMS in the field of intellectual and industrial property. Austrian institute for EMS research, Vienna (2007)
9. Schmidt, T.: An empirical analysis of the effects of patents and secrecy on knowledge spillovers. ZEW discussion Paper, n. 06-048, Mannheim, ftp://ftp.zew.de/pub/zew-docs/dp/dp06048.pdf
10. Shane, S.A.: Cultural influences on national rates of innovation. Journal of Business Venturing 8, 59–73 (1993)
11. Shane, S.A.: Uncertainty avoidance and the preference for innovation championing roles. Journal of International Business Studies 26, 47–68 (1995)
12. Steensma, H.K., Marino, L., Weaver, K.M.: Attitudes toward Cooperative Strategies: A Cross-Cultural Analysis of Entrepreneurs. Journal of International Business Studies 31(4), 591–609 (2000)
13. Triandis, H.: Individualism and Collectivism. Westview Press, Boulder (1995)
14. Zarzeski, M.T.: Spontaneous harmonization effects of culture and market forces on accounting disclosure practices. Accounting Horizons 10(1), 18–37 (1996)

New Forms of Managerial Education in Knowledge Society

Milan Maly

University of Economics, Prague, Department of Management
W. Churchill Sqr. 4, Prague 3, 130 67
maly@vse.cz

Abstract. Paper illustrates the ways of knowledge sharing in transition companies in two main topics: the methods of knowledge management of local managers in joint ventures with foreign partners from free market countries and the development of leadership behavior and decision making styles in privatized companies in the Czech Republic. Lack of previous experience is the main reason for the adoption of different ways of managerial education. Several systems like the tandem, distant learning, mixed, foreign and hired managerial systems are analyzed, and the role of both partners, local and foreign managers, is specified. The analysis of leadership behavior and decision-making styles specifying five levels of participation. Czech managers consider it to be appropriate to use a more autocratic style than a participative one. Only in a few cases, mostly in joint ventures, can we see some elements of the partnership style.

Keywords: Knowledge sharing, Knowledge transfer, Joint ventures, Leadership behavior.

1 Introduction

The transition from a centrally planned to a free market economy raised several new issues in the area of knowledge transfer and managerial education. Government control of the implementation of strategic decisions has been supplanted by private ownership. Strategic decision-making has shifted from the government planning body to the corporate level. Company top management is responsible for new investment decision-making, including advanced technologies and automation. However, company managers have no previous experience in this area.

Transition doesn't mean only privatization, liberalization or parliamentary multiparty political system, but the decisive role is given to transfer of knowledge, including knowledge management. The most powerful mean in this area is just knowledge management. According to Lehaney et al (2004) knowledge is obtained from experts and is based on expert experience, as it requires a higher understanding than information alone. This is exactly what managers from transition countries require predominantly. Our experience shows that the capacity of knowledge management as powerful tool for increasing managerial level is still utilized insufficiently.

Therefore, it is extremely important to create the necessary conditions for high quality management education. Joint ventures and direct foreign investment are the

M.D. Lytras et al. (Eds.): WSKS 2009, CCIS 49, pp. 200–207, 2009.

best means to affect changes in the attitudes, perceptions, values and ways of thinking of these former "centrally planned" managers.

Different knowledge management projects indicated in the literature (Quintas et al, 1999) are implemented in the period of transition in the organizations in Central and Eastern Europe, namely in the form of knowledge sharing, targeted on learning, capturing and sharing knowledge. New business models and opportunities realised by mostly consultancy firms are supplemented in transition economies by foreign partners in joint-ventures, acting as the initiators of capturing the knowledge of the managers of the skilful managers of foreign partner. Together with the other authors (Kyriazopoulos and Samanta-Rounti, 2008) we came to similar findings that resistance by managers and employees was a critically important factor influencing the success of knowledge sharing.

2 Managerial Learning

The joint foreign – East European business ventures have adopted different ways of educating local managers. The method adopted not only depends on the degree of the foreign partner`s desire for dominance; it is also closely related to various technological and economic factors. If the local company cannot attain adequate levels of productivity through economies of scale because of the small size of the domestic market and is, therefore, heavily dependent on the marketing network of its foreign partner (e.g. the automobile joint venture Skoda-Volkswagen), then there is, understandably, very close cooperation under the heavy influence of the foreign partner. This cooperation is enhanced by the local company`s need for investment capital and know-how transfer. The high costs of R&D and the trend towards shortening innovation cycles leads to even closer cooperation between the two partners in the management area. Conversely, if the product is of high quality and the innovation cycle is relatively long, then the main result of cooperation can be productivity enhancement through capital investment. The local management is left with relatively more autonomy.

Our statistical sample consisted of a detailed analysis of 60 joint ventures. The examples are based on the insights and experiences of managing directors as related during personal conversations, through published conference proceedings, or gleaned through student interviews.

2.1 Systems of Local Managers Education

Several different systems of managers` education were identified in our research (Maly, 2001). They are mostly based on knowledge sharing. Knowledge sharing implies learning, since learning is a process of acquiring knowledge. Linking measures enables seekers of information to identify and then contact people with specific knowledge (OECD, 2003).

The description of the experiences with four joint-venture companies and one unsuccessful example of implementation of so called "hired" managerial system follows.

2.1.1 The Tandem System

This method of managerial sharing was implemented by the Skoda-Volkswagen joint venture in the 1993 – 1995 period. The marriage with Volkswagen was actually less a joint venture and more of a take-over. Volkswagen gained only a 31% stake initially but it did receive full managerial responsibility. From the beginning, Volkswagen ended up owning a full 100% stake. The only rationale for the phased sale was the fact that the government could not sell the company outright for political reasons. These reasons can be seen when taking into account Skoda`s contribution to the national economy as a whole. Fully 3.5% of the workforce of the former Czechoslovakia accounted for 5% of all exports. Nevertheless, in a fairly short period of time, Skoda has become totally controlled by Volkswagen. The latter`s marketing department already touts this in their frequently recurring slogan, "Skoda is an integral part of the Volkswagen Group".

The tandem system is organized so that every Czech mid-level manager has a Volkswagen counterpart. They have equal authority and are responsible for the same tasks. Both must work very closely together. The arrangement has its positive and negative aspects, some of which are cited here.

The main advantage is that the tandem system is a very direct learning method. Both partners learn from each other during the actual working process. Learning is not limited to mere facts and figures but is extended to unwritten behavior modes encompassing broader socio-cultural areas. Discrimination against nationalities did not seem to be as prevalent a problem as one may have expected. An informal division of competencies in possible (as in the parenting model where certain things are decided by the mother, others by the father).

Communication problems may turn out to be the largest negative feature. Differences in language and the backgrounds of the two groups may cause confusion both up and down the hierarchy. It could happen that the two partners fail to co-ordinate their actions. As always when people have to work very closely together, the situation has the potential for conflict (which can also be seen as positive in the proper circumstances). However, the current rules may, at times, become offensive to Czech mid-level managers. Their signatures mean almost nothing if they are not confirmed by their foreign counterparts.

2.1.2 The Distant Learning System

This system is being used in one Czech joint venture. Glaverbel is the largest producer of flat glass in Central Europe. Formerly the flat glass division of the state-owned Sklo Union Company, Glaverbel became a subsidiary on 1 January 1991, and a joint stock company created through Czechoslovakia"s privatization program. On 1 April of the same year, a joint venture was created with the Belgian Glaverbel Group (the main flat glass manufacturer in the Benelux). Glaverbel acquired a 40% stake, which increased to 51% in 1991, and to 67% in April of 1992.

Every Czech mid-level manager in Glavarbel Czech has a counterpart (a similarly positioned manager at Glaverbel in Belgium). The Czech partner has the duty to consult with their Belgian counterpart in advance of all important decision or referrals to higher management. The two partners have to reach a consensus and then the local partner can implement the decision or submit a referral. The company thus intends to do everything possible to help its middle management – as well as all other personnel to undertake the gigantic move from a planned economy to a market economy. This is being done in close co-operation with Glaverbel through the mastery of complex advanced technology equipment as well as by training in modern methods, which are applicable to all areas of the company (e.g. sales, finance, management and human resources). Proper management is therefore guaranteed, leading, to the overall success of the venture.

2.1.3 The Mixed System

Cokoladovny was a Czech company that was privatized in 1990. In 1989, Cokoladovny had about a 60% share of the domestic chocolate and biscuit market, indicating a near-monopolistic position. Real changes started only when BSN and Nestle took over the firm in 1992 and Cokoladovny entered into the long-term process of transformation. The final goal of the firm is to change the behavior of all its employees. Currently the French and Swiss managers are aiming to adopt the BSN and Nestle models to the Czech context.

Top management is drawn from both the local and the foreign companies. Both have their responsibilities and competencies. In the case of Cokoladovny, the majority is foreign managers, but both sides have advisors from the other. It is a more indirect method of influence that the Skoda-Volkswagen tandem system, because the local manager still has the power of decision. This means less conflict potential, but less freedom of action for the Czech partner than before.

2.1.4 The Foreign Managerial System

Ralston-Baleria Ltd, a joint venture producing consumer batteries, was established in 1991. The Eveready Battery Company (EBC), the European division of the US-based Ralston Purina Battery Company, is the foreign majority partner and Palaba is the Czech minority partner (Matesova, 1994). Ralston"s main goal was to use joint venture as a means to penetrate the Central and East European markets. Palaba"s main goal was to survive in the increasingly competitive environment by transferring know-how in order to substantially improve the quality of its main product, consumer batteries. Before 1990, Palaba was the monopoly producer in our domestic market. It was well known for its low quality products, packed in paper jackets, which led to frequent leaking, minimal durability and environmental damages. The new joint venture has begun producing new generation consumer batteries which are much more durable, contain no mercury or cadmium, and are wrapped in metal jackets making them environmental friendly.

The foreign partner has imposed a managerial system, which may be called a typical US approach to the turnaround of a Czech company, as the same system exists, albeit with minor variations, in several American acquisitions or joint ventures. The operation of the joint venture is based on an elaborate project and budget. Local management is only authorized to implement the project or to propose changes. The foreign partner leaves very little room for local initiative. The management of the joint venture has no authority to change the budget. Most decisions are made at either the European or US headquarters level. Representatives from the headquarter travel back and forth. However, local people have many opportunities for professional growth. The foreign investor is following its global strategy; the new joint venture must fit into the global picture. The turnaround is made possible through transfers of equipment, cash and know-how from the parent firm, which is financially strong, has the necessary marketing and production expertise, world market share and well-developed distribution channels.

2.1.5 The Hired Managerial System

This method of knowledge sharing can be taken as one of the most innovative ways of knowledge management. This very promising access to the enlargement of knowledge managerial projects (Quintas et al, 1999) finished in 1995 as unsuccessful attempt to implement knowledge management into former centrally planned company.

Tatra is a very famous off-road truck manufacturer in the Czech Republic. The company`s history goes back to the beginning of the previous century. At that time, the company manufactured cars. After the Second World War, Tatra started large off-road truck production.

The company was a very successful exporter to Eastern Bloc countries like the former USSR and China, to many developing countries and to the Czech military. At the beginning of the 1990s, the export markets vanished, as did the military market. Tatra fell into debt. Production decreased substantially. Tatra's management embarked on a novel course of action. Instead of creating a joint venture, they decided to recruit their own top managers from the West. In 1993 they hired Gerald Greenwald, who was famous for the greatest turnaround in automobile manufacturing history, the rescue of Chrysler. The two other managers were David Shelby, a financial expert, and Jack Rutherford, a production and operations specialist. A two-year contract was signed with the trio, the main task being to lift the company out of its red morass and put it into the black. Tatra's Czech managers expected that the Americans would assist them mainly in resolving the financial situation (debt restructuring), finding new markets (Canada, South America) and building up the company's image.

This experiment completely failed. In the middle of 1994, the Americans were forced out. An analysis of this experiment shows the main reason for the failure. It was stated by the members of the Managing Board of Tatra that the Americans had behaved correctly, but there was definitely a lack of responsibility, devotion and commitment on their part.

3 Leadership Behavior

Application of knowledge management in transition institutions follows mostly so called "soft access", or People-Track (Sveiby, 1997) contrary to IT-Track which is for many reasons not suitable for the specific conditions of transition period. The main obstacle for using this IT-Track in transition is primarily the fact that IT development is lagging behind in former centrally planned economy. Parallel to Kyriazopoulos and Samanta-Rounti (2008), among the most important factors of successful implementation of knowledge management in People-Track access are teamwork and leadership activity. Especially proper leadership behavior, for example consulting and group or participative decision making styles create the best preconditions for effective knowledge sharing.

Leadership behavior is taken as one of very important factors of managerial education in knowledge society. It creates the necessary preconditions for the implementation of the managerial knowledge in the process of the real decision making.

Our analysis of leadership behavior is based on the Vroom-Yetton (1973) model. The model comprises three elements which are interconnected in the logic of the contingency theory: There is (1) no leadership strategy (style) which is successful in all situations, (2) therefore the situations have to be diagnosed, and (3) rules have to be found that explain which strategy best matches which situation.

Leadership strategies – according to the model (Reber, Auer – Rizzi, Maly, 2004), a leader can choose from five levels of participation when making a decision (AI, AII, CI, CII, GII). These strategies range from an autocratic decision (AI) to a total group decision (GII). AI represents 0% and GII 100% participation. The assignment of different participation scores for the strategies between the extremes of the scale is based on empirical studies in which managers rate the distances on a 1 to 10 scale. As a result, AII represents 10%, CI 50% and CII 80% participation. "A" stands for autocratic, "C" for consultative and "G" for group decision. "I" stands for the concentration on one person (AI = leader alone, CI = one-on-one consultation with all subordinates who could be affected by the decision), and "II" stands for the inclusion of two or more persons at the same time.

3.1 Decision Strategies

A1: You solve the problem or make the decision yourself using the information available to you at the present time.

AII: You obtain any necessary information from subordinates, and then decide on a solution to the problem yourself. You may or may not tell subordinates the purpose of your questions or give information about the problem or decision you are working on. The input provided by them is clearly in response to your request for specific information. They do not play a role in the definition of the problem or in generating or evaluating alternative solutions.

CI: You share the problem with the relevant subordinates individually, getting their ideas and suggestions without bringing them together as a group. Then you make the decision. This decision may or may not reflect your subordinates" influence.

CII: You share the problem with your subordinates in a group meeting. In their meeting you obtain their ideas and suggestions. Then you make the decision, which may or may not reflect your subordinates" influence.

GII: You share the problem with your subordinates as a group. Together you generate and evaluate alternatives and attempt to teach agreement (consensus) on a solution. Your role is much like that of a chairperson, coordinating the discussion, keeping it focused on the problem and making sure that the critical issues are discussed. You can provide the group with information or ideas that you have, but you do not try to "press" them to adopt "your" solution and you are willing to accept and implement any solution, which has the support of the entire group.

The applied method and data collection were dominated by a clear action orientation. No questionnaire was used and all data were collected by administering a "problem set" in the form of thirty decision-making situations. The thirty cases were selected and rewritten from actual descriptions of real decisions provided to the authors (Vroom/Yetton/Jago, 1976) by hundreds of real managers and were validated with the assistance of trained managers.

The problem set was administered to managers who, at the time of data collection, were unfamiliar with the Vroom/Yetton model. In addition to the cases, they only received the definition of the five strategies and were asked to select one for each case.

The data were collected prior to leadership training programs. In such a training program, the respondents were not providing a "favor" for researchers since their main concern was the improvement of their own leadership behavior. All of the participants received feedback, in which their first reactions to the problem set were compared to a description of the model. Training was provided to assist the participants in using the diagnostic questions and the decision rules for upcoming leadership decisions in their home organizational environment.

Transition managers are typical with higher preferences for autocratic leadership styles, in higher disagreement with the prescriptions of the Vroom/Yetton models and in most of the main effects.

How can these results be explained for the former centrally planned economics, which politically brought about a revolution and a reorganization of its economy from central state planning and state ownership to a market system with a privatization campaign and an opening for international competition? Did more drastic changes remain on the national level and somehow manage not to penetrate the organizational and individual levels? The latter seems to be the reality, in spite of the fact that individual leaders show a high readiness for flexibility with high scores in their standard deviation. Is a "configurationally" view the best approach to explain stability within a change process? In a simplified picture, we could argue that a model of three main levels would bring us closer to an explanation of this paradoxical situation of stability within the flux of change. The change took place on the societal/political level; the population worked and fought for the right to vote, to exercise the right of government participation, to express more individuality, and to support private ownership. At the individual level, these are indicators that similar values and flexibility exist but do not have a place at the organizational level of private enterprises and it does mean that this potential can be tapped. Perhaps a change at this organizational level can only be brought about when the opportunity is administered congruently, and the "whole" and its "parts" can find an optimal (ideal). The existing "values" need the appropriate situational conditions in order to be transformed into "actions".

4 Conclusion

Knowledge sharing, used by almost all joint ventures, when combined with internships in the foreign partner`s subsidiaries is one of the best ways to retrain the local managers quickly and effectively. Many large Western corporations and some small companies have developed their own education, training and retraining programs to compensate for the shortcoming of their official business school systems.

In the issue of leadership behavior the unchanged inner hierarchical governance structure of the many directly or indirectly state-owned companies does not force managers to change their habits. In the leadership seminars, managers stated repeatedly: "I would like to include my subordinates in the decision-making process, but they expect me to make the decisions alone. That way if the decision is wrong, I alone take the blame". Perhaps a communication problem exists (who tells whom first, what is expected in reality) or the leader forgets his/her responsibility as "model" and has to be the front runner when it comes to admitting he/she does not have all of the information and therefore needs help and advice and depends on the commitment of subordinates to get the job done effectively.

Several companies became part of international corporations. In these cases the managers are currently in conflict between the aspirations of the foreign company and their own culturally bound ways of doing things.

An example of a very successful model and partnership is the cooperation between Volkswagen and Skoda. In this situation, above mentioned Tandem System can be seen as a bilateral consensus-seeking program within one company, namely a structure with some elements of the partnership style on the national level.

References

Kyriazopoulos, P., Samanta-Rounti, I.: Approaches to Knowledge Management in Greek Firms. In: Lytras, M.D., Carroll, J.M., Damiani, E., Tennyson, R.D. (eds.) WSKS 2008. LNCS (LNAI), vol. 5288, pp. 286–295. Springer, Heidelberg (2008)

Lehaney, B., Clarke, S., Coakes, E., Gillian, J.: Beyond Knowledge Management. Idea Group Inc. (2004)

Maly, M.: The management strategies of Czech companies in transition. International Journal of Manufacturing Technology and Management 1(4/5), 405–411 (2001)

Matesova, J.: Palaba nad ist joint venture, Case study, 94-021. Czech and Slovak Managemen Center, Celakovice (1994)

OECD: Measuring Knowledge Management in the Business Sector: First Steps, Paris and Ministry of Industry, Canada (2003)

Quintas, P., Jones, J., Demaid, A.: An Introduction to Managing Knowledge, Unit 1 of Open University Business School's Managing Knowledge programme. The Open University, Milton Keynes, UK (1999)

Reber, G., Auer-Rizzi, W., Maly, M.: The behavior of managers in Austria and the Czech Republic: An intercultural comparison based on the Vroom/Yetton Model of leadership and decision making, vol. 9(4). Rainer Hampp Verlag (2004) ISSN 0949-6181

Sveiby, K.E.: The New Organizational Wealth. Managing and Measuring Knowledge-Based Assets. Berret-Koehler, San Francisco (1997)

Vroom, V.H., Yetton, P.W.: Leadership and decision-making. Pittsburgh University Press, Pittsburgh (1973)

Vroom, V.H., Yetton, P.W., Jago, A.G.: Problem Set, vol. 5. Yale University, New Haven (1976)

Managing Project Landscapes in Knowledge-Based Enterprises

Vladimir Stantchev[1,2] and Marc Roman Franke[3]

[1] Berlin Institute of Technology, Berlin, Germany
vladimir.stantchev@tu-berlin.de
[2] Fachhochschule fürOekonomie und Management (FOM), Berlin, Germany
[3] BearingPoint GmbH, Frankfurt am Main, Germany

Abstract. Knowledge-based enterprises are typically conducting a large number of research and development projects simultaneously. This is a particularly challenging task in complex and diverse project landscapes. Project Portfolio Management (PPM) can be a viable framework for knowledge and innovation management in such landscapes. A standardized process with defined functions such as project data repository, project assessment, selection, reporting, and portfolio reevaluation can serve as a starting point. In this work we discuss the benefits a multidimensional evaluation framework can provide for knowledge-based enterprises. Furthermore, we describe a knowledge and learning strategy and process in the context of PPM and evaluate their practical applicability at different stages of the PPM process.

Keywords: knowledge-based enterprises, knowledge and learning strategy, knowledge and learning process, project portfolio management, evaluation, innovation.

1 Introduction

The themes of the different schools of thought in the knowledge society converge to the following five pillars: knowledge and learning objects, processes, strategies, systems, and performance [1]. We regard Project Portfolio Management (PPM) [2] as a promising strategy in this context, together with a set of processes.

This paper assesses the applicability of PPM in knowledge-intensive project landscapes, focusing on product development projects in mechanical and plant engineering. The introduction of PPM in this field is hindered by missing reference processes and methods.

We evaluate an assessment framework that we have proposed in [3]. It serves as a set of strategy and processes for knowledge and learning. The framework extends existing portfolio management approaches and makes them more suitable for knowledge-intensive project landscapes. Furthermore, it also includes a best practice process with business services for project admission, selection and prioritization.

The rest of this work is structured as follows: Section 2 presents the state of the art in the management of project landscapes and the terminology we use. In

M.D. Lytras et al. (Eds.): WSKS 2009, CCIS 49, pp. 208–215, 2009.

Section 3 we describe our assessment framework for such landscapes that serves as a knowledge and learning strategy. In Section 4 we present a best practice knowledge and learning process that implements this strategy. In Section 5 we compare our proposed approach with existing approaches and give key insights from our empirical and experimental evaluation. Section 6 contains a summary of our results and outlook on our future research activities.

2 Preliminaries

The application of portfolio management further to the management of knowledge-intensive project landscapes is a transfer from its origin in the financial theory. It is implemented within the innovation and product development process, the central component of the product lifecycle management (PLM). PLM itself is a complex knowledge-intensive lifecycle approach which often shows deficiencies and inefficiencies.

In this work we discuss PPM in its knowledge and learning-related context of project-, program-, multiproject- and portfolio management. Projects and programs are singular tasks. They are explicitly separated from one another and are the responsibility of their project managers. With the operation of many project initiatives, conflicts occur within the whole project landscape in terms of overlapping knowledge areas, intra-project coordination, resource and competence staffing, as well as budget allocation. A portfolio serves as a precisely defined collection of associated projects and programs. Therefore, the organization and administration of a portfolio plus its overall implementation as a knowledge and learning strategy to achieve the company's strategic and business goals are closely related and interdependent. Overall, we can regard portfolio management as the central management of one or several portfolios in terms of identification, prioritization, authorization, organization, realization and the controlling of its associated projects and programs [3].

There are three general views of portfolio management in knowledge-intensive project landscapes – organizational, process-oriented, and chart-oriented. The first is an organizational view with regards to its hierarchical administration of projects and the central collection and consolidation of project information such as project goal, budget cost, timeline, resource demand and risk class in superior project portfolios [2]. The second process-related view describes the iterative and cyclic process of selecting and prioritizing new requests and active projects [4]. A third understanding of the word "portfolio" is the multi-axis portfolio chart, which is used as a graphical management decision tool [5]. As an own organizational unit, the project and portfolio management office (PMO) governs as a neutral instance and is authorized to audit the accuracy of the process compliance.

3 Knowledge and Learning Strategy for Complex Project Landscapes

We regard the comparative evaluation of requested project proposals with the appropriate criteria to finally select and prioritize valuable projects as the key component of a successful knowledge and learning strategy for complex project landscapes. This evaluation produces a ranking list which serves as an information basis to support the final portfolio decision of accepting or rejecting proposed requests and also to continue or cancel currently running projects. Therefore, an evaluation system should consider all relevant projects, those actively running as well as these prospectively requested. Furthermore, it should cover different project and knowledge categories [6].

3.1 Structuring Knowledge Objectives - Project Requests

An eminent benefit of portfolio management is the consistent collection of project requests in a central information repository. By using a uniform request form for all project types and following a standardized acceptance and review process, we can compare the contributions of projects from different knowledge areas. Beside the common project master data (e.g., name, timeline, milestones, demand manager), we need to consider knowledge-related information about implementation risk and strategic values, and also provide a detailed business case.

3.2 Budget Restrictions and Knowledge Resources

Project budget cost is defined as the projected cost demand that is required while conducting the project during a certain period of time. Compared to an *a priori* estimated planning cost, the once allocated budget is fix and binding. Availability of knowledge workers is particularly limited and, therefore, turns out to be the most likely capacity bottleneck. Therefore, we have to consider human resources, together with their knowledge and competence profiles, during the project's conceptual planning phase.

3.3 Aligning Core Knowledge and Project Landscape

An important characteristic of successful product development projects is their conformity with the overall corporate strategy. This reflects the degree of alignment between enterprise-wide knowledge and learning objectives on the one side, and, on the other side, the currently conducted project activities. Key factors have already been referred to in [4]. By suggesting an even more granular breakdown of the strategic value assessment to several specific knowledge-related driver classes such as business, IT or technical product drivers, our approach goes beyond common evaluation frameworks from existing portfolio models. An exemplary catalog of strategic drivers has evolved based on experience from consulting engagements at leading German companies in the sector industrial products. For the sector industrial products we introduced six specific knowledge-related drivers on the subject of the technical competence (Technical Drivers) in [3]. These drivers are knowledge-related measures to evaluate the future value of technical products and to forecast future prospects of the product success.

3.4 Balancing Knowledge Areas in Complex Project Landscapes

The allocation of resources to projects virtually corresponds to an operationalization of the corporate vision and business strategy, particularly in their relation to knowledge and learning objectives of the enterprise. By classifying projects to a strategic bucket according to their knowledge class or category we are essentially expressing the companies' strategic direction. Another point of high relevance for evaluating projects is their risk. Assessing the impact of knowledge-related risks is best practiced as a survey about hierarchically arranged risk categories. Here we can also apply typical risk considerations, such as excess project costs, or failing adherence to schedule. The main objective is nevertheless an assessment of risks relating to technical feasibility. A total score or risk index will be finally calculated as an aggregated value from these singular risk classes.

4 Knowledge and Learning Processes for Complex Project Landscapes

The following steps typically comprise a product development process: Project proposal acceptance, Requirement specification and preliminary organizational activities, Conceptual knowledge-intensive work on product and process design (e.g., FMEA, QFD, Feasibility studies), The actual product development including quality management and prototyping, Market launch and distribution marketing, Start of series manufacturing (in case of mass products). A so called *Stage Gate Process* was developed to provide a generic pattern to formalize, structure and thus improve the efficiency of these innovation management and product development tasks [7].

Our proposed knowledge and learning process is installed as the gate between the conceptual knowledge-intensive work and the actual product development.

Phases and steps of the proposed knowledge and learning process are shown in Figure 1(adapted from [3]). It aims to provide an equilibrium of knowledge-intensive project landscapes. This model extends existing process models [8,7]

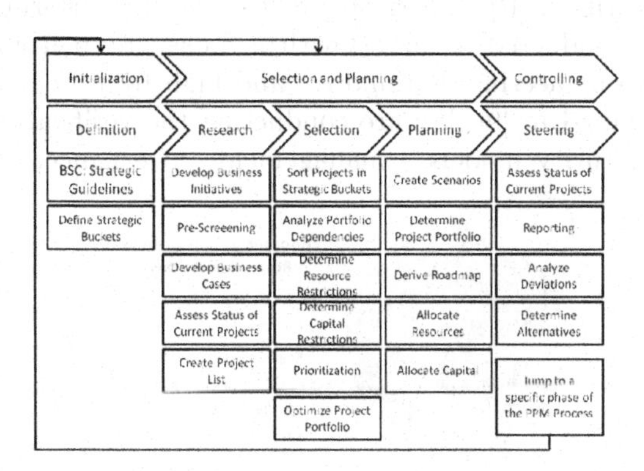

Fig. 1. The Equilibrium of Knowledge-Intensive Project Landscapes

with a more comprehensive evaluation process, particularly suited for knowledge-intensive projects. Furthermore, it specifies discrete business services for each process step with regards to a possible realization of technical services upon these business services. The process begins with the definition of knowledge-related strategic drivers to assess the overall strategic value of a project (as specified in Section 3), followed by the elaboration of a catalog of classification attributes to categorize projects into homogeneous knowledge groups, the so called strategic buckets. After the next step of Pre-Screening (early rejection of obviously unprofitable, invaluable in relation to knowledge objectives, and non-compulsory proposals) the process proceeds further from step to step and finally reaches the controlling phases, where the running portfolio is actively monitored to achieve the equilibrium. During this last and continuously running phase (since the initial start of portfolio management in the company) of status monitoring, new incoming project proposals compete with active projects for budget and resources. Thereby, an optimization of the total portfolio in terms of strategic knowledge and financial value can result in re-prioritization of current projects with a minor benefit. This might cause a delay or a hold-on status, or even a cancellation of running projects ahead of schedule.

5 Approach Assessment and Evaluation

The particular consideration of knowledge-related portfolio risk as a risk class and the multi-dimensional categorization in strategic buckets is a significant step forward as compared to existing PPM approaches.

We evaluated our approach with several empirical and experimental methods. In this section we will describe some key results from one empirical evaluation (design study) and one experimental evaluation.

5.1 Empirical Evaluation

One of the objectives of our empirical evaluation is the verification of our claim that a higher maturity in PPM corresponds to a better performance in project development. We conducted a survey with a focus group from companies in the areas of plant engineering, automotive and high tech. Our survey uses the methodology presented in [9]. We are conducting the verification for every dimension of our maturity models – management, governance, process, system,

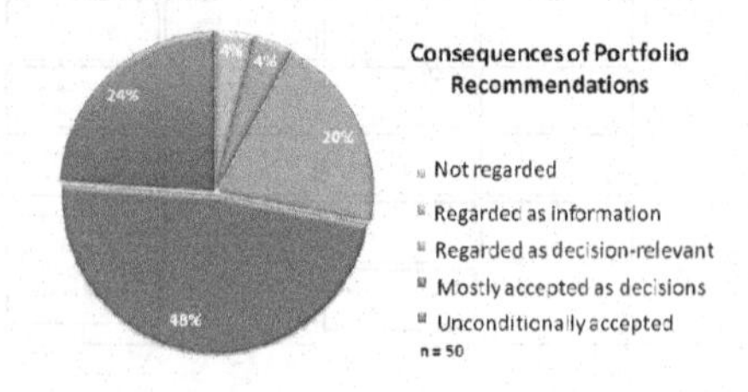

Fig. 2. Acceptance of Results of the Presented PPM Process in the Portfolio Decision

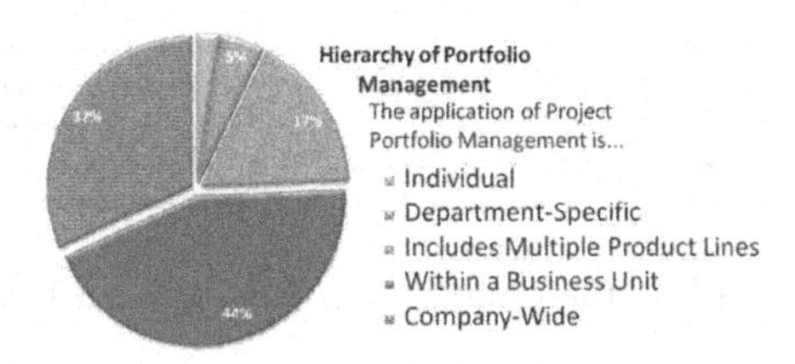

Fig. 3. Hierarchy of PPM

resource management, and social aspects. Thereby we assess quantitative and qualitative parameters during the survey. This allows us to more clearly establish correlations between PPM maturity and project success. Figure 2 shows the relevance of PPM as one aspect of our empirical evaluation – more than 90% of the respondends regard the recommendations of the PPM process as highly relevant to the actual portfolio decision.

Figure 3 shows our results concerning the hierarchy level of PPM application - more than 75% of the respondendts apply PPM at the level of a business unit, or company-wide. This further coroborates our recommendation that a PPM process should consider knowledge and learning aspects from the overall project landscape.

Figure 4 shows the decision criteria for project investments that our respondents most commonly considered. It shows that almost all considered criteria are knowledge and learning-related. Examples are: customer needs, strategy fit, innovation / technology, risk, market attractiveness, as well as assortment fit.

While Return On Investment (ROI) was most commonly used as a decision criterion, it is also related to a knowledge and learning process, as it typically denotes the *expected* ROI from a given project investment.

Another aspect of our evaluation regarded the usage of software tools to support the knowledge and learning-related tasks within PPM. Figure 5 shows an overview of results. Our results show that missing integration and applications that do not cover the process completely are major obstacles for more advanced knowledge and learning applications in the context of PPM.

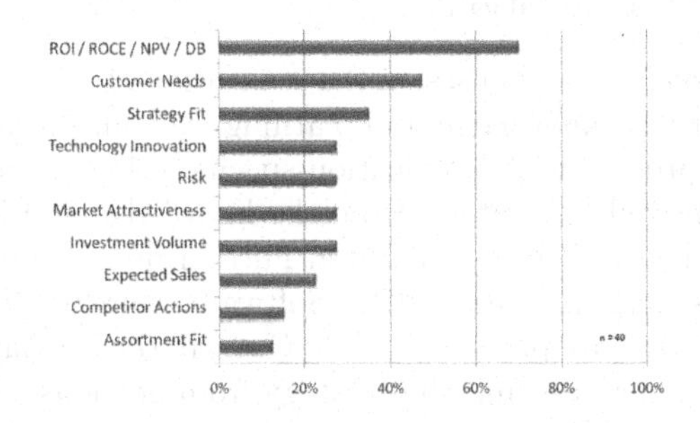

Fig. 4. Decision Criteria for Project Investments

Fig. 5. Software Usage in the Context of Knowledge and Learning in PPM

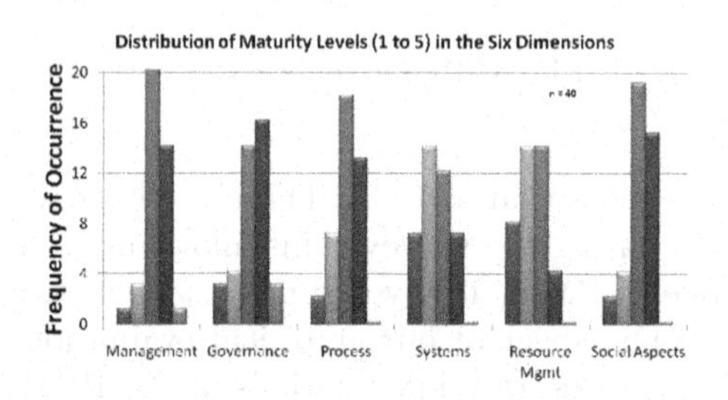

Fig. 6. Maturity Levels of the Respondents (Bars in Every Dimension Denote the Occurrence of a Certain Maturity Level, e.g., 22 Respondents Have a Maturity Level of 3 in the Dimension "Management")

One of our main objectives in the overall project is the definition and evaluation of a holistic maturity model for PPM in knowledge-intensive project landscapes. Figure 6 shows the aggregated view of maturity levels of all respondents in the six different dimensions - management, governance, process, systems, resource management, and social aspects. Of particular relevance are the comparatively low levels in the dimensions *systems* and *resource management*. They reflect a significant gap concerning knowledge and learning within PPM - the high requirements from the areas of management, governance and process are not met from existing software systems. Furthermore, the resource management seldom accounts for competence and knowledge profiles of the personnel resources. This is a major obstacle for the introduction of more advanced knowledge and learning aspects in PPM (e.g., automated consideration of previous project experience and specific qualifications).

5.2 Experimental Evaluation

Our experimental evaluation focuses on the realization of our proposed approach within a software-based knowledge and learning system. We apply a standard software selection process with application-specific criteria, similar to the one used in [10]. We derived the criteria from the knowledge and learning strategy and process as defined in Sections 3 and 4. Furthermore, we used studies from Gartner and Forrester in the area of PPM-Software. More detailed results of our experimental evaluation are presented in [3]. Overall, the evaluated system offers a fair coverage of the process and the strategy from Sections 4 and 3.

6 Conclusion and Outlook

The field of complex knowledge-intensive project landscapes is a suitable environment for the evaluation of knowledge and learning objects, processes, strategies, systems, and performance as defined in [1]. In this work we proposed project portfolio management as a knowledge and learning strategy, together with a knowledge and learning process that brings this strategy to life. Furthermore, we provided some key results of the empirical and experimental evaluation of our approach that confirm its viability and applicability. Our future research activities are focused in the area of suitable systems for knowledge and learning, as well as inter-organizational aspects of project portfolio management and open knowledge aspects in this field.

References

1. Lytras, M., Sicilia, M.: The Knowledge Society: a manifesto for knowledge and learning. International Journal of Knowledge and Learning, 1 1(2), 1–11 (2005)
2. Rajegopal, S., Waller, J., McGuin, P.: Project Portfolio Management: Leading the Corporate Vision. Palgrave Macmillan, Basingstoke (2007)
3. Stantchev, V., Franke, M.R., Discher, A.: Project Portfolio Management Systems: Business Services and Web Services. In: ICIW 2009: Proceedings of the 2009 Fourth International Conference on Internet and Web Applications and Services. IEEE Computer Society Press, Los Alamitos (2009)
4. Cooper, R.G., Edgett, S.J.: Ten ways to make better portfolio and project selection decisions. Technical report, Ancaster, Ontario, Canade (2007)
5. Pepels, W.: Produktmanagement: Produktinnovation, Markenpolitik, Programmplanung, Prozessorganisation. Oldenbourg Wissenschaftsverlag (2006)
6. Kunz, C.: Strategisches Multiprojektmanagement: Konzeption, Methode und Strukturen. Deutscher Universitätsverlag (2007)
7. Cooper, R.G., Edgett, S.J., Kleinschmidt, E.J.: Portfolio Management for New Products. Perseus Books, Cambridge (2001)
8. Schott, E., Campana, C.: Strategisches Projektmanagement. Springer, Heidelberg (2005)
9. Banker, R.D., Bardhan, I., Asdemir, O.: Understanding the impact of collaboration software on product design and development. Info. Sys. Research 17(4), 352–373 (2006)
10. Stantchev, V., Hoang, T.D., Schulz, T., Ratchinski, I.: Optimizing clinical processes with position-sensing. IT Professional 10(2), 31–37 (2008)

Gender Differences in the Continuance of Online Social Networks

Na Shi[1], Christy M.K. Cheung[2], Matthew K.O. Lee[3], and Huaping Chen[4]

[1] University of Science and Technology of China-City University of Hong Kong Joint Advanced Research Center
shina99@student.cityu.edu.hk
[2] Department of Finance and Decision Sciences, Hong Kong Baptist University
ccheung@hkbu.edu.hk
[3] Department of Information Systems, City University of Hong Kong
ismatlee@cityu.edu.hk
[4] School of Management, University of Science and Technology of China
hpchen@ustc.edu.cn

Abstract. Social network sites (SNS) have become increasingly popular in the past few years benefiting from the rapid growth of Web 2.0 applications. However, research on the adoption and usage of SNS is limited. In this study, we attempt to understand users' continuance intention to use SNS and investigate the role of gender. A research model was developed and tested with 213 respondents from an online survey. The results confirm that users' continuance intention to use SNS is strongly determined by satisfaction. The effect of disconfirmation of maintaining offline contacts on satisfaction is more important for women, while the effect of disconfirmation of entertainment is more salient for men. Implications of this study for both researchers and practitioners are discussed.

Keywords: Continuance Intention, Online Social Networks, Gender, Satisfaction, Expectation Disconfirmation Theory.

1 Introduction

Social network sites (SNS) are web-based platforms which allow users to build their own profiles in a bounded system and share connections with their friends within this system [1]. It has become an increasingly popular form of information technology (IT) for individuals to communicate and entertain together due to the rapid growth and popularity of Web 2.0 applications. According to a national survey of teenagers conducted by Pew Internet & American Life Project [2], more than half (55%) of the online American youths use online social networking sites. Facebook, as one of the fastest-growing and best-known SNS, has attracted more than 200 million users.

As a new technology with only a few years of history, there is only very little research on the adoption and diffusion of SNS. Prior research on SNS was mostly associated with its adoption and acceptance [3], [4], [5]. However, researchers are increasingly interested in the post-adoption of IT in the past few years [6], [7], as it

M.D. Lytras et al. (Eds.): WSKS 2009, CCIS 49, pp. 216–225, 2009.

concerns the long-term success of IT adoption. This suggests that research on post-adoption of SNS is needed to better understand SNS usage. In this study, we build and empirically test a theoretical model to understand users' continuance intention to use SNS.

Meanwhile, the moderating effect of gender is another emphasis in this study. The importance of studying gender differences with respect to IT has been demonstrated by a growing body of research investigating these differences in a variety of contexts, including e-mail [8], instant messaging [9], and Internet use [10]. Interesting results have been found from these studies. However, it is necessary to investigate these important differences in new IT innovations (e.g., SNS) to get more complete understandings. In addition, a recent report revealed a gender difference in SNS usage, which showed that girls were more likely to use SNS to strengthen pre-existing friendships [2]. Have these variations in mind, it would be interesting to examine how men and women use SNS differently. Therefore, we incorporate gender into our model and examine its role in understanding the continuance intention to use SNS.

2 Theoretical Background

2.1 Expectation Disconfirmation Theory

The expectation disconfirmation theory (EDT), initially used in the social psychology literature, describes an individual's behavioral process from the initial pre-use expectations of a product to the post-use perceptions of the product [11]. In recent years, the EDT has been widely used to understand the continuance use of an information system in the information systems discipline. Bhattacherjee [7] was one of the very first researchers who introduced the EDT to the IS field. He has argued that continuance intention to use an IS depends on the satisfaction level with initial usage and the perceived usefulness of continued usage. Further, user satisfaction is affected by the discrepancy between the pre-use expectation of the IS and the perceived usefulness (disconfirmation), and the post-use perceived usefulness is influenced by the level of disconfirmation. Studies have confirmed that the EDT is successful in explaining the IS usage behavior from initial acceptance to continuance (discontinuance) [12], [13], [14].

2.2 SNS Motivations

Exploratory empirical studies have identified the motivations of SNS users. For example, Nyland et al. [15] identified five motivations of using SNS: to meet new people, to entertain, to maintain relationships, social events, and media creation. Joinson [4] identified seven motivations: social connection, shared identities, photographs, content, social investigation, social network surfing, and status updates. However, it is too extensive to include all possible motivations in one study. Therefore, we only focus on those that seem to have stronger effects, including maintaining offline contacts and entertainment. Maintaining offline contacts means that people use SNS to connect with their old friends, classmates, colleagues, and other people they know in person [16]. Entertainment refers to the use of SNS for filling up free time, taking break and having fun [16]. People can entertain themselves in their spare time and pass time when they feel bored.

2.3 Gender Issue

Gender differences in using a specific communication technology is connected with the different communication and interaction styles between men and women [17]. Studies have indicated that women are more expressive and likely to engage in social-oriented activities, whereas men focus more on task-oriented activities [18]. Women use the telephone more frequently for maintaining friendships over a distance [19], and men and women hold differences in communication patterns and preferences towards the Internet and computer-mediated communication [10].

Just as the existence of gender differences in communication, men and women also have different attitudes towards technology [20]. Research on gender differences with respect to IT has attracted attention in recent years [8], [21]. Studies in IS research have also identified gender as an important moderating variable [6], [9], [22]. Meanwhile, Boneva et al. [23] have suggested that online gender differences are consistent with the gender differences found in traditional forms of communication such as face-to-face or telephone interactions. Women consider computers more useful for keeping in touch with family and friends than men do [17], and men spend more time using the Internet for entertainment and leisure than women [24].

3 Research Model and Hypotheses

Figure 1 presents the research model of this study. This study integrates two salient motivations of using SNS into expectation disconfirmation theory. Furthermore, we expect gender will moderate the effects of disconfirmations on satisfaction with SNS.

3.1 Satisfaction

Satisfaction reflects a user's psychological or affective state resulting from a cognitive appraisal of disconfirmation [7]. Based on expectation disconfirmation theory, previous research in IS has supported the relationship between satisfaction and IS continuance intention [25], [26]. These post-adoption studies have all demonstrated that

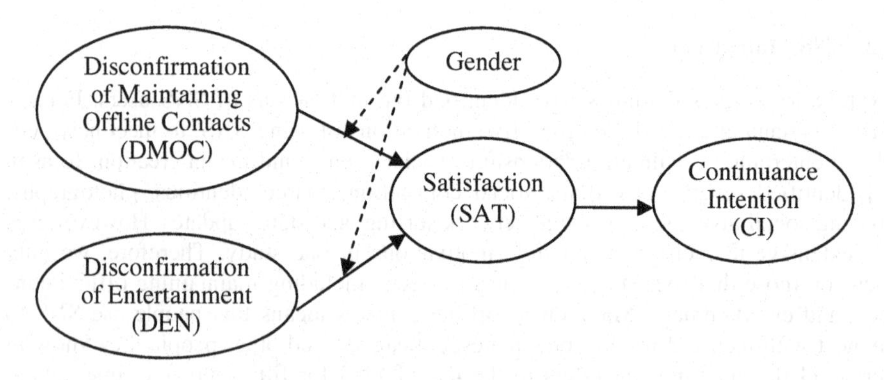

Fig. 1. Research Model

satisfaction is positively associated with IS continuance intention [27], [28]. High level of satisfaction will result in the increase of users' continuance usage intention. Consistent with prior research, we expect that user satisfaction with prior usage experience will positively influence their continuance intention to use SNS. Therefore,

H1: User satisfaction with SNS has a positive effect on users' continuance intention to use SNS.

3.2 Disconfirmations

Prior literature has suggested that it is the discrepancy (disconfirmation) between the expectation and the perceived performance that primarily determines user satisfaction in the post-use stage [7]. As mentioned before, maintaining offline contacts and entertainment are two major motivations of using SNS. Satisfaction is expected to be determined by the disconfirmations between the expectation and the perception of how well these two motivations are achieved. We believe that if the perceived performance of a SNS is better than or the same as the user's pre-use expectation, he/she will be satisfied and will continue to use it. In contrast, if his/her expectation is not met, he/her will be dissatisfied and will give up the use of SNS. Therefore,

H2a: Positive disconfirmation of maintaining offline contacts has a positive effect on user satisfaction with SNS.

H3a: Positive disconfirmation of entertainment has a positive effect on user satisfaction with SNS.

Studies in sociology have argued that women value connection and cooperation more than men [29] and have more extensive social network than men [19], [30]. In addition, prior research has found that men spend more time on the Internet for entertainment and leisure than women [24], where as women prefer using computer technologies to expand their social networks and keep in touch with others [17]. According to these findings, we believe that the effect of disconfirmation of maintaining offline contacts will be more important for women than for men, whereas the effect of disconfirmation of entertainment will be more important for men than for women. Therefore,

H2b: Positive disconfirmation of maintaining offline contacts influences user satisfaction more strongly for women than for men.

H3b: Positive disconfirmation of entertainment influences user satisfaction more strongly for men than for women.

4 Research Method

Empirical data for this study was collected by an online survey among Facebook users. Minor changes on wordings were made according to 30 Facebook users' comments. Then, the hyperlink to the online English questionnaire was posted in several groups on Facebook, which are made up of users in Hong Kong. In order to encourage participation, an incentive of supermarket coupons were offered as lucky draw prizes for participants of the survey. A total of 213 usable questionnaires were collected. Table 1 provides a summary of the demographic characteristics of male (N=93) and female (N=120) respondents.

All measures used in this study were derived from prior studies [7], [16] with minor modifications to fit the specific research context (as shown in Appendix A). In particular, disconfirmations of maintaining offline contact and entertainment used seven-point scales ranging from "Much lower than your expectation" (-3) to "Much higher than your expectation" (3). Satisfaction used seven-point semantic differential scales. Continuance intention used seven-point Likert scales anchored with "strongly disagree" (1) and "strongly agree" (7).

Table 1. Sample Characteristics

Characteristics	Male (N=93) No. (Percentage)	Female (N=120) No. (Percentage)
Age		
18 or below	6 (6.45%)	9 (7.5%)
19-28	69 (74.2%)	92 (76.7%)
29-35	12 (12.9%)	18 (15.0%)
36 or above	6 (6.45%)	1 (0.8%)
Education level		
Secondary or high school	8 (8.6%)	9 (7.5%)
Diploma or equivalent	7 (7.5%)	8 (6.7%)
University or above	78 (83.9%)	103 (85.8%)
Experience with the social network site		
6 months or less	29 (31.2%)	21 (17.5%)
7-11 months	23 (24.7%)	39 (32.5%)
1 year or more	41 (44.1%)	60 (50%)
Number of contacts on the social network site		
50 or less	26 (28.0%)	19 (15.8%)
51-100	20 (21.5%)	23 (19.2%)
101-200	21 (22.6%)	43 (35.8%)
More than 200	26 (27.9%)	35 (29.2%)
Frequency of visiting the social network site		
Once or more per day	56 (60.2%)	80 (66.7%)
Once or more per week	33 (35.5%)	31 (25.8%)
Twice or less per month	4 (4.3%)	9 (7.5%)

5 Data Analysis

Partial Least Squares (PLS) was used to analyze the research model. PLS enables the researchers to assess both the measurement model and the structural model simultaneously. In addition, PLS has the ability to model latent construct under minimal restrictions on measurement scales, residual distribution and sample size [31]. Therefore, we chose PLS to perform data analysis in this study.

5.1 Measurement Model

Convergent validity is assessed by the composite reliability (CR) and the average variance extracted (AVE). A CR of 0.7 or above and an AVE of greater than 0.5 are acceptable [32]. As shown in Table 2, all the measures exceed the recommended thresholds. Meanwhile, to ensure the discriminant validity, the square root of AVE for each construct should be greater than all the correlations between this construct and other constructs [32]. The results in Table 3 suggest an adequate level of discriminant validity of the measures.

5.2 Structural Model

Table 4 presents the results of data analysis. Test of the significance of all paths was performed using the bootstrap resampling procedure. The results show that satisfaction has a strong impact on continuance intention, with a path coefficient of 0.547. In turn, satisfaction is strongly determined by both disconfirmation of maintaining offline contacts and disconfirmation of entertainment, with path coefficients of 0.321 and 0.400 respectively.

To examine the moderating effect of gender, we performed analysis in male group and female group separately. As shown in Table 4, all the direct effects are significantly supported for both two groups, however, different influence strengths have been found between men and women. The significance of difference in path coefficients between two groups was calculated as Keil et al. [33] suggested. The results demonstrate that the effect of disconfirmation of maintaining offline contacts is more important for women, while the effect of disconfirmation of entertainment is more important for men.

Table 2. Item Loadings

Construct	Items	Male	Female		Construct	Items	Male	Female
DMOC	DMOC1	0.816	0.793		SAT	SAT1	0.871	0.837
	DMOC2	0.868	0.643			SAT2	0.913	0.865
	DMOC3	0.778	0.763			SAT3	0.858	0.830
	CR	0.861	0.779			SAT4	0.908	0.839
	AVE	0.675	0.542			CR	0.937	0.908
DEN	DEN1	0.852	0.883			AVE	0.788	0.710
	DEN2	0.921	0.872		CI	CI1	0.913	0.895
	DEN3	0.867	0.929			CI2	0.902	0.818
	CR	0.911	0.923			CR	0.903	0.847
	AVE	0.775	0.801			AVE	0.823	0.735

Table 3. Correlations Matrix of Constructs

Male	DMOC	DEN	SAT	CI
DMOC	**0.822**			
DEN	0.635	**0.880**		
SAT	0.633	0.741	**0.888**	
CI	0.514	0.544	0.624	**0.907**
Female	DMOC	DEN	SAT	CI
DMOC	**0.736**			
DEN	0.549	**0.895**		
SAT	0.492	0.464	**0.843**	
CI	0.292	0.373	0.480	**0.857**

Note: Diagonal elements are the square roots of average variance extracted.

Table 4. Summary of the Results of Data Analysis

	Full (N=213)	Male (N=93)	Female (N=120)	Differences Male vs. Female (t value)
SAT-CI	0.547***	0.624***	0.480***	--
DMOC-SAT	0.321***	0.273***	0.340***	5.359***
DEN-SAT	0.400***	0.568***	0.278***	26.471***
R Square (SAT)	41.4%	59.4%	29.6%	--
R Square (CI)	30.0%	38.9%	23.0%	--

(Note: ** for p<0.01; *** for p<0.001).

6 Discussion and Conclusion

In response to the need of better understanding the post-adoption issues of SNS, this study makes an effort to study users' continuance intention to use SNS, as well as the role of gender in the research model. As the results of this study show, satisfaction is confirmed to have a positive effect on users' continuance intention to use SNS, while satisfaction is significantly determined by the positive disconfirmations of maintaining offline contacts and entertainment. Further, the moderating effect of gender is confirmed on the relationships between disconfirmations and satisfaction. In particular, the disconfirmation of maintaining offline contacts is more important for women, while men are more sensitive to the disconfirmation of entertainment.

This study makes theoretical implications in several ways. First, the current study extends the expectation disconfirmation model by incorporating two major motivations of using SNS, which improves the understanding of the decision process of users' continuance intention to use SNS. Our findings confirm the effect of EDT in predicting users' continuance intention in SNS context. Second, this study also contributes to the SNS research by enriching the limited existing studies on the post-adoption of SNS. Although SNS has been growing explosively and becoming more and more popular in the past few years, research related to SNS in IS field is still rare. More studies are needed to identify influencing factors help to understand SNS users' behavior. Our study on users' continuance intention to use SNS is expected to bring on more insights and inspirations into SNS. Last, the moderating effect of gender is empirically examined in this study and our results also confirm gender differences exist in the use of SNS. Therefore, future research on SNS is suggested to pay attention to gender issue.

This study also provides some practical implications for SNS managers. First, the significance of satisfaction indicates that users' continuance intention to use SNS is positively related to their post-usage satisfaction. SNS service designers should devise strategies that can help to increase user satisfaction which will be conductive to retaining existing users. Meanwhile, these satisfied users may help to bring in new users by word-of-mouth promotions. Second, our results imply that the increased satisfaction through better perceptions of maintaining offline contacts and entertainment will ultimately motivates the users to continue using SNS. SNS designers should develop corresponding features and functions to satisfy users' need in these two aspects better. Third, SNS managers should be aware of gender differences in the use of SNS and adopt different strategies in promoting and maintaining the level of satisfaction for men and women. Accordingly, since the positive disconfirmation of maintaining offline contacts is more salient for women, managers should try to convince the effectiveness and convenience of SNS on keeping relationships to women. In turn, the positive disconfirmation of entertainment is more salient for men, thus managers should highlight and emphasize to men that using SNS can obtain entertaining and interesting.

This study is subject to some limitations. First, the selection of respondents was restricted in Hong Kong, which could restrict the generalizability of this study. The moderating effect of gender may not be inferable to other regions due to cultural differences. More research of gender in SNS continuance behaviors is suggested to compare with the results of this study. Second, the research model explains only 29.6% of the variance in satisfaction for women, 38.9% and 23.0% of the variance in continuance intention for men and women respectively. Future research is suggested to extend this research and identify other important factors, such as social influence, that may help to understand SNS continuance behaviors.

Acknowledgments. The work described in this article was partially supported by a grant from the Research Grant Council of the Hong Kong Special Administrative Region, China (Project No. CityU 145907).

References

1. Boyd, D.M., Ellison, N.B.: Social Network Sites: Definition, History, and Scholarship. Journal of Computer-Mediated Communication 13(1), 210–230 (2008)
2. Lenhart, A., Madden, M.: Social Networking Websites and Teens: An Overview (2007), http://www.pewinternet.org/pdfs/PIP_SNS_Data_Memo_Jan_2007.pdf (accessed on March 5, 2009)
3. Gangadharbatla, H.: Facebook Me: Collective Self-Esteem, Need to Belong, and Internet Self-Efficacy as Predictors of the Igeneration's Attitudes toward Social Networking Sites. Journal of Interactive Advertising 8(2), 1–28 (2008)
4. Joinson, A.N.: 'Looking at', 'Looking up' or 'Keeping up with' People? Motives and Uses of Facebook. In: CHI 2008, Florence, Italy (2008)
5. Raacke, J., Bonds-Raacke, J.: Myspace and Facebook: Applying the Uses and Gratifications Theory to Exploring Friend-Networking Sites. CyberPsychology & Behavior 11(2), 169–174 (2008)
6. Ahuja, M.K., Thatcher, J.B.: Moving Beyond Intention and toward the Theory of Trying: Effects of Work Environment and Gender on Post-Adoption Information Technology Use. MIS Quarterly 29(3), 427–459 (2005)
7. Bhattacherjee, A.: Understanding Information Systems Continuance: An Expectation-Confirmation Model. MIS Quarterly 25(3), 351–370 (2001)
8. Gefen, D., Straub, D.W.: Gender Differences in the Perception and Use of E-Mail: An Extension to the Theory Acceptance Model. MIS Quarterly 21(4), 389–400 (1997)
9. Ilie, V., et al.: Gender Differences in Perceptions and Use of Communication Technologies: A Diffusion of Innovation Approach. Information Resources Management Journal 18(3), 13–31 (2005)
10. Teo, T.S.H.: Demographic and Motivation Variables Associated with Internet Usage Activities. Internet Research 11(2), 125–137 (2001)

11. Oliver, R.L.: A Cognitive Model of the Antecedents and Consequences of Satisfaction Decisions. Journal of Marketing Research 17(4), 460–469 (1980)
12. Au, N., Ngai, E.W.T., Cheng, T.C.E.: A Critical Review of End-User Information System Satisfaction Research and a New Research Framework. Omega 30(6), 451–478 (2002)
13. Doong, H.-S., Lai, H.: Exploring Usage Continuance of E-Negotiation Systems: Expectation and Disconfirmation Approach. Group Decision and Negotiation 17(2), 111–126 (2008)
14. McKinney, V., Yoon, K., Zahedi, F.M.: The Measurement of Web-Customer Satisfaction: An Expectation and Disconfirmation Approach. Information Systems Research 13(3), 296–315 (2002)
15. Nyland, R., Near, C.: Jesus Is My Friend: Religiosity as a Mediating Factor in Internet Social Networking Use. In: The AEJMC Midwinter Conference, Reno, Nevada (2007)
16. Ellison, N., Steinfield, C., Lampe, C.: Spatially Bounded Online Social Networks and Social Capital: The Role of Facebook. In: The Annual Conference of the International Communication Association, Dresden, Germany (2006)
17. Debrand, C.C., Johnson, J.J.: Gender Differences in Email and Instant Messaging: A Study of Undergraduate Business Information Systems Students. Journal of Computer Information Systems 48(3), 20–30 (2008)
18. Wood, W., Rhodes, N.: Sex Differences in Interaction Style in Task Groups. In: Ridgeway, C.L. (ed.) Gender, Interaction, and Inequality, pp. 97–121. Springer, New York (1992)
19. Walker, K.: I'm No Friends the Way She's Friends: Ideological and Behavioral Constructions of Masculinity in Men's Friendships. Masculinities 2, 38–55 (1994)
20. Venkatesh, V., Morris, M.G., Ackerman, P.: A Longitudinal Field Investigation of Gender Difference in Individual Technology Adoption Decision-Making Processes. Organizational Behavior and Human Decision Processes 83(1), 33–60 (2000)
21. Wilson, M.: A Conceptual Framework for Studying Gender in Information Systems Research. Journal of Information Technology 19(1), 81–92 (2004)
22. Venkatesh, V., et al.: User Acceptance of Information Technology: Toward a Unified View. MIS Quarterly 27(3), 425–478 (2003)
23. Boneva, B., Kraut, R., Frohlich, D.: Using E-Mail for Personal Relationships: The Difference Gender Makes. The American Behavioral Scientist 45(3), 530–549 (2001)
24. Weiser, E.B.: Gender Differences in Internet Use Patterns and Internet Application Preferences: A Two-Sample Comparison. CyberPsychology & Behavior 3(2), 167–178 (2000)
25. Liao, C., Chen, J.-L., Yen, D.C.: Theory of Planning Behavior (TPB) and Customer Satisfaction in the Continued Use of E-Service: An Integrated Model. Computers in Human Behavior 23(6), 2804–2822 (2007)
26. Lin, C.S., Wu, S., Tsai, R.J.: Integrating Perceived Playfulness into Expectation-Confirmation Model for Web Portal Context. Information & Management 42(5), 683–693 (2005)
27. Limayem, M., Cheung, C.M.K.: Understanding Information Systems Continuance: The Case of Internet-Based Learning Technologies. Information & Management 45(4), 227–232 (2008)
28. Thong, J.Y.L., Hong, S.-J., Tam, K.Y.: The Effects of Post-Adoption Beliefs on the Expectation-Confirmation Model for Information Technology Continuance. International Journal of Human-Computer Studies 64(9), 799–810 (2006)
29. Meyers, R.A., et al.: Sex Differences and Group Argument: A Theoretical Framework and Empirical Investigation. Communication Studies 48, 19–41 (1997)
30. Wellman, B.: Men in Networks: Private Communities, Domestic Friendships. In: Nardi, O.M. (ed.) Men's Friendships, pp. 74–114. Sage, London (1992)

31. Chin, W.W.: The Partial Least Squares Approach to Structural Equation Modeling. In: Marcoulides, G.A. (ed.) Modern Methods for Business Research, pp. 295–336. Lawrence Erlbaum Associates, Mahwah (1998)
32. Fornell, C., Larcker, D.F.: Evaluating Structural Equation Models with Unobservable Variables and Measurement Error. Journal of Marketing Research 18(1), 39–50 (1981)
33. Keil, M., et al.: A Cross-Cultural Study on Escalation of Commitment Behavior in Software Projects. MIS Quarterly 24(2), 299–325 (2000)

Appendix A

Disconfirmation of Maintaining Offline Contacts (Ellison et al., 2006)

Compared with your pre-expectation, indicate your perception of the experience of using Facebook in performing the following functions:

1. To check out someone you met socially.
2. To learn more about other people in your classes/workplace.
3. To keep in touch with your old friends.

Disconfirmation of Entertainment (Ellison et al., 2006)

Compared with your pre-expectation, indicate your perception of the experience of using Facebook in performing the following functions:

1. To fill up free time.
2. For fun.
3. To take a break from your homework/work.

Satisfaction (Bhattacherjee, 2001)

My overall experience of using Facebook is:

1. Very dissatisfied/ Very satisfied.
2. Very displeased/ Very pleased.
3. Very frustrated/ Very contented.
4. Absolutely terrible/ Absolutely delighted.

Continuance Intention (Bhattacherjee, 2001)

1. I intend to continue using Facebook in the future.
2. I will keep using Facebook as regularly as I do now/more than I do now.

It's Not Only about Technology, It's about People: Interpersonal skills as a Part of the IT Education

Ricardo Colomo-Palacios[1], Cristina Casado-Lumbreras[2],
Ángel García-Crespo[1], and Juan Miguel Gómez-Berbís[1]

[1] Universidad Carlos III de Madrid, Computer Science Department
Av. Universidad 30, Leganés, 28911, Madrid, Spain
{ricardo.colomo,angel.garcia,juanmiguel.gomez}@uc3m.es
Karakorum Servicios Profesionales
C. Zurbarán 5, Pozuelo de Alarcón, 28223 Madrid, Spain
ccasado@karakorumsp.es

Abstract. The importance of what have been termed the "soft skills" for the professional development of IT professionals is beyond any doubt. Taking account of this circumstance, the objective of the current research may be phrased as two separate questions. In the first place, determining the importance which IT related degree students place on these types of competencies for their professional future. In the second place, the importance which the development of the mentioned competencies has been given during their studies. The realization of an empirical study has fulfilled the two objectives described. The results demonstrate, on the one side, the moderate relevance which students assign to interpersonal competencies, especially emotional competencies, in contrast to the international curricular recommendations and studies concerning labor markets. On the other hand, the results indicate the scarce emphasis which lecturers have placed on the development of such competencies.

Keywords: Soft Skills, IT Education, Interpersonal skills, Emotional competences.

1 Introduction

The concept of social or interpersonal abilities refers to the social competencies or capacities which people possess. However, this concept, which is generally perceived as rather simple, has been captured in numerous definitions. In the presence of such an abundance of definitions, Linehan's [1] definition can be selected as a base for the present study. Interpersonal skills are "the complex capacity to exhibit behavior or response patterns which optimize interpersonal influence and resistance to undesired social influence (efficiency in both objectives), which simultaneously maximizes gains and minimizes the losses in the relationship (efficiency in the relationship) and maintains one's own integrity and sense of domain (efficiency with respect to oneself)". Lazarus [2] was one of the first authors to establish the principal response or behavioral dimensions which characterize social skills from a clinical perspective: 1) The capacity to say "no", 2) The capacity to ask for favors and make requests, 3) The capacity to express positive or negative feelings and 4) The capacity to initiate,

M.D. Lytras et al. (Eds.): WSKS 2009, CCIS 49, pp. 226–233, 2009.

maintain, and finish conversations. The response classes outlined above are based on the four category types proposed by Lazarus [2]. In this sense, specifically with respect to the Information, Communication and Technology field, interpersonal skills are considered general and necessary for professionals [3], and employers at international level consider interpersonal skills as a key aspect for recent graduates from the technical area [4].Taking into account the unquestionable importance of interpersonal skills, the concept of competency related to this aspect has been included in curricular initiatives in diverse disciplines. Competences can be defined as an individual's core skills (motives, traits, self-concept, knowledge, and abilities) that are causally related to a specific, effective criterion and/or a superior performance at work [5].

Particularly within the domain of IT related degrees, curricular initiatives are being developed for the set of five fields which includes: Software Engineering [6], Computer Engineering [7], Information Systems [8], Information Technology [9] and Computer Science [10], which as a set have been given the generic name Computing Curricula 2005. These proposals are jointly sponsored by a committee comprised of the IEEE and ACM. In addition to the five volumes mentioned, a study is currently being realized which draws together and compares aspects of the diverse proposals [11]. This publication emphasizes the importance of interpersonal communication, which includes aspects such as written and oral communication, presentation, interaction with clients, sales activities, and their comparison across different disciplines. With this objective in mind, minimum and maximum values from 1–5 on a Likert-type scale have been included for the weight which interpersonal communication should be assigned within the categories mentioned. The values presented are the following: Computer Engineering (3.4), Computing (1.4), Information Systems (3.5), Information Technology (3.4) and Software Engineering (3.4).The scores assigned indicate that considerable importance has been attributed to the interpersonal communication category across all of the distinct disciplines. The most comprehensive analysis of interpersonal skills may be found in the volume dedicated to Information Systems [8].

A second aspect worth mentioning is the inclusion of interpersonal skills within the set of capacities of graduates. In this case, interpersonal skills form part of the category "interpersonal, communication and team skills" and have four associated behaviors: listening, encouragement, motivation and operating in a diverse environment. The specification of interpersonal skills both as part of the body of knowledge as well as the capacities of graduates has initiated, as a third notable characteristic, its inclusion within the programs of various subjects recommended in the curriculum for graduates of the discipline. The study which has been develop which will be described in what follows, consisted of exploring the judgments of young people regarding the importance they give to interpersonal skills in their suture professional career. The particular case under analysis in the current paper is a study directed exclusively towards IT related degrees students about to complete their degree, whose incorporation in the labor market is evident.

Traditionally, it has been assumed that the organizational practice of the IT professional does not require mayor skills in the interpersonal domain. The professional reality has demonstrated that the IT professional requires the development of skills which go beyond the management and application of a series of technical aptitudes, whether he is a coder, team leader or manager.

2 Interpersonal Skills as a Research Field in IT

Today soft skills represent a wide study area in the IT field. In [12], it was affirmed that the most important skills for new IT professionals were soft skills. Other authors have drawn more refined conclusions, and consider that soft skills are more important than technical skills for less experienced IT personnel [13],[14]. In contrast, opposing views to this research [15], [16] state that technical skills are of a higher importance than soft skills. Finally, other works claim that technical and behavioral skills are likely to be complementary [17]. Leaving aside the relative importance of soft skills of professionals at any level in the IT field, the necessity to possess interpersonal skills, one of the soft skills, is without doubt. However, this reality, which IT professionals are familiar with, in confronted by the professional stereotype. This stereotype for IT professionals includes anti-social [18] and solitary [19], to name but a few. However, this stereotype, which is found in many media such as cinema [20] and television [21], is far from being certain.

Examining at study carried out between academics and professionals in the IT field [22], both groups indicate that, behind personal traits, interpersonal skills are the second most important factor for professional practice. It is also interesting to note that IS academics rate interpersonal skills, less important than IS practitioners do, while IS academics deem the variables in the IS technology area as important as or more important than IT practitioners do. This gap was already demonstrated in the 1990s in [23]. Other studies [24] in the same area, however realized among IT students, indicate that technical skills are important but are not sufficient within themselves, and need to be complemented with other skills such as communication and interpersonal skills. Furthermore, other studies completed in High School environments indicate that the opinions of the students with relation to interpersonal skills are closer to the stereotype [25]. Independently of the professional stereotypes of the different IT roles which the IT professional may adopt, the importance of interpersonal skills is highly notable. [26] provides a detailed description of the skills in the context of the different IT roles.

To summarize, some authors indicate that universities must not only include in their curriculum the hard skills of technical expertise, but also the soft skills of interpersonal communication, intrapersonal knowledge, leadership and collaborative skills that lead to a cohesive team [27].

3 The Study

The present study has consisted of the analysis of the importance which students in the final year of a Computer Engineering degree place on soft skills, particularly, interpersonal skills, for their professional future. In order to achieve this objective, a questionnaire has been applied in which the various interpersonal skills have been characterized with 18 associated behaviors. The application of the questionnaire had a double objective. In the first place, to quantitatively determine the opinions of the students with respect to interpersonal skills in the context of professional life, as well as their opinion of the importance which was placed on such skills during their higher education. In the second place, it was aimed to examine whether or not there were statistically significant differences in the responses according to gender. The subsequent sections detail the sample, the questionnaire applied, as well as a discussion of the results obtained.

3.1 Sample and Questionnaire

The present study has been developed based on the participation of 92 students in their final year of a Computing Engineering degree from several universities in Madrid (Politécnica de Madrid, Carlos III de Madrid, Autónoma, Pontificia de Salamanca, Pontificia de Comillas y UNED). The sample was composed of a total of 22 women (24%) and 70 men (76%).This gender difference is caused by the uneven proportion of men versus women enrolled in the course. The average age of the students was established as 25.7 years.

The questionnaire was comprised of 18 questions or items which describe behavior or competencies. These items collectively formed the category "Interpersonal Skills". The items selected for inclusion in the questionnaire were chosen based on the integration of 12 response classes, which are generally accepted as basic social skills in the Psychology field, with other items whose descriptions are more oriented towards a work environment. The respondents were requested to indicate the importance of the skills using a Likert-type scale, on which the value 1 indicated "of no importance", and the value 4 indicated the opinion "very important". Additionally, the students were elicited for their perceived importance of the different skills in the context of their academic career, based on what was communicated to them by their lecturers. The scale described above was also used to indicate scores in this case.

3.2 Results and Discussion

With the objective of determining the scores obtained for each element, an average and standard deviation was calculated for the results obtained in relation to the relative importance of the scores. The results are demonstrated in Table1.

Table 1. Average and Standard Deviation of the perceived importance of elements

Element	Average	Std Dev.
Capacity to accept compliments	1,55	0,58
Capacity to make compliments	1,60	0,54
Capacity to express annoyance, discomfort or anger in a justified way	1,68	0,57
Capacity to recognize the emotions of others	1,84	0,77
Capacity to express appreciation or affection	1,85	0,77
Capacity to reject requests of other people	1,85	0,78
Capacity to perceive the reactions of others to our own actions or opinions	1,88	0,81
Capacity to admit ignorance about something	1,98	0,81
Capacity to request a change in the behavior of another person	2,00	0,80
Capacity to accept the help of others	2,00	0,85
Capacity to take into account the opinions of others	2,05	0,91
Capacity to make one's opinion be heard	2,18	0,95
Capacity to express personal opinions, including disagreement	2,20	0,97
Capacity to make requests in an adequate fashion	2,22	0,84
Capacity to excuse oneself	2,25	0,94
Capacity to accept constructive criticism regarding one's own behavior	2,27	0,96
Be capable of listening with attention	2,46	0,87
Capacity to initiate and maintain conversations	2,48	0,84

Examining the results, it is possible that the stereotype of the IT professional has distorted the importance of the competences for the students in a similar effect to the influence of this stereotype on the rest of society. Several studies have identified the stereotypes of IT Professionals [18], [19], [20], [21]. These stereotypes accentuate negative characteristics of professionals such as anti-social or solitary, to name but a few. Thus, Table 2 below displays the results of the perceived importance by the students of the skills, but in relation to the emphasis placed on such skills during their university career:

Table 2. Average and Standard Deviation of the perceived importance of Interpersonal Skills elements during higher education degree

Element	Average	Std Dev.
Capacity to accept compliments	1,24	0,43
Capacity to make compliments	1,26	0,44
Capacity to express annoyance, discomfort or anger in a justified way.	1,28	0,45
Capacity to recognize the emotions of others	1,14	0,35
Capacity to express appreciation or affection	1,17	0,38
Capacity to reject requests of other people	1,70	0,64
Capacity to perceive the reactions of others to our own actions or opinions	1,33	0,52
Capacity to admit ignorance about something	1,52	0,58
Capacity to request a change in the behavior of another person	1,30	0,46
Capacity to accept the help of others	1,47	0,64
Capacity to take into account the opinions of others	1,90	0,88
Capacity to make one's opinion be heard	1,95	0,87
Capacity to express personal opinions, including disagreement	1,95	0,87
Capacity to make requests in an adequate fashion	1,62	0,74
Capacity to excuse oneself	1,26	0,44
Capacity to accept constructive criticism regarding one's own behavior	1,29	0,46
Be capable of listening with attention	2,18	0,89
Capacity to initiate and maintain conversations	2,39	0,91

An initial analysis of the scores, examining both the importance of the skills in general, as well as the emphasis placed on the skills during university studies, indicates the discrepancy between both analyses. However, it is interesting to note that the "Capacity to initiate and maintain conversations" is the most valued capacity, and additionally, the capacity which students considered was assigned the most importance during their entire third level career. It is evident that almost all of the skills demonstrated low scores in relation to the attention they were given during students' university studies. Consequently, there is a large gap between the importance of the skills for the students, and the perceived importance which was placed on such skills during their academic career.

From the perspective of gender differences, the average displayed by the groups of the relative importance of the skills in Table 1 was almost always higher for the female category. In order to measure the level of statistical significance of this difference, the researchers applied the Students' T test. This statistical method, comparison of two means, was used to carry out one-way between-groups analysis of variance (ANOVA). The level of statistical significance was set at 0.05.Table 3 displays the competencies which demonstrate statistically significant differences between both groups:

Table 3. Significant differences between the importance of elements according to gender

Element	Test
Capacity to recognize the emotions of others	(t(21)= -2,21, p<.05)
Capacity to express appreciation and affection	(t(21)= -2,434, p<.05)
Capacity to reject requests from others	(t(21)= -3,215, p<.05)
Capacity to perceive the reactions of others to one's own acts or opinions	(t(21)= -3,306, p<.05)
Capacity to admit ignorance about something	(t(21)= -2,978, p<.05)
Capacity to elicit a change in the behavior of another person	(t(21)= -2,978, p<.05)
Capacity to accept the help of others	(t(21)= -2,978, p<.05)
Capacity to take into account the opinions of others	(t(21)= -3,071, p<.05)
Capacity to make requests in an adequate fashion	(t(21)= -2,628, p<.05)
Capacity to excuse oneself	(t(21)= -2,090, p<.05)
Capacity to accept constructive criticism about one's own behavior	(t(21)= -2,090, p<.05)
Be capable of listening with attention	(t(21)= -2,731, p<.05)
Capacity to initiate and maintain conversations	(t(21)= -2,731, p<.05)

Examining the table displayed above, it can be seen that female students are more sensitive to the importance of interpersonal skills than male students, as significant differences emerged between both samples. This outcome supports the results of previous studies which indicate that as a general rule, women have abilities superior to those of men in expressing and communicating emotions towards others, as has been demonstrated by numerous authors e.j. [28].

4 Conclusions

Summarizing the results of the current study, the findings indicate that the opinions of students and thus future IT professionals do not correspond to the proposals in curricular initiatives, nor to the increasing importance assigned to such skills in a professional environment. While young people principally minimize the importance of emotional competencies except for those which could be considered communication competencies in a more strict sense, academic and professional literature is attributing increasingly greater importance to emotional competencies. It can be concluded that likewise, the students generally do not perceive the importance placed on such skills during their university studies.

In this environment, institutions and teaching professionals in engineering disciplines should foster the awareness, learning and development of these key competencies for successful professional performance, even prior to the implementation of such recommendations in practice. It can be affirmed that both professionals as well as lecturers are not effectively communicating the professional reality of the IT professional. In order to achieve this, both professional associations as well as those in the academic field should be encouraged to establish effective communication and education measures to obtain the outcome that students, on the one hand, receive improved training in social competencies, and on the other hand, perceive these competencies with high esteem for a professional role.

References

1. Linehan, M.M.: Interpersonal effectiveness in assertive situations. In: Bleechman, E.A. (ed.) Behavior modification with women. Guilford Press, New York (1984)
2. Lazarus, A.A.: On assertive behaviour: A brief note. Behavior Therapy 4, 697–699 (1973)
3. Lee, C.K.: Transferability of Skills over the IT Career Path. In: Proceedings of the 2005 ACM SIGMIS CPR conference on Computer personnel research, pp. 85–93 (2005)
4. Gruba, P., Al-Mahmood, R.: Strategies for Communication Skills Development. In: Proceedings of the sixth conference on Australian computing education, vol. 30, pp. 101–107 (2005)
5. Spencer, L.M., Spencer, S.M.: Competence at Work. Models for Superior Performance. Willey and Sons, New York (1993)
6. Software Engineering. Curriculum Guidelines for Undergraduate Degree Programs in Software Engineering (August 23, 2004)
7. Computer Engineering. Curriculum Guidelines for Undergraduate Degree programs in Computer Engineering, Final Report. IEEE Computer Society & ACM (December 12, 2004)
8. Gorgone, J.T., Davis, G.B., Valacich, J.S., Topi, H., Feinstein, D.L., Longenecker, H.E.: IS 2002, Model Curriculum and Guidelines for Undergraduate Degree Programs in Information Systems (2002)
9. Computing Curricula, Information Technology Volume, Draft (April 2005)
10. Computing Curricula 2001. Computer Science. Final Report, (December 15, 2001)
11. Shackelford, R., Davies, G., Impagliazzo, J., McGettrick (eds.): Computing Curricula 2004. The Overview Report. Draft, (November 22, 2004)
12. McMurtrey, M.E., Downey, J.P., Zeltmann, S.M., Friedman, W.H.: Critical Skill Sets of Entry-Level IT Professionals: An Empirical Examination of Perceptions from Field Personnel. Journal of Information Technology Education 7, 101–120 (2008)
13. Kovacs, P.J., Caputo, D., Turchek, J., Davis, G.A.: A survey to define the skill sets of selected information technology professionals. Issues in Information Systems Journal 7(1), 242–246 (2006)
14. Young, D.: The relative importance of technical and interpersonal skills for new information systems personnel. Journal of Computer Information Systems 36(4), 66–71 (1996)
15. Lee, S., Yen, D., Havelka, D., Koh, S.: Evolution of IS professionals' competency: An exploratory study. Journal of Computer Information Systems 41(4), 21–30 (2001)
16. Koong, K.S., Liu, L.C., Lui, X.: A study of the demand for information technology professionals in selected internet job portals. Journal of Information Systems Education 13(1), 21–28 (2002)
17. Litecky, C., Arnett, K., Prabhakar, B.: The paradox of soft skills versus technical skills in IS hiring. The Journal of Computer Information Systems 45(1), 69–76 (2004)

18. Martin, C.D.: Is Computer Science a profession? ACM SIGCSE Bulletin 30(2), 7–8 (1998)
19. Craig, A., Paradis, R., Turner, E.: A Gendered View of Computer Professionals: Preliminary Results of a Survey. ACM SIGCSE Bulletin 34(2), 101–104 (2002)
20. Colomo-Palacios, R., Gómez-Berbís, J.M., García-Crespo, A.: IT Professionals: The Seventh Art Perspective. Novatica 187, 58–61 (2007)
21. García-Crespo, A., Colomo-Palacios, R., Gómez-Berbís, J.M., Tovar-Caro, E.: IT Crowd, Are we just Stereotypes? IT Professional 10(6), 24–27 (2008)
22. Lee, S., Koh, S., Yen, D., Tang, H.L.: Perception gaps between IS academics and IS practitioners: an exploratory study. Information & Management 40(1), 51–61 (2002)
23. Trauth, E.M., Farwell, D.W., Lee, D.: The IS expectation gap: Industry expectations versus academic preparation. MIS Quarterly 17(3), 293–307 (1993)
24. Medlin, B.D., Dave, D.S., Vannoy, S.A.: Students' views of the importance of technical and non-technical skills for successful IT professionals. The Journal of Computer Information Systems 42(1), 65–69 (2001)
25. García-Crespo, A., Colomo-Palacios, R., Gómez-Berbís, J.M., Tovar-Caro, E.: IT Professionals' Competences: High School Students' Views. Journal of Information Technology Education 8(1), 45–57 (2009)
26. Feeny, D.F., Willcocks, L.P.: Re-designing the IS Function around Core Capabilities. Long Range Planning 31(3), 354–367 (1998)
27. Woodward, B.S., Ashby, S., Litteken, A., Zamora, S.: Student Perceptions of Information Technology Preparedness and Important Job Skills. Information System Journal 6(38) (2008)
28. Schwartz, G.E., Brown, S.L., Ahern, G.L.: Facial muscle patterning and subjective experience during affective imagery: Sex differences. Psychophysiology 17, 75–82 (1980)

Understanding Organizational Learning via Knowledge Management in Government-Link Companies in Malaysia

Asleena Helmi, Zainal Ariffin Ahmad, and Daisy Kee Mui Hung

Swinburne University of Technology, Sarawak, Malaysia
University Sains Malaysia
ahelmi@swinburne.edu.my, {zaba,daisy}@usm.my

Abstract. The knowledge management or KM discipline conjures a host of understanding and impact upon the global business community albeit commercially or socially. Regardless of the different approach to KM, it has inarguably brought about changes in viewing the knowledge capabilities and capacities of organizations. Peter Drucker (1998) argued that knowledge has become the key economic resource and the only source of competitive advantage. Hence organizational learning is an integral part of KM initiatives and has been widely practiced in many large organizations and across nations such as Europe, North America and Asia Pacific. Thus, this paper explores the KM initiatives of government link companies (GLCs) in Malaysia via synergizing knowledge strategy and capabilities in order to achieve competitive advantage.

Keywords: Organizational learning, knowledge management, value proposition, knowledge strategy.

1 Introduction

Snowden (1999) defined KM can be defined as the identification, optimization and active management of intellectual assets either in the form of explicit knowledge held in artifacts or tacit knowledge possessed by individuals and communities. Knowledge can be further categorized as into two forms: tacit knowledge (example: internalized ideas and experience) and explicit knowledge (example: codified text, maps and manuals). Hence Nonaka and Takeuchi (1995) and Davenport and Prusak (2000) posit that KM is required for managing knowledge flow among knowledge workers. The more pertaining question in relation to KM is how best to leverage on knowledge as a strategic resources. The underlying notion of the statement above relates to the necessity for corporations to develop knowledge strategies to sustain competitive advantage. According to Callahan (2004), a knowledge strategy includes: (i) actions that are intended to result in anticipated business outcomes and (ii) actions that emerge as result of the many complex activities undertaken within an organization. It can be argued that these actions based upon organizational knowledge translate into value propositions such as better customer relations and new ways of working. Value

M.D. Lytras et al. (Eds.): WSKS 2009, CCIS 49, pp. 234–247, 2009.

proposition is defined as measures in which KM can increase organizational performance (McManus, Fredericksen, Wilson & Snyder, 2004). Proponents of KM such as Wigg, Malhotra and Porter have suggested knowledge as the strategic resource for competitive advantage. Whereas Zack (2001) has defined knowledgė strategy as competitive strategy built around a firm's intellectual resources and capabilities, then actions it may take to manage gaps or surpluses (e.g. recruiting for particular skills, building online documentary repositories, establishing communities of practice, acquiring firms, licensing technologies, etc.) are guided by a knowledge management strategy. Hence knowledge strategy is oriented toward understanding what knowledge is strategic and why (Snyman and Kruger, 2007).By identifying its competitive positioning (core, advanced or innovative), a firm would be able to ascertain knowledge gap and initiate a more effective knowledge strategy. In other words, the study aims to provide a mapping of knowledge network positioning (see Figure 1) among the GLCs in order to understand inter-organizational learning capabilities across the companies.

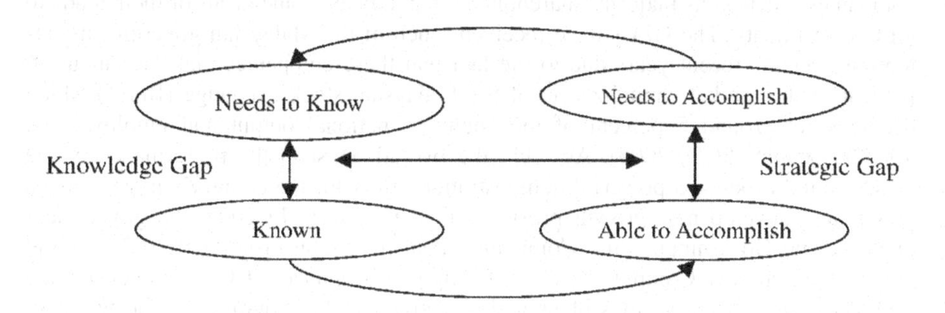

Fig. 1. Knowledge gap in developing KM strategy

According to Quinn, Anderson & Finkelstein (1996), the intellect of a contemporary organization operates at four levels such as the following:

- **Cognitive knowledge or know-what:** refers to the basic mastery of the discipline. It is essential but insufficient for commercial success.
- **Advanced skill or know-how:** refers to the translation of book learning into effective execution.
- **System understanding or know-why:** refers to the knowledge of the cause-and-effect relationship underlying a discipline.
- **Self-motivated creativity or care-why:** refers to the will, motivation and adaptability needed for success. Creative and motivated groups outperform groups with greater financial or physical resources (Khalil, 2000)

However the real issue is how to explicate knowledge. According to Skyrme (1998), many organizations have become so complex that their knowledge is fragmented, difficult to locate and share, and therefore redundant, inconsistent or not used at all. In today's environment of rapid change and technological discontinuity, even knowledge and expertise that can be shared is often quickly made obsolete. Koulopoulos and Frappaolo (1999) purported that, "knowledge management emphasizes the re-use of previous experiences and practices, but its focus is on mapping these to the changing landscapes of the market (Hossain and Ming Yu, 2004). In the broadest context, by identifying core knowledge and developing knowledge maps, companies may be able to explicitly link strategy, knowledge and performance. Nonetheless for the purpose of this study, a knowledge strategy map that was developed by Zack (1999) is important to identify the knowledge gap for the study sample population namely the government linked corporations in Malaysia. As such this study attempts to explore the knowledge strategies deployed by government link corporations (GLCs) in Malaysia via analyzing the linkages between the organizations' knowledge typology and knowledge value propositions.

1.1 Why the GLCs?

The GLCs are defined as companies that have a primary commercial objective and in which the Malaysian Government has a direct controlling stake (Putrajaya Committee, 2005). It is also important to recognize the role of the government linked investment companies (GLICs) as majority shareholders that allocate some or all of their funds to GLC investments. The GLCs have received much of the Malaysian governments' attention over the recent years due to the fact that these companies make up about 36 percent of the market capitalization of the Malaysian stock exchange (Bursa) which accounts for around 5 percent of the country's national output and employ some 400,000 people (PCG, 2005). As such, the overall boost in the performance of the GLCs would produce a powerful demonstration effect for the country's larger private sector, and develop new growth prospects in the country. In 2005, the government launched the Government Link Companies Transformation Programme (GLCT) and created the Putrajaya Commission on GLC High Performance (PCG) was to ensure the implementation of the various plans and objectives of the programme. The two core elements of the GLCT are the value creation plan (VCP) and ten initiatives of the GLCs. The VCP is a long-term plan of how the GLCs can achieve better performance. Consequently it is important to note that among the more important of the ten initiatives is to intensify performance management in all GLCs to ensure focus on the key business priorities-particularly value creation-and to retain knowledge of key experts (PCG, 2005).Under the transformation programme, each GLC is to develop their own VCP based upon their own capabilities and the GLICs' own assessment of the company. The PCG has also defined performance indicators as the complete set of measures derived from the company strategy which reflects the underlying value drivers of a company, and helps diagnose areas indicators for action. Essentially, the GLC's performance indicators and targets should be tightly linked to strategy and address all relevant aspects of value creation within each GLC - both at the company level and also at the lower levels of the organization such as divisions or business units.

2 KM Studies and Initiatives in Malaysia

Over the years, there have been various studies on KM in Malaysia which were conducted by academicians and practioners. The analysis came from various perspective mainly organizational, humanistic (HR), technological and strategic views across different industries such as logistics, natural resources; higher education institutions and medical fields. According to Earl (1997), there are four key factors in establishing knowledge as a strategic competence: knowledge system, network, knowledge worker and organizational learning. According to Syed-Ikhsan and Rowland (2004), all public and private organizations need to manage both tacit and explicit knowledge effectively especially in ensuring that the organization can take full advantage of the organizational knowledge and technologies. Based upon their case study (Syed et.al, 2004), it was found that KM was more advanced in the public sector as organizational knowledge is seen to be much more important in the public sector than the private sector mainly because the employee has long been identified as the key knowledge repository in the public sector. Similarly the study by Kalsom and Syed Noh (2006) discovered that the key performance indicators relating to improved efficiency of people and operations, improved products and services and increased responsiveness to customers' needs are among the top three indicators selected for the underlying factor on how to improve existing knowledge management system in local authorities in Malaysia. Nonetheless, aspects of organizational learning and the dimensions of leadership in the forms of commitment from top management are also crucial to ensure the implementation of KM practices such as knowledge acquisition and sharing (Toh, Ramayah & Jantan, 2006). Thus, drawing upon the various study on KM in Malaysia it can be argued that organizations need to develop competitive intelligence (CI) in order to achieve successfully KM initiatives. Lendrevies and Lindon (1999) defined CI as a strategic process that begins with determining the customer's needs, recognition of competitors strengths and weaknesses, assessment of likely activity and end with the identification of the company's own strength and weaknesses.

In Malaysia, there have been evidences to suggest that the GLCs in Malaysia have been practicing CI since the beginning of their operations (Samsudin and Abdul Kadir, 2006). The study found that the companies practiced CI strategies for decision-making and marketing strategies. Similarly, in a study of global competitiveness among multinationals in Hong Kong by Ng, Tuan and You (2008) discovered that innovations are critical to global competitiveness, with FDI playing a critical role. The findings of the research also implied that innovation factors such as governmental and scientific policies plays an important role in directing the future paths of business innovations. In this research KM is seen as the enabler for competitive advantage for the GLCs and certainly the GLC transformation programme greatly provides the support in developing a knowledge-based society. Hence it can be argued that KM is being practiced among GLCs as CI is a pattern of knowledge initiatives that can be reapplied with adaptation to solve problems (Callahan, 2004). Essentially the study by Badruddin and Shariff (2008), purported that knowledge takes on a number of roles: first, knowledge is, in itself, both a tangible and intangible resource (Hall, 1993); second, having access to knowledge supports any decision making about resources; third, a capability in KM enable those within the organization to leverage the most service from knowledge and other resources; and fourth, effective KM initiatives make contribution to innovation which in turn lead to better performance of Malaysian listed GLC. However, many organizations claim that they are practicing KM but not many were successful mainly due to the fact that there has been a missing link between KM implementation and performance outcomes (Chong, 2006). Therefore it is imperative to identify what are the pertinent value propositions and new business drivers to enable knowledge strategy of the GLCs to be effective and competitive in the long haul.

3 Research Methodology

The study is an exploratory case study that will be applying a multiple-case study approach. A multimethod approach was chosen in order to determine whether converging evidence (triangulation) might be obtained via by applying different methods. The research would include primarily inter alia, case studies and questionnaires. Yin (2003) purported that the former is used to gain insight into causal processes whereas the latter provided an indication of the prevalence of the phenomenon. The strength of case study stems from the three tenets of the qualitative method: describing, understanding, and explaining (Tellis, 1997). According to Bontis (2002), a qualitative approach was appropriate in the study of knowledge (intellectual capital) being a strategic resource as it has an intuitive appeal that provides opportunity for practitioners to collaborate with researchers in understanding the complex inner-workings of KM (Choo and Bontis, 2002). To date, exploratory case studies on KM by Syed et. al (2003), Raja Suzanna (2006), Kalsom & Syed Noh (2006) and Toh et.al (2006) have been proven to be feasible in studying KM in both public and private sectors in Malaysia. Whilst the earlier studies explored the relations of KM to organizational culture, people and innovation, this study would contribute to further proliferation in identifying new performance measurement in relation to KM and organizational performances. The study sample population was 52 GLCs in Malaysia. The unit of analysis is the organization per se. These companies range from providing end-products and trading of knowledge commodities in the form of ideas, experiences and consultancies. The GLCs were chosen for the study as they have been urged by the government to be exemplary of world class standards in organizational performance. More importantly with high investments in recruiting knowledge workers and accountability in implementing multi-million projects, the GLCs need to be better poised in deploying knowledge strategies to remain ahead against competitors. Gilabert, Cruz and Pedro (2009) argued that knowledge workers provide organizations with valuable cognitive skills and expertise. As such the notion of competitive positioning amongst the GLCs need to be identified to address the knowledge strategy gap across the industries. The two primary research instruments were a survey in order to establish a knowledge strategy profile and interviews with middle management whom was involved in company decision-makings. The survey was designed based on a 5-Likert point (Trochim, 2006) of "very high/ very low" scale to indicate the degree of agreement of the statements being asked. Whereas the interviews were based on unstructured questions in order to in order to create rapport and gain confidence from interviewees. The interviewees would include employees deemed as initiators or implementers of KM such as the head of departments in human resource, KM and senior management such as chief executive officers (CEO) and chief operating officers (COO). Verbatim feedback and comments of the interviewees will be included in the study. As such the interview is taken as a follow-up measure of the questionnaire in order to gain in-depth insight and experiences into implementing KM initiatives. **It is our believe that the interviews and developing a conceptual knowledge typology will establish a clearer understanding into the epistemology of knowledge diffusion and collaboration among organizations which will further accelerate Malaysia's knowledge society (i.e: Vision 2020).**

4 Findings

The purpose of this research was to explore the various KM strategies of the GLCs and analyzing organizational performance based on value propositions. Since the research is qualitative case study, a descriptive analysis will be used in order to profile the GLCs to the various categories of knowledge strategies and feedback from the interviews. A total of 28 responses were received within the time limits established. This represented a response rate of 53.8 percent. However, only 26 copies of the survey were exploitable and relevant to the organizational wide and unit KM initiative assessment. The two samples were rejected for two reasons. The main one was based on the fact that one of the organizations was not involved in KM whereas the second one was due to response incompleteness. The following section discusses the data analysis on the GLCs demographic profile this includes types of industry (organization) and types of respondents (knowledge worker). A total of 7 industries ranging from product to services sector participated in the study. From the survey, it was discovered that the majority of the GLCs that responded came from the construction sector 7 (26.9%); next being finance sector 6 (23.1%); automotive at 5 (19.2%); communications at 4 (15.4%); plantations 2 (7.7%) and petrochemicals and medical sector at 1 (%) respectively. Thus, it can be deduced that in Malaysia, the GLCs business orientation are mainly in the developments of heavy industries such as manufacturing and building infrastructures. However from Table 2 above, there is an emerging growth trends across the service sectors (38%) particularly from finance and communications companies. This progression is a positive indication that the companies are placing more efforts in value added provisions of their goods and services to enhance organizational performance.

4.1 Knowledge Strategy Positioning

The GLCs were in a unique position in that they are homogeneous as a state-owned enterprise corporation yet heterogeneous in their business dealings and interactions. In keeping the anonymity of the respondents each company (node) was identified as company 1-26. From the survey, a total of 9 companies were in the innovative position (A); 15 companies within the advanced position (B) and 2 belonged to the core position (C). Each position reflects the company's capability to effectively manage knowledge.

Based upon Figure 2 above, the majority of the GLCs (**57%**) had acquired the **advanced skill** (know-how) and knows why in leveraging on organizational learning and high absorptive capacity in knowledge diffusion. Among the skills acquired were the abilities in information extraction, real-time collaboration, semantic web and expertise management location. From the survey, it was also discovered that **35 percent** (node A) of the GLCs were discovered as being **innovative** companies illustrating high competencies in developing specific function knowledge, incorporating best practices programs and advocating communities of practice across its business relations. While **8 per cent** (node C) of the companies were in **core** stages whereby firms have cognitive knowledge or know-what of knowledge management. Core knowledge essential in KM but it will not be sufficient for successful take-up of knowledge strategy initiatives within the organization.

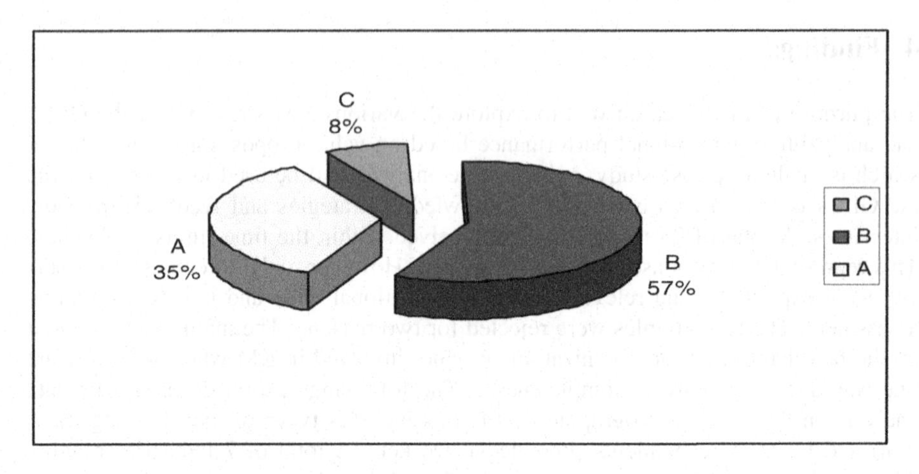

Fig. 2. The GLCs compositions of knowledge strategy

4.2 GLCs Value Propositions

According to McManus et.al (2004), value propositions have been as measures of strategic resources to enhance organizational performance. More importantly, value proposition would depend upon the "positioning" of the company (product or service) which it takes in the market (Brown, 2007). In other words, companies need to decide to be better than their competitors at one specific thing and that provides the differentiation piece of the value propositions.

Table 1. VP results of the GLCs

VP	Scores (%)	Rank
1(better decision –making)	50	1
2 (better customer-relationship)	50	2
3 (Faster response to key business issues	50	1
4 Improved employee skills	58	1
5 Improved productivity	62	1
6 (Increased profits)	62	1
7 (Increased innovation)	46	2
8 (sharing best practices)	46	1
9 (reduced costs)	38	1
10 (new ways of working)	38	1
11 (increased market share)	42	1
12 (create business opportunity)	38	1
13 (product development)	38	1
14 (staff attraction)	35	1
15 (increase share price)	38	0

However for the purposes of the study, the value propositions of the GLCs have been linked to the success indicators of their KM initiatives. In the survey, each of the value propositions (VP) were numerically coded and ranked successful as either to (2) high extent, (1) some extent and (0) little extent (see Table 1). Table 1 below illustrates the level of KM initiatives success based upon the VPs. Each VP was then ranked and their levels of success were translated into accumulated weighted averages in percentages (scores). It was found that the higher scores of VPs was derived from some extent of success of **improved productivity (VP 5) and increased profits (VP 6).** While the majority (50%) of the success indicators came from better decision-making (VP1) and faster response to key business issues (VP 3), both at which were achieved to some extent. While the GLCs achieved high success in developing better customer relationships (VP 2) at a 50% score. However from the survey, it was further discovered that the GLCs had achieved rather **lower levels of success in staff retention (VP 14) at a score of 35%** (see Table 3). Ramayah (2008) argued that the retention of knowledge workers has been an issue due to the decrease in the amount of time employees are given to acquire new knowledge and greater mobility in the workforce. Thus the GLCs and similar organizations need to find alternative and innovative ways to sustain knowledge workers interest and create micro-environments such as knowledge banks that would allow greater cross collaboration and exchange of knowledge within the company in order to tap and encourage the growth of the intellectual capacity of their knowledge workers. In interpreting the VP trend in Table 3, it can be argued that in the context of organizational performance, GLCs achieved better success in relation to knowledge exploitation in namely increased in productivity and profits level. However success in factors of knowledge exploration such as new ways of working, levels of innovation, staff attraction and new business opportunities were comparatively lower than the 50% average score. In summary, the GLCs need to rethink and improve on their positioning in order to enhance organizational performance and achieve value propositions which are more knowledge-based to be sustainable in the long-haul.

4.3 Feedback from Interviews

In the study, a total of 10 interviews were scheduled to be conducted across the 3 knowledge strategy categories of the GLCs namely the innovative (4), advanced (4) and core (2). The general guidelines and topics to be discussed were also sent via email or facsimile to the offices of the senior management. One of the criteria used to select interviewees was that they have to be in position or reviewing, deciding and endorsing organizational policies to some extent. At some of the sessions, there would be two senior management personnel attending. However for the purposes of the study, the target populations are the organizations (GLCs) themselves, hence the interviewees' opinion would be deemed as organizational perception as all the interviewees had adequate experience and involvement which contributed towards organizational performance. The interview sessions will be referred to as case A through to case D in order to facilitate discussion and ensure anonymity. Both cases A and B were from the innovative category whilst cases C and D were from the advanced categories of knowledge strategy. Case A represents a GLC from the petrochemical industry, case B is a GLC from the telecommunications industry, case C represents a

GLC from the automotive industry while case D is from the plantations industry. In general, the respondents from all four cases claimed that their organizations were aspiring to be learning organizations with some at preliminary while others being more advanced with implementing KM initiatives. Nonetheless, some of the interviewees would regard their organizations as moderate performers as some of the KM initiatives implemented had led to an increase in employees' professional competency and increase in profits. Their assertions were verified through their company reports and synomous with the survey findings on value propositions.

4.3.1 How Was KM Being Developed in the GLCs?

The opening question was broad and general question designed to establish rapport and asked the interviewees to share their view or experiences on developing KM initiatives. From the responses, several common themes emerged in the following ways. Firstly, there was a general observation by the interviewees that a learning organization promotes a knowledge sharing culture which enables sharing information and experience in a more open and flexible manner. Interviewee A commented that: "Our company went so far into developing a 'Professional Learning Model' centered around the key value flows or exchanges concerning i) goods and services, ii) knowledge acquired and iii) intangible benefits". Then each level of competency will be linked to the KPI's according to the performance indicator guidelines as provided in the GLCT's Blue Book. Secondly, organizational competencies are centered on knowledge capabilities (example: soft skills and experiences) and resource capacities (such as: number of in-house experts and work processes). Thirdly, KM initiatives were developed on incremental stages across each business units or department. "The idea is to try and loop our learning experiences about new problems and solutions back to the data and knowledge bases so they will be available to all including the support staff in order to be more effective": lamented interviewees B. Finally, there is a shared concern by the respondents that the KM initiatives should be closing the current and future knowledge gaps which is required to transform the GLCs into multinationals in a global and knowledge-based economy. As expressed by interviewees D: "Our company aims to acquire all the ten initiatives and a bonus would be if we achieve it earlier than 2015". However some of the interviewees were skeptical in the light of the recent economic changes and turbulences worldwide such as in the United States would delay the ten year period of the transformation target of the GLCT that would affect plans of international investment and general operational expansion. Drawing from the common themes of the interview responses, it can be said that as that there needs to be match or dynamic link or alignment between organizational competency (knowledge assets) and strategic knowledge management for organizations in order to enhance competitive advantage.

4.3.2 How to Link Value Propositions with Knowledge Strategy?

Traditionally organizational performance for many large corporations lies in bottom-line ratios on financial profits or loses. However this form of measure has changed with the inception of the KPIs by the GLCs to measure new value added propositions linked to various factors of intangible assets such as the level of customer satisfaction, product or service quality, business cycle-time and even market positioning. Hence as expressed by interviewee A, " Our company has developed a knowledge map known as the 'Heat Map' that measures performance across each business unit according to several categories namely; 'below par', 'base', 'leading' and 'pace-setter'. Currently our company is performing at par to other competitors in the resource industry overseas". Interviewee C lamented that " the KPIs are helpful as it provides a well-balanced set of performance measure and they experienced benefits such as increased revenue, cost revenue due to less travelling and faster turnaround time". Although there was a general consensus on the usefulness of the KPIs, questions arises when the managers were trying to determine the level or value of knowledge (i.e: actual value) being spent or incorporated in decision-making process or the creation of end products. As said by interviewee D that, "currently we use the balanced scorecard to measure organizational performance but it is difficult to match the appropriate KPIs to the capabilities of each department". In brief, for the managers, the problem lies in finding out which indicator(s) really matters and effects performance.

4.3.3 What Are the Challenges of KM Initiatives?

Ultimately the role of the CKO is instilling knowledge collaboration across the different business units and requires the support of a 'KM evangelist' group to champion the cause and develops the initiatives across the organization (Debowski, 2006). A high percentage (50%) of the interviewees agreed to this job designation of KM. However interviewees B and C expressed that the challenges lie in cultivating a knowledge culture as the Asian business portfolio approach to decision-making have limited boundaries which caused fewer enterprise performance measures and less opportunity for knowledge sharing and cross collaboration within the company. As commented by interviewees D, "the final decision still lies with top management but without their support, KM would not be feasible". This view illustrates the contradicting roles of top management in making KM initiatives a success whereby it does facilitate KM initiatives within the organization but it would not necessarily be effective if there was a disconnect on knowledge collaboration between the middle and top management. Nonaka and Takeuchi (1995) attributed that Japanese company such as Honda, Canon and Matsushita's ability to tap into the intellectual capital of their workers to create and expand knowledge via a "middle-up-down" approach to management. As commented by case study A, "if we do not link knowledge capabilities to performance appraisal, then a problem of staff retention would arise". This view can be substantiated that the lower score of value proposition for the GLCs was staff attraction. Thus the middle managers play the key role in engaging the cognitive capabilities between the ideals of top management and the chaotic realities of the frontline. In line with this view, interviewee A commented that, "the challenge for organizations is to remove the silo-thinking of employees at large and matching skills and competencies to problem-solving". As such case study A had incorporated an a "KM Away-Day"- an event in which the KM evangelist group and peers alike within the company commune and conducts brainstorm sessions to analyze and identify KM needs and milestones achieved across different business units. These comments clearly summarizes that collaborative effort is needed to ensure the success and implementation of KM initiatives and offers new managerial implications to organizations to adopt a middle-up-down rather than top-down KM approach as the former would provide better linkages of diverse assets into unique capabilities and leverage in pursuit of new opportunities. In other words, the middle-up-down approach would allow for greater flexibility to the organization in which resources would be managed with greater responsiveness (Skyrme, 1998).

5 Conclusions

Based upon the survey conducted and data analysis, it has been found that knowledge and resource dynamics is needed and is existent within the GLCs by their ability to leverage on internal knowledge structure and competence initiatives. Figure 3 below illustrates the various levels of GLCs knowledge strategy positioning between core, advanced and innovative stages. Each numbering represents the 26 GLCs. The knowledge strategy positioning indicates the organization's capabilities in managing knowledge assets in order to achieve competitive advantage. In the study, the GLCs have shown predominance within the advanced stages, thus companies are leveraging on know-how and know-why cognitive competencies. According to the resource based view (RBV) (Prahalad et.al, 1990), on achieving competitive advantage via combining heterogonous resources and collaborative work, it was discovered that the GLCs namely from the construction industry were able to tap into their knowledge capabilities and resource to provide new business opportunities to different industry sectors. By drawing upon the results of the survey and further synergizing into Nonaka's SECI (1995) model and Hansen and Zack (1999) knowledge typology, a knowledge mapping of the GLCS can be derived as shown in Figure 3. From the RBV context, Figure 5 below illustrates that the spiral organizational learning or knowledge movement for the GLCs began developing knowledge capabilities from an explicit knowledge stand-point in which levels of know-what was derived primarily from policies and work manuals. Then the learning process moved outwards to levels of know-how (explicit) when ideas are captured in knowledge banks, then flowed onto tacit knowledge levels of know-why when organizational goals are linked to performance appraisals then ultimately care-why or motivation when employees share a cleared purpose of the KM initiatives. Thus in contrast with the Japanese companies, the local companies are inclined towards an outward approach to organizational learning which begins with an explicit to tacit experiences. This view can be further substantiated when it was discovered in the study that the majority of the GLCs were in advanced stages whereby according to Quinn et.al (1996), advanced skill or know-how refers to the translation of book learning into effective execution. Thus more efforts need to be placed upon knowledge exploration.

Based upon the knowledge typology model (Figure 5), it can be concluded that the GLCs realize the importance of KM which further substantiates the main proposition of the study that high levels of knowledge exploration dynamism of knowledge assets and resources (high levels of know-why and care-why) is needed to achieve competitive advantage. However from the study, it was discovered that a main challenges to knowledge strategy were in discovering ways of embedding knowledge capture in business process and developing interorganizational relations (example: increasing the rate of knowledge absorption of employees). One approach is perhaps to take a less top down approach to imposing rigid KM initiatives but rather in favor of facilitating the idiosyncratic learning requirements of individuals via a "middle-up-down" approach. This approach would allow for an increase of dynamism or spiral flow across the four cognitive thought dimensions within the organization. By drawing upon the background study and feedback from the GLC executives, the GLCT programme is deemed as the impetus or catalyst in developing a KM framework which is better attuned or aligned with the needs of the country to foster and nurture the growth of

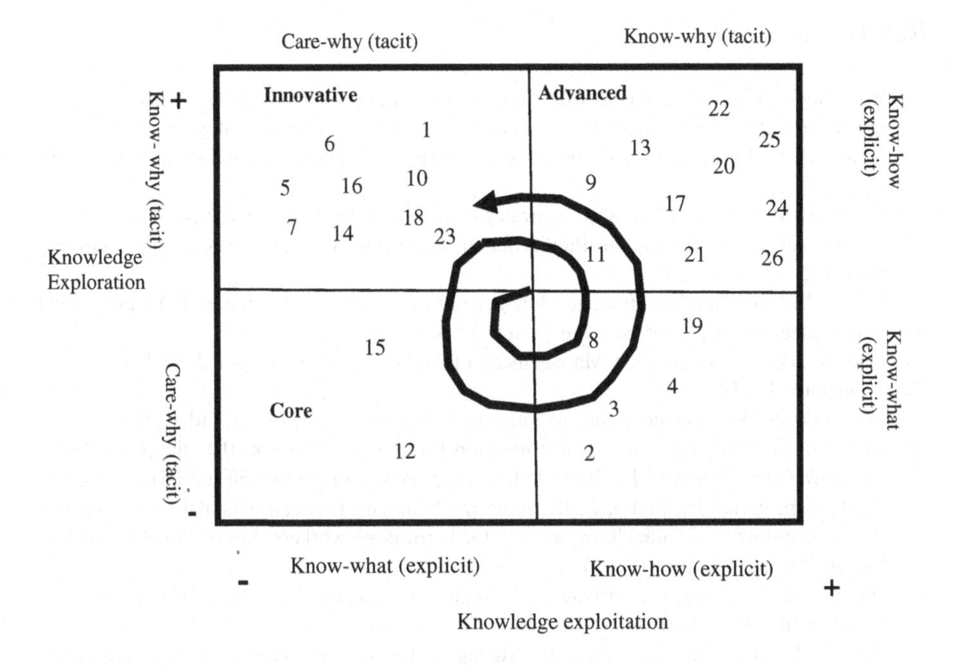

Fig. 3. Knowledge typology/ mapping of the GLCs

more local multinational corporations based upon intellectual capabilities and competencies. Thus the continued support from the government is significant to gain critical mass of KM practices for organizations and corporations alike in Malaysia. Knowledge is thus the fundamental basis of competition and organizational performance.

Competing successfully on knowledge requires either aligning strategy to what the organization knows, or developing the necessary knowledge and capabilities needed to support a desired strategy. Currently the GLCs are in a transition or advanced stage whereby they are focusing more value added knowledge and exploring new capabilities in order to gain new opportunities whilst sustaining competitive advantage. Ultimately a company needs to decide must determine whether its efforts are best focused on longer-term knowledge exploration (example: joint ventures and alliances), shorter-term exploitation (example: assessing strategic gaps), or both and balancing its knowledge processing resources and efforts in anticipation of global markets. **In the scenario of knowledge typology of the GLCs in Malaysia, the business community at large may be able understand the epistemology of organizational knowledge and leverage on the typologies of knowledge collaboration and exchange among companies in the Asia Pacific region. Pertaining to the embodiment of creating a knowledge society vision, this research has sought to highlight the significance of organizations as a source of knowledge and value creating unit.** In conclusion, the findings of this study can further proliferate other areas of potential research of KM-how organizational culture can affect KM initiatives, considering the diversity of knowledge workers and the scope of knowledge work involved.

References

1. Blue Book, Version 2. Intensifying Performance Management: Guiding Principle, Putrajaya Committee on GLC High Performance (PCG), Kuala Lumpur, Malaysia (2005)
2. Bontis, N.: Managing organizational knowledge by diagnosing intellectual capital. In: C.W. (2002)
3. Callahan, S.: Crafting a Knowledge Strategy. Anecdote Pty. Ltd., Australia (2004)
4. Chong, Pandya: KM issues in Public Sector. Electronic Journal of Knowledge Management 1(2), 25–33 (2003)
5. Choo, C.W., Bontis, N.: Strategic Management of Intellectual Capital and Organizational Knowledge. Oxford University Press, New York (2002)
6. Chowdhury, N.: Knowledge Management in Malaysia. In: KM Asia 2007 Conference, Singapore (2007)
7. Debowski, S.: Knowledge Management. John Wiley & Sons/Australia, Ltd. (2006)
8. Government-Link Companies Transformation Programme, "Section II-Policy Guideline". Putrajaya Committee on GLC High Performance (PCG), Kuala Lumpur (2005)
9. Gilabert, E., Cruz, D., Pedro, J.: Breaking the boundary between personal- and work-life skills: parenting as a valuable experience for knowledge workers. International Journal of Knowledge and Learning (2009)
10. Hansen, M., Oetinger, B.: Introducing T-Shaped Managers: KM's Next Generation. Harvard Business Review, 106–116 (March 2001)
11. Housel, T., Bell, A.: Measuring and Managing Knowledge. McGraw-Hill International Edition, New York (2001)
12. Khalil, T.: Management of Technology-The Key to Competitiveness and Wealth Creation. McGraw-Hill International Editions, New York (2000)
13. Malhotra, Y.: Is Knowledge The Ultimate Competitive Advantage? Business Management Asia (2003)
14. Mcmanus, D.J., Fredericksen, D.K., Wilson, L.T., Snyder, C.A.: Business Value of Knowledge Management: Return on Investment of Knowledge Retention Projects (2004)
15. Nonaka, I., Takeuchi, H.: The Knowledge-Creating Company. Oxford University Press, Oxford (1995)
16. Ng, L., Tuan, C., You, A.S.: Gaining metropolis-city competitiveness through innovations: the opinions of multinationals (2001–2006). International Journal of Technology Enhanced Learning 4(6), 553–566 (2008)
17. Prusak, L.: Where did knowledge management come from? IBM Systems Jounral 40(4), 1002–1007 (2000)
18. Prahalad, C.K., Hamel, G.: The Core Competence of the Corporation. Harvard Business Review 68(3), 79–91 (1990)
19. Quinn, J.B., Anderson, P., Finkelstein, S.: Managing Professional Intellect: Making the Most of the Best. Harvard Business Review (March-April 1996)
20. Snowden, D.: Three Metaphors, two stories and a picture - how to build common understanding in Knowledge Management Programmes, Knowledge Management Review. Melcrum Publishing (March/April 1999)
21. Skyrme, D.: Commercializing Organizational Knowledge. ArkGroup Asia.Com (1998)
22. Syed-Ikhsan, S.O.S.b., Rowland, F.: Benchmarking knowledge management in a public organisation in Malaysia. Loughborough University, UK (2003)
23. Tellis, W.: Introduction to Case Study. The Qualitative Report 3(2) (July 1997)

24. Trochim, W.M.K.: Research Methods-Knowledge Base: Likert Scales. Cornell University, USA (2006)
25. Wigg, K.M.: What future KM users may expect. Journal of KM 3(2) (1999)
26. Yin, R.K.: Case study research: Design and Methods, 3rd edn. Sage Publications, Thousand Oaks (2003)
27. Ahmad, Z.A.: Chief Academic Learners as Learners: Adult Learning Patterns within an Organizational Context. Northern Illionois University (1994)
28. Zack, M.H.: Developing a knowledge strategy. California Management Review 41(3), 125–145 (1999)
29. Zack, M.H.: Developing a knowledge strategy- epilogue (2001), http://web.cba.neu.edu/mzack/articles/kstrat2/kstrat2.htm

A Generic Core Knowledge Management Process: Locating Crucial Knowledge

Michel Grundstein

MG Conseil, 4 rue Anquetil, 94130 Nogent sur Marne, France
Paris Dauphine University, LAMSADE, 75016 Paris, France
mgrundstein@mgconseil.fr

Abstract. In the Knowledge Society, the enterprise increasingly develops its activities in a planetary space. The hierarchical Enterprise locked up on its local borders is transformed into an Extended Enterprise without borders, opened and adaptable. In this context, the actors are confronted with new situations that increase their initiatives and responsibilities, whatever their roles and their hierarchical positions are. For their missions, through the Enterprise's Information and Knowledge System, beyond relevant information, they must access to knowledge and individual and collective skills widely distributed in the planetary space of their organization. In such context, the challenge is to well identify and locate "crucial knowledge" that is a set of knowledge, which is essential for the enterprise. This article presents GAMETH®, a specific approach that fits with the "Locating Core KM Process" that constitutes one of the operating elements of the Model for General Knowledge Management within the Enterprise (MGKME).

Keywords: Activity analysis, Crucial Knowledge, GAMETH®, Locating Core KM Process, Interpretative Framework, MGKME, Sensitive process.

1 Introduction

« What makes knowledge valuable to organizations is ultimately to make better the decisions and actions taken on the basis of knowledge [1]. » In the Knowledge Society that is taking place, the enterprise increasingly develops its activities in a planetary space. The hierarchical Enterprise locked up on its local borders is transformed into an Extended Enterprise without borders, opened and adaptable. In such conditions, the range of autonomy of action is increasing for more and more individuals, whatever are their hierarchical levels and roles: they are placed in situations that need to take decisions. They become decision-makers who use and produce more and more knowledge as a basis for their efficiency. By this very fact, Extended Enterprises are more and more concerned with Knowledge Management as a key factor for improving their decision making processes. Notably, they need to locate and identify essential knowledge to capitalize on. Thomas A. Stewart pointed out this issue as early as 1991 [2]. Since that time, Companies launched numerous KM initiatives. Later on, the same author notices the fatal effect of KM initiatives that were not subjected to advisability studies. Stewart states [3] "*One flaw in knowledge management is that it*

M.D. Lytras et al. (Eds.): WSKS 2009, CCIS 49, pp. 248–257, 2009.

often neglects to ask what knowledge to manage and to what end (p.117)." This raises the problem of identifying which knowledge justifies a KM initiative. To deal with this issue, we developed a Global Analysis Methodology so-called GAMETH®, with the aim of identifying and locating "crucial knowledge".

In this article, after having set out the background theories and assumptions, we describe GAMETH®. Finally, we present lessons learned from two case studies.

2 Background Theories and Assumptions

In this section, after having describe the concept of "crucial knowledge", we introduce the basic foundations of GAMETH®.

2.1 The Concept of "Crucial Knowledge"

Crucial knowledge supplies essential resources that are used by an enterprise's support and value-adding processes. Support and Value-adding processes derive from the value chain described by Porter [4] who identifies nine value-adding activities that he classifies into two main categories. The "*primary activities*" are: 1) in-bound logistics, 2) operations, 3) out-bound logistics, 4) marketing & sales, and 5) Services. The "*support activities*" are: 1) business infrastructure, 2) human resource management, 3) technological development, and 4) supplies.

Support and Value-adding processes represent the organizational context for which knowledge is essential factors of performance. It is in this context that is implanted a KM initiative. We should consider KM activities in order to identify knowledge that is essential factor to enable support and value-adding processes to achieve their goals efficiently. This knowledge will be crucial depending of a multi criteria analysis [5]. Notably, knowledge will be "crucial knowledge" depending of its degree of vulnerability, and its impact on the objectives and the durability of the firm.

For example, such is the case for knowledge characterized as follow: knowledge is rare, specific and unique, imperfectly diffused, non- substitutable, difficult to pass down; the cost to develop or purchase that knowledge is very high and the period required getting it is long; and possible loss of that knowledge can cause an unacceptable risk for the strategy and life durability of the firm, by weakening its core competencies, endangering the performances of its business units and reducing its market share. Crucial knowledge can be tacit or explicit as defined by Polanyi [6].

2.2 The GAMETH®'s Foundations

A Brief History of GAMETH®

GAMETH® is one of the results of the CORPUS project initiated and led from 1991 to 1995 into the Framatome Group[1]. The scope of CORPUS was to elaborate a set of concepts, methods and tools aimed at contributing to capitalizing on company's knowledge assets [7].

[1] French Nuclear Power Plant Company, first transformed into Framatome ANP, then integrated into AREVA Group in September 2001.

At the beginning, CORPUS deliverable was a complementary approach to manage the advisability phase of an information project with the aim of integrating knowledge capitalization functionalities into the specifications. As an example, for a quotation improvement project, this approach leads to highlighting a problem that we had decided to call "*knowledge traceability*", that is a generic problem based on the following needs: the need to refer to earlier facts, the need to refer to analogous cases, the need to ask questions about earlier choices, and the need to rely on experience feedback. Beyond a system that helps to prepare quotations, the solution implemented the functionality necessary for *"knowledge traceability"* (pp.144-145).

Later on, we have considered that this approach could be generalized, and since 1997, it has been consolidated as a Global Analysis Methodology, the so-called GAMETH®.

The postulates

GAMETH® rests on three postulates described hereafter.

Knowledge is not an object

Knowledge exists in the interaction between an interpretative Framework (incorporated within the head of an individual, or embedded into an artifact), and data. This postulate comes from the assumption emphasized by Tsuchiya [8] concerning knowledge creation ability. He emphases how organizational knowledge is created through dialogue, and highlighted how "commensurability" of the interpretative frameworks of the organization's members is indispensable for an organization to create organizational knowledge for decision and action. Here, commensurability is the common space of the interpretative frameworks (e.g. cognitive models or mental models) of each member. Tsuchiya states, "*It is important to clearly distinguish between sharing information and sharing knowledge. Information becomes knowledge only when it is sense-read through the interpretative framework of the receiver. Any information inconsistent with his interpretative framework is not perceived in most cases. Therefore, to share individual's knowledge, members' interpretative frameworks commensurability is indispensable.*" (p. 89).

In other words, knowledge that we use to understand a situation, solve a problem and act, results from the sense given, through our interpretative frameworks, to data that we perceive among the information transmitted to us. Consequently, explicit knowledge, codified, stored, and processed in digital information system, is not more than information. We call it "Information source of knowledge for someone". We consider this information as knowledge when members having a large commensurability of their interpretative frameworks commonly understand it. For example, such is the case for members having the same technical or scientific education, or members having the same business culture. In these cases, codified knowledge makes the same sense for each member.

Company's knowledge includes two main categories of knowledge

Within a company, knowledge consists of two main categories. On the one hand, explicit knowledge includes all tangible elements (we call it "know-how") and, on the other hand, tacit knowledge [6], includes intangible elements (we call it "skills"). The tangible elements take the shape of formalized and codified knowledge in a physical format (databases, procedures, plans, models, algorithms, and analysis and synthesis documents), or are embedded in automated management systems, in conception and production systems, and in products. The intangible elements are inherent to the individuals who bear them, either as collective knowledge - the "routines" that are non-written individual or collective action procedures [9], or personal knowledge (skills, crafts, "job secrets", historical and contextual knowledge of environment, clients, competitors, technologies, and socio-economic factors).

Knowledge is linked to the action
From a business perspective, knowledge is essential for the functioning of support, and value-adding processes [4]. Activities contributing to these processes use and create knowledge. Thus, the actions finalize the company's knowledge. This point takes into account the context and the situation, which allow using and creating knowledge. In particular, we must analyze the role of the decision-makers involved with these activities in order to achieve the company's missions. Therefore, knowledge is linked to their decisions, their actions, and their relations with the surrounding systems (people and artifacts).

3 The GAMETH® Description

GAMETH® fits with the "Locating Core KM Process" that constitutes one of the operating elements of the Model for General Knowledge Management within the Enterprise (MGKME) developed by Grundstein [10], [11], [12]. The generic KM processes answer the problem of capitalizing on company's knowledge defined in the following way [7]: "*Capitalizing on company's knowledge means considering certain knowledge used and produced by the company as a storehouse of riches and drawing from these riches interest that contributes to increasing the company's capital*" (p. 141).

Several problems co-exist. They are recurring problems for a company. These problems constitute a general problematic that has been organized in five categories. Each of these categories contains sub-processes aimed to contribute a solution to the set of overall problems (see Figure 1).

The Locating Core KM Process deals with the location of Crucial Knowledge, that is, Knowledge (explicit or tacit) that is essential for decision-making processes and for the progress of the support and value-adding processes [4]. It is necessary to identify it, to locate this knowledge, to characterize it, to make cartographies of it, to estimate its economic value, and to classify it. Thus, GAMETH® provides the elements that lead to identifying the problems, clarifying the needs for knowledge, identifying and locating potential crucial knowledge, specifying the value-based assessment of this knowledge, and finally, determining "crucial knowledge".

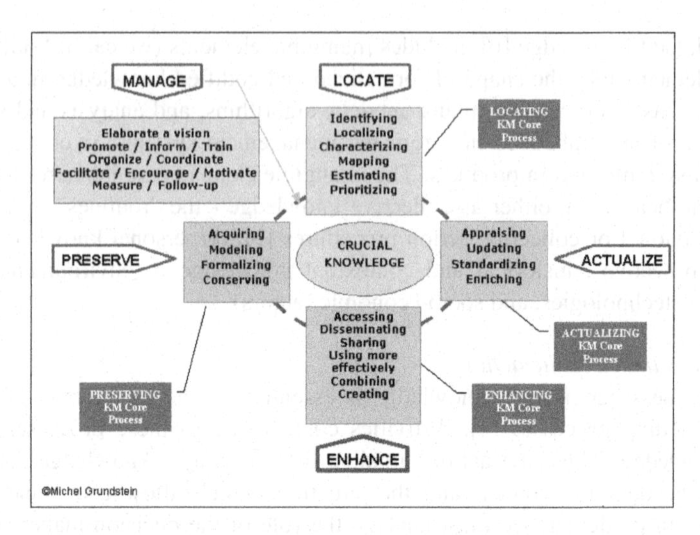

Fig. 1. The Generic Core KM Processes

The approach consists of three main phases conformed to three guiding principles described in the following section.

3.1 The GAMETH® 's Main Phases

In short, GAMETH® approach consists of three main phases gathering the following steps:

Phase 1: Definition Study
The first phase, called "Definition study", aims at constructing the problem space. During this phase we specify the project context, define the domain and the limits of the intervention and determine the process, which is to be subjected to an in-depth analysis. The phase includes four steps: (i) Defining the domain and specifying the context of the operation; (ii) Delimiting operational processes, production processes and organizational entities (operational units, functional services, partners, clients) dealing with the production of goods and services; (iii) Modeling the domain of intervention (functional and structural models of the organizational entities, communication network model); and (iv) Determining sensitive processes.

Phase 2: Identification of the Crucial Knowledge
The second phase, called "Identification of the Crucial Knowledge", aims at distinguishing the problems that weaken the critical activities, i.e. the activities that might endanger the sensitive processes. The phase includes five steps: (i) Modeling sensitive processes; (ii) Assessing the risks to which the sensitive processes are exposed, and determining the critical activities for these processes; (iii) Identifying the constraints and malfunctions that weigh down on these activities; (iv) Distinguishing the determining problems; (v) Locating and characterizing the crucial knowledge.

Phase 3: Determination of the Axis of a Knowledge Management Initiative
The third phase, entitled "Determination of the Axis of a Knowledge Management Initiative", is intended to define, localize and characterize the knowledge to be capitalized. It aims at answering the question: Who utilizes which knowledge during what phase in the sensitive process cycle? The phase includes five steps: (i) Clarifying the knowledge requirements for the resolution of the determining problems; (ii) Localizing and characterizing this knowledge; (iii) Assessing the value of this knowledge and determining the crucial knowledge; (iv) Outlining a project for the improvement of the decision-making and value-adding processes; (v) Determining the axes of a knowledge management initiative.

The approach is finalized by the company's strategic orientation, and the deliverable is an Advisability Analysis Report.

A typical schedule is presented on the figure 2. In this figure, the term "stakeholder", as defined by Roy and Bouyssou [5], refers to *"individuals or groups of individuals who, because of their value system, directly or indirectly influence the decisions, either at first degree because of their intervention or at second degree by the manner in which they use the action of other individuals"*.

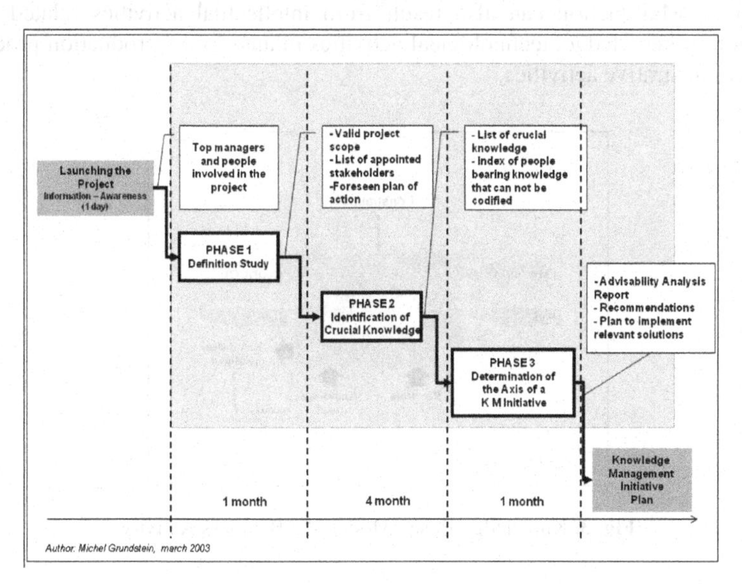

Fig. 2. Typical GAMETH Project Schedule

3.2 The GAMETH®'s Guiding Principles

GAMETH® brings three main principles with respect to the modeling of the company, the knowledge analysis method and the process modeling approach.

The Modeling of the Company
From the point of view of knowledge that she utilizes and creates, Company can be represented as a set of activities that make up the processes that are necessary to achieve the company's mission. The SADT method [13] inspires the activity model,

presented in Figure 3. However, there are two differences. First, it distinguishes two inputs: (i) the material transformed into a product by the activity; (ii) the data that inform on the status of this material and this product. Second, it includes the notions of produced knowledge and used knowledge.

Each activity focuses on the objective to reach. It transforms material into a product. It receives the data required for its well functioning and supplies the data for the functioning of other activities. It consumes financial resources and techniques. The activities use and produce specific knowledge (expertise and skills). They are subjected to constraints. These constraints can either be external to the activity (imposed conditions such as costs, time, quality, specifications to be respected, technical financial resources, human resources, and uncertainties related to delivery and the quality of the input materials), or internal to the activity, resulting from the limits of the admissible scope of the activity (zone of autonomy). The activities can lead to malfunction, that is the gap between the expected and the obtained results. Malfunction is a symptom of either internal sources (directives, procedures, processes, particular action logic that may be maladapted to the situation), or external sources (inadequate materials, unreliable data, badly adapted resources and insufficient or erroneous knowledge). Malfunction can also result from intellectual activities related to the production of knowledge, technological activities related to the production process or purely administrative activities.

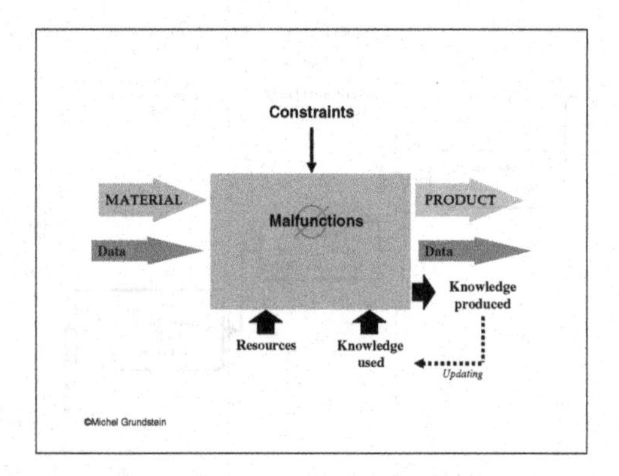

Fig. 3. Knowledge-Based Model of a Business Activity

The Knowledge Analysis Method

The knowledge analysis method focuses on the so-called "sensitive processes". A sensitive process is a process, which represents the important issues, which are collectively acknowledged. These issues concern weaknesses in the process presenting a risk of not being able to meet the cost or time objectives, the required quality for the goods or services produced, obstacles to get over, challenges difficult to reach, goods and services that are strategic assets of the company. Creativity sessions, built upon the knowledge held by the responsible persons within the intervention domain, engender determining "sensitive processes". We describe the analysis method hereafter.

The problems and constraints can weaken the activities and may even endanger the process to which they are supposed to contribute. Therefore, the sensitive processes are submitted to a risk assessment. This assessment helps to determine the "critical activities". The problems related to these activities are called "determining problems". The relaxation of organizational constraints can lead to a rapid removal of these problems. The identification of the remaining determining problems leads to the identification of the knowledge that is required for their resolution. This knowledge can be qualified as "crucial knowledge" depending on its actual value.

Thus, the GAMETH® Framework does not involve a strategic analysis of the business objectives. It rather suggests focusing on the analysis of the knowledge that is relevant for the activities and insures efficiency of processes in concordance with the business missions.

The Process Modeling Approach
Besides the advantages of the process modeling approach highlighted by Kruchten [14], in the GAMETH® Framework the process modeling approach follows constructivist logic. In order to distinguish potential crucial knowledge, the process modeling approach bases on the observation that processes, formalized through numerous procedures that prescribe action rules and operational modes, often differ from how these processes are perceived in actual world. Additionally, we observe that actors are often well aware of their part of the process, but ignorant with respect to the overall process in which this part has to operate.

The process modeling approach comprises formalization, with the stakeholders, of objectives relative to sensitive processes. The approach consists of the co-construction of the process' representation, utilizing the partial knowledge that stakeholders have acquired through the activities that they are supposed to perform. This approach is using a computer and a video projector. The representation of the process is outlined according to the progress of the work, and is shown on a screen in real time. Throughout the analysis, the problems encountered provide the possibilities for the identification of information and communication relations between actors, not recorded in documents, and the identification of the knowledge required for the resolution of these problems. The advantage of this constructivist approach is that it stimulates collective engagement, which is primordial for a successful outcome of a knowledge management initiative.

4 Application Examples

We applied GAMETH® in different contexts, following the typical schedule presented on figure 2. Hereafter we describe some case studies, and lessons we learned.

The first example comes from the French Institute of Petroleum (IFP). The second example comes from the French National Center for Scientific Research (CNRS) Engineering Sciences Department (SPI).

The IFP has applied the GAMETH® Framework in order to set up a pragmatic approach to the capitalization of knowledge within the context of a research and development project. The initiative has been taken by the Quality Direction and was carried out as part of a five-month internship within a M.Sc. program (Research Master) ending in June 2002. The objective of the research was to facilitate the identification of potential crucial knowledge through a selection of the documents, which would contain possibly valuable future assets as part of the final steps of a project. The application at the IFP showed the compatibility of the GAMETH® approach with the ISO 9004 (December 2000) recommendations. Furthermore, the alignment of the knowledge management discourse with the quality management discourse has turned out to be a key factor in the success of the project.

Within the French National Center for Scientific Research (CNRS), the SPI department intended to launch a project in order to capitalize its internal information as well as the information produced by its attached research laboratories. The GAMETH® approach has been applied during a M.Sc research internship (Master research) ending in June 2003. The objective of the study was to facilitate the decision-making process through the identification of potential crucial knowledge (both tangible and tacit) required for the well functioning of a sensitive process within the SPI: the recruitment of engineers and technical personnel. The main objective was to identify the critical activities and knowledge to be capitalized within the process. The application at the CNRS showed that, from a methodological viewpoint, the GAMETH® approach should be limited to one single process and involve at most 10 individual actors in order to be feasible within a six-month period.

Here are some lessons learned:

- The essential conditions for a successful implementation are: (i) include an initiation phase to familiarize the actors with the concepts of knowledge management; (ii) insure the involvement of (an important part of) the management, which is normal in any quality assessment approach; (iii) make sure that the GAMETH® approach is implemented by an individual familiar with the Enterprise.
- The analysis of the results leads to a reasoned and shared vision of the sensitive process by the stakeholders of this process. This emphasizes also the impact of the process being analyzed on different levels of the organizational activities. Several problems result in fact from the interrelation of processes.

5 Conclusions and Future Trends

In the Knowledge Society, Enterprises are concerned with Knowledge Management (KM) as a key factor for improving their efficiency and competitiveness. In this article, referring to the Model for General Knowledge Management within the Enterprise (MGKME), we focused on one of the operating elements suggested by this model: the “Locating Core KM Process” involved by the problem of capitalizing on company’s knowledge. To deal with this issue, we presented the Global Analysis Methodology so-called GAMETH®, and briefly described two case studies.

The case studies showed the relevance of GAMETH® leading to the construction of a “problem space”, to the identification of stakeholders, and to the clarification of knowledge requirements. Because of the constructivist approach logic, the involved actors contribute to the clarification of the problem and the elaboration of the solution. The approach crystallizes a learning process marked by the engagement of the stakeholders to learn together to articulate the problems and to develop the solutions. In this way, the approach acts as a catalyst of change.

However, the applications of GAMETH® are limited to projects that involve no more than 10 to 20 stakeholders. To overcome these limits, Inès Saad in her thesis

[15], presented a generalized method to make GAMETH® usable for any complex project. This method is based on "decision support system" theories. It was conceived and validated in the PSA Peugeot Citroen French Company.

In the future, we will carry on new applications of GAMETH®, and extending its field to organizations and complex projects following the way opened by Ines Saad's thesis.

References

1. Davenport, T.H., Prusak, L.: Working Knowledge. How Organizations Manage What They Know. Harvard Business School Press, Boston (1998)
2. Stewart, T.A.: Brain Power. How Intellectual Capital Is Becoming America's Most Valuable Asset. Fortune (June 1991)
3. Stewart, T.A.: The Wealth of Knowledge: Intellectual Capital and the 21tst century Organization. Currency Doubleday, New York (2002)
4. Porter, M.E.: Competitive Advantage: Creating and Substaining Superior Performance. The Free Press, New York (1985)
5. Roy, B., Bouyssou, D.: Aide multicritères à la décision: méthode et cas, Paris, France (1993)
6. Polanyi, M.: The Tacit Dimension. Routledge & Kegan Paul Ltd., London (1966)
7. Grundstein, M.: "CORPUS", an Approach to Capitalizing Company Knowledge. In: Ein-Dor, P. (ed.) Artificial Intelligence in Economics and Management, pp. 139–152. Kluwer Academic Publisher, Tel-Aviv (1996)
8. Tsuchiya, S.: Improving Knowledge Creation Ability through Organizational Learning. In: ISMICK 1993 Proceedings, International Symposium on the Management of Industrial and Corporate Knowledge, pp. 87–95. University of Compiègne, Compiègne (1993)
9. Nelson, R.R., Winter, S.G.: An Evolutionary Theory of Economic Change. Harvard University Press, Cambridge (1982)
10. Grundstein, M.: MGKME: A Model for Global Knowledge Management within the Enterprise. In: Proceedings of 2nd International Conference on Intellectual Capital, Knowledge Management and Organizational Learning, ICICKM 2005, pp. 201–211. ACL, American University in Dubai (2005)
11. Grundstein, M.: Knowledge Workers as an Integral Component in Global Information System Design. In: Law, W. (ed.) Information Resources Management: Global Challenges, ch. XI, pp. 236–261. Idea Group Inc., Hershey (2007)
12. Grundstein, M.: Assessing the Enterprise's Knowledge Management Maturity Level. Int. J. Knowledge and Learning 4(5), 415–426 (2008)
13. Marca, D.A., McGowan, C.L.: SADT. Structured Analysis and Design Technique. McGraw-Hill, Inc., USA (1988)
14. Kruchten, P.: The Rational Unified Process. An Introduction. Addison Wesley Longman, Inc., Reading (1999) (first printing, November 1998)
15. Saad, I.: Une contribution méthodologique pour l'aide à l'identification et l'évaluation des connaissances nécessitant une opération de capitalisation. Thèse de doctorat. Université Paris-Dauphine, Paris (2005)

Applying IT Governance Concepts and Elements to Knowledge Governance: An Initial Approach

Juan Ignacio Rouyet[1,2] and Luis Joyanes[1]

[1] Universidad Pontificia de Salamanca, Campus de Madrid, Pso. Juan XIII, 3, 28040 Madrid, Spain
[2] Quint Wellington Redwood, Alfonso XII, 38, 28014 Madrid, Spain
i.rouyet@quintgroup.com, luis.joyanes@upsam.net

Abstract. As the era of knowledge-based economy is emerging, the importance of knowledge governance is gradually increasing. The question of how the governance mechanisms influence on the knowledge transactions is becoming increasingly relevant. However, the theoretical approaches have yet to solve outstanding issues, such as how the the micro-level governance mechanisms influence the knowledge processes or what kind of organizational hazard could decrease the benefits form the knowledge processes. Furthermore, the deployment of empirical studies to address the issues mentioned is arguably needed. This paper proposes a knowledge governance framework to assist effectively in the implementation of governance mechanisms for knowledge management processes. Additionally, it shows how this may be implented in a knowledge-intensive firm and proposes specific structures and governance mechanisms.

Keywords: knowledge governance, knowledge management, IT governance, IT service management.

1 Introduction

During the last few years academic researchers on knowledge management have shifted from developing processes for the administration of information to proposing new holistic approaches. Nowadays, knowledge management is less about technology and more about dealing with both people and technology together [1].

In order to cope with the issue of knowledge management, the concept of knowledge governance have emerged into the scenario. The Knowledge Governance Approach (KGA) is a sustained attempt to uncover how knowledge transactions and governance mechanisms are matched, using economic efficiency as the explanatory principle [2]. However, it is necessary to understand the mechanisms of motivation and cognition at the level of individuals [3]. Some attempts have been made to develop models of knowledge governance [4] [5] [6], and define governance structures [7] [8]. Still, the KGA needs to solve some research questions [2]. The ultimate outcome has to be the result of broadening the scope of knowledge management research through a greater integration with other sciences [9].

One of the new disciplines to take into account for the deployment of the KGA could be the research made in the field of the IT Governance. The topic of the IT Governance is under development, though some significant results have been released [10].

M.D. Lytras et al. (Eds.): WSKS 2009, CCIS 49, pp. 258–267, 2009.

The proposal of this paper is to apply the Peterson's model of IT Governance [11] as an initial approach to address the research questions of the KGA. The framework on IT Governance applied to KGA has been deployed in a knowledge-intensive firm in the field of professional services with oustanding results.

This paper is organized as follows:

First of all, a review of the KGA is shown to highlight the existing research questions. Then, we will describe the most important topics on IT Governance, explaining the model of Peterson in IT Governance. Finally we will propose how to apply that model to the KGA and how the implementation in a knowledge-intensive firm has been carried out.

2 Literature Review

2.1 The Knowledge Governance Approach

According to Foss [2], the KGA approach is based on the hypothesis that the knowledge processes can be influenced and directed through the deployment of governance mechanisms. In particular, the formal aspects of organization that can be manipulated by management, such as organization structure, job design, reward systems, information systems, standard operating procedures, accounting systems, and other coordination mechanisms.

The foremost explanatory principle of the KGA is economic efficiency, and the background of the knowledge governance approach is mainly in the transaction cost economics. However, to the extent that the knowledge is held by the "people" [12] it is necessary to understand the mechanisms of motivation and cognition on the level of individuals (micro-mechanisms) [3].

Dealing with the intangible aspects of the knowledge transfer in multinational corporations has brought up the issue of knowledge governance structures [8]. The scholars provided a conceptual typology of three governance structures: "exchange", "entitlement" and "gift". Alternatively, a recent empirical study has been presented by Zhang and Wen [7]. From transaction cost theory and knowledge-based view of a firm they classify the knowledge governance structures into three approaches: market-based, authority-based and consensus-based. The research shows that consensus-based approaches could exert the biggest influence on knowledge integration efficiency.

Moreover the scholar's literature on KGA also shows models for knowledge governance [4] [5] [6]. Nonetheless those models do not propose a satisfactory framework to explain the governance structures and mechanisms. Bueno's model introduces a conceptual scheme showing the fact that knowledge as a resource is converted somehow into capacity at organizational level. The proposal of Mantín and Rodríguez is focused in the relationship between the intellectual capital and the EFQM Excellence Model. Finally, Smits and De Moor's model is closer to knowledge management than to a real knowledge governance approach.

To complete this review of the KGA it is necessary to point out the problems the KGA is set out to solve. The goal of KGA is to link governance mechanisms and knowledge processes. Foss [2] provides some relevant issues regarding that link that have to be addresed: the impact of the motivation (incentives) and governance mechanisms on knowledge processes; conversely, the organizational and exchange hazards of knowledge processes.

The current stage of the development in the KGA field requires empirical researches that can shed light on this crucial question for the KGA. Broadening the scope of the knowledge management research through a greater integration with other sciences [9] is mandatory in order to address successfully the KGA's research questions. This paper proposes an overall knowledge governance framework based on the IT Governance discipline in order to organize the elements of the knowledge governance. Moreover, it shows the first results of its deployment in a knowledge-intensive firm.

2.2 Concepts of IT Governance

In the current economy, Information Technology (IT) has become as pervasive as knowledge management. The fact that the evolution of knowledge management is linked to the technology [13] makes IT and knowledge management have somewhat common concerns. For instance, both IT and knowledge management are struggling to show their business value [14] [15].

One of the most important challenges for the existing IT management is to move from the management of the technology to the management (delivery) of the IT services [16]. To deliver that kind of services and to provide value to the firm, the IT must be aligned to the business [17]. The current academic and practitioners' studies show that the alignment between IT and business has to be extended to the level of the governance [18].

A single definition of IT Governance does not exist [11], which reveals that the concept is evolving. According to the IT Governance Institute [19], IT Governance focuses on the structures of leadership, organization and processes to sustain the corporate governance. From our point of view this approach is worthy because it highlights the relevant point that the most important targert of IT Governance is to support the corporate governance. Therefore, the structures and organization have to encompass both the IT function and the business side of the firm and the relationships to each other. This approach is the underlying basis for the Peterson's IT Governace framework [11]. Peterson's framework places IT Governance structures, processes and relational mechanisms in a comprehensible view.

Some scholars have developed researches based on the Peterson's framework. A significant empirical study was released [10] explaining a real implementation of IT Governance in a major Belgian financial group. The deployment of an IT Governance project could be carried out following the phases proposed by Clementi and Carvalho [20].

This paper proposes that Peterson's framework is a useful scheme to place and analyze the elements of knowledge governance.

3 Knowledge Governance Framework

Our proposal is to translate Peterson's IT Governance framework to the field of knowledge governance.

The new new approach will be called Knowledge Governance Framework (KGF). It allows to place easily the elements of the knowledge governance so that they can be analyzed. It also allows for the comparation of several models of knowledge governance. The representation of the KGF is shown (Fig. 1).

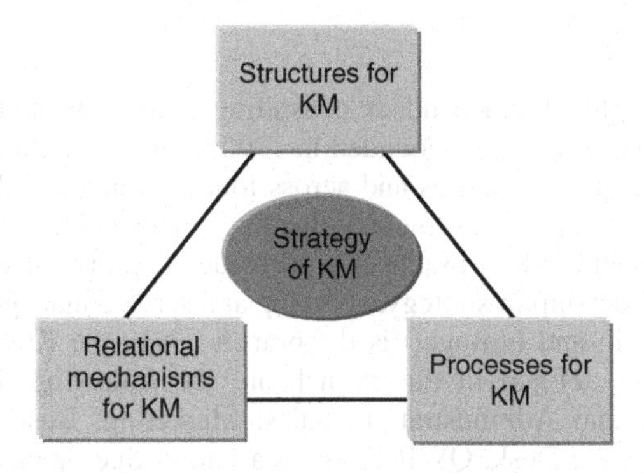

Fig. 1. The Knowledge Governance Framework (KGF)

The lines linking the boxes point out the fact that the processes can be influenced and directed through the deployment of governance mechanisms. At the core of the framework we place the strategy of knowledge management.

The crucial theme of the KGA is how to influence the processes of the knowledge management. However, it is necessary to decide the target of that influence. According to the IT Governance Institute [19], the objective of IT Governance is to sustain the corporate governance. In the same way, the target of the KGA should be to sustain the corporate governance. It means that we think of the existence of two kinds of logic domains: the business domain and the knowledge domain. Both domains do not exist as an actual organization entity, but are embedded in corporate vision and organizational culture. They are related with the concept of hypertext organization of Nonaka and Takeuchi [21].

This point is relevant because focuses on the fact that the knowledge governance's relational mechanisms have an external dimension whereby the alignment between business domain and knowledge management domain needs to be achieved. Therefore, the relational mechanisms have to be defined for enhancing the flow of knowledge inside the knowledge domain and between each other domain. We will come back to this point in the case study.

4 Applying the Framework to a Case

The power of the KGF as a tool for defining and analyzing the knowledge governance's structures can be appreciated in a case study. Therefore, the paper describes the case of the KGF implementation in a knowledge-intensive firm. This work aims to offer an empirical outcome to the development of the KGA [2].

Data was gathered by conducting several face-to-face interviews with business and support representatives (CEO, Business Development manager, Service Delivery manager, Practice Desk manager). The interviews were completed in the period of June 2008 to September 2009. After that, the specific approach of the KGF was implemented and was launched in January 2009.

4.1 The Case: QWR

QWR is a leading global independent consulting firm dedicated to resolving IT-related organizational challenges. Founded in 1992 in the Netherlands, it is nowadays operating in more than 49 countries and across four continents. QWR provides strategy, sourcing and service management to leading organizations from all industries, creating and implementing best practices worldwide. The firm offers thought leadership and practice leadership in strategy, sourcing and service management for IT.

QWR Iberia (Spain and Portugal) is the branch where the KGF was applied initially. The functions defined in this branch are the following: Directors, Human Resources, Finance and Administration, Sales, Marketing, Business Development, Operations, and Practice Desk. QWR Iberia is a knowledge-intensive firm. Business Development has to keep the company in the edge of knowledge, in order to the firm to offer the most innovative services in strategy, sourcing and service management for IT. The services are delivered by the Operations department. The Practice Desk contains an electronic repository to store the QWR's information. The firm's portfolio of services include training and consulting, all integrated across the domains of customer's business and IT.

4.2 The Knowledge Governance Framework in QRW

The Senior Management initiated the project in the Summer of 2008 by asking a consultant to develop a knowledge management system for the Iberia branch. The following were the business drivers: the process of innovation was not successfully implemented due to a constant rework; an inefficient process of resource allocation, causing loss of revenue; and finally, it was found that the firm was not actually aware of the amount of knowledge it possessed, which meant intangible and unaffordable costs for a knowledge-intensive firm. To solve that critical situation it was necessary to redefine the structures and to become more organized. QWR Iberia stablished structures, processes and relational mechanisms, which are described in the following sections.

4.2.1 Strategy

The following underlying principles (KM principles) were established to address the business drivers mentioned above: knowledge management has to move QWR Iberia to action (to adapt); problem recognition and knowledge claim are in the core of knowledge management (QWR Iberia seeks for business problem solving and new business problem recognition); knowledge management has to develop the explicit knowledge dimension in QWR Iberia.

Besides, the governance structures should be created to incentive the following knowledge-oriented culture: openness, whereby mistakes and past failures are openly shared and discussed without the fear of punishment; free time, thereby QWR Iberia has to free up time for their employees to perform knowledge management activities such as knowledge creating or sharing; mutual trust; and collaboration.

4.2.2 Structures

The scope of the knowledge management was established according to the KM principles. The knowledge of the firm was split into two categories: business processes' knowledge (BPK) and support processes' knowledge (SPK). The knowledge management system will just concern to the BPK. Inside the BPK, the following knowledge domains (KD) were defined: sourcing, IT Governance, IT Service Management and IT Service Management Regulations. The knowledge domains cover the 80% of the business activity of QWR. Therefore, the two first KM principles were addressed. Any other BPK out of the KD is considered grey literature. This made the knowledge management system cost-efficient. The knowledge domains are divided into knowledge areas (KA) to go down to other levels of knowledge and pursuit all pieces of knowledge.

With the aim to create the knowledge-oriented culture mentioned above, the core structure of the knowledge management system is the Community of Practice (CoP) [22]. Particularly, there is a CoP per knowledge domain and the roles of CoP Leader (CoPL) and CoP Member (CoPM) were defined to carry out with the activity of the every CoP. The CoPL is the owner of a domain of knowledge and acts as the product manager of the products and services related to his domain of knowledge. Thereby, the CoPL manages the knowledge of his KD applying a business vision. Additionally, the role of the Chief Knowledge Officer (CKO) [23] was named and the Practice Desk role was redefined to support the CoP and to manage the grey literature.

Table 1. Committees representing business and knowledge management system

Committee	Business representatives	KM system representatives
KM Steering Committee	Members of Directors; Members of the management committee	Chief Knowledge Officer (CKO); Community of Practice Leaders (CoPL)
Knowledge Committee for Business Development	Business Development manager; Sales manager	Community of Practice Leaders (CoPL); Practice Desk
Knowledge Committee for Business Operation	Project leaders	Community of Practice Member (CoPM); Practice Desk

Table 1 shows the committees where the business and the roles of the knowledge management system are represented. This kind of committees ensures the alignment of the knowledge management with the business needs thereby the corporate governance is sustained by the knowledge governance framework. The committees define the hypertext organization [21] covering both business and knowledge management activity.

4.2.3 Processes

QWR Iberia has got implemented a quality management system based on the definition of business and support processes. It was decided to implement only a few number of knowledge management processes in ordet not to overload the quality management system. The following were defined: knowledge strategy, to identify the knowledge needs and to state the proper knowledge management's strategy; knowledge generation, to create new knowledge or to pull the existing knowledge (tacit and explicit) which is spread all around the firm's branch; knowledge sharing, to share the knowledge created, pulled or updated; knowledge application, to convert the knowledge into business actions for the firm's branch.

Two decisive knowledge tools were defined to make effective the knowledge management processes. They are the Knowledge-map (K-map) and the Practice Green Book (PGB). The K-map is based on Gupta, Iyer and Aronson [24] and categorizes the knowledge existing in QWR Iberia. The K-map is linked to two databases: a database with information about the library of the QWR (explicit knowledge) and a database with information about the knowledge of the QWR's employees (tacit knowledge).

The Practice Green Book (PGB) is an open document describing the state-of-the-arte knowledge in QWR Iberia for every KD. The target is to make most tacit knowledge, which has been gained from the live experience of the company, explicit. The knowledge is extracted mainly from the project's deliverables.

4.2.4 Relational Mechanisms

Relational mechanisms are the governance mechanisms whereby an "organization" is established. "Organization" is used in the sense of Jones [25], that is, the formal and informal allocation of decision rights and the mechanisms that enforce such rights. In addition, we include the mechanisms to monitor the decisions [26].

In QWR the consultants' activity is monitored through a web-based tool. To enforce and monitor the knowledge management activity, every QWR's employee must reflect the time spent in any process of knowledge management in the tool. Furthermore, the consultants have personal objectives related to knowledge management from knowledge generation to knowledge sharing and application. The information collected with this tool is used later to obtain metrics of the activity spent in the knowledge management processes, which will be underlying data to business performance metrics.

One of the targets of the knowledge management system is to offer knowledge services to the Business Development function. The services were defined in a concrete set of deliverables, such as white papers, brochures and press articles. The business development deliverables and the updating of the K-map and PGB are monitored on a quarterly-basis.

5 Conclusions

As the era of knowledge economy is emerging, the importance of knowledge governance is gradually increasing. The question of how the governance mechanisms influence on the knowledge transactions is becoming increasingly relevant. However, the theoretical approaches have yet to solve outstanding issues. Furthermore, at the present stage of development, what is arguably most needed is the deployment of empirical works to address the issues mentioned above.

The main contribution of this paper lies on the proposal of a unique relevant framework for knowledge governance. Besides, the framework has been deployed in a knowledge-intensive firm and the main findings can be described as follows:

1. Usability of the framework. The Knowledge Governance Framework represents quite a useful framework to adress the concepts linked to the KGA. The structures and decisions rights can be easyly defined; topics concerning the knowledge processes, such as unit of analysis or knowledge transactions dimensionalisation, are conveniently analyzed.
2. The framework allows us to deploy the concept of hypertext organization. Virtual structures can be defined and implemented over the functions constituiting the organization. To the extent that the knowledge is pervasive in a knowledge-intensive firm. The governance structures for the knowledge management can be separately built in. These are to be interwoven with the business structures, and the framework shows the relationships to each other.
3. The paper shows a categorization of the knowledge and a definition of a unit of analysis. The latter is particularly important because it defines identifiable "pieces" of knowledge. The "pieces" are related to the business activity of the firm and split into two levels: Knowledge Domains and Knowledge Areas. Moreover, the transactions of the "pieces" are well defined as services for the Business Development function within a concrete set of deliverables (i.e. white papers, brochures or press articles).

The proposed framework has been applied in a knowledge-intensive firm. However, it can be used in other scenarios related to the emerging knowledge society:

1. Inter-enterprises scenario: the KGF can be deployed externally, extending the limits of the business domain. Knowledge management not only is a matter inside the enterprise's boundaries, but both inside and outside together [27]. While companies share knowledge efficiently and effectively with the environment, they can become both revenue increasing and social responsible. Thereby, enterprises have to create appropriate structures and relational mechanisms between companies (KGF approach) to share knowledge, according to the stakeholders' interests.
2. NGO scenario: We will build a real knowledge society when not only information but knowledge flows steady to all parts of the society [28]. Whilst knowledge management relies increasingly on technology (i.e. Internet), we know that knowledge management is about dealing with both people and technology together. Non Governmental – Non for Profit Organizations are the key factors to achieve social benefits from knowledge management. Therefore, NGOs have to create knowledge structures and relational mechanisms (KGF) interwoven with the social networks to spread knowledge as the future's resource.

References

1. Lee, H., Choi, B.: Knowledge Management Enablers, Processes, and Organizational Performance: An Integrative View and Empirical Examination. Journal of Management Information Systems 20(1), 179–228 (2003)
2. Foss, N.: The Emerging Knowledge Governance Approach: Challenges and Characteristics. DRUID Working Paper No. 06-10 (2006)

3. Grandori, A.: Governance Structures, Coordination Mechanisms and Cognitive Models. Journal of Management and Governance 1, 29–42 (1997)
4. Bueno, E.: Fundamentos Epistemológicos de Dirección del Conocimiento Organizativo: Desarrollo, Medición y Gestión de Intangibles. Economía Industrial 357, 13–26 (2005)
5. Smits, M., De Moor, A.: Measuring Knowledge Management Effectiveness in Communities of Practice. In: Proceedings of the Proceedings of the 37th Annual Hawaii International Conference on System Sciences (HICSS 2004) Track 8, p. 8 (2004)
6. Martín, J., Rodríguez, O.: EFQM Model: Knowledge Governance and Competitive Advantage. Journal of Intellectual Capital 9(1), 133–156 (2008)
7. Zhang, P., Wen, P.: An Empirical Study on the Knowledge Governance Choices and Knowledge Integration Efficiency in Work Teams. In: 4th International Conference on Wireless Communications, Networking and Mobile Computing, WiCOM (2008)
8. Chenga, P., Ju, C., Ibrahim, T.: Governance Structures of Socially Complex Knowledge Flows: Exchange, Entitlement and Gifts. The Social Science Journal 43, 653–657 (2006)
9. Choi, C., Cheng, P., Hilton, B., Russell, E.: Knowledge Governance. Journal of Knowledge Management 9(6), 67–75 (2005)
10. De Haes, S., Grembergen, W.: IT Governance Structures, Processes and Relational Mechanisms: Achieving IT/Business Alignment in a Major Belgian Financial Group. In: Proceedings of the 38th Hawaii International Conference on System Sciences (2005)
11. Peterson, R.R.: Information Strategies and Tactics for Information Technology Governance. In: Van Grembergen, W. (ed.) Strategies for Information Technology Governance. Idea Group Publishing, Hershey (2003)
12. Davenport, T., Prusak, L.: Working Knowledge: How Organizations Manage What They Know. Harvard Business School Press, Cambridge (1998)
13. Prusak, L.: Where Did Knowledge Management Come From? IBM Systems Journal 40(4), 1002–1006 (2001)
14. Brynjolfsson, E., Hitt, L.M.: Computing Productivity: Firm-Level Evidence. MIT Sloan Working Paper 4210-01 (June 2003)
15. Kalling, T.: Knowledge Management and the Occasional Links with Performance. Journal of Knowledge Management 7(3), 67–81 (2003)
16. Davenport, T.: Putting the I in IT. In: Marchand, D.Y., Davenport, T. (eds.) Mastering Information Management. Prentice Hall, Englewood Cliffs (2000)
17. Henderson, J., Venkatraman, N.: Strategic Alignment: Leveraging Information Technology for Transforming Organizations. IBM System Journal 38(2&3) (1993)
18. De Haes, S., Grembergen, W.: Analysing the Relationship between IT Governance and Business/IT Alignment Maturity. In: Proceedings of the 41st Hawaii International Conference on System Sciences (2008)
19. IT Governance Institute: Cobit 4.0. Rolling Meadows: IT Governance Institute (2005)
20. Clementi, S., Carvalho, T.: Methodology for IT Governance Assessment and Design. In: Suomi, R., Cabral, R., Hampe, Felix, J., Heikkilä, A., Järveläinen, J., Koskivaara, E. (eds.) Project E-Society. Buildmg Bricks. IFIP International Federation for Information Processing, vol. 226, pp. 189–202. Springer, Boston (2006)
21. Nonaka, I., Takeuchi, H.: The Knowledge-Creating Company: How Japanese Companies Create the Dynamics of Innovation. Oxford University Press, New York (1995)
22. Wenger, E.: Communities of Practice. Learning, Meaning and Identity. Cambridge University Press, Cambridge (1998)
23. Guns, B.: The Chief Knowledge Officer's Role: Challenges and Competencies. Journal of Knowledge Management 1(4), 315–319 (1997)

24. Gupta, B., Iyer, L., Aronson, J.: Knowledge Management: Practices and Challenges. Industrial Management & Data Systems 100(1), 17–21 (2000)
25. Jones, G.: Transaction Costs, Property Rights, and Organizational Culture: An Exchange Perspective. Administrative Science Quarterly 28, 454–467 (1983)
26. Weill, P., Ross, J.W.: IT Governance: How Top Performers Manage IT Decision Rights for Superior Results. Harvard Business School Press, Boston (2004)
27. Verago, R., Chatwin, F., d'Alessi, F., Zanette, A.: Eye Knowledge Network: A Social Network for the Eye Care Community. In: Lytras, M.D., Carroll, J.M., Damiani, E., Tennyson, R.D. (eds.) WSKS 2008. LNCS (LNAI), vol. 5288, pp. 22–30. Springer, Heidelberg (2008)
28. Georgiadou, K., Kekkeris, G., Kalantzis, M.: Inclusion in the Information Society for the "Excluded" Women in Greek Thrace. In: Lytras, M.D., Carroll, J.M., Damiani, E., Tennyson, R.D. (eds.) WSKS 2008. LNCS (LNAI), vol. 5288, pp. 460–468. Springer, Heidelberg (2008)

The Concept of Embodied Knowledge for Understanding Organisational Knowledge Creation

Yoshito Matsudaira and Tsutomu Fujinami

Japan Advanced Institute of Science and Technology,
Graduate School of Knowledge Science
1-1 Asahidai, Nomi, Ishikawa, 923-1292, Japan
s0660016@jaist.ac.jp, fuji@jaist.ac.jp

Abstract. Our goal in this paper is to understand, in the light of intuition and emotion, the problem-finding and value judgments by organisational members that are part of organisational knowledge creation. In doing so, we emphasise the importance of embodied knowledge of organisations as an explanatory concept. We propose ways of approaching intuition and sense of value as these are posited as objects of research. Approaches from the first, second, and third-person viewpoints result in a deeper grasp of embodied knowledge of organisations. Important in organisational knowledge creation is embodied knowledge of organisations, which has a bearing on problem-finding before any problem-solving or decision making takes place, and on value judgments about the importance of problems that have been found. This article proposes the concept of embodied knowledge, and, by introducing it, gives a profound understanding of that facet of organisational knowledge creation characterised by tacit knowledge held by organisational individuals.

Keywords: embodied knowledge, organisational knowledge creation, distributed cognition, tacit knowledge.

1 Introduction

Business administration is a branch of social science. Its subject of research is the activities of firms and other organisations. Knowledge-based management is a branch of business administration. Within this domain, knowledge-based management is the study of the activities of organisations, focused on the creation, utilisation, and maintenance of knowledge. In contrast, cognitive science is a branch of behavioural science. Its subject of research is the workings of the human mind as well as cognition and behaviour. These two branches of science – business administration and cognitive science – each pursue their own path of development. Carefully looking into the past, however, there was a time when both were closely linked. It is Simon who is most noteworthy regarding this point.

Simon [1] was very interested in the process of decision-making by people and organisations. Decision-making may be defined as acting to find ways to solve a given problem by applying the human and material that are resources available. According to

M.D. Lytras et al. (Eds.): WSKS 2009, CCIS 49, pp. 268–278, 2009.

Simon [1], decision-making skills are constrained by cognitive limits. Assuming that the decision-making is objective and rational, the selection of the unique best alternative comes only after knowing all conditions and investigating all of the consequences resulting from the selection. In fact, however, the decision-maker is subject to the constraint of "bounded rationality" because of factors such as incompleteness of knowledge, difficulties of anticipation, and the scope of behavioural possibilities. Therefore, it may be assumed that completely rational decision-making is not feasible. As a result, it is thought that human beings created organisations to overcome such constraints.

Simon [1] divides decision-making into two categories: (1) decision-making based on value premises incorporating the ambitions and enthusiasms of senior management, which is the process of drawing conclusions from a wide range of premises involved in the process of decision-making by organisations, and (2) decision-making from factual premises that can be verified objectively. Barnard [2] attached great importance to the former (1), whereas Simon eliminated value judgments, assuming that the important thing was to base decision-making on factual premises. This technique was called logical positivism. However, as Flyvbjerg [3] argued, social science is fundamentally different from natural science in terms of its need to address issues of subjectivity, such as values, context, and power. As humans are both subjects and objects of research at the same time, research in social science cannot totally be free from subjective factors.

According to Simon, both individuals and organisations are information-processing machines that take and process information from the environment to solve a problem and adapt to the environment based on a given goal. There is a limit to human information-processing skills. However, it is entirely feasible to make assessments incorporating objective rationality within the scope of certain bounds, and organisations are mechanisms for determining the boundaries of this scope. Organisations gather information efficiently by concentrating and centralising the information secured by individuals. This system enables better decision-making. The structure of organisations is explainable from the viewpoint of efficient information-processing. The science of decision-making developed by Simon has been applied to both business administration and cognitive science, and it has pursued its own path of development in each field. However, both fields share a common point in that they retain the viewpoint of information-processing with respect to people and organisations. In terms of business, the development of database and decision-support systems in connection with computerisation, and their introduction into organisations, have improved profitability. In fact, both types of systems are part of the same stream of development, in the sense that both were developed on the foundations of the information-processing paradigm. However, there are two constraints on the information-processing paradigm:

- The process of "problem-finding" prior to problem-solving is not stated clearly.
- The human sense of values is not taken into consideration.

This is the necessary conclusion stemming from the emphasis placed on human logic in the information-processing model.

The key to overcoming this constraint is the concept of embodied knowledge. By introducing the concept of embodied knowledge, this article attempts to identify the skills required by the individual members in the creation of organisational knowledge, and to understand the best way to cooperate with organisational members who have outstanding qualifications. The former topic is regarded as belonging to the field of cognitive science, whereas the latter is considered to be a part of the domain of business. In order to address this issue, it is necessary to adopt an integrated approach that combines both viewpoints. In addition, this paper proposes an approach which clarifies the embodied knowledge of organisations.

2 Logical Positivism and the Present

There are evident differences if we compare the present situation with that which prevailed during the period from the 1960s to the 1970s, when Simon was playing an active role. At that time, there were implicit premises that enabled the immediate identification of the kind of problems that were present. Today's vague search situation, in which the question is first of all whether problem exists or not, and what kind of problem it is, assuming that the problem does exist, was not envisioned. In contrast, present era is an era in which one cannot determine at a glance whether a problem exists or not and what kind of problem it is.

Utilising the concept of decision-making, logical positivism explains the actions of human beings in an organisation on the basis of empirical, observable facts. However, Simon's decision-making approach is powerless in a situation in which problems are not visible. Decision-making commences with the definition of a problem. Insofar as the problem is described clearly, countermeasures may be determined clearly too, and the members of an organisation may all understand and share the same content. However, not all of these factors have been established in today's world. We do not know what it is best to do because we do not understand what the problem is. That is the situation in which we find ourselves now.The situation exposes the limits of Logical Positivism.

3 Necessity for Spontaneousness

While Simon was playing an active role from the 1960s to the 1970s, it was possible to grasp both the activities of organisations and individual cognition in the light of decision-making. It was a time when problems were clear. After examining a problem closely, it was possible to make plans and execute them as the pace of the world was relatively slow. At present, however, it is difficult or impossible to see the whole because of distribution. It has become a situation in which addressing problems cannot keep up with sudden changes.

An example may help to explain the matter. The format of games is shifting from the American football format to football (soccer). In American football, the head coach determines the strategy, and the players perform the tasks at coach's command. It is more or less decided that given tasks are performed in given situations, depending on the team. In football (soccer), in contrast, the coach determines the overall strategy, but in practice the particulars of the game are left to the individual player. Why is the way of playing the game left up to the individual player? This is because match-ups against opponents take place frequently at the same time, for the situation is always changing, and instructions for individual players are not given.

During the past half-century, we have witnessed business changing from the format of American football to that of football (soccer). The change of format of the game puzzles many organisations and individuals, who cannot seem to find neither the kind of game that is suited to them nor the right way to play.

The key to the research presented below is the concept of embodied knowledge. Embodied knowledge exists at the individual and organisational levels. Returning to the example of football, we should pay attention to the national team of Brazil. We cannot but admire their exquisite play, while at the same time they prominently display individual skills. In the moment at which victory has been decided, there is a scene in which all team members leap shoulder to shoulder, and we have a vivid impression that the team itself has emerged as a total entity. It is at such times that we can see the embodied knowledge of organisations.

Here, in the prominent display of individual skills and the extraordinary teamwork that highlight the way in which individuals work together as a team, we can sense the dynamic exhilaration about which we have been talking. The above example shows the course of action that organisations should aim at achieving in the future. At this point in time, we should pay extra attention to the following items: organisational strategy has meaning only when the individual skills of the team members are good. Attempting to use organisational skills to make up for the inferior skills of individuals does not work. A top-down methodology likewise does not work. This is a methodology in which an excellent leader makes a good plan and the personnel tries to perform the allotted task by following his/her decisions.

4 Explicit Knowledge and Embodied Knowledge of Organisations

4.1 Explicit Knowledge of Organisations

The explicit knowledge of organisations involves to thinking by logical inference. On the other hand, the embodied knowledge of organisations refers to intuitive understanding and empirical judgment. Bureaucratic organisations are the foundations of the explicit knowledge of organisations [4].

According to Weber [4], in short, bureaucratic organisations are characterised by (1) a system of rules, and the members of the organisation are obliged to always follow those rules; (2) there are systems of hierarchical official authority and top-down command; (3) proposals and final decisions as well as a wide range of commands are formalised and anchored in writing; and (4) members of the bureaucracy get special training in performing their duties.

Weber compared bureaucratic organisations with such distinctive characteristics to precision instruments. They have structural characteristics designed and optimised in advance, like clocks, and they function in a regular, repetitive manner. All members of bureaucratic organisations utilise thinking by logical inference to perform their tasks in a repetitive manner. Moreover, there are few changes in the situation and environment surrounding bureaucratic organisations. Therefore, in bureaucratic organisations, there is no basic need for members of the bureaucracy to have embodied knowledge, including intuitive understanding and empirical judgment.

4.2 Embodied Knowledge of Organisations: Distributed Cognition and Collaboration

Let us examine the embodied knowledge of organisations. In attempting to balance the opposing pair between the organisation and the individual, the key ideas are distributed cognition [5] [6] [7] [8] and collaboration [6] [8]. Distributed cognition is defined as incorporating the feelings and thoughts experienced by individuals in various situations into the team with the team drawing a total image of the situation into which it is has been placed. Collaboration is defined as the individual displaying his skills while collaborating with others after having recognised the overall image. The two terms, distributed cognition and collaboration, reflect the position of phenomena viewed at the organisational level. With our selection of terms, we adopt the standpoint of visualising phenomena through the prism of the organisation. According to Nonaka and Takeuchi [8], socialisation and internalisation correspond to distributed cognition, and externalisation and combination correspond to collaboration, visualised in the light of the relationship with the SECI model. This paper confirms the SECI model [9] [10] [11] [12] [13] (pp.9-10).

Socialisation: Socialisation is the process of converting new tacit knowledge through shared experiences. Tacit knowledge can be acquired only through shared experience, such as spending time together or living in the same environment. Socialisation typically occurs in a traditional apprenticeship, where apprentices learn the tacit knowledge needed in their craft through hands-on experience, rather than written textbooks or manuals. Socialisation may also occur in informal social meetings outside of the workplace, where tacit knowledge such as world views, mental models and mutual trust can be created and shared.

Externalisation: Externalisation is the process of articulating tacit knowledge into explicit knowledge. When tacit knowledge is made explicit, knowledge is crystallised, thus allowing it to be shared by others, and it becomes the basis of new knowledge. Concept creation in new product development is an example of this conversion process.

Combination: Combination is the process of converting explicit knowledge into more complex and systematic sets of explicit knowledge. Explicit knowledge is collected from inside or outside the organisation and then combined, edited or processed to form new knowledge. The new explicit knowledge is then disseminated among the members of the organisation. The combination mode of knowledge conversion can also include the 'breakdown' of concepts. Breaking down a concept such as a corporate vision into operationalised business or product concepts also creates systematic, explicit knowledge.

Internalisation: Internalisation is the process of embodying explicit knowledge into tacit knowledge. Through internalisation, explicit knowledge created is shared throughout an organisation and converted into tacit knowledge by individuals. Internalisation is closely related to 'learning by doing'. Explicit knowledge, such as product concepts or manufacturing procedures, has to be actualised through action and practice.

One distinctive characteristic of the SECI model is its emphasis on tacit knowledge. Under the theory of knowledge creation, tacit knowledge is primarily related to socialisation and internalisation. In contrast, explicit knowledge is primarily related to externalisation and combination. Socialisation and internalisation are crucial from the standpoint of emphasising tacit knowledge. Under our framework, they are related to distributed cognition. Therefore, an examination of distributed cognition is conducted below.

4.3 Socialisation as Distributed Cognition

Socialisation emphasises feeling, that is, direct experience conveyed via the body, and intuition. Both feeling and intuition are individual skills. How are they used? And how are they shared with other persons?

Things derived from feeling and things derived via intuition pose something of a problem. That which is derived from feeling is reality-related raw information. The importance of real, on-the-spot experience is often emphasised because it yields information that cannot be obtained simply by reading data collected by others. Information derived from reality contains many items that cannot be converted into data. Some measurement techniques are necessary in order to generate data; however, no measurement techniques exist for many phenomena. When we consider phenomena that we cannot measure, our only recourse is to gather information with our own body.

Things that are grasped through feeling pose something of a problem in many cases. The issue is not problem-solving but rather problem-finding. Why is problem-finding important? This is because we try to solve the problems we have envisioned without taking reality into account. In that case, will we be able to visualise problems if we let feeling act as much as possible and retrieve information from reality? The important point is that we cannot visualise problems by giving free rein to feeling. This is because true problems are hidden behind phenomena. We have to depend upon intuition in order to investigate hidden causes. Intuition enables us to find the true, hidden causes.

It may be that our acceptance or non-acceptance of the working of intuition affects whether we accept the management of tacit knowledge. This is because things derived from intuition are often impossible to verify by means of quantitative data. With quantitative data, proof might be to some extent feasible. However, this means only enhanced credibility, not conclusive evidence. Therefore, it is easy to deny intuition if one's standpoint attaches great importance to quantitative data and proof. Intuition can be easily eliminated by invoking rationality. Should we eliminate intuition or put trust in it? This question highlights a major crossroad in knowledge-based management.

4.4 Internalisation as Distributed Cognition

Emotion has been avoided in the realities of business. It surfaces into awareness only in the theme of motivation. Why is emotion avoided? This is because of the antagonism between reason, rationality, logical thinking and emotion. Reason, rationality, and logical thinking form the basis of the modern sense of values. The themes of problem-solving and decision-making were established under the modern sense of values. In contrast, emotion is irrational and often conflicts with reason. It is often neglected because it leads to wrong decisions. From the standpoint of reason, the issue is to what extent emotion has been tamed. However, we take the standpoint that emotion is a separate kind of intelligence that is different from logical thinking. Emotion is intellectual in the light of its own content. Emotion does not need a build-up of inferences to arrive at a conclusion. The conclusion of liking or disliking, being favourable or unfavourable, appears immediately. It is important to arrive at a conclusion immediately; the distinctive factor here is that assessments made by emotion are faster than relying on logical thinking.

In internalisation, tacit knowledge gains results in assessments that go beyond reason, that is to say, in empirical judgments (sense of values). In other words, this is the skill of distinguishing what one likes and what one dislikes. Likes and dislikes are generated by experience. Good experiences lead to liking the factors involved, and bad experiences lead to disliking the things related to the experience. Excellent empirical judgments are cultivated by the repetition of high-quality experiences.

An individual forms his/her own sense of values via experience. A sense of values gives criteria to assess what is important, and this may differ depending on the person. Therefore, the criteria of value judgments gained through experience are distributed in organisations, and it may be said that a sense of values is also one form of distributed cognition.

5 Capturing the Embodied Knowledge of Organisations

5.1 Approaches from the First, Second, and Third-Person Viewpoints

When the subject of research is intuitive understanding and empirical judgments (sense of values), the question is how they should be handled. The intuition and emotion of individuals is a domain where even cognitive science has made little progress actually. The focus of interest is the issue that the intuition and emotion of organisations are even more difficult to address than the intuition and emotion of individuals.

Intuition and emotion are subjective phenomena. They are not easy to handle as it is difficult to understand other individuals. The only way to proceed is to listen to the reports of the parties concerned regarding the workings of the intuition and emotion of individuals. Therefore, it is necessary to interview the parties concerned. Also group interviews are desirable when the goal is to research the embodied knowledge of organisations. This is because projects are borne by groups. When inquiring about the experience of groups, it is better to pose inquiries to the group as a whole rather than to individuals. This way may be the best we have at our disposal, even though it is not easy to gain admittance to the hearts of the parties concerned. We do have the skill of sympathy. By making comparisons with our own experience, we can assess the accuracy of the intuition of others and share their enjoyment and grief. We can not objectively describe the intuition and emotion of others from the standpoint of the observer. However, sympathy allows us to talk about other mind.

Furthermore, circumstantial evidence is not entirely meaningless; it can provide supporting facts. The presence of subjective reports and of objective facts with consistency enhances the accuracy of subjective reports. Objective facts consist of products and services as results as well as sketches or memos that were produced at the various stages of development of the products and services. What we seek are descriptions of the internal world backed by objective facts. Explanations based on cause and effect are impossible in the case of objective facts as they cannot act as the causes of subjective reality. However, objective facts can become the results of subjective reality. Therefore, subjective reality can be envisioned by backtracking to cause and effect.

There is the subjective reality of the parties concerned, understanding via the sympathy of neighbouring persons, and supporting evidence provided by objective facts. These are thought to be subjective phenomena. As such, they are three different ways of approaching intuition and emotion. It is thought that approaching from three directions at the same time may enable research of intuitive understanding and empirical judgment (sense of values). The three methods of approach may be named as follows:

- the first-person's viewpoint: the subjective reality of the parties concerned
- the second-person's viewpoint: understanding via the sympathy of neighbouring persons
- the third-person's viewpoint: supporting evidence provided by objective facts

It is necessary to observe and analyse phenomena in an integrated manner from three different viewpoints in order to understand human beings and organisations. Varela and Shear [14] pointed out the importance not only of a third-person viewpoint, but of first- and second-person viewpoints, too. Regarding this matter, they take the same standpoint as we do. They assert that, in addition to the third-person viewpoint, any approach must also include a first- and second- person viewpoint if it is to elucidate consciousness. We, however, use this method to arrive at an understanding of the knowledge-creation process and of the abilities of people who work according to that process. At first glance, a similar methodology should be applied to neuroscience when viewed through the prism of science. This is because even though the technology to measure brain functions has made dramatic progress, and it has become possible to measure the state of brain activities more or less concurrently, in real time, only the subject of the examination can judge whether the state of the brain measured corresponds to specific mental states. An experienced brain scientist may guess the mental states of the subject from the activity status of the brain. However, knowing whether the guess is correct is limited to instances in which the brain scientist himself/herself understands the relationship between the activity status of the brain and mental states. Intuition and emotion cannot be traced to other constituents and physical traces (active state of the brain). Intrinsically, they have to be handled raw, and thus are not considered science but are treated in combination with the second-person and the third-person's viewpoint.

5.2 Ways of Representing Viewpoints

We believe that we can basically organise the actuality of the creative activities that we have approached from three different directions by following the SECI model. In the SECI model, knowledge-creation process is regarded as consisting of the following four phases: socialisation, externalisation, combination, and internalisation. From the viewpoint of the embodied knowledge of organisations, what we are interested in is socialisation and internalisation. Therefore, we apply the three directions to summarise what took place in each phase.

In socialisation, we want to ascertain what the problem is by means of our five senses. In the case of quest, the quest is assigned a direction under a sense of values, that is, under the influence of having determined what is important. When this phase is examined from the three viewpoints, it is necessary to carefully observe the following points:

- The first-person's viewpoint: The points that should be carefully observed are those places which explained the discovery of the issues that needed to be addressed. Other places to be carefully observed are those places that talk about the experiences that became opportunities, including the motivation of individuals and similar motivations, which form the backdrop of such issues.
- The second-person's viewpoint: If the leading role is assigned to the parties concerned, focus on the points that show how and when those people in supporting roles accepted the motivation and goals of the parties concerned.
- The third-person's viewpoint: Carefully observe the wide range of attempts and trials that were undertaken in the process of problem-finding, for example, whether attempts were made by the group to visit those places where customers live.

In the case of internalisation, we organise whether and how a sense of values was obtained or verified. Clarify the differences in the sense of values before the launch of the project and after it ended.

- The first-person's viewpoint: things considered to be important (for example, customer satisfaction). Is there satisfaction with the results of the work?
- The second-person's viewpoint: Opinions about the optimal organisational culture and human relations of organisations, the goals of the organisation, and its course of action.
- The third-person's viewpoint: Statement regarding the social role and contribution, and opinions regarding environmental problems and the international situation.

6 Conclusion

In this paper we have discussed what kinds of approaches are feasible towards organisational knowledge creation from the viewpoint of the embodied knowledge of organisations. When we investigate creative activities, such as the development of new products and services, there are many items that cannot be understood only in the light of problem-solving and decision-making. The things that are more important in organisational knowledge creation take place before the stages of problem-solving and decision-making. In other words, this refers to noticing problems that no-one paid attention to previously, and determining whether the problem is worth addressing. Problem-finding and value judgments cannot be explained in the light of reason and rationality. This article explains problem-finding and value judgments in the light of the working of intuition and emotion by using the concept of embodied knowledge.

Compared to half-a-century ago, the world of today is much more fluid and full of changes. Half-a-century ago, individuals and organisations were in situations that were more static than nowadays. Firms had time to work out adequate plans in advance and to examine alternative plans. Today, however, they are in situations in which they have to undertake actions before their thinking has solidified. Even with conduct guidelines, individuals and organisations have to rely upon experience and hunches when time is a serious consideration. The question is why certain people or organisations can manoeuvre well when pushed into a corner, while other people or organisations are clumsy. The standpoint of this paper is that this question may be explained in the light of the concept of embodied knowledge.

As Toffler [15] said, we are now living in a 'knowledge-based society', where knowledge is the source of the highest quality power. Knowledge is the most important resource, replacing the traditional management resources of land, capital, and labour [16]. In knowledge society, competition based on tangible assets such as land, capital, and labour shifts to time managing knowledge based on intangible assets such as embodied knowledge, which are distributed intuition and sense of values held by organisational individuals. Embodied knowledge of organisations is intangible assets. In knowledge society, intangible assets mean knowledge assets. Knowledge assets are the inputs and outputs of knowledge creating process. Knowledge assets have to be created and utilised internally to realise their value, since they cannot be easily bought and sold. Knowledge assets are continuously evolving, since they are both inputs and outputs of knowledge creating activities by organisational members. Today, knowledge assets are sustainable competitive advantage in knowledge society.

Lastly, this paper touches on the contribution to knowledge society vision of the research into embodied knowledge of organisations. First, studying embodied knowledge of organisations can give knowledge workers more profound understanding of knowledge assets, which has a bearing on problem-finding before any decision making takes place, and on value judgments about the importance of problems that have been found. In addition, approaches from the first, second, and third-person viewpoints can classify knowledge assets into the individual level, team level, and organisation level and the classification is very practical for knowledge workers to accumulate or utilise knowledge assets.

Second, by the study, understanding how organisational members find problems when they try to create knowledge and what kinds of sense of values are created after knowledge creating process results in enabling the next knowledge creating process.

References

1. Simon, H.A.: Administrative Behavior. Macmillan, New York (1945)
2. Barnard, C.I.: The Functions of the Executive. Harvard University Press, Cambridge (1938)
3. Fiyvbjerg, B.: Making Social Science Matter: Why social Science Fails and How it Can Succeed Again. Cambridge University Press, Cambridge (2001)
4. Weber, M.: Wirtschaft und Gesellschaft Grunderiss der Sozialokonomik 3, Tuebingen (1947)
5. Suchman, L.: Plans and Situated Actions: the Problem of Human-Machine Communication. Cambridge University Press, New York (1987)
6. Hutchins, E.: The Technology of Team Navigation. In: Galegher, Kraut, J.R., Egido, C. (eds.) Intellectual Teamwork: Social and Technical Bases of Cooperative Work, pp. 191–220. Lawrence Erlbaum Assoc., Hillsdale (1990)
7. Lave, J., Wenger, E.: Situated Learning: Legitimate Peripheral Participation. Cambridge University Press, Cambridge (1991)
8. Hutchins, E.: Cognition in the Wild. MIT Press, Cambridge (1995)

9. Nonaka, I.: Chishiki-Souzou no Keiei [A Theory of Organisational Knowledge Creation]. Nihon Keizai Shimbun-Sha: Tokyo (1990) (in Japanese)
10. Nonaka, I.: The knowledge-creating company. Harvard Business Review, 96–104 (1991)
11. Nonaka, I.: A dynamic theory of organizational knowledge creation. Organization Science 5(1), 14–37 (1994)
12. Nonaka, I., Takeuchi, H.: The Knowledge-creating Company. Oxford University Press, New York (1995)
13. Nonaka, I., Toyama, R., Konno, N.: SECI, Ba and Leadership: a Unified Model of Dynamic Knowledge Creation. Long Range Planning 33, 5–34 (2000)
14. Varela, F.J., Shear, J.: First-person Methodologies: What, Why, How? Journal of Consciousness Studies 6(2-3), 1–14 (1999)
15. Toffler, A.: Powershift: Knowledge, Wealth and Violence at the Edge of the 21st Century. Bantam Books, New York (1990)
16. Drucker, P.F.: Post-Capitalist Society. Butterworth Heinemann, Oxford (1993)

An Extended Model of Knowledge Governance

Laszlo Z. Karvalics[1] and Nikunj Dalal[2]

[1] Szeged University, 6722 Szeged, Egyetem tér 2. , Hungary
zkl@hung.u-szeged.hu
[2] Oklahoma State University, Stillwater, Oklahoma 74078, USA
nik@okstate.edu

Abstract. In current times, we are seeing the emergence of a new paradigm to describe, understand, and analyze the expanding "knowledge domain". This overarching framework – called knowledge governance – draws from and builds upon knowledge management and may be seen as a kind of meta-layer of knowledge management. The emerging knowledge governance approach deals with issues that lie at the intersection of organization and knowledge processes. Knowledge governance has two main interpretation levels in the literature: the company- (micro-) and the national (macro-) level. We propose a three-layer model instead of the previous two-layer version, adding a layer of "global" knowledge governance. Analyzing and separating the main issues in this way, we can re-formulate the focus of knowledge governance research and practice in all layers.

Keywords: knowledge management, knowledge governance, management cybernetics, knowledge networks, knowledge ecology, communities of practice, wisdom management.

1 Introduction

Starting in the mid-1990s, there has been an explosion of ideas and research around the concept of knowledge management. Basically, knowledge management (KM) is concerned with methods, strategies, and technologies to effectively create, represent, and distribute both tacit and explicit knowledge in an organization. Though there are many schools of thought discussed under the broad umbrella of knowledge management and there are wide disagreements about definitions and approaches, broadly speaking, the core components of KM include people, processes, technology, culture and structure [1]. The focus of **early efforts in knowledge management was on technologies to store, capture, and distribute knowledge**. In the next wave, the focus shifted to **behavior, culture and tacit knowledge**, but the development was more conceptual than practical. The recent wave is epitomized by **communities of practice** [2] that emphasize practicality. These emerging waves reflect the new reality that the main methodologies and approaches of classical knowledge management, distilled to cook book definitions and consultant practices, are becoming inadequate and inefficient to address the growing complexity of information and knowledge flows.

M.D. Lytras et al. (Eds.): WSKS 2009, CCIS 49, pp. 279–287, 2009.

In current times, we are seeing the emergence of a new paradigm to describe, understand, and analyze the **expanding "knowledge domain"**. This overarching framework called knowledge governance draws from and builds upon knowledge management and may be seen as a kind of meta-layer of knowledge management. The emerging knowledge governance approach deals with issues that lie at the intersection of organization and knowledge processes.

In this paper, we first outline a few forerunner schools and related directions within knowledge management. Next, we provide a brief survey of the moderately short history of the emerging knowledge governance approach. Finally, we propose an extended model for future knowledge governance research.

2 Related Approaches

Many approaches have attempted to address the intersection of organization and knowledge processes in different ways. To illustrate, we briefly describe a few of many historic to recent ideas and schools, not all of which are at the same level of abstraction.

2.1 Management Cybernetics

Taking birth during the systems movement, Cybernetics is a broad interdisciplinary field that examines the design and function of all types of complex mechanical, biological, and social systems involving processes of communication, feedback and control. The goal of cybernetics is to make goal-oriented systems more efficient and effective. Interestingly, "cybernetics" shares the same root as the word "governance".

Stafford Beer was the first to apply cybernetics to management in the 1960s, calling it the "science of effective organization". Management cybernetics focuses on the study of **organizational design, and the regulation and self-regulation of organizations from a systems theory perspective**. Beer's viable system model [3] (VSM) and Ashby's concept of variety [4] can be used to study different aspects of knowledge management in an individual, organization or network and to model knowledge processes dynamically over time with the goal of improving the organizational systems [5]. Management cybernetic approaches have led to the transformation of organizations particularly of public bodies such as governments and the advancement of new forms of governance.

2.2 Communities of Practice

Arising from the field of organizational learning, the notion of communities of practice refers loosely to interest groups that get together at work and in social settings into which newcomers can enter and learn the sociocultural practices of the community. A community of practice is defined as "a unique combination of three elements: a *domain* of knowledge, which defines a set of issues; a *community* of people who care about this domain; and the shared *practice* that they are developing to be effective in their domain [2, p27]". From our perspective, a community of practice within an organization can be seen as a **practical way to connect people, share existing tacit knowledge, and create new knowledge**. Hence companies are beginning to recognize the importance of communities of practice in dealing with difficult knowledge challenges. From an organizational design standpoint, knowledge initiatives should be designed to leverage and build upon existing "alive" communities of practice. The notion of community of practice need not be restricted to an organization; it extends beyond formal institutional boundaries to distributed networks that we describe next.

2.3 Knowledge Networks

In recent decades, networks have become a ubiquitous form of organization after hierarchies and markets, no doubt helped by advances in information and communications technologies. Networks are neither solely organized like a market nor do they have official hierarchic regulatory structures. Rather, while they may share some characteristics with markets and hierarchies, they are more likely to have informal practices of coordination, common goals or interests, and transaction mechanisms based on attributes such as trust and recommendations rather than prices or administrative orders [6].

The Internet abounds with social networks, gaming and entertainment networks, and knowledge networks. From the perspective of this paper, **knowledge networks provide an effective coordination mechanism for creating, sharing, and distributing knowledge within and across organizations** as well as in specialized domains such as cancer and climate change. Knowledge networks can be seen as being larger, more diffuse and distributed, and less cohesive and practice-oriented than communities of practice.

2.4 Knowledge Markets

In the early knowledge management literature, the "knowledge market" was generally described as a mechanism for **distributing knowledge resources between providers and users**. It was Simard [7] who developed a "*cyclic end-to-end knowledge-market model*". The model is based on nine stages: generate, transform, manage, use internally, transfer, add value, use professionally, use personally, and evaluate. The latest (third generation) vision of knowledge markets is even more ambitious: It views knowledge markets as "formal or informal community contexts, platforms, or environments (real or virtual) used to promote knowledge commerce, trade and exchange, demand and supply, between knowledge buyers and sellers. They are used to organize, coordinate, aggregate, facilitate, communicate, broker, and network flows and exchanges of knowledge between knowledge seekers and knowledge providers [8]".

2.5 Learning Organizations

According to the influential vision of Peter Senge [9, p 3], learning organizations are: "organizations where people continually expand their capacity to create the results they truly desire, where new and expansive patterns of thinking are nurtured, where collective aspiration is set free, and where people are continually learning to see the whole together." From the perspective of learning organizations, the focus of knowledge creation should begin with individuals and **in helping them to learn using tools such as: systems thinking, personal mastery, mental models, building a shared**

vision, and team learning [9]. The prescription for organizational design in this perspective is to build a learning organization using the methods of Senge and other practitioners who have built on this approach.

2.6 Knowledge Ecology

Closely allied to the notion of learning organizations, the concept and practices of "knowledge ecology" recognize the **systemic and holistic nature of knowledge and aspire to go beyond knowledge management to develop shared intelligence and collective wisdom.** Knowledge ecology, in contrast to command and control hierarchies, aims to unleash the full potential of its participants in order to design and support self-organizing knowledge ecosystems, whereby information, knowledge, intelligence, and wisdom can cross-fertilize and feed on one other [10]. The practices of knowledge ecology are in some ways similar to communities of practice though there seems to be a more formal organizing direction in this approach.

2.7 Wisdom Management

Many recent approaches have begun to recognize that the focus on mere knowledge is not enough. Profound organizational and societal crises are *crises* not necessarily because of a lack of information, knowledge or other resources but because of greed, lack of values, and a dearth of wisdom. While wisdom has been a focus of philosophical and religious traditions since antiquity, only in recent times are we seeing attempts to understand wisdom from an organizational science perspective. The data-information-knowledge-wisdom hierarchy [11] [12] is a cornerstone in the knowledge management literature. Although wisdom is at the pinnacle of this hierarchy, few attempts have been made to define wisdom, which is seen as a multifaceted construct. The task is so challenging that the University of Chicago has recently launched a $2 million research program on the nature and benefits of wisdom. Combining the notions of wisdom, communities of practice, and networks, Dalal [13] has proposed the vision of wisdom networks as communities that aim to actualize and inculcate wisdom in specific domains. Wisdom networks are involved in **inquiry of key issues in a domain, the creation and dissemination of wisdom-based learning, counseling, participation in community initiatives, and in building linkages with other wisdom networks.**

3 Knowledge Governance

Thanks to Nicolai G. Foss, the corporate economy "guru", the concept of knowledge governance is expanding and gaining popularity. It is not Foss who coined and introduced the term "knowledge governance", but he provided the biggest contribution to its meaningful enrichment with his Italian colleague, Anna Grandori [14] to lay the disciplinary foundation.

In Foss's theoretical works, knowledge governance is a distinctive approach, having many cross-connections with knowledge management [15]. At first, he refers to only the cross-points of general management, strategic issues and human resource management [16] and defines knowledge governance as follows: "The 'knowledge governance approach' is characterized as a distinctive, emerging approach that cuts across the fields of knowledge management, organization studies, strategy, and human resource management. Knowledge governance is taken up with how the deployment of governance mechanisms influences knowledge processes, such as sharing, retaining and creating knowledge. It insists on clear micro (behavioral) foundations, adopts an economizing perspective, and examines the links between knowledge-based units of analysis with diverse characteristics and governance mechanisms with diverse capabilities of handling these transactions." But over the next two years, Foss gradually broadened the scope of knowledge governance to connect with the **management of intellectual capital, innovation theory, technology strategy, and the international business** itself [17]. In the most recent vocabulary of Foss, knowledge governance "refers to choosing structures and mechanisms that can influence the processes of sharing and creating knowledge [17].

Knowledge governance has two main interpretation levels in the early literature: the company- (micro-) and the national (macro-) level. For example, knowledge governance has been discussed as a profitability issue at the company level and as an effectiveness issue at the government level in the research project series of the University of Bonn, The Center for Development Research (ZEF, Zentrum für Entwicklungsforschung) started in 2000.

Whitley [18] conceptually classified knowledge governance as:

- *entrepreneurial knowledge governance* based upon knowledge codification and privatization, and the organizational methods of generation and usage of new corporate knowledge, and
- *associative knowledge governance*, which addresses the macro-level distribution of the complex forms of knowledge.

They are simultaneously evolving narratives sharing many similarities such as the inclusion of holistic approaches and high-level planning and control functions. For instance, Smits és Moor [19] composed an indicator system to measure the effectivity of corporate knowledge management, dubbing it the "Knowledge Governance Framework", while Mariussen [20] used it to address the integration of the knowledge system and managing on a nation-state level. Both directions are spreading: Vale and Caldeira [21] used the knowledge governance approach for proximity-centered research in localized production systems such as the footwear industry in the north region of Portugal whereas the latest 2009 publications reflect the research focus on macro-level regulation dimensions such as copyright and patent issues.

4 An Extended Framework

We propose a three-layer model instead of the previous two-layer versions, adding a layer of **"global knowledge governance"**. Separating the main issues in this way, we can **re-formulate the focus of knowledge governance research and practice** in all layers.

In the Enterpreneurial layer *(Table 1)*, the main task is to exactly define the relation between knowledge management and knowledge governance: we identify four determining directions (dynamic relations) as shown in the table. These pertain to

performance measurement, reengineering, design, and fusion of local practices outside its boundaries.

On the nation-state level, the knowledge governance approach strongly overlaps with the latest knowledge-related narratives, refreshing some multi-contextual traditional policy fields (*Table 2*).

Table 1. Enterpreneurial knowledge governance: Four directions of penetration at the company level

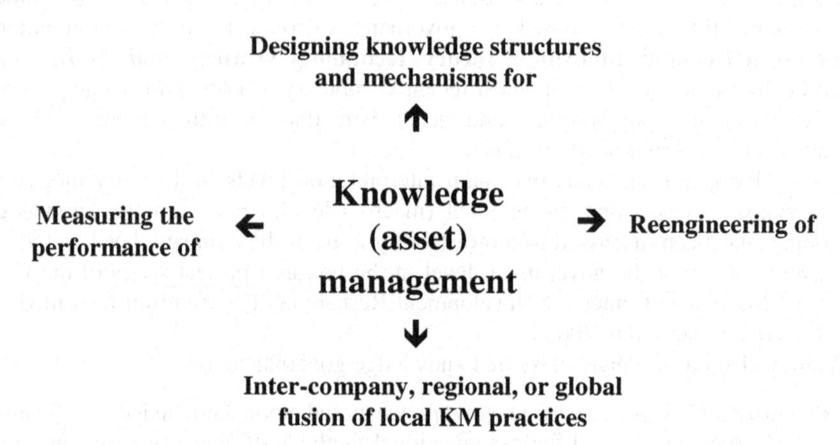

Table 2. Fields of knowledge governance at the nation -state level

"Traditional" policy fields	New type of narratives and interventions
Innovation policy	Data and Knowledge asset policy, "national crowdsourcing" models
Science policy	Planning the structure and resource map of Natural-, Life-, Technical Sciences and Humanities
Education policy and literacy	Information literacy, talent management, lifelong learning schemes
Media and dissemination of scientific information	Nation-state reactions on current Brain Drain, Brain Gain, Brain Sharing issues
Knowledge Industry Development (Fostering attractivity and visibility)	Competition in creative industries, talent hunting
Copyright, patent issues	Indigenous knowledge management, copyleft

There have been a few improvements on shaping global information and knowledge flow since the UN's 1974 Declaration on New International Information Order (NIIO). The radically changing knowledge environment and information infrastructure arrogate concerted and scientifically substantiated efforts to establish a systematic knowledge background. Knowledge governance is not only a likely direction, but an umbrella-like promising paradigm to synthesize and re-formulate similar approaches.

Table 3. Knowledge governance at a global level

"Phenomena" to reflect	Scientific domain	Development/ Policy/ Planning issues
International cooperation in the fields of education, science, and communication	Cultural and Communication politics	Reengineering of UNESCO-type global coordination
Collaborative Research Megaprojects	Sociology of Science	New generation workflow tools, Regulation challenges
Knowledge readiness (knowledge development indicators)	General politics	Narrowing the gap between the developed and under-developed nations and regions
Global Conference and Publication Industry	Knowledge Management	Re-thinking of the channels of distribution of knowledge
Circulation of Brains	(Im)migration, Demography, Sociology	Equation mechanisms, regulation, monitoring, reducing the digital divide
Globalized higher education, virtual universities	Pedagogy, Economy	Quality management, equivalence and interoperability issues, learning management

5 Linkage to Knowledge Society Vision

In the history of knowledge management, it has always been deemed important and useful for the public sphere (i.e. the transaction-rich government institutions) to **adapt and implement the best corporate knowledge management methods and solutions into their everyday management practice**. (In the vocabulary of the Open Research and Open Knowledge Society these standardized methods can be described as "hard aspects" of the knowledge society.)

As the importance of the knowledge governance paradigm grows over the years, there is no doubt that this adaptation process will expand to many global inter-stakeholder policy fields. The resulting professionalization can refresh and revolutionize international multi-agent cooperation forms such as inter-school exchange programmes and collaborative citizen science projects, energizing millions of professional amateurs (ProAms) in the process of knowledge generation and distribution. The knowledge governance approach presents a potentially innovative way to form radically new knowledge producing megamachines [22] that combine and channel the creative energies of the scientific elite, the teachers, and the students towards a sustainable future.

6 Future Research Issues

Based upon our analysis, we propose a few research issues in the emerging knowledge governance approach. One research issue that applies to all levels is: How does an organization, nation, or global group determine what is valid knowledge? Knowledge governance policies address the creation and sharing of knowledge. But what if the created knowledge is invalid or erroneous? This has cost and reputation implications.

Surely, before such knowledge is disseminated, there must be ways to verify and validate the knowledge. We believe that the knowledge governance approach will be strengthened in its normative intent if it incorporates **knowledge verification processes** to ensure valid, reliable, relevant, current and well-founded knowledge.

Other future research issues include: How can an organization, nation, or global community support the creation of **higher forms of knowledge and related attributes such as intelligence, sensitivity, and wisdom**? How can we deal with the inevitable **politics** of knowledge governance, which is likely to be a greater issue at a global level, but can be expected at all three levels? And finally: what is the anatomy of trespassing and cross-connections between the three layers? How can the **global knowledge governance actors and interventions influence the nation-state and company layers** – and vice versa? These are just a few of many issues that need attention in the emerging field of knowledge governance. Clearly, knowledge governance has major implications for organizations, nations and states, and the global society as a whole.

References

1. Spender, J.-C., Scherer, A.G.: The Philosophical Foundations of Knowledge Management: Editors' Introduction. Organization 14(1), 5–28 (2007)
2. Wenger, E., McDermott, R., Snyder, W.M.: Cultivating communities of practice: A guide to managing knowledge. Harvard Business School Press, Boston (2002)
3. Beer, S.: Diagnosing the System for Organizations. John Wiley & Sons, Chichester (1985)
4. Ashby, W.R.: Introduction to Cybernetics. Meuthen, London (1964)
5. Leonard, A.: The viable system model and knowledge management. Kybernetes 29(5/6), 710 (2000)
6. Thomson, G.: Between Hierarchies & Markets: the logic and limits of network forms of organization. Oxford University Press, Oxford (2003)
7. Simard, A.: Knowledge markets: More than Providers and Users. IPSI BgD Internet Research Society Transactions 2(2), 4–9 (2006)
8. Davis, B.: Harnessing Knowledge Markets Research Program Kaieteur Institute for Knowledge Management Toronto 3, 4 (2007)
9. Senge, P.: The Fifth Discipline: the Art and Practice of the Learning Organization. Doubleday, New York (1990)
10. Por, G.: Nurturing systemic wisdom through knowledge ecology. The Systems Thinker 11(8), 1–5 (2000)
11. Ackoff, R.L.: From Data to Wisdom. Journal of Applied Systems Analysis 16, 3–9 (1989)
12. Zeleny, M.: Management support systems: towards integrated knowledge management. Human Systems Management 7(1), 59–70 (1987)
13. Dalal, N.: Wisdom Networks: Towards a Wisdom-Based Society. In: Lytras, M., Carol, J. (eds.) The Open Knowledge Society: A Computer Science and Information Systems Manifesto. Book Series: Communications in Computer and Information Science, vol. 19. Springer, Heidelberg (2008)
14. Grandori, A.: Neither hierarchy nor identity: knowledge governance mechanism and the theory of the firm. Journal of Management and Governance 5(3-4), 381–399 (2001)
15. Foss, N.J.: The Knowledge Governance Approach. Copenhagen Business School Center for Strategic Management and Globalization Working Paper Series (2005), http://ssrn.com/abstract=981353 (February 11, 2009)

16. Foss, N.J.: The Emerging Knowledge Governance Approach: Challenges and Characteristics Knowledge Governance Primer Organization 14, 29–52 (2007), http://organizationsandmarkets.com/2007/02/05/knowledge-governance-primer/ (February 11, 2009)
17. Foss, N.J., Michailova, S. (eds.): Knowledge Governance. Processes and Perspectives. Oxford University Press, Oxford (2009)
18. Whitley, R.D.: The Institutional Structuring of Innovation Strategies: Business Systems, Firm Types and Patterns of Technical Change is Different Market Economies. Organizational Studies 21, 855–886 (2000)
19. Smits, M., Moor, A.D.: Measuring Knowledge Management Effectiveness in Communities of Practice. In: Proceedings of the 37th Hawaii International Conference on System Sciences, pp. 236–244 (2004)
20. Mariussen, A.: New forms of knowledge governance. In: Basic outline of a social system approach to innovation policy DRUID Summer Conference: Creating, Sharing and Transferring Knowledge Copenhagen, June 12-14 (2003), http://www.druid.dk/uploads/tx_picturedb/ds2003-832.pdf (February 20, 2009)
21. Vale, M., Caldeira, J.: Proximity and knowledge governance in localized production systems: the footwear industry in the north region of Portugal European Planning Studies (15/4), 531–548 (2007)
22. Karvalics, Z.L.: The Biggest Human GRID-s of the Future: Hybridization of Science and Public Education. In: Lytras, M.-D., Carroll, J.M., Damiani, E., Tennyson, R.D., Avison, E., Vossen, G., Ordóñez de Pablos, P. (eds.) The Open Knowlege Society. A Computer Science and Information Systems Manifesto, First World Summit on the Knowledge Society, WSKS 2008, Athens, Greece, September 24-26. Proceedings. Communications in Computer and Information Science, vol. 19, pp. 53–56. Springer, Heidelberg (2008)

Assessing the Value Dimensions for Customers in Knowledge Intensive Business Services

Francesco Sole[1], Daniela Carlucci[1], and Giovanni Schiuma[1,2]

[1] CVM-DAPIT, Università degli Studi della Basilicata
Via dell'Ateneo Lucano, 10,
85100 Potenza, Italy

[2] Center for Business Performance, Cranfield School of Management,
Cranfield, Bedfordshire
MK43 0AL, UK

{francesco.sole,daniela.carlucci,giovanni.schiuma}@unibas.it

Abstract. With the rise of knowledge economy, the importance of Knowledge Intensive Business Services (KIBS) has gradually increased as well as their overall impact on economy. However, in comparison with the manufacturing sectors, KIBS sector remains poorly studied. Especially little prior research has been done about the description and assessment of service value dimensions from a business customer's perspective. This paper, on the basis of a literature review, identifies and describes eight different value dimensions. These dimensions represent the building blocks of a conceptual model proposed for assessing the value created by KIBS for business customers. Moreover, the paper describes an application of the model in a real case.

Keywords: KIBS, service value, customer, model, case example.

1 Introduction

Nowadays Knowledge Intensive Business Services (KIBS) hold a key place in the current research agenda. This is because with the rise of knowledge society, the importance of these business services has gradually increased as well as their overall impact on economy [19], [25]. This is mainly due to the fact that KIBS contribute to create a fertile soil for innovation and diffusion of new knowledge and practices between firms and other organizations, as well as across the industries [6], [33]. The increasing importance of KIBS has meant a growing research attention towards these particular businesses. However, in comparison to the manufacturing sectors, KIBS remain poorly studied and their future development has rarely been considered in terms of policies and roles in their respective innovation and productive systems [28]. This paper attempts to enrich the existing theoretical literature on KIBS, by providing some valuable insights, theoretically founded, concerning an area where little prior research has been done, i.e. the description and assessment of the service value dimensions from a business customer's perspective. This represents an outstanding research topic, because of both the phenomenal growth of KIBS sector and the increasing attention towards the value embedded and delivered by knowledge intensive

M.D. Lytras et al. (Eds.): WSKS 2009, CCIS 49, pp. 288–297, 2009.

services. Especially, the paper proposes a conceptual model for identifying and assessing the value produced by KIBS from a business customer's perspective. The model is grounded on the theoretical insights obtained from a literature review regarding the nature and the determinants of value of a knowledge intensive service, as perceived by customers. In particular, starting from the analysis of the management literature, eight different value dimensions are identified and described. These dimensions also represent the building blocks for assessing the value created by KIBS for business customer. Moreover, the paper describes an application of the model in a real case. The paper is structured as follows. In the first section, the context of the research is briefly described. In particular, some main features characterizing KIBS are illustrated. In the second section, the results of a literature review regarding the service value and its determinants are presented. In the third section, starting from the theoretical insights emerging from the literature review, a conceptual model for identifying and assessing service value dimensions for business customer is introduced. In the fourth section, a description of the application of the model is provided. Finally, in the last section, conclusions and suggestions for future research are provided.

2 Knowledge Intensive Business Services

In the last decades, with the rise of knowledge economy, the importance of KIBS has gradually increased as well as their overall impact on economy [19], [25]. This is because these particular companies play a pivotal role in facilitating innovation and economic growth across the sectors [33]. KIBS are believed to serve as 'bridges' for knowledge flow between firms and other organizations, as well as across the industries [6]. Moreover KIBS represent one of the fastest growing areas of the knowledge economy [10], [21]. There are several reasons underpinning this growth such as, for example, the increasing demand for knowledge inputs from organizations, the rise of "new" organisational strategies and management thinking, such as the 'outsourcing' and the focus on core competencies, as well as the increasing attention on service and intangible elements of production and products. The appreciation of the importance of KIBS has also meant a growing research attention towards these particular businesses. Especially, since the last decade, knowledge-intensive services hold a key place in the research and policy agenda. In this regard [21] identifies five reasons founding the increasing interest towards KIBS: i) the rapid growth of the sector; ii) the evidence of the important role of KIBS in enabling upgrading and innovation in firms; iii) the role of KIBS in improving the innovation and export performance of SMEs; iv) the role of KIBS as intermediaries between public sector research organizations and business; v) the importance of KIBS in assisting the formation and survival of new firms that are exploiting technological or market-based opportunities. The emphasis on KIBS is also revealed from the great amount of conceptualizations proposed for these businesses. In particular, looking at definitions of KIBS provided in the management literature, it rises that there is not a common accepted definition [35] and the precise connotation of KIBS remains debatable. However, it seems possible to identity some core elements that typify KIBS. As knowledge intensive service (see e.g. [20], [23]), KIBS are characterized as follows: i) knowledge is an important input of services; ii) services are significantly based on professional competence and knowledge and, therefore, on human capital (e.g. [1], [25]); iii) services themselves are sources of knowledge for a customer or services are used as an input for developing a customer's own knowledge; iv) there is an intensive interaction between a customer and a service provider. It provides a possibility for distribution and creation of new knowledge. Moreover as business service companies, KIBS act like private service companies which sell their services to other companies or to the public sector [26]. Summarizing, as argued by [27] and by [15], KIBS are characterized by: i) knowledge intensiveness of services (which separates them from other services), ii) consultancy-function (i.e. solving function), and iii) service production in close interaction with the client (client orientation). These private companies acquire, analyze, generate and transfer knowledge, embedded in services or products, to public or private organizations or companies, which are not able or don't intend to develop by themselves that knowledge. Indeed, the core activity of KIBS is contribution to the knowledge processes of their clients, through the exploitation of a significant number of highly skilled experts from different scientific branches. Since KIBS offer mainly information and knowledge to their clients and there is a huge variety of knowledge, the structure of the KIBS sector is fragmented. The NACE Rev. 2 classifies KIBS into the following subsector: computer and related services, research and development, legal services, accounting, auditing etc. services, marketing services, technical services, management consultancy, labour recruitment services, training in the private sector. Therefore, knowledge represents the central driver of KIBS activities. Especially the effective and efficient development and exploitation of knowledge play a key role in their value creation process. In fact, as underlined by [11], the provision of service is not an instant purchase, but a process of knowledge transfer, requiring reciprocal learning. Especially, as above highlighted, for these businesses, providing service consists mainly in interacting with the client and finding, during a co production process, a solution to a problem together utilizing knowledge as the most important and critical resource [7].

3 Business Services and Value for Customer: A Literature Review

Service value appears to be a concept which is worthy of increased attention in order to better understand service delivery systems and processes necessary to analyse the customers' wants and needs as well as to supply managers greater insights regarding the decision-making process of customers [5], [8], [36]. Furthermore, service value is a fundamental concept in order to analyse service productivity. As [14] has argued, internal efficiency cannot be managed separately from effectiveness or external efficiency, which can be defined as the firm's capability to produce a certain level of customer value with a given resources structure. On the other hand, from the customer's viewpoint, obtaining value is a fundamental purchase goal and pivotal to all successful exchange transactions [18]. In turn, as highlighted by [17] in their investigation about the service-profit chain, value created for customer means customer's satisfaction and loyalty and, finally, provider's profit and growth. Therefore, the above quotes leave little doubt that service value represents a key factor of interest to service provider companies. In management literature, service value, from a customer's perspective, is widely described as a ratio of total perceived benefits and total perceived costs by a customer [36], or conceptualized as a comparison of the main give and get components [36]. In particular, [8] stated that service value is a cognitive trade-off between customer's perceptions of quality and sacrifice. In this regard, [16] identify service quality as the main "get" characteristic while [9] recognize the sacrifice made to acquire as the salient "gives" component. As a consequence, from the analysis of literature it rises that value is the key linkage between perceived quality and perceived sacrifices. Therefore, as argued by [4], service value is a richer measure of customers' overall evaluation of a service than perceived service quality and, for this reason, according to [5], it could be considered as a function of customers' sacrifice and service quality. Regarding customers' sacrifice, it represents the main "gives" component of service value and, according to [5] and [36], it appears to be perceived by customers as involving both monetary attributes and such non-monetary factors as time and efforts. In addition, [32] suggested that perceived risk should also be identified as a sacrifice component, due to risk is an implicit part of the cost of the acquisition and use of any service. To summarise, according to [5] customers' sacrifice, in business services, can be considered as a composite of perceived monetary price, perceived non-monetary price, and perceived risk. About service quality, several studies have stated that service quality involves more than an outcome dimension; it also includes the way or manner in which the service is delivered during interactions between provider and consumer. In particular, [12] argued about two types of service quality: technical quality, which involves what customers actually receive from the service (i.e. the outcome of the service) and functional quality, which involves the manner in which customers receive the service (i.e. the process of service delivery). Analogously, [29], in their Servqual tool, describe "outcome" and "process" as the two main dimensions of a service. [22] have identified two quality dimensions for the customer: "process quality" and "output quality". The relative importance of these two dimensions has been differently evaluated by scholars. For example, with reference to functional (or process) dimension, [24] described "client relations" as a fundamental activity in the delivery of business services. Otherwise, starting from the consideration that a business service is an intermediate input which is used to achieve the strategic objectives of the customer firm, [34] argued that technical/outcome dimension is of the highest priority than the functional/process dimension. Moreover, with regard to the value of quality dimensions for customers, it seems also important to outline that it depends on specific business service sector. In particular, [2] have stated: "the use by clients of perception and intermediate process criteria to evaluate their advisors may be particularly significant for services with a high technical content, where clients may not have the skills or expertise adequately to evaluate the outputs" (p.257). They conclude that the evaluation of the overall outcomes of a business service is a complex activity which usually requires both objective and subjective criteria.

4 Customer's Value Dimensions in KIBS: A Conceptual Model

Despite both the phenomenal growth of KIBS sector and the relevance of service value as a fundamental measure of customers' overall evaluation of a service, only few academic researchers have attempted to define and model the service value in KIBS. In such a prospect, the study aims to extend the body of literature concerning service value, focusing the attention on KIBS sector. Especially, the paper proposes a conceptual model for identifying and measuring customers' value dimensions in KIBS. The model is grounded on the theoretical insights obtained from the literature review regarding nature and determinants of service value as perceived by customers. In line with statements of [5], the foundation of the theoretical model is that service value embraces two main components: service quality and customers' sacrifices. Especially, starting from the insights revealed by the literature review, the main value dimensions which constitute service quality and customers' sacrifices in KIBS have been identified They are: monetary price, non-monetary price, risk, output, outcomes, impact, relationship and methodology (see Figure 1).

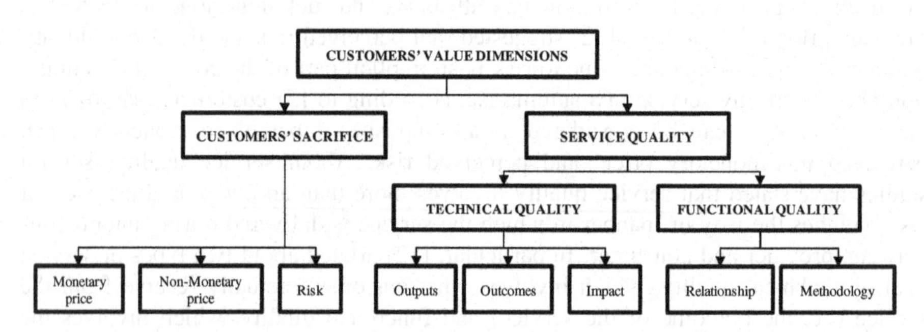

Fig. 1. Customers' Value Dimensions in KIBS

The model includes, according to [5], three main value dimensions of customers' sacrifices: monetary price, non-monetary price and risk. While monetary price dimension is the most obvious factor, otherwise non-monetary price dimension such as time and effort require a further explanation. In particular, we have recognised time and effort as key sacrifice factors, due to the strong involvement of customers during the process of knowledge production and delivery in KIBS. Finally, according to [32], we consider risk dimension as an inherent part of the cost of the acquisition, due to the strong intangible component of a KIBS. Especially, according to [5], several risk components such as financial risk, performance risk and social risk could be placed within the more general risk dimension. In addition, due to the features of the co-production process, which is characterized of a client's situational knowledge sharing with provider, a subsequent customer's risk is strictly closed to the professionalism of the provider related to the protection of customer's private information. With reference to service quality dimensions, in line with the statements of [30] and [17] our assumption is that service quality evaluation is not made exclusively on service outcome; it also involves evaluation of the process of service delivery. In particular, we have split service quality in two main dimensions, as suggested by [13]: technical quality ("what" is delivered) and functional quality ("how" the service is delivered during the service encounter). As previously highlighted, some authors have associated technical quality dimension to service outcomes [31] or output [22], to briefly describe the results for which providers were commissioned. More specifically, [31] have argued that technical quality was clearly the most significant dimension in explaining customers' value. Therefore, starting from the above studies, we have introduced in our model three main technical quality dimensions which well describe all the results provided by a KIBS to its customers, respectively in short, medium and long term, i.e. output, outcomes and impact. At the basis of our choice there is the assumption that services provided by KIBS are intermediate inputs for other firms' production processes, thus customer uses them to achieve several performance objectives with regard both at operations and strategy level. In such a prospect, we define outputs as the direct results of services activities. KIBS outputs, for example, might be the materials and tools. On the other hand, outcomes are the results in the medium-long term. Especially we have distinguished outcomes strictly closed to the customer's business processes from the outcomes related to an individual level. With reference to business processes, we have started from the assumption of [3] regarding the process improvement. They sustain that the process improvement could be generated in the following ways: 1) by improving the execution of specific job steps; 2) eliminating the need for particular inputs or outputs; 3) removing an entire step from the responsibility of the customer; 4) addressing an overlooked step; 5) re-sequencing the steps; 6) enabling steps to be completed in new locations or at different times. Therefore, we have introduced the above six outcomes dimensions to identify the results of a KIBS with regard to a process level. With regard to the individual level outcomes, we refer to precise changes occurred to customers in terms of attitudes, behaviors, knowledge, skills, status, or level of functioning and caused by the received services. Finally, we have introduced the concept of impacts, which are related to organizational, community, and/or system level changes in the long term. They are a logic consequence of the above mentioned outcomes, and they might include improved conditions, increased capacity, and/or changes in the policy arena. Regarding to the functional quality in KIBS, i.e. the manner through which the provider delivers the technical service component, our model provides, in line with [31], two main dimensions, i.e. methodology and relationship. Especially, the methodology dimension reflects the provider's modus operandi; it includes attributes such as problem solving ability, reliability in meeting deadlines, responsiveness and general professionalism displayed. While relationship reflects the relationships developed between client and service provider. As previously mentioned, many scholars have stated that for KIBS the quality of relationships between customer and provider, is a key factor for customers' satisfaction, since it strongly affects the knowledge generation and transfer. We have identified, within relationship dimension, some specific factors such as the involvement intensity of customers in knowledge co-production process and the provider's willingness to share and transfer own knowledge. The identified value dimensions can properly drive the assessment of the value created by a KIBS for its customers. In this regard several authors (e.g. [4], [12], [17], [31]) have argued value assessed by customers is always relative because it is based both on perceptions of the way a service is delivered and on initial customer expectations. In such a prospect customers' value assessment is the result of a trade off between customers' expectations and perceived performance. Moreover we assume that for comprehensively assessing the value created for customer, it is necessary to take account of weight assigned from a customer to each item included in the value dimensions of the proposed conceptual model. This is because considering the weight factor refines the understanding both of value created, as a whole, and its determiners. Therefore, the value created for a business customer by a KIBS has to be assessed by taking account of customer's *expectation* and *performance perception* as well as of the *weight* assigned from the customer to each item characterizing the value dimensions of the model. From an operational point of view, the assessment process may involve different stages. For example, starting from the value dimensions shown in Figure 1, provider may investigate the main items of each value dimension for the business customer. Then, the customer may associate a weight to each item and formulate an expectation, for example on the basis of its current wants and needs, past experiences and provider's promises. Finally, both during knowledge co-production process and after service delivery, customer may evaluate service results, related to each value dimension, as a trade off between priority, expectation and perceived performance. Through the assessment process, KIBS providers may better understand how customer assesses the delivered service. In particular, business provider may assess what are the service value items/dimensions that have been or should be accomplished and the related importance from a customer's point of view.

5 An Application of the Model

The model has been implemented within a company which provides expert advice and a range of bespoke consulting services specifically designed to deepen and enhance business engagement with culture. The company also supports the corporate clients in order to unlock the creativity in their business and talent in their people, by using arts as a tool for transforming and changing lives on both an individual and corporate level. In the paper the company's name is not mentioned for confidential reasons. The model has been implemented with the main aim to identify those items, describing the eight dimensions of the model, particularly valuable for assessing value created for the company's customers. The service analysed in this study was a consulting service aimed at improvement of worker's public speaking and communication skills. The service is delivered through the development and implementation of tips and techniques able to develop and increase people's expertise and confidence. The identification of the items has required a close interaction between managers and researchers. In particular, during a focus group, researchers have guided the development work by providing consultation about the meaning of the model's dimensions and their use. As a result, 24 items, differently spread on the model's dimensions, have been identified. They are: *Monetary Price*: fee paid; *Non-Monetary Price*: personal sacrifice; *Output*: giving time and space to prepare and rehearse (i.e. mental preparation); communicating key messages; confidence in giving a presentation; engaging the audience to create an impact; inspiring emotion and transferring energy; using voice effectively; using eye contact; using physical presence; using breathing technique; *Outcomes*: personal change; professional development; *Impact*: impact on working activities performance; *Methodology*: material and information received before and during the initiative; competence and professionalism of the facilitators; training methodology; structure and duration of the course; location and facilities; *Relationships*: relationships with facilitators; relationships with organisers; atmosphere in the group; personal level of engagement.

Starting from the identified items, a questionnaire aimed to assess customers' priority and expectations has been formulated. This assessment is part of an ongoing project. At the moment the evaluation has involved thirty people.

6 Conclusions

The knowledge society has brought a competitiveness revolution that requires organizations to re-examine the way in which they operate and create value. In particular knowledge has become for organisations the key source for value generation and competitiveness. In such a context, KIBS are becoming increasingly important. This is because KIBS are valuable vectors and producers of knowledge. They essentially turn information and tacit knowledge into practical business tailored solutions for private and public organisations. They base both their own success and that of their clients mainly on the acquisition, development, and management of knowledge. In doing this, KIBS act as valuable "bridges" for knowledge flow between firms and other organizations, as well as across the industries. Despite the growing importance of KIBS into knowledge society, there is yet considerable work to be done for better understanding their role. Especially, more knowledge is required about the value that they generate and how this value can be evaluated. This paper proposes a conceptual model for evaluating the value generated by KIBS according to a customer's point of view. Both providers and customers may take advantage by using the model. Provider may use the model both in order to better manage the value creation process for a single business client and to define and develop the company's strategy. In fact, provider may define and share with the client all the main value dimensions connected to the service provision and ranking them according to the customer's priority and expectations. This, in turn, may help provider to sell a credible and shared promise to client, to explain why and how the service gives value, to better deliver the promised value as well as to learn from the process provision in order to improve future activities. On the other hand, by using the model for investigating the value dimensions for business customers, provider may obtain useful insights for defining and developing a successful strategy. In fact the model could be also useful for provider in order to investigate what are the current value gaps for their business customers. As previously mentioned, also business customers may take advantage by using our model. In fact it represents useful tool both to "negotiate" with the provider the key value dimensions to be accomplished and to compare fairly results obtained and sacrifices suffered. Further research is needed in order to test, first of all, the conceptual model, by examining the nature and robustness of the consistency between service value dimensions as evaluated in practice by customers and the value dimensions suggested in the model. There is a need to develop an operative tool to measure, consistently with the proposed model, the value dimensions considered from a business customer in KIBS sector. Moreover, it would be useful to investigate how business customers' priority and expectations differ across several subsectors of KIBS. Finally, the key role of KIBS in knowledge society calls for an enlargement of the evaluation point of view to remaining KIBS stakeholders.

Acknowledgements

The authors wish to thank the Italian Minister of University and Scientific Research for financing this research, which is part of the National Scientific Research Programs (PRIN 2007).

References

1. Alvesson, M.: Management of Knowledge-Intensive Companies. Gruyter, Berlin (1995)
2. Bennett, R.J., Robson, P.J.A.: The Advisor-SME Client Relationship: Impact, Satisfaction and Commitment. Small Business Economics 25, 255–271 (2005)
3. Bettencourt, L., Ulwick, A.: The Customer-Centered Innovation Map. Harvard Business Review (2008), http://www.hbr.org
4. Bolton, R.N., Drew, J.H.: A Multistage Model of Customers' Assessments of Service Quality and Value. Journal of Consumer Research 17, 375–384 (1991)
5. Cronin, J., Brady, M., Brand, R., Hightower, R., Shemwell, D.: A cross-sectional test of the effect and conceptualization of service value. The Journal of Services Marketing 11(6), 375–391 (1997)
6. Czarnitzki, D., Spielkamp, A.: Business services in Germany: bridges for innovation. The Service Industries Journal 23(2), 1–31 (2003)
7. Den Hertog, P.: Knowledge-intensive business services as co-producers of innovation. International Journal of Innovation Management 4(4), 491–528 (2000)
8. Dodds, W.B., Monroe, K.B., Grewal, D.: Effects of price, brand and store information on buyers' product evaluations. Journal of Marketing Research 28, 307–319 (1991)
9. Drew, J.H., Bolton, R.N.: Service value and its measurement: local telephone service. In: Suprenant, C. (ed.) Add Value to Your Service: 6th Annual Services Marketing Proceedings. American Marketing Association, Chicago (1987)
10. European Foundation for the Improvement of Living and Working Conditions. The knowledge-intensive business services sector content (2006), http://www.eurofound.europa.eu/emcc/source/eu05011a.htm
11. Gadrey, J., Gallouj, F.: The Provider-Customer Interface in Business and Professional Services. The Service Industries Journal 18(2), 1–15 (1998)
12. Grönroos, C.: Strategic Management and Marketing in the Service Sector. Research Reports No. 8 (1982)
13. Grönroos, C.: A service quality model and its marketing implications. European Journal of Marketing 18(4), 36–44 (1984)
14. Grönroos, C.: Service management and marketing, A customer relationship management approach. Wiley, Chichester (2000)
15. Haataja, M., Okkonen, J.: Competitiveness of Knowledge Intensive Services. In: Proceedings of eBRF eBusiness Research Forum conference, Tampere (2005)
16. Heskett, J.L., Sasser Jr., W.E., Hart, C.W.L.: Service Breakthroughs: Changing the Rules of the Game. The Free Press, New York (1990)
17. Heskett, J., Jones, T., Loveman, G., Sasser, W.E., Schlesinger, L.: Putting the Service-Profit Chain to Work. Harvard Business Review, pp. 164–174 (March-April 1994)
18. Holbrook, M.B.: The nature of customer value. In: Rust, R.T., Oliver, R.L. (eds.) Service Quality, New Directions in Theory and Practice, pp. 21–71. Sage Publications, London (1994)
19. Illeris, S.: Location of Services in a Service Society. In: Huib, E., Meir, V. (eds.) Regional Development and Contemporary Industrial Response, pp. 91–109. Belhaven Press, London (1991)
20. Kautonen, M., Schienstock, G., Sjöholm, H., Huuhka, P.: Knowledge-Intensive Business Services in Tampere urban region. Working Reports of the Work Research Centre at the University of Tampere (1998)
21. Kemmis, S.: Knowledge Intensive Business Services (KIBS) (2006), http://www.servicesaustralia.org.au/pdfFilesResearch/DITRKnowledgeServices.pdf

22. Lehtinen, U., Lehtinen, J.: Two Approaches to Service Quality Dimensions. The Service Industries Journal 11(3), 287–303 (1991)
23. Løwendahl, B.R.: Strategic management of professional service firms. Copenhagen Business School Press, Copenhagen (1997)
24. Maister, D.H.: Balancing the professional service firm. Sloan Management Review 24(1), 15–29 (1982)
25. Miles, I., Kastrinos, N., Bilderbeek, R., Den Hertog, P.: KIBS: their role as users, carriers and sources of innovation. Report to the EC DG XIII Sprint EIMS Programme, Luxemburg (1995)
26. Miles, I.: Knowledge-intensive services and innovation. In: Bryson, J., Daniels, P. (eds.) The handbook of service industries. Edward Elgar, Aldershot (2005)
27. Muller, E., Zenker, A.: Business services as actors of knowledge transformation and diffusion: Some empirical findings on the role of KIBS in regional and national innovation systems (2001); Working Papers Firms and Region No. R2/2001. Fraunhofer ISI, Karlsruhe
28. Muller, E., Doloreux, D.: The key dimensions of knowledge-intensive business services (KIBS) analysis: a decade of evolution (2007); Working Papers Firms and Region No. U1/2007 Fraunhofer ISI
29. Parasuraman, A., Zeithaml, V., Berry, L.: A conceptual model of service quality and its implications for future research. Journal of Marketing 49, 41–50 (1985)
30. Parasuraman, A.: Customer service in business-to-business markets: an agenda for research. Journal of Business & Industrial Marketing 13(4/5), 309–321 (1998)
31. Patterson, P., Spreng, R.: Modelling the relationship between perceived value, satisfaction and repurchase intentions in a business-to-business, service context: an empirical examination. International Journal of Service Industry Management 8(5), 414–434 (1997)
32. Peterson, R.A., Wilson, W.R.: Perceived risk and price-reliance schema as price perceived quality mediators. In: Jacoby, J., Olson, J. (eds.) Perceived Quality. Lexington Books, Lexington (1985)
33. Rodriguez, M., Camacho, J.A.: Are KIBS more than intermediate inputs? An examination into their R&D diffuser role in Europe. International Journal of Services Technology and Management 10(2/3/4), 254–272 (2008)
34. Viitamo, E.: Productivity of Business Services - Towards A New Taxonomy (2007), https://oa.doria.fi/bitstream/handle/10024/43346/isbn9789522144560.pdf?sequence=1
35. Wood, P.: Consultancy and Innovation: The Business Service Revolution in Europe. Routledge, London (2002)
36. Zeithaml, V.A.: Consumer perceptions of price, quality, and value: a means-end model and synthesis of evidence. Journal of Marketing 52, 2–22 (1988)

An Interactive Medical Knowledge Assistant*

Bogdan D. Czejdo[1] and Mikolaj Baszun[2]

[1] Department of Mathematics and Computer Science, Fayetteville State University, Fayetteville, NC 28301, USA
[2] Department of Electronics and Information Technology, Warsaw University of Technology, ul. Koszykowa 75, Warsaw 00-662, Poland
bczejdo@uncfsu.edu, mjbaszun@gmail.com

Abstract. This paper describes an interactive medical knowledge assistant that can help a doctor or a patient in making important health related decisions. The system is Web based and consists of several modules, including a medical knowledge base, a doctor interface module, patient interface module and a the main module of the medical knowledge assistant. The medical assistant is designed to help interpret the fuzzy data using rough sets approach. The patient interface includes sub-system for real time monitoring of patients' health parameters and sending them to the main module of the medical knowledge assistant.

Keywords: Medical Knowledge Bases, Medical Software, Rough Sets.

1 Introduction

There is a dramatic increase of medical knowledge that allows doctors to make better diagnostic and treatment decisions. The same time, however, the doctors are facing with the difficulty of accessing and updating this knowledge. The doctors need a computing system to process such knowledge quickly and to assist them in making optimal decisions. For many years Computer Science and Artificial Intelligence disciplines were providing various techniques of knowledge acquisition, modeling, integration, and transformation that could be used to support medical decisions. The initial and many other attempts [4, 5] for creating a useful electronic medical knowledge produced some results but they still have not provided the complete solutions.

In the past two decades a wide variety of information systems have been developed to facilitate medical diagnosis, better known as medical expert systems [7, 8]. These systems have used various methodologies but most of them faced the difficulty of extracting the knowledge from the source text because of complexities of natural language. An additional problem is related with the imprecise character of the medical knowledge. The imprecise knowledge might require special interpretation or additional processing making it even harder to use it appropriately.

The goal of our project was to develop the methodology and the software to provide support for doctors and patients based on specially created medical knowledge bases.

* This work was partly supported by founds on science in 2007 – 2010 as Ordered Research Project of Polish Ministry of Science and Higher Educations.

M.D. Lytras et al. (Eds.): WSKS 2009, CCIS 49, pp. 298–304, 2009.

There are several aspects of our project that made our efforts different that the others and alleviate partially the described problems [6, 9]. First, the methodology and the software can perform analysis of actual medical knowledge base in real time so that we can provide effective interactive assistance for a doctor. Second, we allow knowledge bases to be individualized so that different doctors can use different component in a different way. Third, we allow the remote communications of the physicians with the patients. Fourth, we are creating data for our knowledge bases automatically, by developing a sub-system for real time monitoring of patients health parameters. The remote health monitoring is based on several measurements by devices attached to patients who may be physically distant from the doctor. Presently we are developing automatic measurement, monitoring, and the data transfer for such patients health parameters as glucose level (based on the device Optium Xido / Abbott Laboratories), blood pressure and hearth pulse frequency (based on the device MPX5050), pulse and blood saturation measurements devices (optical pulsoksymeters). Fifth, the guidance can be quickly provided to patients through the interactive interfaces. The guidance can originate directly from the medical assistant software based on the patient's monitored health parameters, or can originate from the doctor alerted by the medical assistant software that the patient might need some intervention.

2 Medical Knowledge and the Rough Set Model

The rough set model [1, 2, 3] is well accepted in the modern Computer Science/Artificial Intelligence disciplines as a method to deal with imprecise categories and approximations. The model assumes that knowledge is closely related to classifications of any objects (real things, states, abstract concepts, processes). The model defines knowledge as a structure that consists of a family of various classifications of objects. As a result the knowledge is manifested in the ability to classify objects. That model suits very well the medical area where the process of classification of patients is of the primary interest with the goal to identify the proper medical treatment. A significant body of the modern medical knowledge is based on patients' classifications and related with these classifications risk factors [10]. Our methodology is based on integrating rough set theory with the practical techniques of knowledge base creation based on ontologies[11, 12, 13, 14, 15, 16].

Let us start with simple classification (partition) of patients as shown in Figure 1. In this figure we used an extended UML (Universal Modeling Language) notation [16, 17,18]. Each rectangle describes a class (set of objects). Alternative names are listed in the left part. We use alternative names to allow the choice between short one character mathematical notation and a full self-descriptive name for the set. In our example Patients are classified as the hospital Admitted Patients and the Non-Admitted Patients. Admitted Patients and the Non-Admitted Patients are represented by two rectangles connected with the Patients rectangle by two lines with an arrow (triangle). To differentiate between the whole set and an element of the set, the right side of each rectangular indicates a variable used for one example instance. Alternative names are again provided. The right bottom part of each rectangle stores examples of specific instances of the set.

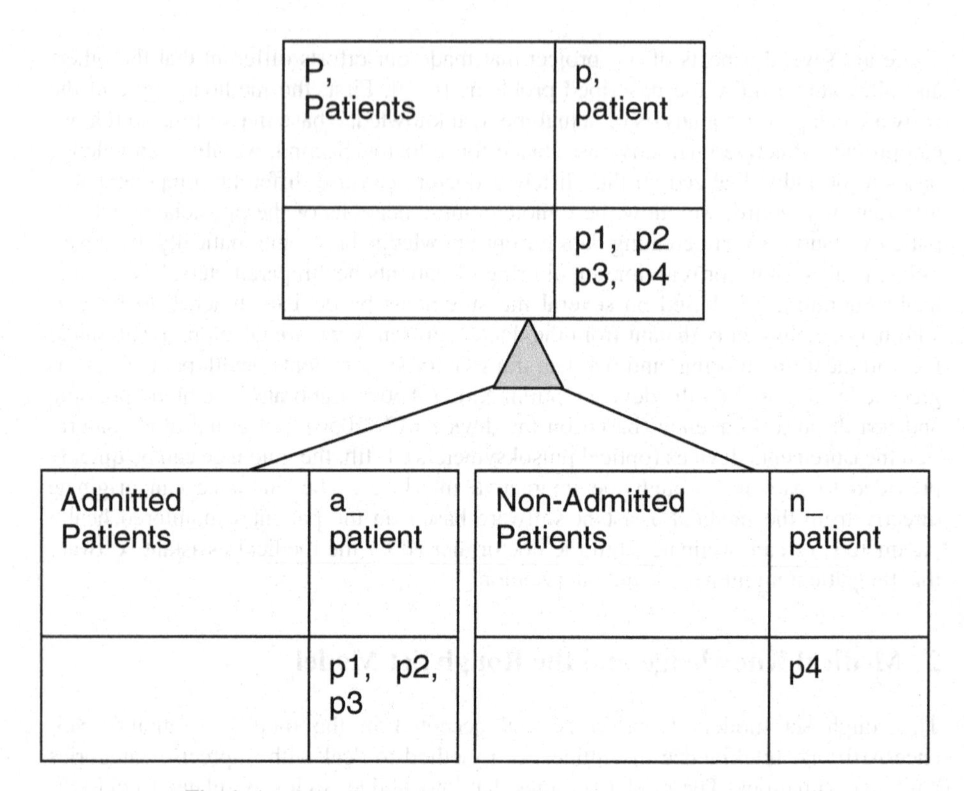

Fig. 1. An example of a simple classification of medical patients

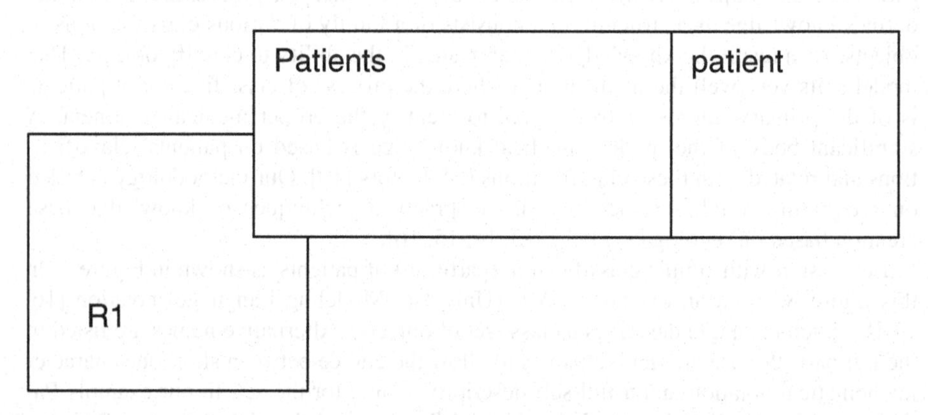

Fig. 2. Classification into hospital Admitted Patients and Non-Admitted Patients using a binary equivalence relationship

In general we can have many classifications. Therefore, our Medical Knowledge Base can be defined as simply a Family of Classifications where each classification consists of Family of Concepts. Each concept is related with several Patients.

The traditional Rough set approach uses equivalence relationship which can be represented graphically by a recursive relationship linking Patients set with itself as

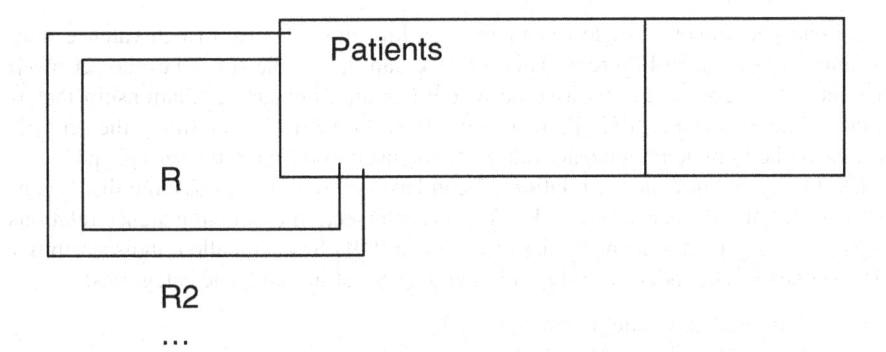

Fig. 3. An example of a Simple Patient Knowledge Base

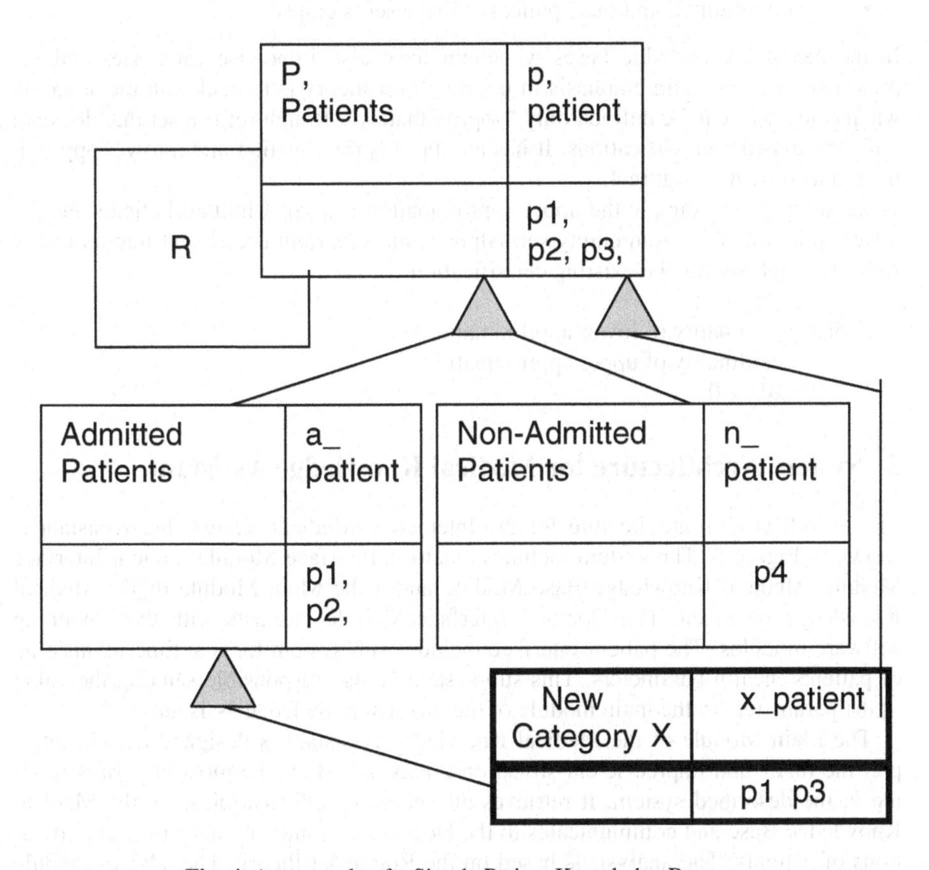

Fig. 4. An example of a Simple Patient Knowledge Base

shown in Figure 2. R1 is an example equivalence relationship based on the hospital admittance i.e. Patients Admitted are related to themselves defining the set {p1, p2, p3}. Also, the Non-Admitted patients are related to themselves defining the set {p4}.

We can generalize a single binary relationship into a family of equivalence relationships, as shown in Figure 3. This way we can define the set R i.e. the set of all elementary categories. Let us assume that R2 is an equivalence relationship that is based on the gender i.e. Male Patients are related to themselves defining the set {p1, p2}. Also, the Female Patients are related to themselves defining the set {p3, p4}.

The family of equivalence relations R can have subsets. Let us assume that S consisting of R1 and R2 is a subset of R. Then the intersection of all equivalence relations for S is also an equivalence relation denoted IND(S) and called indiscernibility relation over S. The indiscernibility relation over S defines all basic categories:

- Admitted and male patients {p1, p2}
- Admitted and female patients {p3}
- Non-admitted and female patients {4}
- Non-admitted and male patients { } (the set is empty)

In the medical knowledge bases we might have also imprecise categories and approximations. The main emphasis in the rough set theory is to deal with the situation when categories can be defined only "approximately". Rough set is a set that does not "fit" any existing classifications. It has an upper approximation and a lower approximation as shown in Figure 4.

In our specific example the upper approximation is a set Admitted Patients and the lower approximation is an empty set. Alpha is an important coefficient that tells how well the rough set fits the existing classification

Alpha = cardinality of lower approximation/
cardinality of upper approximation
= 0/3 = 0

3 System Architecture for Medical Knowledge Assistant

The overall system architecture for the Interactive Medical Knowledge Assistant is shown in Figure 5. The system includes Doctor's Interface Module, Patient Interface Module, Medical Knowledge Base Module and a the Main Module of the Medical Knowledge Assistant. The Doctors' Interface Module interacts with the remaining software modules. The patient interface includes sub-system for real time monitoring of patients' health parameters. This sub-system is also responsible sending the value of the parameters to the main module of the medical knowledge assistant.

The Main Module of the Medical Knowledge Assistant is designed to help interpret the fuzzy and imprecise classifications. This module is the most important module in the described system. It retrieves all necessary information from the Medical Knowledge Base and communicates to the Doctor the results of analysis of classifications of patients. The analysis is based on the Rough set theory. The advisor module computes the upper and lower approximation set and all required coefficients including alpha coefficient. One of the special functions of the Medical Advisor if allowing to do what-if analysis using rough set approximations. The doctor can integrate his/her own experience by modifying existing classifications and the considered set of patients. This way he can quickly get access to all necessary information to make critical decisions.

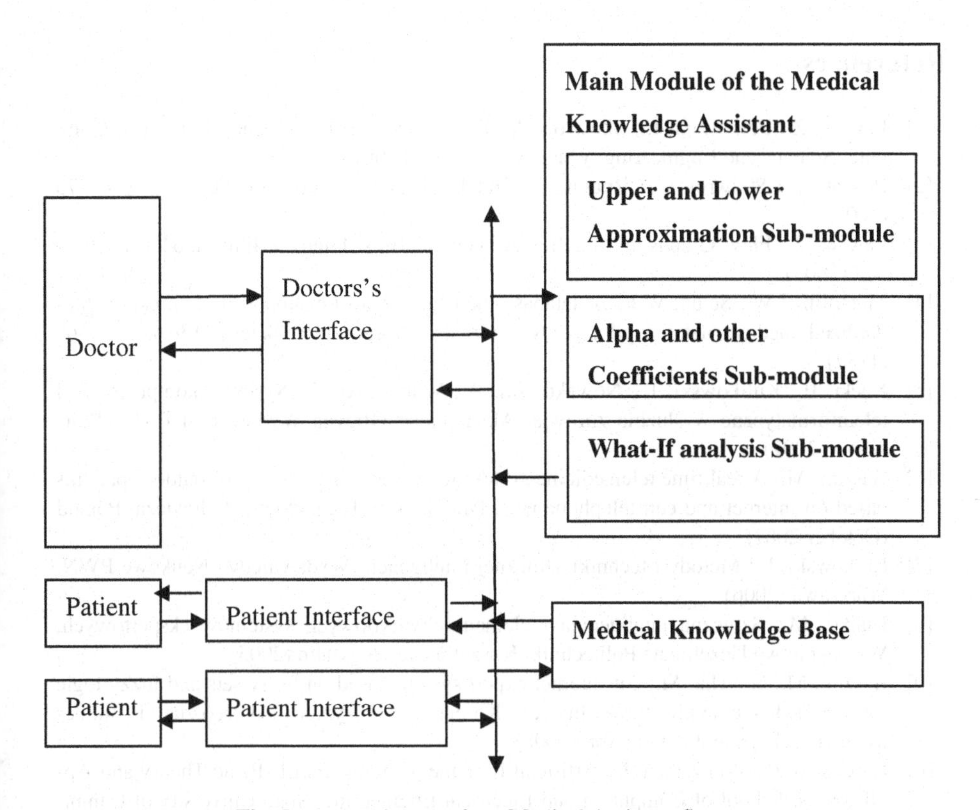

Fig. 5. An Architecture for a Medical Assistant System

The patient Web based interface can also provide a patient with a various health assistance. The patient can communicate with doctor in real time or send a message. The patient can obtain from the medical assistant Web server history of the patient's monitored health parameters. The doctor alerted by the medical assistant software or the software itself can send the patient some guidance. One simple example of guidance for a patient can be to see the doctor immediately.

4 Summary

The paper discussed methodology and software for the medical analysis and medical knowledge base processing. We described the architecture of entire system with an emphasis on the main module of medical assistant. The medical knowledge base module contains all information about the patients classifications required by main module. The doctor and a patient can use an assistant in different modes. One of the important modes is what if analysis when the doctor can change some classifications to see the potential results.

References

[1] Pawlak, Z., Polkowski, L., Skowron, A.: Rough Set Theory. In: Encyclopedia of Computer Science and Engineering. Wiley, Chichester (2008)
[2] Pawlak, Z., Skowron, A.: Rough sets and Boolean reasoning. Inf. Sci. 177(1), 41–73 (2007)
[3] Pawlak, Z.: Flow Graphs and Intelligent Data Analysis. Fundam. Inform. 64(1-4), 369–377 (2005)
[4] Meredith, J.W., Selen, W.J.: A database medical diagnostic support system using standardized medical data: a pilot study. Journal of Information Science 13(6), 353–360 (1987)
[5] Kacki, E., Kulikowski, L., Nowakowski, A., Waniewski, E.: Systemy komputerowe i teleinformatyczne w sluzbie zdrowia, Akademicka Oficyna Wydawnicza EXIT, Warszawa (2002)
[6] Baszun, M.: A real time telemedicine system for wireless monitoring of outdoor patients based on internet and cell telephone nets, Dni Nauki I Technologii, Bialowieza, Poland (October 2007)
[7] Rutkowski, L.: Metody i techniki sztucznej inteligencji, Wydawnictwo Naukowe PWN, Warszawa (2006)
[8] Bialko, M.: Sztuczna inteligencja i elementy hybrydowych systemów ekspertowych, Wydawnictwo Uczelniane Politechniki Koszalińskiej, Koszalin (2005)
[9] Baszun, M., Glówka, M.: An Internet expert system based on fuzzy sets and fuzzy logic for telemedicine applications. In: XV Konferencja Inzynierii Akustycznej i Biomedycznej, Zakopane, Poland (April 2008)
[10] De Castro, L., Von Zuben, F.: Artificial Immune Systems: Part I -Basic Theory and Applications, School of Computing and Electrical Engineering, State University of Campinas, Brazil (1999)
[11] Lytras, M., Naeve, A., Pouloudi, A.: Knowledge Management as a Reference Theory for E-learning: A conceptual and technological perspective. International Journal of Distance Education Technologies 3(2), 66–73 (2005b)
[12] Lytras, M., Pouloudi, N.: Towards the development of a novel taxonomy of knowledge management systems from a learning perspective. Journal of Knowledge Management 10(6), 64–80 (2006)
[13] Sicilia, M., Lytras, M., Rodríguez, E., García-Barriocanal, E.: Integrating descriptions of knowledge management learning activities into large ontological structures: A case study. Data & Knowledge Engineering 57(2), 111–121 (2006)
[14] Vargas-Vera, M., Motta, E.: AQUA - Ontology-based Question Answering System. In: Monroy, R., Arroyo-Figueroa, G., Sucar, L.E., Sossa, H. (eds.) MICAI 2004. LNCS (LNAI), vol. 2972, pp. 468–477. Springer, Heidelberg (2004)
[15] Weng, S., Tsai, H., Liu, S., Hsu, C.: Ontology construction for information classification. Expert Systems with Applications 31(1), 1–2 (2006)
[16] Czejdo, B., Mappus, R., Messa, K.: The Impact of UML Diagrams on Knowledge Modeling, Discovery and Presentations. Journal of Information Technology Impact 3(1), 25–38 (2003)
[17] Booch, G., Rumbaugh, J., Jacobson, I.: The Unified Modeling Language User Guide. Addison Wesley, Reading (1999)
[18] Myers, M.: The field of IS has always been about Relationships not things in themselves. Int. J. Teaching and Case Studies 1(1/2), 15–22 (2007) National Center for Biotechnology Information, http://www.ncbi.nlm.nih.gov

Elaborating a Knowledge Management Plan: A Multi-actor Decision Process

Thierno Tounkara

TELECOM & Management SudParis, 9 rue Charles Fourier,
91011 Evry cedex, France
Thierno.Tounkara@it-sudparis.eu

Abstract. The elaboration of a knowledge management plan should take into account the evolution of the firm' strategy: it is the so-called "strategic alignment". However, existing approaches focus essentially on operational actors' point of view: what about the executive level which defines the firm' strategy? We propose and illustrate, in this article, an approach and cartographic methods to elaborate knowledge management actions aligned with the strategy of the firm. We explore Multicriteria Decision-Analysis field to refine the robustness of our approach.

Keywords: Knowledge Management, Knowledge Management Plan, Strategic Alignment, Strategy, Knowledge Mapping and Muticriteria Decision-Analysis.

1 Introduction

For many years, research in knowledge management produced, for the operational level, methods, tools and approaches for locating, formalizing, sharing and enriching corporate knowledge. This orientation can be explained by the fact that knowledge management appeared as a field to solve operational problems as massive retirements or employees' transfer for example.

Recently, knowledge management field has evolved towards strategic problematics which can be grouped into four questions:

How can we create strategic alignment between the firm' strategy and the knowledge management plan?

1. How can we evaluate the Return On investment (ROI) when implementing a knowledge management plan?
2. What can be the added-value of knowledge management in elaborating a strategic vision of the firm?
3. Which monitoring tools can be used to measure the performance of an implemented knowledge management system?

In this paper, we bring a partial answer to the first two questions. We postulate that strategic alignment of knowledge management plan determines, in part, the value of intangible assets: human capital (skills, talent and knowledge), information capital (databases, information systems, networks and technology infrastructure) and organizational capital (culture, leadership and knowledge management).

M.D. Lytras et al. (Eds.): WSKS 2009, CCIS 49, pp. 305–318, 2009.

We propose a cartographic approach to interact with executive teams to determine their vision of a knowledge management plan aligned with the firm' strategy. Following an empirical methodology we will describe in our paper, we cross two maps:

- a strategy map (in reference to Kaplan and Norton) which is a powerful tool to provide a uniform and consistent way to describe and represent strategy of the firm;
- a business processes map which is a visual representation of the internal processes and activities of the firm.

The result of this cross analysis between strategy and processes is completed by a second analysis with operational teams to elaborate a final knowledge management plan taking into account two different levels of perception (executive and operational levels).

The paper begins by giving an overview of existing knowledge mapping methods. After, we define the problematic of strategic alignment when elaborating a knowledge management plan. We propose an approach and tools to create strategic alignment.

Then, we present a case study performed in Chronopost International Company (in France). Our goal is, using this case, to describe and to illustrate our methodology for strategic alignment of knowledge management actions.

In the last part of the paper, we discuss the robustness of our approach and propose research perspectives in link with the "Multicriteria Decision-Analysis" field.

2 Knowledge Mapping Methods, a Quick Overview

Knowledge mapping (or knowledge cartography) allows the value of the firm's critical knowledge to be enhanced [13] [19]. It is a step to be performed before any operation of knowledge management. Knowledge cartography is an identification of the corporate knowledge. We refer to the definition of knowledge cartography given by [14]: "knowledge mapping is defined as the process, methods and tools for analyzing knowledge areas in order to discover features or meaning and to visualize them in a comprehensive, transparent form such that the business-relevant features are clearly highlighted". It is a new research field in knowledge management and there are few academic papers.

Companies wishing to manage their corporate knowledge must make a precise analysis in order to determine the knowledge they must preserve, develop, abandon etc. Thus, cartography becomes a decision support tool. To this end, there is a need to establish specific criteria in order to evaluate, in the cartography, the most critical knowledge for the company. This is the so-called "cartography of critical knowledge".

Knowledge mapping methods can be categorized into two approaches [3] [17]:

- A "Process" oriented approach

This approach deals with knowledge cartography methods which use modeling, description and analysis of business processes to determine critical knowledge.

As examples, we can cite:

- o GAMETH (Global Analysis METHodology) which is an approach focusing on business processes that connect knowledge to action [4].

  - o Tseng and Huang methodology [18] which is "process" oriented and is guided by problems. It is based on a quantitative analysis of collected information while interviewing some experts.
- A "Domain" oriented approach

In this approach, we try to make an analysis from a mass of information in order to organize it in logic different from the functional approach. In fact, the goal is to ignore the functional structure of the firm, grouping activities into knowledge domains. This task demands an important capacity of analysis because it's not a natural process.

As examples, we can cite:

  - o The knowledge trees of Authier and Levy [2] which are the expression and the consequence, evolving in real time, of training courses and experiences of all members of a given community. The underlying principles of their development are mathematical, philosophical and sociological.
  - o The M3C methodology is the result of various experiences and issues developed in a working group of the Knowledge Management Club in France (www.club-gc.asso.fr) [2] [16]. M3C relies on robust models which have been performed in industrial research centers and also in industrial operational units (Hydro Quebec, GTIE group, Schindler, DGA, PSA Peugeot Citroen, etc.).

    The cartography and the evaluation of knowledge domains are based on knowledge acquisition from experts. Thus, M3C is also a knowledge engineering method and it completes other methods used for the modeling of descriptive and operational knowledge of an expert.

3 Strategic Alignment Problematic

We can synthesize the implementation of existing knowledge mapping methods and approaches into three steps:

- identification of knowledge domains interviewing actors in operational units;
- evaluation of the criticality of knowledge;
- Identification of a set of knowledge management actions which could reduce the criticality of identified domains.

3.1 Limits of Existing Knowledge Mapping Methods

When analyzing these cartographic methods, we can point out two limits:

- **A truncated/fragmented vision**

Existing knowledge mapping methods and approaches lead to a fragmented vision of what should be the knowledge management plan because they only focus on « operational actors » (in business units) points of view. This can be explained by the fact that the actors of operational units are the main actors of knowledge accumulation process.

- **The strategy of the firm is not explicitly taken into account**

It is very difficult, using these approaches, to evaluate the impact of the firm's strategy on business knowledge evolution. Regarding to the strategy of the firm, which knowledge domains should be sustained or should emerge?

So, there is a great risk to have a misalignment between the strategy of the firm and the knowledge management plan. This can lead to investments in non pertinent knowledge management actions that will not deliver the expected return on investment.

For example, investing in a knowledge management program in CRM (Customer Relationship management) would create higher value for a company following a total customer solution strategy than for a company following a low total cost strategy.

3.2 Linking Knowledge to Strategy

In the Resource Based View (RBV) theory [5] [9] [15], Organization is seen as a portfolio of resources (tangible and intangible assets). Following this theory, firms should focus on particular resources which can help it create distinctive and sustainable value to be more competitive. So, the goal of the strategy is to engage the firm in a dynamic process to maintain his corporate resources and to acquire new resources important for its development.

Following the RBV theory, we can define the principle of strategic alignment of knowledge management plan by two assertions:

- The strategy of the firm strongly relies on cognitive processes (organizational learning, knowledge creation, knowledge transfer, innovation, etc.). The firm, given what it knows, must identify the best product and market opportunities for exploiting that knowledge [21]. *So, the evolution of strategy must necessarily lean on the existing corporate knowledge which is a factor for competitiveness.*
- The evolution of the firm' strategy has an impact on the evolution of competencies/corporate knowledge. The firm must identify the knowledge required to execute its intended strategy. *So, the elaboration of a knowledge management plan should take into account this evolution.*

This problematic of strategic alignment is well known in IT research field [6] [20].

To sum up, a knowledge management plan cannot be elaborated without taking into account strongly and precisely the strategy of the firm and, so, the point of view of executive level (executive actors).

4 A Method for Strategic Alignment of the Knowledge Management Plan

4.1 A "Middle-Top-Down" Approach

We propose a "Middle-Top-Down" approach (cf. Figure 1), in reference to the terminology of Nonaka and Takeuchi [10]. Our approach confronts the vision of executive actors (holders of the strategy of the firm) to the vision of operational actors.

First, leaning on the strategy, we identify impacts on business processes ("Top-Down" approach): how competencies and knowledge domains will evolve (criticality, level of maturity)? Then, from identified knowledge domains in operational units, we evaluate the criticality of competencies ("bottom up" approach). In The last step, we confront the two points of view (operational and executive level points of view) and we elaborate a knowledge management plan to be implemented.

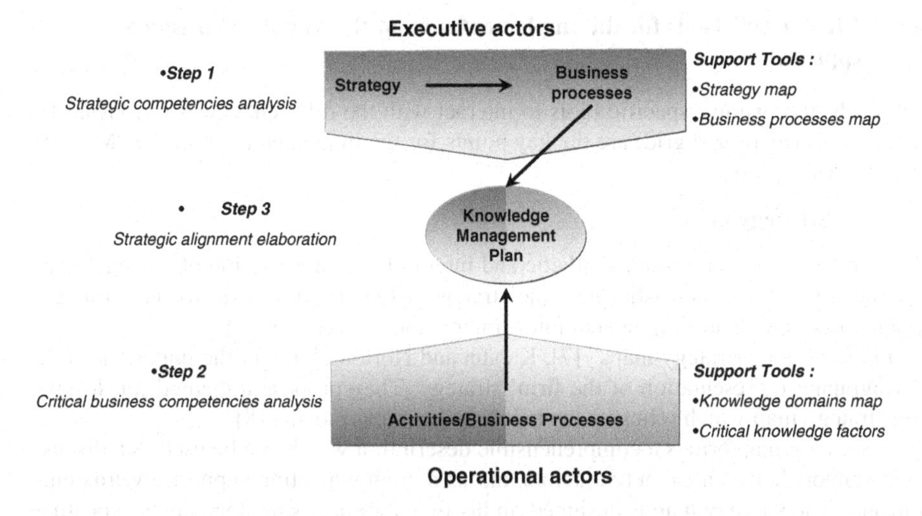

Fig. 1. A Middle-Top-Down approach for strategic alignment of the knowledge management plan

This approach has a double interest:

- guaranteeing the pertinence of the knowledge management actions to be implemented
- making easier acceptance of knowledge management plan by the different actors (operational and executive actors)

Concretely, it seems interesting to use existing knowledge mapping methods (cf. **§2**) for implementing this approach. For the point of view of the operational level, the applicability of these methods has been tested and validated.

However, we cannot use these methods when interacting with the executive level because they are designed for very detailed analysis of competencies. Executive actors have a less technical and more global vision of competencies. Furthermore, in these methods, criteria for knowledge domains and competencies evaluation do not take into account the firm' strategy as It is defined by the executive level.

For analyzing the point of view of executive actors ("top-down" approach), we propose a new and appropriated cartographic method. For the operational level ("bottom-up" approach), we use the M3C cartographic method we briefly describe on **§2**.

These two cartographic methods are the support of a three steps approach (cf. Figure 1):

- **Step 1:** strategic competencies analysis (executive level point of view); the result of this step is a strategic competencies map.
- **Step 2:** critical business competencies analysis (operational level point of view); the result of this step is a critical business competencies map.
- **Step 3:** strategic alignment elaboration; the result of this last step is the definition of the knowledge management plan.

4.2 "Mediation" Tools for the Implementation of the Middle-Top-Down Approach

For each step, we use specific tools to interact with the different actors (cf. Figure 1). These tools (maps and grid) are the key points for the implementation of the "Middle-Top-Down" approach.

- **Strategy map**

The strategy map is a visual, synthetic and intelligible representation of strategic axes for the firm. In this heuristic map, the strategic axes are broken down into strategic orientations which are fragmented into strategic themes (cf. Figure 2).

In their book "strategy maps" [7], Kaplan and Norton highlight the importance of a cartographic representation of the firm' strategy. They propose a method for designing strategy map which relies on the balanced Scorecard model [8].

A strategy map brings a comprehensible description which can be used as a discussion support between executive actors and as a communication support towards employees. The strategy map is designed, in his first version, using documents evocating the firm' strategy (if they exist): middle term plan, synthesis… This first version is then completed and validated by few executive actors which, ideally, are involved in the strategy elaboration.

- **Business processes map**

The business processes map is a visual representation of the processes in the firm (cf. Figure 3). Going from the centre of the map, processes are broken down into activities and activities into sub activities.

This map is designed, in his first version, using official documents as quality referential or description of department activities for example. Then, this version is completed and validated by executive actors or by managers in charge of processes (when they exist).

- **Knowledge domains map**

It is a visual representation by operational actors of knowledge domains they consider essential for their activities.

In the M3C methodology, a knowledge domain is defined as a field of activity of a group of people from whom information and knowledge can be gathered.

The central point of the cartography is the core activity or "core knowledge" which corresponds to its fundamental mission. Around this central point are the knowledge axes, which define the strategic domains of knowledge.

The final knowledge domains in the classification are grouped according to a common finality on the same theme of knowledge, along the knowledge axes. According to the precision required, a domain can be divided into sub-domains and a theme into sub-themes.

- **The critical knowledge Factors grid (CKF)**

The criticality of a domain is an evaluation of risks/opportunities. It may be for example risks of loss of knowledge that can have harmful consequences; interest to develop a domain to obtain advantages for the firm (productivity gains, new market shares…).

We need to define what may be "objectively" the criticality of knowledge and to give a model of evaluation for identifying the most critical knowledge domains in the cartography.

The Knowledge Management Club, in France, has elaborated a grid of generic evaluation called CKF (Critical Knowledge Factors) that is available for the members of the club. This grid has been performed and validated in many French and Foreign companies.

The CKF grid contains 20 criteria regrouped in 4 thematic axes (Rarity, Utility, Difficulty to capture knowledge, Nature of knowledge). That's this grid which is used in the M3C method.

Each criterion is evaluated according to a scale composed of 4 levels, representing the degree of realization of the criterion. Each evaluation of a criterion is based on one question. Each level is expressed by a clear and synthetic sentence by avoiding the vague terms and which lead to confusion ("rating description")

5 Chronopost International Case study

Chronopost International, a transportation and logistics company is a subsidiary of the "La Poste group" in France, the third largest parcel and logistics operator in Europe. With a foothold in the United States since 1997, Chronopost International has established itself as a leading delivery service for shipments originating from the United States.

5.1 The Context of the Study

The strong concurrence and the prices decrease have increased the pressure in the express messaging market.

Furthermore, new regulations require important adaptations: speed limitation to 90km/h for heavy trucks, reduction of flights during night, reinforcement of airport safety and access conditions in town centre.

Chronopost has a real expertise in express messaging and wants to stay a main actor in this market. They contact our research team to realize a pilot project to design a performing system for knowledge and competencies management.

The Chronopost study is part (pilot project) of a bigger project « Competencies Management Process » leaded by the Human Resources Department. The aim of this pilot project is to help the firm having a prospective vision on competencies regarding to the new established strategy plan "Defis stratégiques 2007".

These were two objectives:

- Identifying and evaluating strategic competencies for the future of Chronopost international
- elaborating and implementing a knowledge management plan

The study was performed in 2005. It lasted 6 months. We performed our "Middle-Top-Down" approach to achieve objectives.

- **Step 1: strategic competencies analysis**

The strategic competencies analysis was performed with five members of the executive board of Chronopost International. This first analysis leaded to a very detailed and argued description of competencies Chronopost should focus on to implement its new strategy. The final output was synthesized in a strategic competencies map.

- **Step 2: critical business competencies analysis**

For the second step analysis, a pilot business unity was chosen: "International Express Treatment" at Roissy Airport in France. Using the M3C method (cf. §2.), we interviewed four managers of the business unity.

The results were:

- a knowledge domain map representing their majors competencies areas of the Express International activity at Chronopost
- Threats and opportunities about identified competencies
- a critical business competencies map of the business unity "International Express Treatment"

- **Step 3 : strategic alignment elaboration**

The strategic alignment has linked the critical business competencies analysis with the strategy of Chronopost International. So we could identify knowledge domains for the business unity "International Express Treatment" affected by strategic competencies identified during the analysis with members of executive board.

5.2 Strategic Competencies Analysis

We performed this step in three stages: "elaboration of strategy map", "Elaboration of a business processes map" and "Identification and evaluation of strategic competencies".

- **Elaboration of a strategy map**

In the project « Competencies Management Process », we have reformulated the strategic plan "Défis stratégiques 2007" to design a strategic map (cf. Figure 2) which was validated by the five members of the executive board.

- **Elaboration of a business processes map**

The first version of this map was designed using three official documents of Chronopost International:

- the quality" manual",
- the competencies referential
- a document about competencies management process "REPERES"

Then, the final business processes map was completed and validated by the members of the executive board.

- **Identification and evaluation of strategic competencies**

In this stage, we identified, for each business process, competencies impacted by the new strategy of Chronopost International.

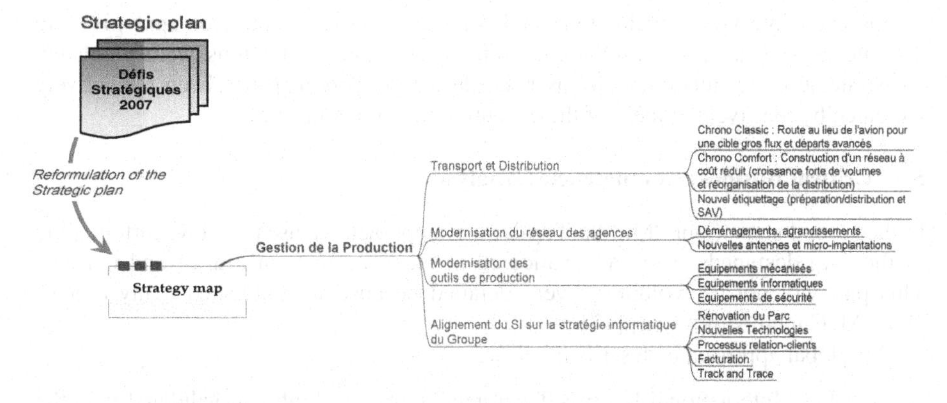

Fig. 2. Partial view of the strategy map elaborated for Chronopost project

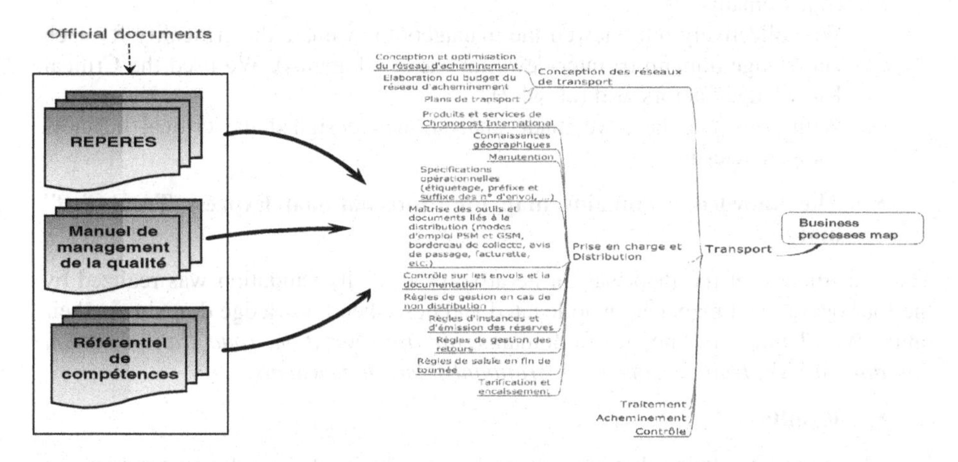

Fig. 3. Partial view of the business processes map elaborated for Chronopost project

We interviewed each member of the executive board. Each interview lasted 1h30min and was realized in three stages:

- Strategy maps and business processes map are presented to the executive actor to let him appropriate them.
- After having read the maps, the executive actor identifies for each strategic axe its impacts on needed competencies to perform business processes: which are the competencies Chronopost International should make emerged or should develop to implement its new strategy?
- Last, we characterize each identified competency evaluating its:
    - o maturity level (is it a competence which must be created, transformed or abandoned?)
    - o criticality level (using the CKF grid described in **§4.2.**, is the competence "non critical, critical or very critical ?).

For each interview, evaluations and arguments were synthesized and given to the interviewed actor for validation. When the whole evaluations were validated, the strategic competencies map has been designed. This map has been collectively validated by the five members of the executive board in a meeting.

5.3 Critical Business Competencies Analysis

In the second step of our "Middle-Top-Down" approach we use the M3C cartographic method we designed, tested and validated in many projects in France and Quebec. This part will not be, voluntary, very detailed because we published many articles about M3C methodology [3] [17].

Our global approach is described below:

- For "International Express Treatment" unity, we built and validated with the four managers (representing operational actors) the cartography of knowledge domains.
- We collectively interviewed the managers to evaluate the criticality for each knowledge domain (6 interviews for the six domains). We used the Critical Knowledge Factors grid (cf. §4.2).
- With collected data, we made analysis and designed the critical business competencies map.

- **The knowledge domains map for "International Express Treatment" unity**

The construction of the map was an iterative process. Its validation was realized by the four operational managers who located and described knowledge domains in their unity. We identify six knowledge domains: *air transportation, road transportation, operational SAV, Hub Treatment, Operational safety and customs.*

- **Results**

First, for each knowledge domain, we made a synthesis of the collective perception (the four operational managers) about the knowledge domain criticality. It is the result of a qualitative (collected arguments) and quantitative analysis (score in each criterion of the CKF grid and the average score for the knowledge domain).

Then, we elaborate the critical business competencies map. It is a visualization of competencies grouped by domain with a colour code indicating the level of criticality.

Last, we list in a table, knowledge domains concerned by specificities it could be interesting to highlight when considering the operational actors points of view: domains with great expertise, domains to be valorised, very vulnerable domains or domains that need to improve/adapt methods for training courses and knowledge transfer. This table is a basis for a more refined analysis and for identification of suitable knowledge management actions.

5.4 Strategic Alignment Elaboration

In this step, we have identified cognitive distortion in the perception operational actors had when considering the strategy of Chronopost International.

We performed the strategic alignment in three stages:

- Building an influence matrix to appreciate the influence level of strategic competencies on knowledge domains identified for "International Express Treatment" business unity.
- Classifying knowledge domains and strategic competencies
- Elaborating the knowledge management plan (knowledge management actions to be set up).

- **Building the influence matrix**

The goal is to identify the potential influence of strategic vision on the business vision and, inversely, the potential influence of the business vision on the strategic vision.

The previous maps (strategic competencies map and critical business competencies map) gave us two complementary points of view.

Using these two maps, we identified knowledge domains (for "International Express Treatment" business unity) affected by strategic competencies identified by the executive actors. We have built a double entry table (the influence matrix) in which interactions between business knowledge domains and strategic competencies are indicated (cf. Table 1).

In this table:

- Knowledge domains for "International Express Treatment" unity are represented in column; the criticality coefficients range from 1 to 4, in reference of the CKF grid.
- Strategic competencies are represented in line; the criticality coefficient range from 1 to 3 (non critical, critical, very critical).
- The last line gives an influence score for each strategic competence. It represents the influence of "International Express Treatment" unity on each strategic competence. We call it the business influence score
- The last column gives an influence score for each business knowledge domain. It represents the influence of the strategy on each knowledge domain of the "International Express Treatment" unity. We call it the strategic influence score.

- **Classifying knowledge domains and strategic competencies**

In this stage, using the strategic influence scores, we have classified knowledge domains for "International Express Treatment" unity (from the higher to the lower score). In the same logic, using the business influence score, we have classified strategic competencies.

Arguments collected during interviews with executive and operational actors were a basis for the validation of these classifications. Classifications are very useful if we want to give a priority level for each knowledge management action in the knowledge management plan.

- **Elaborating the knowledge management plan**

For each business knowledge domain and strategic competencies, we have elaborated recommendations for knowledge management actions using:

- Interviews synthesis and collected arguments
- Identified recurrent elements which characterized the criticality of knowledge domain or competency (unsuitability of the training system, inexistence of knowledge capitalization system, need of a sharing knowledge system, etc.)
- Classifications of knowledge domains and strategic competencies

For a better visibility, the knowledge management plan was structured into themes:

- "*Organization*" when it was managerial actions
- "*Training-Recruitment*" when actions were dealing with learning systems, recruitment for new competencies not available in Chronopost International
- "*Capitalization-transfer*" when it was actions for acquisition, preservation, sharing or documentation of competencies
- "*Innovation*" when actions were dealing with creativity, environment scanning, etc.

Table 1. Example of influence Matrix

strategic Competencies / Knowledge domains		Competence a	Competence b	Competence c	Competence d	Competence e	Competence f	**Strategic influence score**
Criticality		3	3	1	1	3	3	
Domain A	2,7	X	X	X	X	X		**29,7**
Domain B	2,7	X	X	X	X	X	X	**37,8**
Domain C	2,5		X				X	**15**
Domain D	2,6	X	X					**15,6**
Domain E	2,8	X	X	X				**19,6**
Domain F	2,5	X	X	X		X		**25**
Business Influence Score		**39,9**	**47,4**	**10,7**	**5,4**	**23,7**	**15,6**	

6 Discussion : Multicriteria Decision Models and Tools to Refine the Middle-Top-Down Approach

Elaborating a Knowledge Management Plan can be reformulated as a Decision-Analysis problem with two objectives:

- identifying and defining a set of potential knowledge management actions;
- helping Decision-Makers to choose an appropriate sub-set of knowledge management actions to be implemented (the Knowledge Management Plan).

It is a specific Decision-Analysis problematic which can be put in the class of strategic problems as "defining a diplomatic strategy" or "defining a military strategy in a crisis situation" for example [11] [12]. For this class of problems (at the opposite of operational problems as management of production for example), decision theories provide few adapted models and tools because, here, information are qualitative and there are multiple actors with rationalities different.

In this article, we have proposed an approach and tools to define potential knowledge management actions aligned with the strategy of the firm; we consider it is an answer for the first objective of the decision-analysis problem.

For the second objective of the problem (the choice of an appropriate sub-set of knowledge management actions to be implemented), we thing that we can refine our approach and make it more robust using the framework of prescriptive decision theory.

There are two questions:

- How can we define robust and stable criteria, with respect to the Decision Makers' beliefs and preferences, for the evaluation of identified potential knowledge management actions?

  Actually, we only consider the dimension "knowledge criticality" (rarity, utility, difficulty to capture knowledge, nature of knowledge, and maturity of knowledge). Or, we know that Decision Makers' preferences integrate other dimensions as the "economic dimension" (investment costs, Return on Investment, etc.).
- Which multicriteria decision models and tools (examples: ELECTRE, PROMETHEE or TACTIC) can we use to aggregate defined criteria and compare knowledge actions performance?

So, a real research perspective is to explore models and tools from the "Multicriteria Decision-Analysis" field to refine and make more robust our approach.

7 Conclusion

The Scientific community interest for knowledge management is increasing and lessons learned form companies show that the elaboration of a knowledge management plan should take into account three dimensions: strategic, organizational and technological dimensions. The approach and method, we propose here, are designed in this "spirit".

Efficiency and Return on Investment for knowledge management projects must not be evaluated in the short term but in the middle/long term with qualitative criteria. Quantitative evaluation remains a theoretical and methodological problem: usual criteria (cost, quality, duration) are unsuitable to evaluate the impact on organizational performance.

References

1. Aubertin, G.: Knowledge mapping: a strategic entry point to knowledge management. Trends in Enterprise Knowledge Management, Edition Hermes Penton Science (2006)
2. Authier, M., Lévy, P.: Les arbres de connaissances, Editions la découverte (1992)
3. Ermine, J.-L., Boughzala, I., Tounkara, T.: Critical Knowledge Map as a Decision Tool for Knowledge Transfer Actions. The Electronic Journal of Knowledge Management 4(2), 129–140 (2006), http://www.ejkm.com
4. Grundstein, M., Rosenthal-Sabroux, C.: A Way to Highlight Crucial Knowledge For Extended Company's employees. Annales du Lamsade Paris (2003)

5. Hamel, G., Prahalad, C.K.: Competing for the Future. Breakthrough Strategies for Seizing Control of Your Industry and Creating the Markets of Tomorrow. Harvard Business School Press, Cambridge (1994); trad. par Larry Cohen, La conquête du futur. Stratégies audacieuses pour prendre en main le devenir de votre secteur et créer les marchés de demain, Paris, ERPI (1995)
6. Henderson, J.C., Venkatraman, N.: Strategic Alignment: Leveraging information technology for transforming organizations. IBM Systems Journal 32(1), 4–16 (1999)
7. Kaplan, R.S., Norton, D.P.: Strategy Maps: concerting intangible assets into tangible outcomes. Harvard Business School Press, Boston (2004)
8. Kaplan, R.S., Norton, D.P.: The Balanced Scorecard: Translating Strategy into Action. Harvard Business School Press, Boston (1996)
9. Kogut, B., Zander, U.: What firms do? Coordination, identity, and learning. Organization Science 7, 502–518 (1996)
10. Nonaka, I., Takeuchi: The Knowledge Creating Company. Oxford University Press, Oxford (1995)
11. Roy, B.: Multicriteria methodology for Decision Aiding. Kluwer Academic Publishers, Dordrecht (1996)
12. Roy, B., Bouyssou, D.: Aide Multicritère à la Décsion: Méthodes et Cas, Economica, Paris (1993)
13. Saad, I., Grundstein, M., Rosenthal-Sabroux, C.: Locating The Company's Crucial knowledge to Specify Corporate Memory: A Case Study in an Automotive Company. In: Workshop Knowledge Management and Organisational Memory, IJCAI 2003, International Joint Conference on Artificial Intelligence, Acapulco, August 9-16 (2003)
14. Speel, P.H., Shadbolt, N., De Vries, W., Van Dam, P.H., O'hara, K.: Knowledge Mapping for industrial purpose. In: Conférence KAW 1999, Banff, Canada (October 1999)
15. Teece, D., Pisano, J.G., Shuen, A.: Dynamic capabilities and strategic management. Strategic Management Journal 7(18), 509–533 (1997)
16. Tounkara, T., Boughzala, I., Ermine, J.-L.: M3C: Une Méthodologie de Cartographie des Connaissances Critiques dans l'entreprise. In: IBIMA 2005, Caire, December 13-15 (2005)
17. Tounkara, T., Ermine, J.-L.: Méthodes de cartographie pour l'alignement stratégique de la gestion des connaissances, dans Management et ingénierie des connaissances: modèles et méthodes, Lavoisier, février (2008)
18. Tseng, B., Huang, C.: Capitalizing on Knowledge: A Novel Approach to Crucial Knowledge Determination. IEEE Transactions on Systems, Man, and Cybernetics Part A: Systems and Humans (2005)
19. Tsuchiya, S.: Improving Knowledge Creation Ability through Organizational Learning. In: Proceedings International Symposium on the Management of Industrial and Corporate Knowledge, UTC, Compiègne (1993)
20. Venkatraman, N., Camillus, J.C.: Exploring the concept "fit" in strategic management. Academy of Management Review 9(3), 513–525 (1984)
21. Zack, M.: Developing a knowledge strategy. California Management Review 41(3), 125–144 (1999)

Scouting for Drivers of the European Knowledge Society: The Role of Social Participation

Daniele Vidoni[1,*], Massimiliano Mascherini[2,*], and Anna Rita Manca[2,*]

[1] Italian National Institute for Educational Evaluation, Villa Falconieri, via Borromini 5 – Frascati, RM, Italy
[2] European Commission – Joint Research Centre, Via E. Fermi 2749, Ispra, VA, Italy

Abstract. Knowledge is a relational good. The flow of interactions among the individuals of a group provide the necessary opportunities to share the existing knowledge and use it to further accumulate the (human) capital, which is the main productive input for the development of any knowledge economy. In this sense, the opportunities of social interaction are per se a resource and a dimension of knowledge. The analysis of individual civil participation in different kinds of civil formal organizations across Europe further confirms the existence of different European social models and hints to a possible positive parallelism between the effectiveness of economic policies and social policies aimed at fostering social participation.

Keywords: Knowledge Society, Political Participation, Social Capital.

1 Introduction

Knowledge is the intangible asset that drives the growth of any knowledge society. In a narrow sense, knowledge could be identified merely with the stock of resources that individuals possess and use as production resources in their jobs and for the growth of society. Yet, the dynamics of the present society are such that knowledge is a relational good, object to rapid obsolescence and valuable only if shared. Thus, the available stock of knowledge is intrinsically tied to the flow of interactions among the individuals of a group, which are necessary to further accumulate (human) capital. In this sense, the opportunities of social interaction are per se a resource and a dimension of knowledge. The accumulation of such resources builds the stock of social capital which, as Putnam [7] posits, is the resource that "…enhances the benefits of investment in physical and human capital".

The European Commission communication on "Working together for growth and jobs" indeed acknowledges the need of both dimensions and states that "making growth and jobs the immediate target goes hand in hand with promoting social ... objectives." Hence, in the last few years the European Commission, has tried to integrate social policies into their core business to create an equal balance between the

* The analysis contained in this report is personal to the authors and does not necessarily reflect the views of the INVALSI or of European Commission.

M.D. Lytras et al. (Eds.): WSKS 2009, CCIS 49, pp. 319–327, 2009.

economic and social policies. This new logic is based on the conventional wisdom that greater economic prosperity drives jobs growth that in turn drives wellbeing.

This paper focuses on the interactive (social capital) component of knowledge. The detailed analysis of individual civil participation in different kinds of civil formal organizations across Europe hints to a possible positive parallelism between the effectiveness of economic policies and social policies aimed at fostering social participation. The existence of different European social models with different structures and rules [9] drives the implementation of the analysis in two macro areas: the North and Mediterranean Europe, where differences are more evident and a heterogeneous snapshot of Europe's social reality is highlighted.

Our paper is divided up in four main parts. In section 2 we present the general framework of the relations between the individual participation and the perception of different social reality. Section 3 presents the European social model to participation where are investigated the characteristics of the shares of individuals that are more likely to be involved with different kinds of formal organizations. In section 4 we discussed the results and section 5 concludes.

2 Individual Participation and Contribution to Social Capital

The scope of our analysis is to identify the characteristics of the shares of individuals that are more likely to be involved with different kinds of formal organizations and to investigate how each of these groups perceives the political, economical, and relational spheres of social reality.

As Putnam [6], [7] and other researchers [11] suggest, not all participation is conducive to positive results in terms of attentiveness to social and economic challenges. The classical Putnam's axiom makes the distinction between bridging (or inclusive) and bonding (or exclusive) social capital, where the first identifies the outward looking and encompass people across diverse social cleavages like, for instance the civil rights movement, while the second is characterized by inward looking and tend to reinforce exclusive identities and homogeneous groups, like for instance country clubs.

Previous work [2] focuses on the distinction between bridging and bonding social capital, eventually to categorize the various networks and organizations to which people belong.

Canois, Lerais, Mascherini, Saltelli and Vidoni [3] support the image of an Europe's social reality predominated by a bonding social capital linkage where the intensity of change follows a North-South axis in which Scandinavian countries are leader towards any form of social cohesion while the Mediterranean countries are centred on old conformities schemes of family, class and religion. The same North-South axis is identified by Hoskins and Mascherini [5] through the Active Citizenship Composite Indicator which measure the level of active citizenship in Europe. This composite indicator shows a heterogeneous Europe where Nordic countries lead and southern European countries follow. However, the indicator cannot tell us much with respect to the role of the different types of social capital for the progress of European societies because the structure of the indicator considers positively all kinds of participation. Still, as suggested earlier, the prevalence of different forms of participation may imply a trade-off between behaviours aimed at maintaining the status quo or at fostering progress. Hence, it is necessary to investigate whether such trade off does exist and whether the behaviour follows a mere North-South axis, or whether more articulated structures are required.

3 Participation and European Social Models

Favouring the fine-tuning of the process of institutional responsiveness requires identifying the objective characteristics and the perceptions of the actively engaged individuals.

To achieve this aim, we investigated the European Social Survey which ran a specific module on citizenship in 2002.

The European Social Survey (the ESS) is an academically-driven social survey designed to chart and explain the interaction between Europe's changing institutions and the attitudes, the beliefs, and the behavioural patterns of its diverse populations.

The European Social Survey (ESS) aims to be representative of all residents among the population aged 15 years and above in each participating country. The size and the quality of the sample make the European country coverage in the ESS data reasonably good, with 18 EU Member States providing data of sufficient quality. Among the 18 investigated Member States, Austria, France, Hungary, and Ireland have been excluded from the analysis because some variables were completely missing in the dataset as some questions were not asked in the national versions of the questionnaire. Although data on Poland has revealed anomalous patterns we have decided to keep the country in the dataset for the analysis.

The dataset under investigation is therefore composed by 26.491 observations. The sample is representative at the national level for the 14 European countries considered, which are: Belgium, Germany, Denmark, Spain, Finland, United Kingdom, Greece, Italy, Luxemburg, the Netherlands, Poland, Portugal, Slovenia and Sweden. The number of observations was subsequently reduced to 24.023 to exclude people below 18 or above 80 years of age.

In the European Social Survey 2002, individuals were asked about their *ways of participating* to several types of organizations (sport, cultural, trade union, business, human rights, environment/peace, religious, politics, social, teacher/parents organizations), and they could choose among four different behaviours (membership, donating money, participation and voluntary work) recorded as dichotomous variables (values 0/1). For each type of organization, the *strength* of individual participation amounts to the sum of the scores that the individual has obtained with respect to the foresaid possible behaviours.

Specifically, we created a new variable equal to the sum of the four binary variables describing the possible action of engagements in that organization.

$$Y_h = \Sigma_{i=1}^{4} X_{h,1}. \tag{1}$$

where $h=1..10$, is the type of organization and $i=1..4$, are the four different ways of engagement. Each variable can assume a value from 0 (no action taken) to 4 (the person is engaged in all possible ways).

On the basis of the definitions of *bridging* and *bonding* social capital, we divided individual participation into two categories:

1. *Social engagement*, individuals participating to organizations that are outward looking and aim at improving the society at large (Cultural, Human Rights, Social, Religious, Environmental/Peace organization).
2. *Private engagement*, encompasses the organizations that work closer to the private interest of the respondent (Sport, Trade Union, Business, Teacher/Parents, Political Party).

A factor analysis using polychoric correlations for ordered-category data was carried out to validate the consistency of this theoretical grouping. The results of the analysis confirm the hypothesis of the existence of two main groups (social engagement and private engagement) and explain approximately 40% of the variance.

The two groups were used as the basis for constructing two variables – SOCIAL ENGAGEMENT and PRIVATE ENGAGEMENT – given respectively by the aggregation of the variables reporting higher factor loadings in factor 1 (social engagement) and in factor 2 (private engagement), respectively. On the basis of the 10 categories of organizations previously identified (sport, cultural, trade union, business, human rights, environment/peace, religious, politics, social, teacher/parents organizations), the variables *social engagement* and *private engagement* ended up having the following structure:

$$\text{SOCIAL ENGAGEMENT} = Y_{Cultural} + Y_{Social} + Y_{Env/Peace}\ Y_{HumanRights} + Y_{Religious}$$
$$\text{PRIVATE ENGAGEMENT} = Y_{Sport} + Y_{TradeUnion} + Y_{Business} + Y_{Teacher/Parents} + Y_{Political}$$

Subsequently, we created the variable ENGAGEMENT that will constitute the basis of our analysis. ENGAGEMENT is a multinomial variable that takes values from 0 to 3 defined as follows:

ENGAGEMENT = 0 - individuals not participating in any organization;
ENGAGEMENT = 1 - individuals participating just in social organizations;
ENGAGEMENT = 2 - individuals participating just in private organizations;
ENGAGEMENT =3 - individuals engaged in both social and private organizations.

To characterize the factors that can determine the decision of an individual to participate in formal organizations, the variable ENGAGEMENT was considered as the dependent variable, and a set of objective variables reported in the European Social Survey was included as explanatory variables. Explanation of the various patterns of participation

The variable objects of the analysis are qualitative, hence the application of classical regression models (OLS) is not allowed due to the violation of the basic assumptions underlying the model. Modelling qualitative variables with two or more categories requires the use of models based on different assumptions that can produce reliable estimates and allow for the correct application of standard statistical techniques. If the adoption of Logistic Regression could solve the problem of modelling dichotomous variables, in the case of variables that can assume more than two categories the use of the Ordinal Logit or the Multinomial Logistic regression model can face the challenge of providing correct and reliable estimation of the parameters.

The model we decided to adopt to analyze the variable ENGAGEMENT is the Multinomial Logit (mLogit). The mLogit is a straightforward extension of the classical Logistic model and it is a strategy often used in the literature when categories are unordered (as the case of the variable ENGAGEMENT). In the mLogit model a category (Non Participant, in our case) is designated as reference category. The probability of membership in other categories (Social, Private and Full Engagement) is compared to the probability of membership in the reference category. In general for a dependent variable with M categories, the mLogit requires the calculation of M-1 equations, one for each category relative to the reference category to describe the relationship between the dependent variable and the explanatory variables. In our case, three different equations have been computed.

Although possible violations of the Independence from Irrelevant Alternatives (IIA) assumptions need to be further investigated, similar results have been obtained by repeating the analysis and applying several logistic regressions that perform pairwise comparisons between the reference category and the other categories.

4 Results

An initial analysis of the participatory patterns at country level, as shown in figure 1, suggests that individuals in all Nordic countries are preponderantly engaged in both private and social organizations, while more than 50% of the individuals in Mediterranean are not engaged. Greece is an interesting outlier because it shows a strikingly high 45% of people that are involved in private organizations, a result partly due to the peculiar status of Greek business organizations [2].

The analysis of the results of the mLogit model shows an interesting picture, highlighting different profiles of the participants. The result of the analysis presented in this paper compared the Nordic cluster, which includes Sweden, Denmark, Finland and The Netherlands, and the Southern cluster made by Portugal, Italy, Spain and Greece. The results of all the other clusters are in Canois et al. [3]. The picture of Europe which emerges is very heterogeneous because characterized by different dynamics of social participation in each geographical area. This suggests the absence of

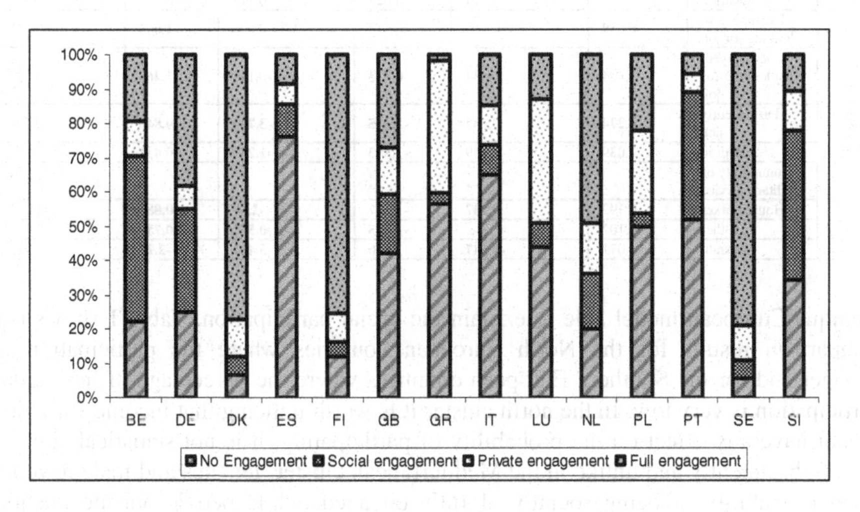

Fig. 1. "Engagement": how is it distributed among European countries?

Table 1. Parameter estimates for the model of Nordic and Mediterranean countries

		Comparison between Non Engaged and Socially Engaged		*Comparison between Non Engaged and Privately Engaged*		*Comparison between Non Engaged and Fully Engaged*	
		Nordic Countries	*Mediterranenan Countries*	*Nordic Countries*	*Mediterranenan Countries*	*Nordic Countries*	*Mediterranenan Countries*
	Explanatory Variables	Coef.	Coef.	Coef.	Coef.	Coef.	Coef.
Household Income	>1800Euros	-0,065	-1,435	-1,489	-1,788	-0,609	1,02
	>3600Euros	0,522	**-3,047**	0,485	**-1,994**	0,151	**-1,919**
	>6000Euros	-0,312	-1,341	-0,366	-1,017	-1,482	0,001
	>12000Euros	0,547	-1,037	-0,346	-0,405	-0,589	-0,725
	>18000Euros	0,725	-1,047	0,184	-0,652	-0,037	-0,18
	>24000Euros	0,735	-0,026	0,434	0,073	-0,13	-0,097
	>30000Euros	0,503	-0,993	0,478	-0,428	0,388	-0,521
	>36000Euros	0,523	-0,504	0,324	**-1,254**	0,575	0,244
	>60000Euros	0,741	-1,398	1,287	0,251	1,434	0,295
	>90000Euros	0,889	...	0,188	...	0,198	...
	>120000Euros	0,019	...	1,205	...	0,891	...
Age	(18-24)	**-1,022**	**-1,025**	0,13	-0,858	-0,004	-0,961
	(24-35)	**-1,136**	**-0,921**	0,203	-0,303	**1,251**	-0,829
	(35-44)	**-1,134**	**-1,294**	**0,807**	0,692	**0,866**	-0,282
	(44-55)	-0,544	**-0,778**	0,511	-0,072	0,542	-0,406
	(55-65)	**-0,779**	**-0,647**	**0,696**	0,395	**0,806**	0,064
	Gender	0,002	0,347	**-0,489**	**-0,909**	-0,113	**-0,614**
Media variables	Watching TV	**-0,103**	-0,07	-0,036	0,02	**-0,194**	-0,095
	Listening to the Radio	0,018	-0,039	0,046	0,028	**0,08**	**0,083**
	Reading Newspaper	0,045	0,036	**0,21**	0,06	**0,346**	**0,196**
	Surfing the Web	0,008	-0,051	0,008	0,019	0,04	**0,026**
	Individual Human Capital	**0,099**	**0,12**	0,03	**0,102**	**0,123**	**0,127**
	Number of persons	-0,054	-0,04	**-0,162**	0,067	**-0,257**	0,084
Domicile	Suburbs of a big city	0,405	**1,039**	-0,022	0,476	**0,853**	0,139
	Small town	**0,492**	**0,994**	0,212	0,545	**0,556**	**0,769**
	Country village	**0,52**	0,564	0,211	0,273	**0,954**	0,483
	Home in the countryside	0,026	0,482	-0,095	0,526	**1,308**	-0,864
	Citizenship	0,462	0,968	**0,883**	0,618	**1,171**	0,632
	Declared Religious	**0,842**	0,094	-0,159	0,162	**0,982**	0,455
Main Activity	In education	0,592	0,014	0,024	0,717	0,298	0,47
	Unemployed, looking for job	-0,058	0,469	-0,617	-0,696	**-1,065**	-0,887
	Unemployed, not looking for job	-0,404	-0,34	**-0,933**	-0,199	**-1,465**	0,334
	Permanently sick	-0,274	0,685	**-1,315**	**-3,734**	**-0,843**	0,009
	Retired	-0,039	-0,169	**-1,593**	**-1,259**	-0,431	-0,35
	Community or military service	...	...	...	...	...	-1,413
	Houseworker	-0,192	0,237	**-1,237**	**-2,56**	**-0,887**	**-1,667**
	Other	0,107	...	**-1,805**	**0,572**	**-0,785**	0,429
	Constant	-2,071	**-3,547**	-1,106	**-3,424**	**-2,26**	**-3,915**

a unique European model able to explain the social participation. Table 1 shows the comparison results for the North European countries where the participation is stronger and for the Southern European countries where the percentage of individual participation is very low. In the north cluster it is worth noticing that income variables do not have any effect on the probability of participating, it is not statistical significant. Subsequently and unlike in the Mediterranean cluster, females and male have an equal probability of being socially of fully engaged, while people outside the job market are less likely to be privately or fully engaged.

This scenario changes completely in the Mediterranean cluster. The first important result concerns the limited number of significant coefficients coming out from the model suggests that, in these area, participation is driven by variables not considered in the model and not collected through the survey. Moreover, the variables used for the analysis are *self reported* and doubts persist on the reliability of the variable *income*. Although the number of common significant variables between the two clusters is reduced, it is still possible perform a comparison in order to highlight differences and common trends. In particular older people are more likely to participate in the socially engaged group in the North as in the Mediterranean cluster and the effect of years of education is positive in both clusters. While in the Mediterranean countries the gender variable play an important role in favor of male, people which are more likely to be privately and fully engaged. The effect of watching TV is negative on the engagement in Northern countries, while is not significant in the Mediterranean cluster. Concerning the other media variable the same effect in both clusters are recorded. Analyzing the effect of the place of domicile in the north countries; individuals that live outside big cities but not completely isolated in the countryside or in small cities have a higher probability to be part of the fully engaged group. This variable is not significant in the Mediterranean cluster. The main activity in the previous week variable has a strong effect in the north countries where non-active participation to the job market are associated with a substantial reduction of the probability of being privately and fully engaged. This effect is not significant in the Mediterranean cluster. Considering all the different described in the two clusters, the adoption of a different model exploring new dimensions and variables appear necessary for modeling the participation in the Mediterranean European countries, while the model represents accurately the northern European cluster.

Yet, the analysis offered already hits to the existence of substantial differences across Europe.

Recalling the initial argument, the level of social participation suggests the existence of a higher stock of interaction opportunities, which facilitates the accumulation of human capital and the economic prosperity of the country. Although it is far from being conclusive, this evidence is further supported by looking at the parallelism between participation and a set of indicators that sketch the country status in terms of economical and social development.

Indeed, limiting the analysis to European OECD countries, we can notice that the intensity of the investments tends to co-vary with welfare state characteristics, especially on a North-South axis. For instance, high welfare spending in the Scandinavian countries goes hand in hand with a relatively higher level of wealth (GDP), a relatively smaller income inequality. Low welfare spending in the Mediterranean countries goes together with a lower level of wealth, a larger income inequality (fig. 2). The picture does not change when looking at investments in R&D, Higher Education, and ICT (merged as "Total investment in knowledge") whereby Finland and Sweden invest more than 5% of their GDP with a specific focus on R&D. Southern European countries, on the other hand stop half way through at about 2%.

Indeed, limiting the analysis to European OECD countries, we can notice that the intensity of the investments tends to co-vary with welfare state characteristics, especially on a North-South axis. For instance, high welfare spending in the Scandinavian countries goes hand in hand with a relatively higher level of wealth (GDP), a relatively smaller income inequality. Low welfare spending in the Mediterranean countries goes together with a lower level of wealth, a larger income inequality (fig. 2). The picture does not change when looking at investments in R&D, Higher Education, and ICT (merged as "Total investment in knowledge") whereby Finland and Sweden invest more than 5% of their GDP with a specific focus on R&D. Southern European countries, on the other hand stop half way through at about 2%.

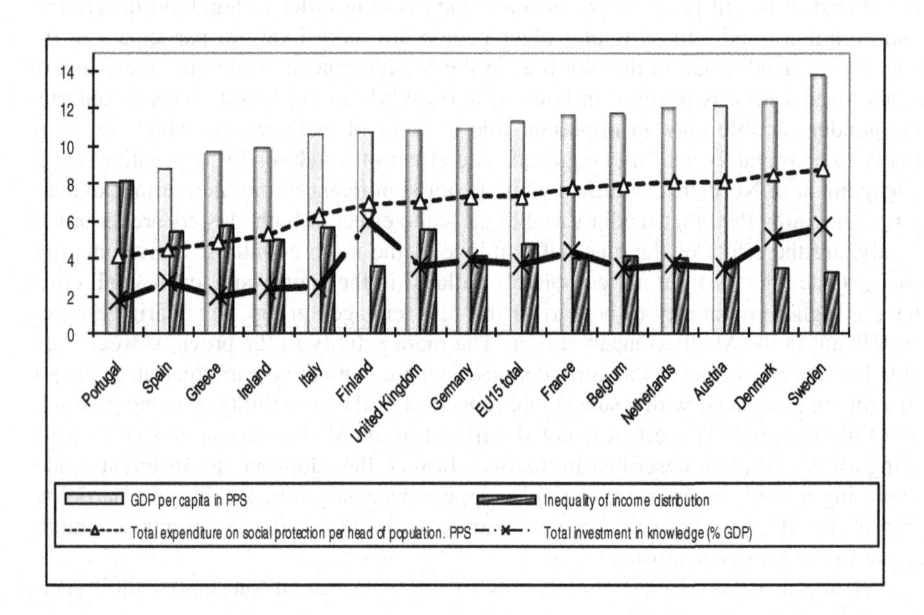

Fig. 2. GDP per capita, social protection, investment in knowledge and inequality of income distribution: an European overview[1]

5 Conclusions

The paper shows that new realities call for a better involvement of social dimensions to foster economic dynamism in two different clusters: the North and the Mediterranean.

The empirical analysis of participation clearly identifies specific patterns according to the distinction between private and public participation. The analysis shows that involvement of citizens is key to social and economic performances, but different models of participation coexist. The variety of participation patterns reflects the way in which different traditions and cultures affect the models of coordination between the State and the stakeholders.

The key element is that participation by itself, in whichever form – is a crucial input for economic and social performance.

One point to highlight is the consistently positive effect of education for all kinds of participation in all the clusters. Such result identifies for looking at education as one of

[1] Data source: Eurostat (Expenditure in social protection and GDP per capita relate to 2004, Inequality of income relates to 2005), OECD [3].

the possibly most transversal policy actions to undertake and suggests the importance of working on the quality of education systems both at basic and higher level.

On a certain level, our empirical analysis suggests that the full development of a knowledge society requires economical and social policies to go hand in hand. Indeed, there is a striking parallelism between the direction of crucial economic indicators (such as GDP per capita, social protection, investment in knowledge and inequality of income distribution) and the level of individual participation to formal organizations. Thus, as recognized by the European Commission's communication on "Working together for growth and jobs," boosting economic growth cannot but go hand in hand with the development of European social policies aiming at integration and social cohesion.

Still, it is also apparent that strong structural differences exist in the approach to participation in different parts of Europe and especially on a North-South axis. These differences impede the application of blanket policies and suggest the need of cooking *ad hoc* recipes closely agreed upon with the stakeholders, who take on the independent and legitimate role of interpreting social needs and of launching cooperation strategies at territorial level.

In this paradigm, favouring individual participation to the public-value creation process is a key element to the development of wellbeing in the future Europe.

References

1. Aranitou, V.: The strengthening of the employers' organizations representation and social dialogue. In: Proceedings of the S. Karagiorgas Conference on Social Change in Contemporary, Greece, Athens (2003)
2. Beugelsdijk, S., Smulders, S.: Bridging and Bonding Social Capital: which type is good for economic growth? In: ERSA conference papers, ersa 2003, p. 517. European Regional Science Association (2003)
3. Canoy, M., Lerais, F., Mascherini, M., Saltelli, A., Vidoni, D.: The Importance of Social Reality for the European Economy: An Application to Civil Participation. In: OECD: Statistics, Knowledge and Policy 2007: Measuring and Fostering the Progress of Societies. OECD Publishing, Paris (2008)
4. Dasgupta, P.: Social Capital and Economic Performance: Analytics. In: Ostrom, E., Ahn, T.K. (eds.) Critical Writings in Economic Institutions: Foundations of Social Capital. Edward Elgar, Cheltenham (2003)
5. Hoskins, B.L., Mascherini, M.: Measuring Active Citizenship through the development of a Composite Indicator. Soc. Indic. Res. 90, 459–488 (2009)
6. Putnam, R.D.: Making Democracy Work: Civic Traditions in Modern. Princeton University Press, Princeton (1993)
7. Putnam, R.D.: The Prosperous Community: Social Capital and Public Life. American Prospect 13, 35–42 (1993)
8. Putnam, R.D.: Bowling Alone: The Collapse and Revival of American Community. New Simon and Schuster, York (2000)
9. Sapir, A.: Globalisation and the Reform of European Social Models. Bruegel policy brief, Bruxelles (2005)
10. Völker, B., Derk, H.: Creation and Returns of Social Capital: A New Research Program. Routledge, London (2004)
11. Woolcock, M.: Social capital and economic development: towards a synthesis and policy framework. Theory and Society 27, 151–208 (1998)

Operationalising the Sustainable Knowledge Society Concept through a Multi-dimensional Scorecard

Horatiu Dragomirescu[1] and Ravi S. Sharma[2]

[1] Bucharest University of Economics - ASE, 6 Piata Romana, Bucharest 010374, Romania
[2] Wee Kim Wee School of Communication & Information, Nanyang Technological University, 31 Nanyang Link, Singapore 637718

Abstract. Since the early 21st Century, building a Knowledge Society represents an aspiration not only for the developed countries, but for the developing ones too. There is an increasing concern worldwide for rendering this process manageable towards a sustainable, equitable and ethically sound societal system. As proper management, including at the societal level, requires both wisdom and measurement, the operationalisation of the Knowledge Society concept encompasses a qualitative side, related to vision-building, and a quantitative one, pertaining to designing and using dedicated metrics. The endeavour of enabling policy-makers mapping, steering and monitoring the sustainable development of the Knowledge Society at national level, in a world increasingly based on creativity, learning and open communication, led researchers to devising a wide range of composite indexes. However, as such indexes are generated through weighting and aggregation, their usefulness is limited to retrospectively assessing and comparing levels and states already attained; therefore, to better serve policy-making purposes, composite indexes should be complemented by other instruments. Complexification, inspired by the systemic paradigm, allows obtaining "rich pictures" of the Knowledge Society; to this end, a multi-dimensional scorecard of the Knowledge Society development is hereby suggested, that seeks a more contextual orientation towards sustainability. It is assumed that, in the case of the Knowledge Society, the sustainability condition goes well beyond the "greening" desideratum and should be of a higher order, relying upon the conversion of natural and productive life-cycles into virtuous circles of self-sustainability.

Keywords: composite indexes, knowledge society, sustainability, multi-dimensional scorecard, policy-making.

1 Introduction

The operationalisation of the Knowledge Society concept becomes increasingly important, as a way for rendering stakeholders (including citizens) aware and responsible of what is at stake for each of them, proactive in accomplishing their role, and open to concertation in vision and in action. Operationalising the respective concept enables the evolution towards Knowledge Society becoming manageable through vision-building, steering and monitoring the dynamics of the respective societal system.

M.D. Lytras et al. (Eds.): WSKS 2009, CCIS 49, pp. 328–337, 2009.

In normal science, conceptualization should precede operationalisation, as the latter is aptly applicable onto already crystallized concepts. Atypically, in the case of the Knowledge Society, operationalisation had been initiated before the completion of the conceptualization phase.

The United Nations ad-hoc Expert Group on Knowledge Systems for Development raised a series of basic questions pertaining to the operationalisation of the Knowledge Society concept: "*What is knowledge, and what is different about knowledge today? What are the antecedents to the knowledge society? What are the contours and the values of the knowledge society? What are the cultural shifts that must take place?*" [1].

The issue of operationalising the Knowledge Society concept under the sustainability condition is hereby considered as a key goal for policy-making, in the vision-building phase, but also in the subsequent ones: policy implementation, monitoring and assessment. Adequate tooling of such an enterprise also requires keeping the balance between the comparative and the contextual approaches, as defined by Barzilai-Nahon [2]. The use of composite indexes pertains to the former, while the latter privileges synoptic frameworks.

Hence, it is intended that the multi-dimensional scorecard advocated in this paper will serve researchers and policy-makers with interests in knowledge management for development in several scenarios. The first would be to perform an ex-post review of a society's growth and development in order to learn and prescribe best practices for others to adopt. Another scenario would be to conduct longitudinal investigations tracking a particular society's progress along the path towards a developed knowledge society. Yet another may be to first identify and then harmonise disparate levels of development within a community such as the European Union or ASEAN. Given the vast amount of data available from the United Nations, the World Bank and NGOs such as the World Economic Forum, the above scenarios would be valid and practical.

The remainder of this paper is structured as follows. In section 2, the importance of overcoming the reductionism fallacy is evidenced with respect to both Knowledge Society and sustainable development. In recent literature reviewed, the former is still predominantly considered from the technological determinism perspective, and the latter - from the environmental one. In section 3, composite indexes used for assessing info-states are reviewed and the inherently limited informativeness, due to their aggregate pattern, is examined, *vis-à-vis* the example of the multi-dimensional synopsis proposed by Spangenberg [3]. In section 4, a multi-dimensional scorecard is proposed as a contextual instrument devoted to the sustainability assessment of the Knowledge Society at national/regional level. The respective scorecard is a re-structured and upgraded version of the policy analysis framework devised by Sharma and his co-workers [4]. The final section concludes.

2 Knowledge Society under the Sustainability Challenge

A core issue regarding the operationalisation of the Knowledge Society concept consists of having it subjected to the condition of sustainability.

Society's path towards sustainability can be described as a transition process, subjected to major challenges taking the form of the need for:

- more accurate models, metaphors, and measures to describe the human enterprise relative to the biosphere;
- substantial improvements regarding citizenship and governance;
- major enhancement in public awareness, along with provision of the education needed for the transition to sustainability;
- tackling sustainability as a series of divergent problems formed out of the tensions between competing perspectives *"that can only be resolved by higher forces of wisdom, love, compassion, understanding, and empathy"* [5].

Traditionally, at the macro-economic level, the sustainability condition used to be applied to the process of development, while, at the micro-economic level, this condition was imposed to the competitive advantage sought by business enterprises.

The 21st Century brought about a higher pressure with the twin imperatives of raising public concern and political engagement with a view to progressing towards both Knowledge Society and sustainable development. However, the perspective originally predominant in understanding sustainability was the environmental one, hardly applicable to the Knowledge Society. It is so because environmental endowments and constraints are of a tangible, ontological kind, while, by contrast, the Knowledge Society involves an emphasis on the intellectually grounded, intangible aspects.

During the last two decades, research inquiries devoted to the Knowledge Society and sustainable economic development concepts unfolded in parallel, but rather independently to one another [6]. Thus, the cross-fertilisation of the two respective lines of inquiry was delayed. A comparison between them allows identifying some common features:

- both concepts are of a similar, high level, generality and complexity;
- in both cases, reflection of a rather explorative kind was first performed at a global scale, while subsequent political commitments and initiatives were taken at national and regional levels;
- both cases were originally tackled from a single dominant perspective, that eventually proved to be reductionist: the technological determinism perspective for the Knowledge Society [7], and the "greening" (environmental) one for the sustainable development [8].

Although an increasing knowledge orientation in tackling the sustainability issue was signaled in the recent literature [9], most of the underlying approaches - such as considering innovation, science, education, or Information & Communications Technology (ICT) as key factors of sustainable development - are selective in focus and difficult to articulate systemically; measurement and analysis tools generated on such grounds are, in their turn, disparate and partial.

Approaches of an integrated type occurred more recently, but, ironically, many of these are lacking a distinct knowledge orientation. For example, the Sustainable Society Index [10] has a comprehensive, multi-dimensional format, but the knowledge orientation is barely present at the individual level only, being limited to the "education opportunities" parameter. In turn, the nascent "Science of sustainability" is claimed to be concerned with *"the dynamic interactions between nature and society"* [11], thus exhibiting a predominant environmental orientation, while the knowledge-oriented aspects are still pending consideration.

As a way to embedding the knowledge dimension into a holistic sustainability mindset, Gallopín and Vessuri [12] stated the necessity of defining a research agenda for the inter- and trans-disciplinary articulation of knowledge, including not only scientific knowledge, but also local, traditional and even lay persons' knowledge.

The sustainable Knowledge Society concept was proposed as an alternative to the propensity for extrapolating the New Economy model, that would lead, in the long run, to unsustainability *"not so much in economic competitiveness, as in relation to social problems (loss of cohesion, growing disparities) and to environmental problems"* [13].

Spangenberg pointed out that Knowledge Society can be sustainable only if developed in a strategically oriented context [3].

Fuchs proposed a broad, multi-dimensional notion of Information Society sustainability, that encompasses ecological, technological, economic, political, and cultural aspects and problems [14]. Knowledge Society's sustainability involves seeking to diminishing risks that affect the societal system, while capitalising on its strengths and opportunities that could be revealed through a knowledge SWOT analysis [15].

According to Bender and Simonovic, consensus can be considered as a measure of sustainability, an idea which fully makes sense if applied to Knowledge Society [16], understood as "*a society [. . .] endowed with the ability and capacity to generate and capture new knowledge and to access, absorb and use effectively information and ICTs*" [17].

At the extent to which the development of the Knowledge Society can be steered, the core qualitative instrument hereby applicable is an overall vision at national level, consensually agreed by its relevant stakeholders. In the same vein, Wiig proposed the concept of "balanced society" [6], that seems consistent with the idea of a "new pact" between science and society in the 21st Century.

The assumption hereby derived asserts that the Knowledge Society involves a sustainability of a higher-order, which is self-sustainability, at national, regional and international levels. This is achievable through turning traditional life cycles of the *"from cradle to grave"* type, currently usual in environmental and productive settings, into virtuous circles, such as, for instance, the *"virtuous circle of development"* [13].

3 Operationalisation through Measurement: From Comparative to Contextual Instruments

There are considerable difficulties with operationalising the Knowledge Society concept, which is a reality in evolution, difficult to apprehend synoptically. Such difficulties foster the scepticism towards the Knowledge Society, up to the limit of considering it a futuristic ideal, or just a metaphor in a rhetoric.

Quantitative approaches aiming at operationalising the Knowledge Society concept preceded the qualitative one, and also outweighed them in terms of popularity. The former strand adopted the methodological and data background from the previous Information Society phase; however, while, in the context of the Information Society, the emphases was placed on capacities and access to ICT, in the Knowledge Society the focus is on use and effects of ICTs in their beneficiary systems.

A high level of attention was devoted lately to composite indexes, that enable ranking and clustering countries or regions, with the possibility of subsequent gap analysis. The range of the composite indexes most widely applied to the Knowledge Society is reviewed in Table 1; however, it should be mentioned that developing procedures for the ranking of communities or societies is not a purpose of this paper.

Table 1. Composite indexes used for Knowledge Society assessment

Index	Objective and scope
ICT Opportunity Index	- estimate of a country's ICT readiness, depending on degree of info-density and info-use - combines 10 quantitative and qualitative indicators from 4 subcategories: networks, skills, uptake, and intensity of use
Digital Divide Index (DIDIX)	- measures, at aggregate level, the inequalities in ICT diffusion, to benchmark national digital divides within EU member states - covers the following 4 disadvantaged, "risk" groups: people aged 50 years and over, women, low-education and low-income groups
Digital Opportunity Index (DOI)	- measures ICT diffusion, using a range of indicators relevant to the profile of a forward-looking Information Society - combines 11 ICT indicators, grouped in three categories: opportunity, infrastructure and use
e-Readiness Index	- estimates a country's ability to make use of digital channels, provided by ICT industry and Internet services development, in order to benefit citizens, business, and government - combines about 100 quantitative and qualitative criteria, organised into 6 groups, among which: connectivity environment, government investment and policy, social and cultural attitudes related to Internet adoption
Knowledge Competitive-ness Index	- benchmarks the knowledge capacity, capability and sustainability of a country/region, reflecting the degree of knowledge use for generating economic value and citizens' wealth - combines selected variables grouped into five categories: human capital, financial capital, knowledge capital, regional economy outputs and knowledge sustainability
Knowledge Economy Index (KEI)	- quantifies the overall level of development of a country/region towards the Knowledge Economy - is calculated as an average score of a country/region, based upon the key performance variables on all Knowledge Economy pillars specified by the World Bank: economic incentives and institutional regime, education and human resources, innovation system, ICT
Network Readiness Index (NRI)	- quantifies a country's/community's propensity for fructifying opportunities provided by ICT - covers three aspects: the environment provided by a given country/community for leveraging ICT, the readiness of the community's key stakeholders to use ICT, and the usage of ICT amongst these stakeholders

Sources: [18], [19], [20], [21], [22], [23], and [24].

Composite indexes should be used for policy-making purposes while taking into account the limitations caused by their particular design:

- their construction, inspired by the technological determinism approach of the Knowledge Society:
- their limited informativeness, due to procedures of calculus used, through which diverse sets of raw data are refined into compact end-constructs;
- their retrospective orientation, that allows comparing states already attained, implicitly assuming the emergent character of the Knowledge Society advance;
- the current scarcity of models that allow correlating diverse composite indexes;
- the fact that composite indexes are used in designing policies of a catching-up, type, attributing the status of a *de facto* standard to the entities upper positioned in rankings, their success recipe being then imposed to those lagging behind on the respective scales.

Ignoring these limitations could lead to misjudgements, which are avoidable through complementing composite indexes with other assessment instruments. A notable attempt to conceiving assessment tools that have a contextual orientation is the set of indicators developed by Spangenberg [3], a synopsis of which is presented in Table 2. It overcomes the kind of limitations above described and is also consistent with the complexity of the Knowledge Society.

Table 2. Synopsis of Spangenberg's Knowledge Society sustainability indicators

Dimension	Knowledge-oriented indicators	ICT-oriented indicators
	I. Mono-dimensional indicators	
Economic	- output of the education sector - velocity of invention and innovation diffusion	- ICT accessibility quotient - ICT contribution to labour productivity growth
Social	- literacy rate - availability of cultural hotspots	- home Internet penetration - access to telecommunications
Environmental	priority of the "greening" concern in the conduct of citizens and organisations	- ICT energy consumption - ICT waste recycling
Institutional	- education expenditure - consumer information standards	ability to use ICT infrastructure and content
	II. Trans-dimensional indicators	
Socio-economic	- demographic distribution of education and training levels - knowledge intensity of production	- weight of ICT expenses in households' income - end-user market demand for electronics and ICT goods
Economic-institutional	education accessibility and affordability	- structure of media ownership - fiscal regime in e-commerce
Socio-institutional	educational attainment in science, informatics, and engineering	employment and social security status of ICT workers

Source: adapted from [3].

The key contributions brought by Spangenberg's system of indicators can be summarised as follows:

- the applicability of the sustainability condition to Knowledge Society is explicitly stipulated;
- the balance between the knowledge orientation and the ICT orientation of the set of indicators is kept;
- the multi-dimensional character is ensured through including several axis of analysis: economic, social, environmental and institutional; moreover, three of these axes are combined into mixed frameworks (socio-economic, economic-institutional and socio-institutional), with a view to going beyond their separate consideration as parallel attributes.

It can be observed that such a multi-dimensional approach facilitates the twinning of the Knowledge Society and sustainable development concepts, and their operationalisation for policy-making purposes.

4 A Knowledge Society Multi-dimensional Scorecard

Given its perceived complexity, Knowledge Society exhibits multiple facets, corresponding to the diversity of perspectives under which it can be approached or to that of the stakeholders involved. Therefore, it was portrayed in diverse ways, through distinguishing a core feature that would most pertinently reveal its essence; the hypostases most frequently invoked are: network society learning society, innovation society, communications society, creative society etc. The opinions according to which those features are "parallel concepts" are contrasting to those pleading for their integration into a multi-dimensional representation.

The complexity of the Knowledge Society justifies the adoption of complexification method for systemic cognition purposes, with a view to obtaining a "rich picture" of its shape and dynamics; the respective method was proposed by Morin and Le Moigne, as underlined by Eriksson [25].

Table 3 above summarised the fundamental dimensions of a scorecard that is claimed to represent comprehensively and parsimoniously a knowledge society.

The original version of this scorecard was developed from a Delphi survey, further subjected to expert confirmation [4].

For the purposes of this paper, the respective framework was expanded and restructured. In this respect, the individual dimensions, identified through literature review, were grouped under the four pillars set up by the World Bank's Knowledge Assessment Methodology [23]. Moreover, the list of proxy indicators was expanded, all of them being regularly released from reputable sources such as the United Nations and the World Bank.

The usability of the proposed scorecard for policy-making purposes is under testing in a field exercise, the foci of the analysis being Singapore, Nigeria, the United States and the United Arab Emirates.

Table 3. A multi-dimensional scorecard of the Knowledge Society at national level

Pillars	Dimensions	Proxy indicators
Education and training	Higher education	public spending on higher education; tertiary education attainment; tertiary graduates in mathematics, science and technology; GDP share invested in higher education
	Training	lifelong learning participation
Innovation systems	Research & development	patent intensity; citation impact of country's scientific output; scientific publications highly cited in patents; receipts of royalty and license fees; R&D employment; Creativity Index
	Net knowledge inflows	international trade in core cultural goods; international trade in ICT goods; technology balance of payments; share of trade in high-tech products; international mobility flows of foreign tertiary students; net migration of skills
	Knowledge networks	composition of telecentre networks; university-industry R&D centres; academic spin-offs; R&D consortia; research sub-contracting; patent citations
	Shared spaces for knowledge creation	co-patents and co-publications; fairs, exhibitions digitised cultural heritage; household expenditure on civic amenities (culture, entertainment)
Information and Communications Infrastructure	ICT accessibility	Network Readiness Index; B2B and B2C sales in e-commerce; broadband Internet subscribership; hosts and websites on the Internet; Internet domain name registrations
	Role of mass media	entertainment and media market: voice & accountability; press freedom
Economic and institutional regime	Rule of law consistent with international norms	Corruption Perceptions Index; Global Peace Index
	Political vision & strategy	Country's Project Maturity Index; political stability; regulatory quality; government effectiveness
	Human rights & freedom	Human Development Index; Index of Personal and Economic Freedom; EIU Democracy Index
	Intellectual property	cyberlaw coverage; rate of piracy in digital intellectual goods
	Business environment that rewards innovation	high-tech companies benefiting from early-stage venture capital investment; venture capital investment for private R&D; Index of Economic Freedom; Business Competitiveness Index

Source: developed by the authors based upon [4].

5 Conclusions

The Knowledge Society development is increasingly recognised as a top priority for all stakeholders concerned, among which policy-makers play a key role in vision building and steering the respective evolution under the sustainability condition. To adequately deal with complexity, comparative measurement tools, such as composite indexes pertaining to the info-state, should be complemented by contextual ones, of the kind of assessment indicators assembled into synoptic frameworks. In the latter respect, a multi-dimensional scorecard of the sustainable Knowledge Society is proposed, based upon the preliminary results of a research in progress that will further proceed to field testing of the usability of this tool.

The authors make no claims about the suitability or adaptability of the multi-dimensional scorecard presented in this paper for the expressed purpose of ranking or comparing societies in terms of the sustainability of their knowledge base. Such a use would be unseemingly futile, as rankings risk to fuel rhetoric and envy among communities that ought to cooperate for the betterment of all. It is reiterated that a multi-dimensional scorecard attests to the strengths, weaknesses, opportunities and threats inherent in a Knowledge Society, with the view of benchmarking and the transfer of best practices among nations, Such practices, it is submitted, reushers in an era of international cooperation and development for all.

References

1. U.N. Department of Economic of Economic and Social Affairs: Expanding Public Space for the Development of the Knowledge Society. Report of the Ad Hoc Expert Group Meeting on Knowledge Systems for Development, September 4-5, United Nations, New York (2003), http://unpan1.un.org/intradoc/groups/public/documents/UN/UNPAN014138.pdf
2. Barzilai-Nahon, K.: Gaps and bits: conceptualizing measurements for digital divide/s. The Information Society 22, 269–278 (2006)
3. Spangenberg, J.H.: Will the information society be sustainable?: Towards criteria and indicators for a sustainable knowledge society. International J. of Innovation and Sustainable Development 1, 85–102 (2005)
4. Sharma, R.S., Ng, E.W.J., Dharmawirya, M., Lee, C.K.: Beyond the digital divide: A conceptual framework for analyzing knowledge societies. J. of Knowledge Management 12, 151–164 (2008)
5. Orr, D.R.: Four Challenges of Sustainability. School of Natural Resources, University of Vermont (2003), http://www.ratical.org/co-globalize/4CofS.pdf
6. Wiig, K.M.: Effective societal knowledge management. J. of Knowledge Management 11, 141–156 (2007)
7. Joux, A.: Theoretical approaches of the knowledge society at UNESCO. Communications 31, 193–214 (2002)
8. Loveridge, D.: On sustainability. "Ideas in Progress" series, paper no. 14, PREST, University of Manchester, Manchester (1999), http://www.personal.mbs.ac.uk/dloveridge/documents/on-sustainability_wp14.pdf

9. Cam, C.N.: A conceptual framework for socio-techno-centric approach to sustainable development. International J. of Technology Management and Sustainable Development 3, 59–66 (2004)
10. Sustainable Society Foundation: Description of the Sustainable Society Index, SSI (2008), http://www.sustainablesocietyindex.com/ssi-description.htm
11. Clark, W.C., Dickson, N.M.: Sustainability science: The emerging research program. Proceedings of the National Academy of Sciences 100, 8059–8061 (2003)
12. Gallopín, G., Vessuri, H.: Science for Sustainable Development: Articulating Knowledges. In: Guimarães Pereira, A., et al. (eds.) Interfaces between Science and Society, pp. 35–51. Greenleaf Publishing, Sheffield (2006)
13. Fontela, E.: Foresighting the new technology wave. EC-HLEG Key Technologies, Brussels (2005), ftp://ftp.cordis.europa.eu/pub/foresight/docs/kte_toward_sks.pdf
14. Fuchs, C.: Sustainability and the Information Society. In: Berleur, J., Nurminen, M., Impagliazzo, J. (eds.) Social Informatics: An Information Society for all? In Remembrance of Rob Kling, pp. 219–230. Springer, Heidelberg (2009)
15. Sharma, R.S., Wirawan, A., Foong, S.G.C., Win, Y.M.: Bridging the knowledge gap in developing societies: A field study of Indonesia, Vietnam and Myanmar. The International J. of Technology, Knowledge and Society (to appear, 2009)
16. Bender, M.J., Simonovic, S.P.: Consensus as the measure of sustainability. Hydrological Sciences 42, 493–500 (1997)
17. D'Orville, H.: Towards the global Knowledge and Information Society – The challenges for development cooperation, INFO 21, United Nations Development Program (2000), http://www.undp.org/info21/public/pb-challenge.html
18. ITU: Measuring the Information Society. The ICT Development Index, ITU, Geneva (2009)
19. Hüsing, T., Selhofer, H.: DIDIX: A digital divide index for measuring inequality in IT diffusion. IT & Society 1, 21–38 (2004)
20. ITU & UNCTAD: World Information Society Report 2007. Beyond WSIS. ITU, Geneva (2007)
21. Economist Intelligence Unit & IBM Institute for Business Value: E-readiness rankings 2008. Maintaining momentum. Economist Intelligence Unit, London, New York, Hong Kong (2008)
22. Huggins, R., Izushi, H., Davies, W., Shougui, L.: World Knowledge Competitiveness Index 2008. Centre for International Competitiveness, Cardiff School of Management, University of Wales Institute, Cardiff (2008)
23. World Bank Institute: Measuring Knowledge in the World's Economies, Knowledge Assessment Methodology and Knowledge Economy Index. The World Bank, Washington, D.C (2008)
24. Dutta, S., Mia, I.: The Global Information Technology Report 2008 - 2009. Mobility in a Networked World. INSEAD and World Economic Forum (2009)
25. Eriksson, D.: A principal exposition of Jean-Louis Le Moigne's systemic theory. Cybernetics and Human Knowing 4, 35–77 (1997)

Extending Static Knowledge Diagrams to Include Dynamic Knowledge*

Bogdan D. Czejdo[1] and Thompson Cummings[2]

[1] Department of Mathematics and Computer Science, Fayetteville State University, Fayetteville, NC 28301, USA
[2] School of Arts and Sciences, St. George's University, Grenada
bczejdo@uncfsu.edu, tcummings@sgu.edu

Abstract. UML class diagrams are typically used for representing a static knowledge about objects, their properties and relationships. In this paper we will demonstrate how inference capabilities can be added to static UML graphs resulting in knowledge diagrams. Knowledge diagrams can be built to gain a deeper understanding of a subject area, to prepare better presentations about the subject, to guide knowledge discovery, or to support inference. In this paper we introduce knowledge diagrams that allow for a controlled incompleteness and inconsistency.

Keywords: UML, Knowledge, Static and Dynamic Knowledge, Inference.

1 Introduction

The Universal Modeling Language (UML) includes class diagrams that mainly involve static modeling. The class diagrams describe class properties and classifications that are based on a subclass hierarchy, an aggregation hierarchy, an association relationship, or any combination of these [2, 3, 4, 8].

There have been many projects that showed the usefulness of UML for modeling of a domain specific knowledge [4, 5]. There are also UML models such as state diagrams to capture dynamic knowledge [3]. In this paper, we use a different approach and extend static diagrams to include rules. By including rules in class diagrams we can model knowledge much more effectively and include a dynamic component of the knowledge. UML models with rules can be built to gain a deeper understanding of a subject area, to prepare better business presentations about the subject (including Web presentations), or to serve as a basis for teaching support systems [6, 7, 8, 9, 10, 11] by allowing students to issue queries and requests for explanations of the data.

We assume that the application specific knowledge needs to be acquired from documents and converted into knowledge diagrams. Because of this need we focus not only on the diagrams with rules but also on the development of methodology and knowledge acquisition tools that extract application-specific knowledge from natural language documents. We are currently developing tools and methodologies to support

* This work was partly supported by Belk Foundation.

M.D. Lytras et al. (Eds.): WSKS 2009, CCIS 49, pp. 338–345, 2009.

the activity of employing diagrams for knowledge modeling. A guided knowledge discovery process is supported in which new information is requested and added to the diagram in a controlled manner. This paper centers on the integration of rule-based reasoning that is commonly found in knowledge-based and expert systems [9] within a data diagram framework. In particular, we introduce the concept of rule inheritance with the possibility of the overriding a rule by another rule that has been placed lower in the subclass hierarchy. We also introduce other mechanisms to combine evidence that originates from different rules. The resulting knowledge diagrams need to be verified and checked for inconsistency. Obviously, many conflicting rules need to be either removed or reconciled. The uniqueness of our approach is that we allow for some controlled inconsistency and interpret it as fuzzy information in the knowledge diagrams.

Once the knowledge diagram is obtained it can serve as a basis for support of different types of inferences. The proposed knowledge diagrams with rules can be extremely important for "what-if" analysis extending the power of "what-if" spreadsheet analysis to the power of "what-if" reasoning. This can be accomplished because changes in any section of the diagram will be able to “propagate” to the remaining sections by means of the rules used to guide knowledge discovery and knowledge evolution.

2 Knowledge Diagrams with Data and Rules

We extended UML diagrams to model a domain specific knowledge containing data and rules. We called these diagrams knowledge diagrams. The modeling using knowledge diagrams is object oriented, meaning that whatever system is being modeled, its components become abstract objects that have some properties (also called attributes). A class is a collection of these abstract objects. Additionally there exist relationships between objects that graphically are typically represented as links between classes.

Generally, there are several graphical constructs that can be useful for knowledge modeling. Classes with their name and properties are graphically represented as boxes as shown on the example class *Programming Language* with the properties *name*, *tedious*, *hard to debug* shown in Figure 1. The domain of each attribute is specified between brackets. For the name attribute, the “PCdata” domain identifies any string. For the *tedious* attribute of the class *Programming Language* there are two values, "Yes" and "No". Similarly, there are two values for the attribute *hard_to_debug*. This indicates that the programming language can either be *tedious* or not, *hard to debug* or not. If no domain is specified it means that it contains only one value, "Yes".

The instances of a class are typically hidden or represented in a separate diagram. We feel that combining a class diagram with the instance diagram gives the user a better frame of reference to enter data and to see these two concepts much closer connected. As a result in our system we assumed a combined representation the class *Programming Language* and its instances as shown in Figure 1.

Let us assume that there are two instances of the class *Programming Language*, namely *C++* and *XML*. Let us also assume that *C++* is *tedious* language. If we want to capture this limited knowledge in our diagram we can create several (two in our case) staggered windows each corresponding to one class instance and attach it to the bottom of the class diagram as shown in Figure 1. The question mark identifies the incomplete knowledge. The user can easily switch between instance windows to see properties of each instance.

Programming Language	
name[PCdata] tedious[Yes, No] hard_to_debug [Yes, No]	
C++	XML
name = C++ tedious = Yes hard_to_debug = ?	

Fig. 1. An example of knowledge diagram for the class *Programming Language* with no rules

Rules can augment the knowledge diagrams and help specify values for some properties. In our case, we assume the rule that if programming language is *tedious* then it is *hard to debug*. It can be represented by rule R1:

tedious[= Yes] → hard_to_debug[= Yes] {R1}

Since we use "Yes" as a default value therefore we can improve readability and re-write the rule as

tedious → hard_to_debug {R1 - alternative notation}

Rules in our knowledge diagrams are placed in Class Rules and Instance Rules boxes as shown in Figure 2. Since rule {R1} has "hard_to_debug" property as a right hand side slot, it will be associated with "hard_to_debug" attribute in the class *Programming Language* (on the same line) as shown in Figure 2. Generally, the rules are

Programming Language		Class Rules
name[PCdata] tedious[Yes, No] hard_to_debug [Yes, No]		 tedious → hard_to_debug {R1}
C++	XML	Instance Rules
name = **C++** tedious = **Yes** hard_to_debug = <u>Yes</u>		 tedious → hard_to_debug{R1}

Fig. 2. An example of a knowledge diagram for the Class Programming Language with rules

applicable to all instances, but there are some exceptions. Therefore, a separate box Instance Rules was created not only to list the rules that are applied but also to record the exceptions. Once the rules are specified, the incomplete information for the property "hard_to_debug" can be generated.

In general, in our knowledge diagrams, each instance property can:

- be unknown which is represented by a question mark,
- have a value provided explicitly (e.g. *tedious* for C++) and this is represented by bold characters,
- be derived based on rules (e.g. hard_to_debug for C++) and this is represented by underlined characters.
- have the same value provided explicitly and derived and this is represented by underlined bold characters.
- have different values provided explicitly and derived (this will be discussed later)

3 Propagation of Rules

Let us consider knowledge diagrams with several classes. In Figure 3 there is shown an example of a subclass relationship. Other diagrams can also use aggregation or named association. Subclass relationships are especially important because they allow inheritance of properties and rules.

Let us look at the class "Programmer" as shown in Figure 3. Its attributes are "skill_level" "project" and "salary". These attributes have domains which illustrate the various levels at which a programmer may work. According to the diagram, a programmer's salary may fall between $40,000 and $100,000 a year. A programmer's skill level may be low, average, or high. Additionally, a programmer may work on either a simple or complex projects.

Let us also consider the subclass C++ Programmer as shown in Figure 3. It inherits all attributes "skill_level" "project" and "salary". Let us now discuss rules for the class Programmer as shown in Figure 3. For example, if a programmer works on a complex project, then his or her salary will generally be more than $55,000. Also, if his or her skill level is high, then his or her salary will be above $50,000. These rules provide a brief method of notation for the variability in this profession. The subclass hierarchy may cause inheritance of rules. In our example, we see that the range of values for salary can be restricted for all C++ Programmers by applying rule R4.

Rules can also be defined based on relationships. For example let us consider a diagram that connects the class *Programming Languages* from Figure 2 with the class *Programmer* from Figure 3 using the *programs_in* relationship. For such diagram we can define a new rule based on the relationship e.g. rule R5:

programs_in.hard_to_debug -> salary[> 70,000]} {R5}

Assuming that John programs in C++, the presence of this rule will cause modification of John salary to the new range salary[70,000 – 100,000].

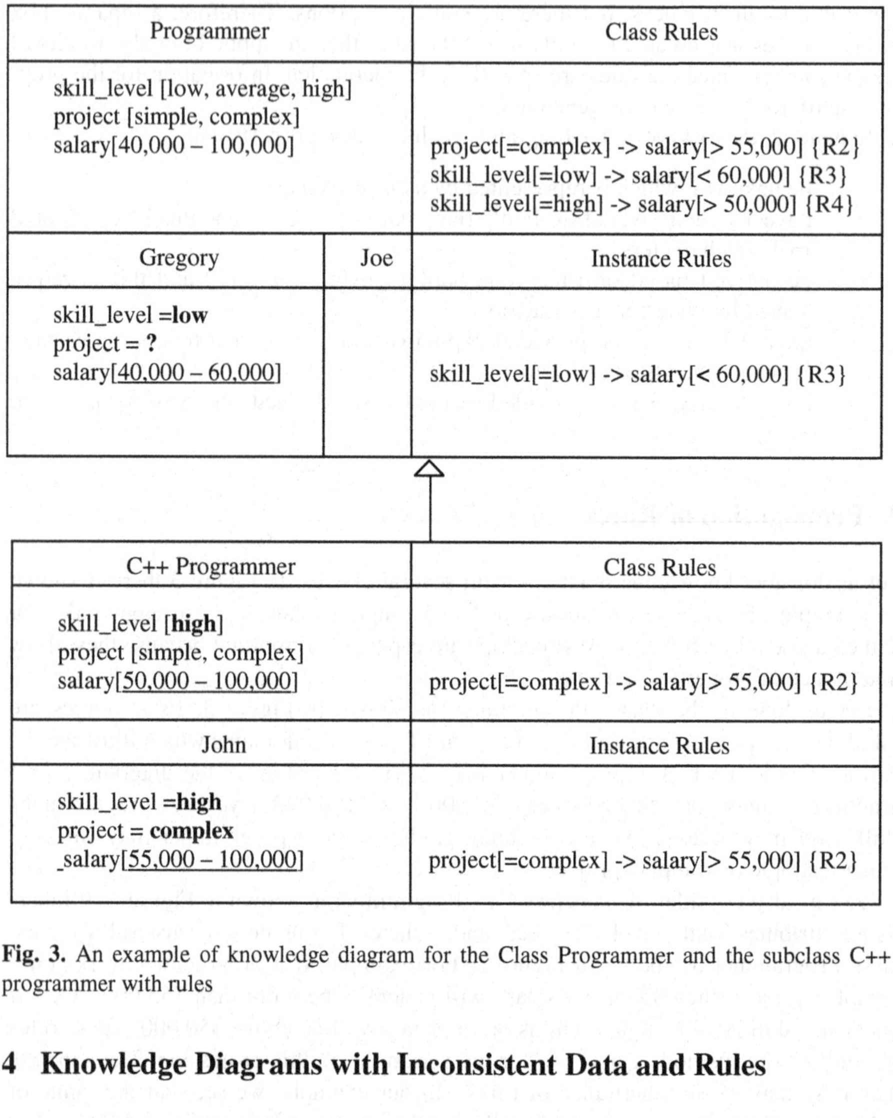

Programmer		Class Rules
skill_level [low, average, high] project [simple, complex] salary[40,000 – 100,000]		project[=complex] -> salary[> 55,000] {R2} skill_level[=low] -> salary[< 60,000] {R3} skill_level[=high] -> salary[> 50,000] {R4}
Gregory	**Joe**	**Instance Rules**
skill_level =**low** project = **?** salary[40,000 – 60,000]		skill_level[=low] -> salary[< 60,000] {R3}

C++ Programmer	Class Rules
skill_level [**high**] project [simple, complex] salary[50,000 – 100,000]	project[=complex] -> salary[> 55,000] {R2}
John	**Instance Rules**
skill_level =**high** project = **complex** salary[55,000 – 100,000]	project[=complex] -> salary[> 55,000] {R2}

Fig. 3. An example of knowledge diagram for the Class Programmer and the subclass C++ programmer with rules

4 Knowledge Diagrams with Inconsistent Data and Rules

As we acquire new knowledge it should be continuously integrated with the existing knowledge diagrams. In this process we should be able to expand our knowledge diagrams and include new facts and rules. Typically, we would be able to insert new explicit and generated values without causing any inconsistencies. Some of the operations on knowledge diagrams, however, can introduce inconsistency.

Some of these inconsistencies can be real contradictions that should be resolved and as a result the knowledge diagram can be significantly improved. Many inconsistencies, however, cannot be simply resolved. In our approach, it means that in our previous knowledge diagram we were too confident about some facts and now we are admitting some fuzziness. We need to maintain this information to assist us in continuous integration of knowledge diagrams with the new knowledge.

Let us analyze three situations that have to be distinguished with respect to the processing of multiple rules: all rules derive the same conclusion, rules derive different conclusions, but the conclusions are not contradictory, or rules derive different conclusions, some of which are contradictory.

The first situation does not pose any particular problem; we simply draw the conclusion that is shared by all rules.

When the second situation occurs, we can choose the most general hypothesis that is consistent with the rules' conclusions. For example, if one rule suggests that the value of an attribute A should be either 1 or 2 or 3 and a second rule suggests that the value of A should be 2 or 3 or 4 we would infer that the value of A should be 2 or 3. Such a solution is recommended but can be overridden by the user. He/she can choose to introduce fuzziness and keep these two alternatives stored.

In the third situation at least two of the rules' conclusions are inconsistent; e.g. one rule might suggest that the value of attribute A should be 1, whereas the other rule suggests that the value should be 2. In this case, we use the specificity of the involved rules with respect to the class hierarchy to determine which conclusion should be selected. In our approach, specificity is defined as follows:

Rules that are associated with more specific classes (with respect to the class hierarchy) take precedence over less specific rules and rules with specific left hand side conditions take precedence over less specific ones.

In the case that class specificity is not sufficient to determine which rule to choose, we do not draw any conclusions with respect to the involved attributes.

Let us consider another example that would lead us to inconsistency between data and rules as shown in Figure 4.

<table>
<tr><td colspan="2">Programming Language</td><td>Class Rules</td></tr>
<tr><td colspan="2">name[PCdata]
tedious[Yes, No]
hard_to_debug [Yes, No]</td><td>tedious → hard_to_debug {R1}</td></tr>
<tr><td>XML</td><td>C++</td><td>Applied Rules</td></tr>
<tr><td>name = XML
tedious = Yes
hard_to_debug = No
user-defined
flexibility</td><td></td><td>tedious → hard_to_debug[=Yes]

user-defined →flexibility {R6}</td></tr>
</table>

Fig. 4. An example of knowledge diagram for the class Programming Language with the instance XML

The rule processing results in the following contradictory conclusions:

1. Due to the fact that XML is a programming language and is tedious, we infer that XML is hard to debug based on the rule in the programming language class {R1}.
2. On the other hand, in the specification in the XML class we recorded that XML is not hard to debug.

The user can choose one of these alternatives e.g. cancel the rule R1 for XML. The user can also introduce fuzziness and keep these two alternatives stored as shown in Figure 4.

The two conclusions are contradictory. Let us first discuss an option of automatic resolution of this conflict. Due to the fact that the rule that infers XML is not *hard to debug* is associated with the class *Programming Language*, the conclusion of the instance overrides the specification of the more general class, and we therefore infer that XML is not hard to debug.

Let us also look at an option of a dialog based resolution. In general, the user has three options. As the first option the user can choose the data value and reject the generated value. In this case the rule can be rejected for the specific instance or for the whole class or a subclass. The second option for the user is to choose the generated value and reject the data value. The third option for the user is to ask the system to keep the knowledge about inconsistency. Practically it means that the user admits that values are fuzzy. Therefore, by creating an inconsistency aware system we introduce and allow fuzziness of data. Such inconsistency aware system, in turn, can provide an important assistance for the user i.e. the user can be informed about other inconsistencies and can make a more knowledgeable decision.

5 Conclusions

In this paper, we showed how to use knowledge diagrams to model knowledge that can include the rules. Based on this knowledge diagrams the users can issue queries or requests for explanations. Our approach allows for the incremental addition and evolution of knowledge. This is very useful for "what-if" analysis extending the power of "what-if" spreadsheets analysis to the power of "what-if" reasoning. This can be accomplished because changes in any place of the diagram are able to "propagate" to other places in the diagram by means of the rules.

References

[1] Antill, L.: Towards active case based learning in IS. Int. J. Teaching and Case Studies 1(1/2), 146–158 (2007)
[2] Bollen, P.: How to overcome pitfalls of (E)ER and UML in Knowledge Management education. Int. J. Teaching and Case Studies 1(3), 200–223 (2008)
[3] Booch, G., Rumbaugh, J., Jacobson, I.: The Unified Modeling Language User Guide. Addison Wesley, Reading (1999)
[4] Czejdo, B., Mappus, R., Messa, K.: The Impact of UML Diagrams on Knowledge Modeling, Discovery and Presentations. Journal of Information Technology Impact 3(1), 25–38 (2003)

[5] Czejdo, B., Morzy, T.: Knowledge, Knowledge Security, and Meta-Knowledge. In: Proceedings of 1st World Summit on the Knowledge Society (WSKS 2008). Communications in Computer and Information Science (CCIS) Series. Springer, Heidelberg (2008)
[6] Lytras, M., Naeve, A., Pouloudi, A.: Knowledge Management as a Reference Theory for E-learning: A conceptual and technological perspective. International Journal of Distance Education Technologies 3(2), 66–73 (2005b)
[7] Lytras, M., Pouloudi, N.: Towards the development of a novel taxonomy of knowledge management systems from a learning perspective. Journal of Knowledge Management 10(6), 64–80 (2006)
[8] Myers, M.: The field of IS has always been about Relationships not things in themselves. Int. J. Teaching and Case Studies 1(1/2), 15–22 (2007); National Center for Biotechnology Information, http://www.ncbi.nlm.nih.gov
[9] Sicilia, M., Lytras, M., Rodríguez, E., García-Barriocanal, E.: Integrating descriptions of knowledge management learning activities into large ontological structures: A case study. Data & Knowledge Engineering 57(2), 111–121 (2006)
[10] Vargas-Vera, M., Motta, E.: AQUA - Ontology-based Question Answering System. In: Monroy, R., Arroyo-Figueroa, G., Sucar, L.E., Sossa, H. (eds.) MICAI 2004. LNCS (LNAI), vol. 2972, pp. 468–477. Springer, Heidelberg (2004)
[11] Weng, S., Tsai, H., Liu, S., Hsu, C.: Ontology construction for information classification. Expert Systems with Applications 31(1), 1–2 (2006)

Against the Reign of Ignorance

Carel S. De Beer

Department of Information Science, University of Pretoria, Pretoria, South Africa

Abstract. If we are serious about "a better life for everyone" in a knowledge society, the most obvious and most probably the ultimate condition to achieve this would be to ensure genuine quality of knowledge against the reign of ignorance in whatever format it may emerge and to involve everybody in a situation of collective intelligence for the achievement of this dream. Not the mere acceptance of knowledge, but the certainty that we are talking of "the right kind of knowledge" is at issue.

Serious reflection is required about true knowledge in its dynamic complexity and the obstacles in the way of achieving it of which ignorance is its biggest enemy. It is indicated to what extent an alternative territory of thought may be the answer since this will enable humans to develop an insight into the threats of ignorance and the sources thereof, to contemplate a genuine knowledge of knowledge, to engage in collective intelligence activities that will help to overcome the threats and will contribute to the building of an authentic knowledge society for the benefit of all.

Keywords: Reign of ignorance, knowledge of knowledge, territories of thought, collective intelligence, informatisation.

1 Introduction

In view of the dynamics and complexity of the notion of knowledge and its immense history of thousands of years any contemplation and discussion of a knowledge society as if it is a new invention should include reflection on these aspects as well and not only any facet of it that may be of secondary importance. Given this assumption this paper has been prepared from a philosophical rather than a technological point of view. In this respect ignorance features prominently and should be put high on the agenda.

Why is it important to take such a view point against ignorance to thematise it for a summit on the knowledge society? Is it not too negative? Is ignorance really so prominent and such a threat in a so- called knowledge society? Is knowledge not the only thing that really matters? Why bother with ignorance? The assumption is that ignorance, especially in a camouflaged sense, very often under the pretension of knowledgeability, is very prominent, even where nobody would expect its presence, and that great damage can be done not only to the true quality of knowledge, but also to the integrity of the true knowledge society. As a matter of fact, in camouflaged form its popularity is so predominant that it takes over the scene quite easily, hence the reign of ignorance.

M.D. Lytras et al. (Eds.): WSKS 2009, CCIS 49, pp. 346–356, 2009.

Ignorance in this sense poses a vital problem, perhaps more so because the urge to fight against it and oppose it by encouraging and promoting thinking, learning and cultivation, is very often and generally speaking, objected to by the remark: "Why, what is the use of learning, reading, knowledge, what is the use of it all? We cope well without this kind of effort." (Cf Steiner 1999). This is a demonstration of its reign albeit quite a cynical move. On the other hand, it should be realised what the implications may be of falling into the trap of ignorance as a reigning force: barbarism, de-spiritualisation, the total loss of meaning, reductionism, general shallowness and emptiness of concepts, and forgetfulness of evil, are some of the implications brought to the fore by a number of intellectuals. The societal implications are really vast and comprehensive and deserve thorough attention.

If we are indeed and really serious about "a better life for everyone", or, about "quality of life for everyone", the most obvious and most probably the ultimate condition to achieve this would be to ensure genuine quality of knowledge against the reign of ignorance in whatever format it may emerge... This is our challenge. Not the mere acceptance of knowledge, but the certainty that we are talking of "the right kind of knowledge". Serious reflection is required about knowledge and its qualities. An "epistemo-critique" may be required; it may even be necessary to go beyond the notion of critique towards an a-critical disposition. (Cf Serres 1995:125-166).

2 Imagine a World without Knowledge

Let us imagine for a moment that ignorance, as the absence of knowledge, or as the presence of distorted, illusionary knowledge, takes over. Let us try for a moment to imagine what the world would be like without knowledge. No science, no culture, no civilization, no progress. No increase in life span, no decrease in child mortality, no technical developments. Not to mention anything about the utter dependence of wealth and affluence on knowledge.

But we have to be careful: Let us not try for one moment to think that everything mentioned regarding the knowledge society and the threat of ignorance can be ascribed to computers en technical developments. Do we have a knowledge society because of the internet, World Wide Web, Web 2 or 3, multimedia, computers, nanotechnology, e-mail, etc., like many people would think or would like us to believe? What a misleading assumption would that be! What these developments bring to the knowledge society is that it may certainly facilitate knowledge in many respects, its use, its development, dissemination, application and invention, etc. It may certainly be of immense and substantial support. On the other hand it may also be introducing a kind of dubious knowledge, knowledge as commodity, as something marketable, something for sale, something like a manageable tool, but not something dynamic and complex, something that can save humans, human lives and human civilizations. All of them built indeed on some kind of knowledge but then of a kind different from knowledge in that age old sense of the word.

But then, what knowledge are we talking about?

3 A Determination of Conceptions of Knowledge

Edgar Morin (1990:18-21) warns in this regard against "the pathology of knowledge" which consists in disjunction, reduction and abstraction that together constitutes "a paradigm of simplification". Disjunction indicates the radical isolation of the three major scientific fields namely the physical, life and human sciences. The effort to remedy this disjunction is another simplification: the reduction of the complex to the simple. Such knowledge finds its rigour and its operationality in the measurable and the calculable. In this way we arrive at what he calls "a blind intelligence" that destroys togetherness and totalities and when that happens what emerges is a "new, massive and stupendous ignorance". This leads to infinite tragedies before we end up in a supreme tragedy. (Cf. in this regard also Morin 1986).

We have to assess our conceptions of knowledge carefully that they do not bring us close to ignorance and on the edge of the abyss – the downfall of humankind. Beware of the ignorance neatly packaged under the label of knowledge; that is the most dangerous, really fatal kind of ignorance, because it is so misleading and filled with snares. Many strategic and ideological tactics are very often put in place, consciously or unconsciously, to empty our conceptual instruments of meaning or to make it mediocre.

Compare, for example, the notion of knowledge and a knowledge society developed by Stiegler et al (2006:163-173) and their views on Unesco's notion of knowledge and a knowledge society as it has been articulated in the Tunis Conference documents of 2005. A careful analysis will clearly show to what extent knowledge can feature prominently in different approaches immaterial of deep discrepancies. Certain fundamental issues should be addressed at an occasion like this and that did not happen according to the authors. What are these issues? "The principal question not posed by the Tunis Conference of Unesco is the question of the relationship between information, knowledge, technology, industry, and society as the undertaking of an international politics of transformation of the actual capitalism and which will enable it to go out of the impasse in which it became essentially a capitalism of consumption, which produce [not knowledge but] no more than deafening, de-sublimation and, lets say the word, mental, mortal, intellectual, spiritual and aesthetic regressions in all domains of what Valery and Arndt called "the life of the spirit".

The failure to ask this question leads unavoidably to a situation of de-spiritualisation. Facing this spiritual misery humans feel that they have an irreducible need of spirit." But spirit here is not to be understood in the exclusively religious sense of a hope of salvation and a life hereafter. The life of the spirit cannot be reduced to the religious world. Here it can be added that the life of the spirit should also not be reduced to the world of machines. "Contrary to this", Stiegler et al. (2006) proposes "that the life of the spirit is first of all to be found in everyday work, in the relation between individuals, in the instruments of communication, and singularly through those that affect the great public spaces and times that are television audiences and users of mobile phones, which become multimedia, the users of internet and the web, and those of all these cultural and cognitive technologies of information and communication that deploy themselves today with generalised numerisation and which constitute the principal motor of actual economy" (Stiegler, et al., 2006:163-4).

In view of the above they want to mobilise collective intelligence on a world wide scale as the major effort that contributes to the re-invention of spirituality, the posing of the principal question, and the generator of sound knowledges. Pierre Lévy (1997) before them was already extensively engaged in a similar project of creating intelligent communities on the basis of collective intelligence as an effort to reach a condition of true and authentic knowledge.

We must also take heed, however, of the warning by Edgar Morin (1986:10) that we must be careful that we do not ignore and destroy the treasures of knowledge in our energetic struggle against ignorance. Keep in mind what close connection there exists between knowledge, truths, and thought and spirituality. For this reason it will pay to be very thoughtful about the notion of knowledge that we subscribe to.

4 Considering the Notion of Ignorance

Why is it important to get to know and understand ignorance? Pynchon (1984:15-16) states it as follows: "Everybody gets told to write about what they know. The trouble with many of us is that at earlier stages of life we think we know everything – or to put it more usefully, we are often unaware of the scope and structure of our ignorance. Ignorance is not just a blank space on a person's mental map. It has contours and coherence, and for all I know rules of operation as well. So as a corollary to writing about what we know, may be we should add getting familiar with our ignorance, and the possibilities therein for ruining a good story."

What is ignorance about? To ignore means to pretend that there is nothing. In this regard ignorance is about the lack or, absence even, of knowledge, especially necessary knowledge for certain situations. But also of knowledge in general that is very necessary to cope with life and its challenges – suffering, death, lack, hatred, violence and killing. Related terms, each with its own history and dynamics, that amplify the notion of ignorance are illusion and stupidity. Both earned an enormously negative reputation in certain intellectual and cultural circles.

The situation of ignorance can happen by accident or because of circumstances, or by deprivation of some kind, or by deliberate decision not to know because of the effort that is required (due to a kind of laziness), or due to part of the human make-up that does not want to know, or that prefers not to know like Lacan's "will to ignorance" or "passion for ignorance" (Lacan 1975:110) and Heidegger's "urge to blindness" (Heidegger 1971:19), or it can also be ascribed to human fallibility – the human being's essential inability, even of the brightest amongst us, to know to the full, or by ideological and other forms of distortion, deception and illusion.

It seems as if a dominant element of ignorance happened to be woven into the fibres and structures of societal systems: cultural, scientific, technical, economic and professional. Only careful analyses demonstrate to what extent ignorance prevails and emerges everywhere with fatal consequences, despite all the pretensions in these systems regarding knowledge and commitments to knowledge. These consequences have been identified and analysed by a number of thinkers. Steiner (1999) relates the consequences to barbarism, cruelty and abuse; Stiegler (2001) is explicit about the fact that it creates nothingness, malaise and nihilism on a grand scale; Ronel (2003) does not hesitate to connect it to the term stupidity. All three identify these fatal consequences for individuals and for societies, even cultures and civilizations.

Ignorance is never, or not very often, at least, the total absence of knowledge. It is much rather a distortion of knowledge, or an illusion regarding knowledge, or a lack of complete knowledge. In no case is it absent. As Ronel (2003:29) puts it: "Ignorance is not just a blank space". It has got structure, direction and coherence. There is something like "the will to ignorance" as emphasised by Lacan (1975) and

Felman (1987). In this sense it is not a hopeless case at all. Ronel(2003:29) emphasises that "ignorance holds out some hope, you can get to know it, maybe move on". Irrespective of its dangers it can be turned around into constructive achievements as worked out by Felman (1987:77-81). In the case of "the blind urge" Heidegger (1971) warns against it; "we must guard against it". The reflection on the question "What is called thinking?" must enable us to renounce access to this urge for, this natural inclination towards blindness. This blindness, The blind intelligence of Morin (1990), the equivalence of ignorance, would permit us "to snatch a quick answer in the form of a formula", a quick answer that would be easily graspable. This strategy of "a quick answer", and "the form of a formula" may sound impressive, may lead to some actions, but it takes us away from the real question namely "what is called thinking". Ignorance is the enemy of thinking. And real thinking assists us to resist and to give up "quick answers" and ready-made "formulas". Although they look effective they accommodate all possible dangers of ignorance. (CF Ronel 989:26-45) for extensive commentary on this, although with another context in mind).

The above arguments call for decisiveness and careful analyses of the structural accommodation of ignorance in all sectors of society. This must clearly and honestly be identified; the dangers must be spelled out, and be dealt with in a thorough and consistent way. What are the dangers or consequences?

Some of the serious dangers or consequences should be looked at and kept in mind.

On the other hand, it should be realised what the implications may be of falling into the trap of ignorance as a reigning force: Barbarism is indicated by Steiner (1999) as an inevitable consequence; de-spiritualisation and the total loss of meaning is emphasised by Stiegler (2001); Morin (1986) shows to what extent reductionism as an example of ignorance leads to barbaric manifestations; the general shallowness that leads to emptiness of concepts or to the taking of a certain limited meaning in language for granted, has been highlighted by Blair (2006) and Steiner (2001); "the forgetfulness of evil" as a substantial form of ignorance against which Dupuy (2002) warns us. The dangers of ignorance lie in the fact that ignorance is very often camouflaged by distorted views of knowledge. Ignorance is more a matter of lack of complete knowledge and the unwillingness to achieve complete knowledge. The societal implications of these issues are really vast and comprehensive and deserve thorough attention. All these contribute to what Jean-Jacques Salomon (2006) calls "a civilisation at high risk" and to what Dominique Lecourt (2009) describes as "the age of fear". Humanity has got a central responsibility to fight against this fatal reign of ignorance.

When ignorance is understood in this sense it must be realised that there are many sources to which its emergence can be ascribed. The following are some of the most important ones: Paradigmatic reductionisms (Morin); ideological distortions (Habermas); socio-cultural blockages (absolutisms, especially culturalisms); anthropological limitations (Lazslo, Felman/Lacan; Heidegger/Ronel); linguistic depletion (Steiner, Granger, Blair); libidinal economic destructions (Stiegler, Passet). All these factors contribute substantially to the destruction of our ability to know to the full and hence to the emergence and the thriving of ignorance with all the negative consequences pertaining to it.

Each one of these identified issues play a major role in the skewed construction of what is called knowledge and deserves special attention because of the implications of the distortions and illusions created by each of them in its own way. Reductionisms, distortions, blockages, limitations, depletions and destructions are all making contributions to the impact they have on our images of what knowledge is supposed to mean and to achieve and what the knowledge society is about.

5 Firm Weapons against Ignorance for the Sake of Knowledge and Knowledgeability

The dangerous thing, but natural in the context of the Enlightenment dynamics, is to expect adaptation as an answer and to be persistent about the fact that "there is no alternative" (TINA) (The Chomskian logic as Stiegler calls it), since there is no other solution than to adapt. The problem is that this attitude provokes the opposite of what is intended, namely it constitutes the fatal creator of ignorance. The model of adaptation to the status quo leads to entropy, to the rejection of all vision to the future, to a certain blindness, to the annihilation of all differences (to nihilism) and that all this implies that there is no chance for an alternative. The human brain, in this process, becomes a merchandise, standardise as perfectly as possible, that is a total absence of thought, a systematic organisation of the downfall, the loss, of the value of the spiritual. Adaptation, which is also conformation, leads to this absence of consciousness that constitutes the renouncing of all capacity to decide that it may also be possible to think otherwise, especially to think otherwise in the sense of Ronel who suggests a move into another "territory of thought".

This brings us back to the concern of Steiner: Why is science, music, art unable to protect humans against the radically inhuman? Perhaps it is because all happens within the same territory of thought – the "thought drug" which Ronel refers to. Only when we move beyond these limitations into another territory of thought it becomes possible to think otherwise, to think multidimensional rather than one- dimensional, to re-invent spirituality, to move against ignorance, to contemplate alternatives. The rest of this essay will try to briefly explore some of the aspects related to the point of departure sketched above, or, how to proceed in the context of an alternative territory of thought.

5.1 The Human Being's Unique Ability to Know

It is important to remember in this context that a special feature of being human is the ability to know, to think, to understand, to pursue and to invent. These qualities make humans the unique beings they are. As Stiegler (2006:164) puts it: "Knowledge is something that, by way of structure, has always been defined as being owned by, or belonging to humans". Knowledge is a human issue. Any effort to inhibit or obstruct these abilities, these special qualities, must be resisted and forcefully opposed. Any direct or even indirect celebration of ignorance falls in this category of issues that deserves to be vehemently opposed and resisted. It leads inevitably to the degradation of humans and opens the road to barbarism, brutality and abuse.

Whatever means that can be employed for humans to achieve their ultimate in this regard should be accepted without for one moment accepting that by means of these assisting materials humans are set free from their responsibility to know to the ultimate: it is our moral duty to know to the maximum (Gadamer). Knowledge is, in other words, a matter of ethics, a moral duty, especially with regard to the intimate link between ethics and thought and ethics and understanding (Morin 2004). He emphasises that the barbaric evil of human against human is driven by hatred, lies and lack of understanding, in other words, by ignorance. The fight against the reign of ignorance thus becomes an ethical issue.

This is, however, clearly not at all a matter of quantity of knowledge that is at stake in the *first* place, but much rather a matter of insight, understanding and wisdom – quality of knowledge in the ultimate sense.

This unique quality should not be taken away from people, but should much rather be cultivated and enhanced so that the full intellectual potential of humans can flourish. For this reason, amongst others, knowledge should be seen as a collective effort of the human race. On this same basis collective intelligence should be pursued on a grand scale.

5.2 The Need for an Alternative Territory of Thought

We must immediately be clear about one thing: The answer against the reign of ignorance does not lie in the mere accumulation of knowledge, the better distribution of knowledge, the more prolific management of knowledge, the connections with knowledge networks, better thinking strategies and more rational approaches, more education and learning. History has shown that knowledge, art, music, thought, insight are not necessarily able to protect us, the human race, against the barbarisms, cruelties, abuses and many more that humans are capable of. George Steiner (1999) was clear about this and did not have a clear answer. Ronel (2003) is faced by the same problem. Her suggestion is to emphasise that "the territory of thought" that we frequent for our answers, namely the Enlightenment tradition with its "thought drug" is perhaps not the adequate solution. "At this point in our shared experience of history it may be time to contemplate getting off the thought drug, powerful and tempting as it is, that allows equivalences to be made between education and decency, humanism and justice.... [T]here always exists the danger that Enlightenment remains related to hallucination, to favourable forms of comforting deception."(Ronel 2003:24).

For this reason we have to move out of this territory of thought and seriously contemplate an alternative. This is not easy. We have to compel ourselves "to confront every mask of good conscience to which commitments have been urgently made" (Ibid). She continues: "But the violence to which the world succumbed is of understanding: understanding is itself at issue. There is a frontier beyond which the Enlightenment cannot go in order to lend its support or illuminate the poverty of being. This in part is why I take the route through the other German tradition, that which opens the dossier on unintelligibility and non-understanding. Cutting through another territory of thought, the tradition initiated by Schlegel's reflections on that which cannot be adapted to human understanding uncompromisingly searches out a language, marked by the crisis of permanent parabasis, that would be capable of answering to the punishing blows of an indecent, unassimilable historical injury." (Ibid). Clearly more understanding, better management, new research along the same lines will not provide answers.

We need an alternative approach, another territory of thought, which will take us beyond the inadequate possibilities offered by the traditional mode of understanding and thinking. We have to move away from the dogmatic image of thought, understanding and truth to a new image of thought which will introduce us to new ways of understanding and different notions of truths. The weapons against the reign of ignorance, that will briefly be brought forward, should be equipped with this different quality of thought, should emphasise the unique capacity of humans to go beyond the "standard" approaches to thought, truth, knowledge, and understanding .

All of a sudden the views of Steiner make sense when he concludes his *Barbarism of ignorance* with four crucial points: The human spirit is totally indestructible, poetry can save humans even in the face of the impossible, language and reality have a close relationship. Perhaps because of these three points it is in the fourth place necessary to be joyful.

5.3 The Knowledge of Knowledge

As counter move any effort to stimulate, cultivate, motivate and enhance the full quality of knowledge against ignorance in the knowledge society must be embraced, supported and pursued. What is needed is "the knowledge of knowledge" to use the terminology of Edgar Morin (1986). It is strange that in the knowledge society the question of the knowledge of knowledge is not much of an issue. A certain conception of knowledge is simply taken for granted. The idea of knowledge as a complex and dynamic phenomenon is never really addressed. The search for a proper understanding of knowledge is hardly pursued while one would expect this should be a central issue in a knowledge society.

The struggle against the reign of ignorance can hardly be undertaken without this understanding. The matter of knowledge is never straightforward. There are many definitions of knowledge and many views on what represents real and authentic knowledge. For some it is best represented by so-called "indigenous knowledge"; for others scientific knowledge in terms of a specific and well defined definition of science represents the only true knowledge; others claim religious convictions and beliefs as true knowledge, and so we can go on. In the context of knowledge management knowledge is most of the time in the last analysis understood as a manageable entity or substance that can be sold, consumed and used. Any dynamic and complex aspect poses a threat and should be avoided. There are convincing efforts by many to move beyond this perspective, but most of the time without abandoning the substantialist perspective, mainly because not doing this creates tension and uncertainty.

5.4 The Real Means with Which to Counter Ignorance

All means that can assist in this regard should be involved, like the capacity to embrace complex, multiple thinking, the courage to work towards the reinvention of human spirituality, the rediscovery of the powerful function of language together with the acknowledgement of the necessity of linguistic and poetic abilities in a knowledge society, the reconfiguring of the "the will to ignorance" as well as "the urge to blindness" in view of possible positive and creative outcomes to the benefit of the knowledge society and all its peoples. (See once again the views of Lacan (1975, 1982) and Felman(1987), Heidegger (1971) and Ronel (1989) referred to above in this regard). A critical issue here would be how to reconcile Stiegler's keenness to resist ignorance with the Freudian and Lacanian acceptance of ignorance as a significant dimension of being human? Then there is "the uncertain quest" emphasised by Salomon (1999) who claims that scientific knowledge, in terms of a certain understanding of science, can never be the only form of knowledge or absolute knowledge. The sciences do not create gods. Other ways of understanding scientific dynamics highlight the problematic issues pertaining to scientific knowledge and the claims related to this kind of knowledge. Numerous examples by scientists of world renown exist but there is no time or space to pursue this here.

5.5 Collective Intelligence

In order to come to terms with the idea of a knowledge society, different from the Unesco strategy where knowledge is considered along the following lines, we have to take the protest against the force of ignorance to a different level. We have to "mobilise collective intelligence on a world scale" as articulated respectively by Stiegler et al. (2006:166, 163-173) and Lévy (1997). We need to re-invent foundations for this initiative that differ from those of Unesco (the traditional currently accepted bases that go through under terms like knowledge societies, knowledge economy, cognitive capitalism, society of intelligence and revolution of intelligence (Stiegler et al.:163). The [quantitative] increase of knowledge and intelligence is all that counts). In this regard intellectual property becomes a main element in economic development and this is according to Stiegler et al. is highly problematic in view of the fact that all this involves alienation, expropriation, and hegemony, and, moreover, that the question of property is also the question of the proper (the own), of appropriation, and of the conditions of this appropriation. Knowledge is something that, by way of structure, has always been defined as being owned by, or belonging to humans, as a human as well as individual genre, belonging to nobody and constitutes the common good of humanity that every human being must be able to appropriate. (Op. cit.:164).

5.6 A Multidimensional Notion of Informatisation

Informatisation considered here not as a mechanical concept but as a concept working with the multifaceted notion of information in the sense of creating form and bringing light but then form and light in the most comprehensive sense possible. All the dynamic and complex dimensions of knowledge, all possible territories of thought and the multidimensional development of truths should be brought into the centre of the picture of the knowledge society. Information as knowledge for action should be introduced in order thereby to regenerate societies and communities into intelligent communities on the basis of collective intelligence. Noology or "the science of the knowing mind" (Morin) must be revived and be made active to such an extent that the noosphere, or the sphere of ideas, will flourish.

6 Conclusion: A Society of Ignorance or a Society of Knowledge?

It is more a matter of opposing the reign of ignorance, in other words, of fighting a society of ignorance, rather than taking it for granted that we are living in a society of knowledge. The main question is then how to oppose, how to protest, against the reign of ignorance and its fatal consequences in such a way that we can bring back into the central position the reign of knowledge? Knowledge then understood in the sense that it is emphasised all the time by Bernard Stiegler namely "know what to do", "know how to live", and "know how to theorise". How to introduce it where the taste for it is lost? It calls for a comprehensive, all-incusive approach. What is manifested is a kind of malady of the socius and what is then required is nothing less than a kind of socio-therapy. For this to be achieved a number of things are required: the re-invention of human spirituality (against the mechanisation, industrialisation and instrumentalisation of humans); the re-invention of togetherness, collectivity, the "us" for con-naissance to flourish (against all forms of de-individuation and de- socialisation) - a "no" to dissociated and de-spiritualised individuals, but a "yes" to the establishment of intelligent communities on the basis of collective intelligence; the re-invention of the depths of cultural wealth and expression (against all forms of consumerism; knowing science, art and livelihood); the re-invention of a mode of ir-reduction (the refusal of reductionism of whatever kind) and the emergence of the noosphere, the world of ideas; the re-invention of language in its powerful poetic fullness (against the depletion and impoverishment of language).

The protest against ignorance is a decisive "no" to the creation of degenerated consumers and an emphatic "yes" to humans as well-integrated corpse-psyche-logos-nous-totalities – fully re-spiritualised beings, interested in the invention of knowledges in order to ensure the future and well-being of all of us.

References

1. Blair, D.: Wittgenstein, language and information: 'Back to the Rough Ground'. Springer, Heidelberg (2006)
2. Dupuy, J.-P.: Avion-nous oublie le mal? Bayard, Paris (2002)
3. Felman, S.: Jacques Lacan and the adventure of insight: psychoanalysis in contemporary culture. Harvard University Press, Cambridge (1987)
4. Heidegger, M.: Wass heisst Denken? Max Niemeyer Verlag, Tubingen (1971)
5. Lacan, J.: Le Séminaire, Livre XX: Encore. Seuil, Paris (1975)
6. Lacan, J.: The subversion of the subject and the dialectic of desire in the Freudian unconscious, in Ecrits: a selection. Tavistock Publications, London (1982)
7. Lévy, P.: Collective intelligence: mankind's emerging world in cyberspace. Plenum Trade, New York (1997)
8. Morin, E.: La Méthode 3: Connaissance de la connaissance. Seui, Paris (1986)
9. Morin, E.: Introduction à la pensée complexe. ESF Éditeur, Paris (1990)
10. Morin, E.: La Méthode 4: Les idées. Seuil, Paris (1991)
11. Morin, E.: La Méthode 6: Éthique. Seuil, Paris (2004)
12. Ronel, A.: The telephone book: technology, schizophrenia, electric speech. The University of Nebraska Press, Lincoln (1989)
13. Ronel, A.: Stupidity. The MIT Press, Cambridge (2003)

14. Salomon, J.-J.: Survivre à la science:une certaine idée du futur. Albin Michel, Paris (1999)
15. Salomon, J.-J.: Une civilisation à hauts risques. Éditions Charles Léopold Mayer, Paris (2006)
16. Serres, M.: Conversations on science, culture and time. The University of Minnesota Press, Ann Arbor (1995)
17. Steiner, G.: Barbarie de l'ignorance. Le Bord de L'Eau, Bordeaux (1999)
18. Steiner, G.: Grammars of creation. Faber and Faber, London (2001)
19. Stiegler, B.: La technique et le temps 3. Le temps du cinéma et la question du mal-être. Galilée, Paris (2001)
20. Stiegler, B.: Ars Industrialis. Réenchanter le monde: le valeur esprit contre le populisme industriel. Flammarion, Paris (2006)

Italian University Students and Digital Technologies: Some Results from a Field Research

Paolo Ferri[1,2], Nicola Cavalli[2], Elisabetta Costa[2], Andrea Mangiatordi[2], Stefano Mizzella[2], Andrea Pozzali[2], and Francesca Scenini[2]

[1] Department of Educational Sciences "Riccardo Massa", University of Milan-Bicocca, Piazza dell'Ateneo Nuovo, 1 - 20126, Milan (italy)
[2] Observatory on New Media "Numedia Bios", University of Milan-Bicocca, Piazza dell'Ateneo Nuovo, 1- 20126, Milan (italy)
{paolo.ferri,andrea.pozzali}@unimib.it, {nicola.cavalli,elisabettacosta1,andrea.mangiatordi, stefano.mizzella,francesca.scenini}@gmail.com

Abstract. Developments in information and communication technologies have raised the issue of how a kind of intergenerational digital divide can take place between "digital natives" and "digital immigrants". This can in turn have important consequences for the organization of educative systems. In this paper we present the result of a research performed during the course of 2008 to study how university students in Italy make use of digital technologies. The methodology was based on a mix of quantitative and qualitative approaches. A survey research was done, on a sample of 1186 students of the University of Milan-Bicocca, based on a questionnaire administrated through the Intranet of the University. A series of focus groups and in depth interviews with students, parents, and new media experts was furthermore performed. The results are consistent with the presence of a strong intergenerational divide. The implications of the results for the future organization of educative systems are discussed in the paper.

Keywords: digital natives, digital immigrants, social networks, education.

1 Introduction

The development of the so-called "Knowledge Society" is strictly linked to the increasing pervasiveness of information and communication technologies in all aspects of everyday life. Even fields that are traditionally quite resistant to changes need to adjust to the strong transformative power of these technologies. The case of education represents a paradigmatic example of how technological advancements can determine the possibility to introduce significant innovations, that have anyway to take into consideration the need to find a correct balance with traditional procedures and organizational practices; old models and processes of knowledge transmission through generations have to be somehow adapted to a new situation: *"Contrarily to what happened previously to older generations when radio and, particularly, television emerged, digital technologies, and the services associated with them, convey something completely*

M.D. Lytras et al. (Eds.): WSKS 2009, CCIS 49, pp. 357–365, 2009.

new: they modify not only the speed at which people deal with and manage information but also how they eventually transform it into knowledge" [1].

The issues at stake are manifold, and will probably grow in relevance in the next future. As far as the organization of educational systems is concerned, one of the point that deserves a specific attention is linked to the potential development of a sort of intergenerational digital divide, as the one described by Papert [2] in *The Connected Family*. Papert underlined the fact that, given their style of "enlarged communication" [3] and their strong technological alphabetization, the Digital Kids (i.e. children that grow up in societies where Internet connections, mobile phones and videogame consoles are readily available) were likely to develop communicative practices and attitudes radically different from those of their parents and teachers. Making reference to a well known classification developed by Prensky [4], this intergenerational divide is based on the contraposition between "digital natives" and "digital immigrants": while "digital natives" show a growing enthusiasm for computers and digital technology, this enthusiasm often scares teachers, parents and scholars.

Digital natives, says Prensky, communicate and learn making an extensive use of digital tools, such as computers, video games or online encyclopedias, and this "extended digital environment" often represents their natural learning environment [5, 6]. These tendencies are further reinforced by the diffusion of social and communication applications as the ones that characterize the so-called Web 2.0 [1]. On the other side, digital immigrants on average still make use of more traditional tools such as books and libraries and can therefore have some difficulties in fully understanding the potential of digital technologies in the training field. This not only can result in a diminished efficiency of training processes overall, but can also be negative on a specific motivational ground, as long as there is the risk of creating a discrepancy between the ways in which learning and communication take place in ordinary life and the ways in which learning and communication take place in formal training environments. This does not hold only for kids and pupils, but it concerns every level of education, as long as also University students have to be considered as digital natives [7].

Many different empirical studies have, especially in the last few years, brought evidence in support of Prensky's thesis. A recent research of the U.S. National School Boards Association (NSBA) [8] shows that the number of hours spent at the computer by pupils and college students has now equaled the amount of time spent at watching television; moreover, a significant proportion of this time is not used only for video gaming and purely recreational ends, but also for educational activities such as "studying" or "creating and sharing content". According to NSBA data, the 59% of children and college students interviewed uses the Internet to download or search for texts and educational content and to find information or news related to teaching, while the 50% uses the network as an extension of the group work done at school: to "do the homework", to connect to virtual classes, to realize collective on line works, to receive tutorship and assistance from teachers. Digital natives are more and more using the web also to socialize and as a way of self expression: more than 37% of them update their site every week, the 30% have a blog and the 17% post a new content in it at least once in a week.

It must be underlined that this increasing rate of usage of digital technologies can lead to the development of learning attitudes that can be quite different from the traditional, "analogic", ones. Some authors point to the existence of a sort of "anthropological" difference [9] that can specifically characterize the digital natives students of the multicultural and globalized informational society [10,11]. A new generation of adolescents and students is growing, who not only show a strong cultural hybridization, but are also "structural technological symbionts" [9], as long as they are getting used to consider digital technologies as a natural extension of their physical abilities.

It seems quite clear that this mutation can have deep social consequences. As was the case with other previous major technological *and* cognitive revolutions - think for example to the invention of printing [12] - the co-evolution between humans and technology can determine the emergence of new cognitive styles and learning attitudes [13]. Digital natives in fact already seem to show a series of peculiar behaviors [14,15]: in particular, they learn more through screens, icons, sounds, games, and by "surfing" into virtual environments than through words and texts; the transmission of information and knowledge, moreover, takes place more through the constant contact with the network of "peers" rather than in more hierarchical ways.

Multitasking is another significant feature that seems to characterize the way in which digital natives approach digital technologies [16]: it is quite frequent, for example, for young people to chat and listen to music while studying, and at the same time remaining in contact with the group of peers through Messenger or other instrument for social networking. Indeed, digital natives can actually make use of a large amount of social communication tools, that represent the well known features of the so called Web 2.0: from Habbo to My Space, from Facebook to Twitter, from MSN Messenger to FriendFeed, from Slideshare to Linkedin, from YouTube to Wikipedia, not mentioning the blogs.

Unfortunately, there seems to be a lack of systematic data on the possible mid and long term consequences of such practices on learning performances, even if some preliminary evidence is already available. The New Millenium Learner (NML) Project [1], carried out by OECD/CERI on a representative sample of students (age 6-15) of the OECD countries aims to analyze in detail how digital natives learn by using digital technologies. The research, based on a questionnaire that is administered in addition to the questionnaire used for the PISA research, shows how the use of ICTs can significantly and positively influence the educational attainment of young students.

Among other things, the NML research seems to provide evidence of the fact that access to ICT indeed has an impact on learning performances. On average, students who have access or own a computer in fact get 506 PISA points, while the ones who do not own a computer or are unable to access it from home get 478 points (it must be underlined that on the entire sample the average score is 482 and that scores that are above 500 points are considered as good scores). What is even more interesting is that the best scores are obtained by those students who make an intense use of technologies at home and who make a moderate use at school: *"Probably the most important conclusion of all is that the correlation between home use and academic attainment is greater than in the case of school use in most countries, even when allowances are made for the effects of different socio-economic contexts. In particular, students who do not have access to a computer at home tend to be lower achievers than the others and, secondly, it would also seem to be the case that students using computers at home less often had below-average results"* [1, pag.15].

This kind of evidence not only casts some doubts on the actual efficiency of the ways in which ICTs are currently employed at school, but it also raises the need of exploring more in depth the variables that can play a role in producing such results. The research we present in this paper starts exactly from this point; more specifically, it assumes that, when assessing the impact of ICTs on learning processes, a mere quantitative assessment of access opportunity or frequency of use is not enough; it is necessary to take a more in depth look at the way in which technologies are actually used. Digital natives and digital immigrants may indeed present radically different approaches to information and communication technologies. It is certainly possible to trace back these differences to the fact that natives use these technologies a lot more than immigrants, but this will represent only a superficial analysis of the phenomenon. What is more important is the fact that styles of communication and of technological appropriation may be radically different between natives and immigrants. The theoretical basis of our research hypothesis assumes that digital natives may develop specific communication practices, that can in turn have considerable effects on the overall processes of communication at a social level and particularly on the configuration of education systems.

2 The Research

2.1 Methodology

During the course of 2008 we performed a research on the "medial diet", and more specifically on styles of media consumption and usage of university students (18-22 years old). The decision of focusing on this particular population was driven by two basic reasons. While there are, as already mentioned, some evidences available concerning the relationships between ICTs and pupils and adolescents, there seem to be quite a lack of data focusing on university students. This holds in particular for the Italian situation. The second reason is linked to the fact that one of the variables on which we would like to focus our attention was the diffusion and the profiles of use of social networks tools and sites such as Facebook, MySpace, Twitter and so on that can be considered as one of the most interesting features of the current phase of transition from the "old" Internet to the so-called Web 2.0 [17]. As it is well known, and as many statistics available on the Net seem to confirm, on average the use of this type of tools is more diffused among people from 18 to 25 years old.

The methodology of our research was based on a mix of quantitative and qualitative approaches. A survey research was done, on a sample of students of the University of Milan-Bicocca, based on a questionnaire that was accessed through the Intranet of the University. We restricted our analysis to students frequenting First Level Degree Courses: this gave us a total population of 21054 students. To avoid selection sample biases, we choose to administrate the questionnaire when students accessed the Intranet in order to complete their test on informatics, a compulsory examination that all students need to pass if they want to go on in the course of their study (also for this reason, we choose first level students, as older students might already have completed the test). As the students' registration number was recorded, we were able to avoid the possibility of double answers. Moreover, we also controlled if the sample obtained was statistically significant as long as the distribution of students in different faculty was concerned, and we corrected the biases with a second, more focused, administration of the questionnaire. This was done in order to assure that our final sample of 1186 students was representative of the overall population of students of the University of Milan-Bicocca.

Some of the students were also involved in a series of focus groups and in-depth interviews, that were performed in order to collect more information on the motivations that make people connect to the Net, the diffusion of different instruments of social networking, the ways in which digital media are replacing traditional ones and so on. To address the theme of the intergenerational digital divide, we also realized a series of focus groups with some students' parents. A different series of focus groups was furthermore performed, involving a set of qualified experts and practitioners operating in the new media sector, that was instrumental in helping us to understand the point of view of people that are currently involved "from the inside" into the current wave of developments.

2.2 Results

The questionnaire was divided in three main parts: a preliminary section, focused on general data concerning the use of digital technologies and the access to the Internet (What kind of technology do students use? Where? When? For how much time, etc.). The second part was specifically focused on the "medial diet": we collected data concerning the usage of different media, both in the "analogic" (books, newspapers, television, radio) and in the "digital" version (e-books, on-line news, web-tv and web-radio, etc.). In the last part of the questionnaire, we specifically focused on a series of tools and platforms for social networking, trying to analyze their diffusion among the population of students and the reasons that lie at the basis of their use.

By comparing the quantitative results drawn from the questionnaire with the evidence arising from the focus groups and the interviews, we can underscore a few relevant points. First of all, as was largely expected, university students use the Internet a lot more than their parents. If we consider the subjects who never or rarely (less than hour a week) use the Internet we find that only 6,8% of the students fall in this category, compared with almost 40% of the fathers and almost 60% of the mothers. For students, it seems that the Internet is replacing other, more traditional, media: while the 68,7% of our sample connects to the Internet more than 5 hours a week (with more than a student out of four that connects to the Internet for more than 20 hours and another 24,6% who connects between 10 and 20 hours a week), the rate of usage of television and radio is far lower. Almost three students out of four listen to the radio less than five hours a week, with the 31,7% of students listening less than one hour and only the 2,6% listening more than 20 hours a week. The same seems to hold for what concerns television, even if in this case the rates of use are a little bit higher: the 53,8% of the students watch TV less than five hours a week, with more than a student out of ten watching less than one hour and only 4% watching more than 20 hours a week. For what concerns reading, finally, the 13,7% of our students never read a book (except the ones required for studying) and almost a student out of two reads less than 5 books in a year.

Taken together, these results seem to confirm that, even for university students, the computer and the Internet are quickly becoming the preferred media. Evidences coming from the focus groups and the interviews clearly confirmed this point, also adding some more qualifications, in particular for what concerns the digital divide between generations; for example, we think that this excerpt from an interview is highly representative of the type of relationship that some students are developing with their PC: "I think that my all life could be easily confined within a 4x4 square meter room, with a bed, a WC, a little kitchen and a computer... I wouldn't need anything else". On the contrary, many parents have confessed the great difficulties and discomfort encountered when they have to revert, for reasons mainly linked to working necessities, to the use of computers. Quite curiously, one of the points in which discrepancies between students and parents appear more evident is linked to the different use of e-mail and instant messaging: while parents still largely prefer the e-mail, students are more and more shifting towards IM. Indeed, only a student out of four uses the e-mail every day, while more than half of the sample (the 57%) uses IM every day.

Apart from verifying different frequency of use between generations, our research aimed also, as already said, to ascertain what kind of use do students actually do of digital technologies, and for what reasons. For this reason, we performed a cluster analysis on the data of our questionnaire, seeking to identify some specific sub-groups of student that can be characterized by having the same type of attitude and behaviors. This analysis allowed us to identify three main clusters[1]; the first cluster (26,3% of our sample) gathers together those subjects that we could classify as "active and creative users" of the Internet and of the new technologies. These subjects have high levels of technology and media consumption in all the categories considered. Furthermore, what characterizes them the most, in comparison with the other two clusters, is the high propensity to create content in an active way and to upload new and original contributions on the Net, in particular for what concerns the use of MySpace, the active participation to communities such as YouTube and Wikipedia[2] and the use of blogs.

A second cluster (19,6% of the sample) is made up of those subjects who present a low level of media and technology consumption overall. These subjects tend to use the Internet much less than the other two groups (the weekly hours of connection to the network varies on average between 1 and 5), are more inclined to define themselves as "basic users" or "beginners" and are not so involved in the use of IM. It must also be underscored that these subjects have, on average, a lower level of media consumption, in particular as long as reading books and newspapers is concerned.

---

[1] It was not possible to include all the case in the cluster analysis due to the high level of "don't know/don't answer" responses to some specific questions. For this reason, the percentages presented in the text do not sum up to 100%.

[2] From our data, it seems that the percentage of subjects connecting to other type of sites, such as Facebook, Twitter, Fickr and Slideshare and so on are quite lower than could have been expected, if looking to similar research performed in other countries or by comparison with data related to other ages (in particular 14-18 years old). It must be taken into consideration, however, that at the time our research started (February 2008) the overall popularity of some of these sites, in Italy, still had to reach its peak and was actually quite low: this holds in particular for Facebook. It is likely that performing the same research now can offer quite different results: actually, we are working at an extension of our initial research design, by involving also students coming from other Universities.

The majority of our sample (41,5%) falls anyway into the third cluster, that gathers those subjects who, while showing a high level of media and technology consumption overall (in some cases even significantly higher than those of the first group), have a very low propensity towards the active creation of contents. These subjects show a very intense use of the Internet, especially for what concerns instant messaging services, and show also a strong willingness to take part in online initiatives[3]: what helps to differentiate them with respect to the subjects of the first group is the fact that they show a very low level of creative involvement with the Net. It should be stressed in fact that, for this group, the propensity to upload new content on YouTube and Wikipedia is very low (even lower than that of the subjects included in the second group, characterized as already said by low levels of media consumption and of Internet usage). The most striking difference, anyway, is to be found in relation to the propensity to create new contents on MySpace: while this propensity is shown by nine subjects out of ten, among those included in the first group, the same holds only for the 4% of subjects included in this last cluster.

2.3 Comments

Taken together, the results of our research seem to highlight a few relevant points. First of all, and this was largely expected, there is indeed a strong digital divide between university students and parents for what concerns the frequency and the type of use of digital technologies and of the Internet. This adds to the already available evidence on digital kids and pupils and confirms that the transition from the Gutenberg generation to the New Millenium Learner might be considered in term of a one-shot discontinuity rather than a continuous and graded transition, as all the younger generations seem to be involved in the same way, even with some specific differences, in this process.

The second point to be underlined is linked to the division of our sample in three groups, that has come up from the cluster analysis, and can shed some more lights on the debate concerning the possible consequences of such a pronounced usages of the Net by young people. In particular, the questions at stake here seems to connect with the fact that an increased use of the computer can hamper the creativity and imagination of youngster. Data coming from our research seem to counter this type of arguments, as long as a significant proportion of our sample seem to be actively involved in a creative use of the Internet. This should not lead us to undervalue, anyway, that the majority of subjects still shows a low level of active involvement in the creation and sharing of new contents. The fact that in some cases these subject are also those that spend most of their time connected to the Internet confirm that the most important variable to look at is not the mere frequency of use, but rather the reasons that motivate people to connect.

3 Conclusions

The results of our research show how Italian university students (18-22) have to be structurally considered as digital natives, as long as they strongly prefer going digital

[3] With "online initiatives" we mean for example the involvement in online questionnaire or market analysis, the collection of signatures to support specific causes, the organization of public events and so on.

to communicate, search for information, listen to music, and also to study and cooperate together. This is clearly displayed, among other things, by the capillary diffusion of instant messaging and other similar tools and by the rising popularity of social networking.

What are the possible consequences of this type of results, for the future of formal education systems? And more in general, how this can be related to the process of development of the so-called "Knowledge Society"? First of all, it appears quite clear that the traditional face to face and "books based" style of teaching risks to become more and more inadequate, especially if it is not complemented also by other, more advanced, practices and methods. Formal education, primary schools, colleges and University will probably have to massively introduce digital tools in their curricula, if they want to escape the risk of incurring in a complete loss of interest in formal education by digital natives. This will be instrumental also in helping to close the gap between the learning styles of the natives and the training styles of immigrants teachers and professors. This kind of efforts will surely require massive investments in digital infrastructure (Internet connections, video projectors, interactive whiteboards, etc.) but the hardware alone is not enough.

If we want to grasp the real training potential of digital technologies a more in depth reflection on teaching methodologies is strongly needed. It is not enough to simply substitute blackboard and chalk with e-books and the Web, if we don't change our pedagogical models accordingly. It is quite evident that the problem is systemic and requires for a comprehensive approach. In the Northern Europe the digitalization of teaching practices is on-going and is progressing somehow faster than in other European countries, as some international reports seem to show [18]; some steps have been taken also in Italy, as long as the digitalization of colleges and University education is concerned, with the identification of a set of steps as the following ones [7]:

a. definition of a global plan to make the formal education go digital through the massive use of open source Learning and Content Management System, to handle the needs of on-line communication of the natives;
b. adoption of international standard (SCORM) for projecting and building Learning Objects (LO) on the different subjects;
c. allocation of financial resources in a plan for teachers and professors training in education technology;
d. setting up of open content repository of LOs and digital curricula, following good practices in this field as the ones defined by the Open University and by the MIT Open Archive Initiative.

Taken together, all this can be interpreted as a sign of a certain kind of dynamism that is finally manifesting itself even into a system, like the Italian one, that is still characterized by much diffidence and that is, as a consequence, lagging behind in comparison with other systems more prone to innovate in this field. Anyway, we strongly believe that the kind of problems we are talking about are global, as they impact on a global level, that is fully independent from national boundaries between States. For this reason, they cannot be left to the individual initiatives of each single Country – what is needed is a global, European solution, that can define a lists of common priorities and guidelines.

References

1. OECD: New Millennium Learners: a Project in Progress. Optimising Learning: Implications of Learning Sciences Research, http://www.oecd.org/dataoecd/39/51/40554230.pdf
2. Papert, S.: The Connected Family: Bridging the Digital Generation Gap. Longstreet Press, Atlanta (1996)
3. Boldizzoni, D., Nacamulli, R.C.D.: Oltre l'Aula: Strategie di Formazione nell'Economia della Conoscenza. Apogeo, Milano (2004)
4. Prensky, M.: Digital Natives, Digital Immigrants. On the Horizon, IX (2001)
5. Ferri, P.: La Scuola Digitale: Come le Nuove Tecnologie Cambiano la Formazione. Bruno Mondadori, Milano (2008)
6. Ferri, P., Mantovani, S. (eds.): Bambini e Computer. Alla Scoperta delle Nuove Tecnologie a Scuola e in Famiglia. Etas, Milano (2006)
7. Ferri, P.: E-learning. Didattica, Comunicazione e Tecnologie Digitali. Le Monnier, Firenze (2005)
8. National School Boards Association: Creating and Connecting. Research and Guidelines on Online Social — and Educational — Networking, http://www.nsba.org
9. Longo, G.O.: Il Simbionte. Prove di Umanità Futura. Meltemi, Roma (2003)
10. Castells, M.: The Information Age. Economy, Society and Culture. The Rise of Network Society, vol. I. Blackwell Publishers, Malden (1996)
11. Castells, M.: Flows, Networks, Identities. In: McLaren, P. (ed.) Critical Education in the New Information Age, pp. 37–64. Rowman & Littlefield, New York (1999)
12. Eisenstein, E.L.: The Printing Revolution in Early Modern Europe. Cambridge University Press, Cambridge (1983)
13. De Kerckhove, D.: Brainframes. Technology, Mind and Business. Bosch & Keaning, Utrecht (1991)
14. Veen, W., Vrakking, B.: Homo Zappiens. Growing up in a Digital Age. Network Continuum Education, London (2006)
15. Veen, W.: A New Force for Change. Homo Zappiens. The Learning Citizen 7, 5–7 (2003)
16. Foehr, U.G.: Media Multitasking among American Youth. Prevalence, Predictors and Pairings. Menlo Park, The Henry J. Kaiser Family Foundation (2006)
17. O'Reilly, T.: What Is Web 2.0, Safari Books Online, http://www.oreillynet.com/pub/a/oreilly/tim/news/2005/09/30/what-is-web-20.html
18. E-learning Nordic 2006: Uncovering the Impact of ICT on Education in the Nordic Countries, http://itforpedagoger.skolverket.se/digitalAssets/177565_English_eLearningNordic2006.pdf

Success Factors in IT-Innovative Product Companies: A Conceptual Framework

Enric Serradell-López[1], Ana Isabel Jiménez-Zarco[1], and Maria Pilar Martínez-Ruiz[2]

[1] Department of Business Sciences, Open University of Catalonia
eserradell@uoc.edu, ajimenezz@uoc.edu
[2] Marketing and Market Research Department, Universidad de Castilla-La Mancha
MariaPilar.Martinez@uclm.es

Abstract. The purpose of this paper is to explore the role that Information Technologies (ITs) play in product innovation processes, both as an element that strengthens marketing processes making possible consumer integration within the innovation framework, as well as the development of the organizational learning process. There are two main uses of ITs in marketing processes. First, they increase the likelihood and efficiency of communication and cooperative relationships between different agents, both inside and outside the organization. Second, they manage market data correctly, prioritizing its acquisition, storage and dissemination throughout the company, thereby developing market intelligence that can be used in the decision-making process.

Keywords: product innovation process, consumer integration, cooperative relationships.

1 Introduction

In a new era in the business environment, organizations must respond quickly and efficiently to market fluctuations. In order to increase their productivity and competitiveness in the long term [20], [34] organizations are characterized by: i) the development of an organizational culture oriented towards the customer, learning, and innovation; ii) the intensive use of Information and Communication Technologies (ITs); iii) the establishment of dynamic and flexible structures; iv) specialization in core business activities; and v) the integration of different agents in the value chain – through the creation of multiple, close cooperative relationships [31].

In this new competitive landscape, marketing acquires a special relevance [1]. So, as a a global and interactive process that extends to all departments and functions in the company, marketing makes possible: (i) the construction and maintenance of stable relationships with different agents –both within and outwith the organization [9], and (ii) the correct management of market information favouring its diffusion and use in the decision making process [11].

In line with the factors highlighted above, Wei and Morgan [37], recognise product innovation as being the third key factor for organizational success. The development of new products allows the company to access new markets, or strengthen their competitive position in existing ones – with focus on increasing customer satisfaction and loyalty [4].

M.D. Lytras et al. (Eds.): WSKS 2009, CCIS 49, pp. 366–376, 2009.

Yet in today's competitive markets companies are forced to develop new products that accomplish several objectives simultaneously if they want to be successful. Products should be competitive in global markets, offer good value to customers, be environmentally friendly, enhance the strategic position of the company and be introduced at the right time. To meet this formidable set of objectives, companies have to concentrate their efforts and resources on the development of a complex technical process that leads them to create radical and/or incremental product innovations. Further, they need to consider the importance of marketing in the innovation process, as well as to embrace new tools and management techniques, applications, and devices based on the widespread use of ITs [36]. In this way, they may ensure that the new product is totally adapted to market needs and requirements, as well as enhance company performance.

In an attempt to develop the ideas noted above, the research described in this article aims to detail the relationship between those factors discussed in the preceding paragraphs, and the ways in which they ensure the success of an organization and new products in the market. In the same way, this article will analyse the key role that IT technologies play in marketing activity, specifically in the development and maintenance of these factors and in the promotion of their relationship.

The paper is structured as follows: in section 2 we present an analysis of the product innovation process. The importance of the development of new products and services for the company necessitates a review of the marketing state of the art, which leads to the identification of specific success factors. Customer cooperation, identified as one of the key factors for performance of the product innovation process, and its market effects, is tackled in the third part of the paper. Capable of offering important benefits during the new product development (NPD) process, client cooperation is most effective when the level of integration with the company is high. The fourth section of this article analyses the concept of organizational learning, as one of the other key elements of product innovation process performance. Product innovation is defined as a learning process where knowledge is both input and output – the extent to which the organization has a sound knowledge of the market, as well as the different processes that affect NPD, will greatly affect the efficiency and effectiveness of the innovation process and resultant market performance. In section 5 we analyze the role of Information and Communication Technologies (ITs) –and their intensive use in marketing processes– within the product innovation process. The development of an effective marketing function favours the creation of close, long-lasting relationships between the company and the client, which in turn become a strong indicator of success. In the same way, marketing also guarantees the development of both internal and external processes that build knowledge. Furthermore, if the usage of ITs in marketing activities is high, the probability of producing effective marketing processes is increased. Finally, we present the main conclusions of our work.

2 The New Product Development Process

Traditionally, the concept of product innovation has been tied to the technological change process. However, in the last years the literature has identified an important relationship between the product innovation process and the presence of intangible assets, such as investment in R+D+I (Research and Development and Innovation), or in human capital [32]. Thus, from a marketing viewpoint, product innovation has to be considered as a learning process, with the basic resource as knowledge, although simultaneously it is also the main result. So, Affected by a wide range of internal company and environmental factors, and benefited by both competition and cooperation (both internally and externally with other companies), innovation provides both radical technological changes and small, cumulative improvements, that increase the performance of existing technologies [32].

Different fields have shown an increasing interest in determining the success criteria for new products. Recent studies on NPD have discovered a number of traits that contribute to success or failure. Identified success factors, although different in each research work, show a consensus on the importance of cooperative relationships, the organizational learning process and the use of ITs In this sense, Griffin and Hauser [8] show that NPD activities and outcomes as being strongly influenced by the firm's ability to generate disseminate and use market information On the other hand, Tatikonda and Stock [28] among others, suggest that ITs create two main competitive advantages that enhance NPD and organizational performance. First, by positively supporting, through seamless integration within the process, time to market, quality, innovativeness and market share. Second, by improving teamwork, marketing competences and outcomes.

3 The Product Innovation Process and Consumer Integration in the Value Chain

Nowadays, NPD can be characterized as a complex and uncertain process where speed has become a key competitive element. In order to reduce NPD uncertainty and

Table 1. Benefits of cooperation
Source: Own elaboration

Benefits of cooperation
The establishment of workteams made up of experts in different functional fields who adopt flat, highly adaptable structures, wherein decisions are taken in a decentralised way
The supply of economic, human and technological resources to reduce complexity, cost and duration of the process (Littler *et al.*, 1995). The improvement of communication and information exchange processes.
The improvement of relationships between agents in the environment, internalizing the project and promoting the development of working relationships in which members actively participate.
The transfer of information, experience and new technologies that help to identify, and resolve quickly and efficiently, any problems that might arise (ooperation guarantees the circulation of information among agents and its use in the innovation process, thereby improving the activities of investigation and NPD.
Positive support for developed products that meet new needs and demands, and the development of a more efficient innovation process that incorporates the "voice of the customer" together with the experience and know-how of other agents
A reduction of the uncertainty surrounding the product's future (, and its dependence on the timing of product launch, while improving on the results obtained, and ensuring a favourable response from the market.

time dependence, and therefore improve their results, organizations value the establishment of cooperative relationships ([7], [14]. The type of established relationship varies, from a simple transaction or one-off exchange to total consolidation, with the integration of the members of a complete organisation or team making for an authentic alliance [7], [14], [35]. The relationship partner may take the form of universities and research centres [22], competitors or distributors and consumers [3], [33].

There is empirical evidence to show the existence of a positive relationship between cooperation and a successful innovation process [7]. Cooperation offers important benefits to the companies, which are summarized in Table 1.

Interaction, cooperation, credibility, confidence and commitment are recognised in the literature as elements that strengthen the relationship between the organization and consumers, making them more stable with time. Their importance is such that when these elements are dramatically increased, the parties involved begin to share the same principles, culture and values, committing resources with the aim of achieving a common strategic objective [1]. In this situation, as noted by Battacharya and Sen [5], the consumer becomes the key part of the value chain, achieving a high level of integration within the organization.A close relationship with the consumer offers the organization important benefits, to such an extent that they must be considered internal strategic assets to be exploited. In fact, with activities like product innovation, client collaboration is essential. As commented by, Battacharya and Sen [5] and Thomke and Von Hippel [29], the integration of the consumer in the company increases their levels of confidence, commitment, satisfaction and involvement. This increases the quantity and quality of the information that they contribute to the organization, and allows them to adopt a more active role in the development and success of certain strategic activities.

Others authors suggest that the role of consumer should be active and intimate to the design and development process, and come much earlier in the process. They consider product innovation as an iterative, learning process where consumers are often useful contributors of new ideas and feedback, and play an active role in new product testing and launch. Moreover, Pitta and Franzak [18] suggest that consumers' integration in the team enhances the chances of NPD success. Consumer value comes from their product usage in day-to-day life making their experience and knowledge useful in several ways, including the increase in market acceptance of new products, product improvement, an early identification of market trends, and a source of new product ideas.

4 Organizational Learning: Relationship and the Information Management Process

In the current environment, information constitutes a strategic asset [17], [19], [30]. Information is the basis of organizational knowledge[1]. It is also the element on which most firm´s abilities to respond quickly and efficiently to customer demands, create new markets, develop new products or control new technologies [17], [23]. In fact

[1] Some authors use the concept of *organizational knowledge* [23], others the concept of *organizational memory* or *marketing intelligence* [15]. Consistent in all concepts is the means by which knowledge is stored for the future.

learning faster than competitors may be the only source of sustainable competitive advantage in dynamic and turbulent markets. Thus, the more critical challenges for any business are to create the combination of culture and climate that maximizes organizational learning and creates superior consumer value. Slater and Narver [25] suggest that for a business to maximize its ability to learn about markets, creating a market orientation is only the start. A market-oriented culture can achieve maximum effectiveness only if is complemented by a spirit of entrepreneurship and an appropriate organizational climate, namely structures, processes and incentives for implementing the cultural values [37]. Some authors suggest that an organizational structure comprised of multiple close relationships and which considers the consumer as an internal agent, has the necessary elements for developing an organizational learning process [1], [12],[15]. On the other hand, the integration of the client in the organization facilitates their participation and total collaboration in the development of new products and other risky, complex activities. Finally, if the team has a strong degree of integration, its members will show heightened levels of involvement and cooperation, in turn creating a climate which favours collaboration and the interchange and effective use of information [18]. Several works within the area of information processing and market orientation confirm this fact, suggesting that internal relationships are an essential element in the organizational learning process.

Organizational learning is the development of new knowledge or insights that have the potential to influence behaviour [18], [23]. The processing of information leads to knowledge and facilitates behaviour change by the organization that leads to better performance. Thus, all businesses competing in dynamic and turbulent environments must pursue the processes of learning behaviour change and performance improvement [16].

Organizational learning involves an organizational, cultural and structural change, as well as the development of new activities. However, large established organizations are complex social entities for which change is difficult [15]. In order to simplify and reduce these complexities into, for example, lack of knowledge, resistance to change or bad procedures, companies organize routine activities or processes in a collective way. Indeed, this complex collective action is not possible without shared routines that can be reproduced easily, since routines minimize the need to redefine each situation anew. Marketers have begun to recognize that for market orientation, implementing learning organization concepts can be a means of increasing the competitive performance of the company [16] as well as the competitiveness and effectiveness of marketing [18]. Through changing organizational behaviour, learning can be translated into a competitive advantage and superior performer. However, Maltz and Kohli [15] note that the superior position lies in a firm's ability to *use* knowledge, not simply in its access. Slater and Narver [25] describe three ways that learning use can influence organizational behaviour: action-oriented, knowledge-enhancing and affective use (see Figure 1). These three types of knowledge-use form a continuum, form direct to indirect, of the effects of learning use on behavioural change.

An organization has a foundation for sustained competitive advantage when it possesses skills and resources that provide a superior value to the consumer, are difficult to imitate, and are capable of multiple applications [25]. On the one hand, a company provides superior value to the consumer when its culture and climate foster behaviour that leads to improvements in effectiveness or efficiency, which in turn provides additional benefits or lower prices for consumers [9]. On the other, imperfect imitation might be the product of a socially complex organizational environment that is difficult for competitors to understand or emulate [25]. Finally, the company is capable of multiple applications when an organizational system provides unique insight into opportunities in new or existing markets [10].

Organizational learning is valuable to firms in this context because it focuses on understanding and effectively satisfying their expressed and latent needs through new products, services, and ways of doing business [23]. This should lead directly to superior outcomes, such as more new product success, superior consumer retention, higher consumer-defined quality, and ultimately, superior growth and/or profitability. According to Weerawardena [36], few issues have been characterised by as much agreement among organizational researchers as the importance of product innovation to organizational competitiveness and effectiveness. Innovation is viewed as a central concept in the search for a consumer value-based differentiation strategy, and NPD has emerged as one of the critical strategic concerns for firms [24].

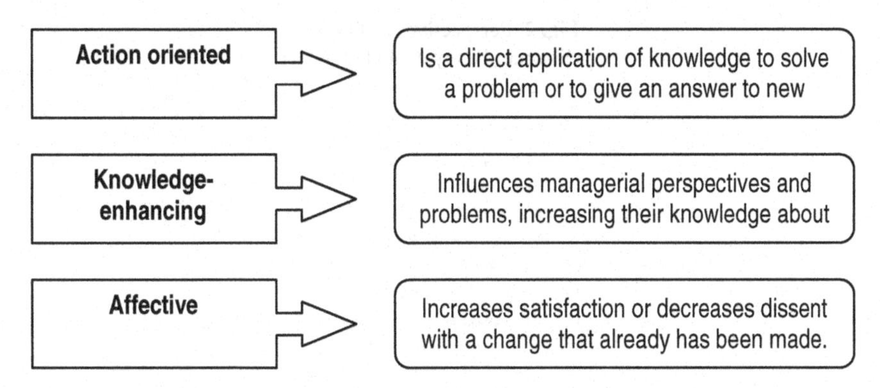

Fig. 1. Ways that learning use can influence organizational behaviour
Source: Slater and Narver [25]

5 ITs Use in Marketing: Enhancing NPD Processes

Traditionally, the literature suggests that there are many different perspectives, or aspects of ITs that must be considered [6] (see Figure 2).

Today, ITs must be conceived broadly to encompass the information that businesses create and use, as well as the wide spectrum of increasingly convergent and linked technologies that process that information. Therefore, ITs can be viewed as a collective term for a wide range of software, hardware, telecommunications and information management techniques, applications and devices [6], [19].

The literature suggests that ITs enhance the degree of companies' internal integration allowing the creation of more flexible and adaptive organizational structures, fully market and consumer-oriented [35]. As Porter and Millar [19] suggest one concept that highlights the role of ITs in competition is the value chain.Nevertheless, these changes are especially important in marketing activities, where ITs play and important role in enhancing the productivity and effectiveness of certain activities or functions [6], [35]. So, ITs, provides ready access to a vast array of global information resources and facilitate the gathering of valuable competitive knowledge and

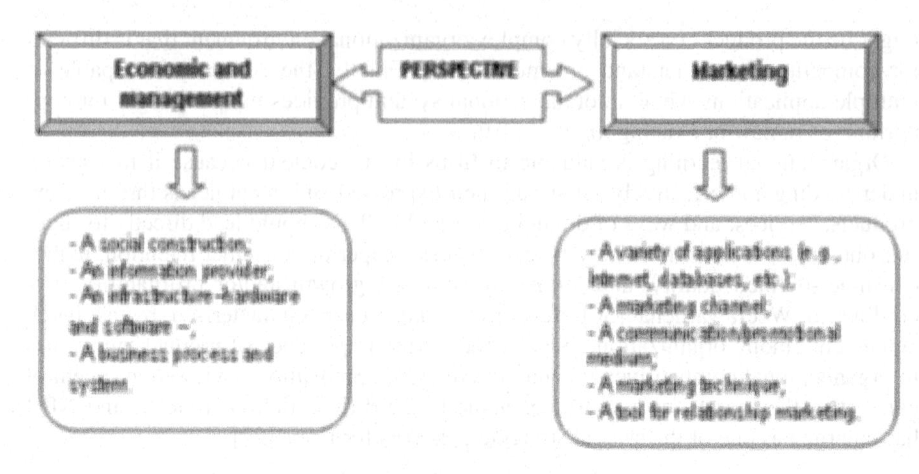

Fig. 2. Perspectives of ITs
Source: Brady et al.[11]

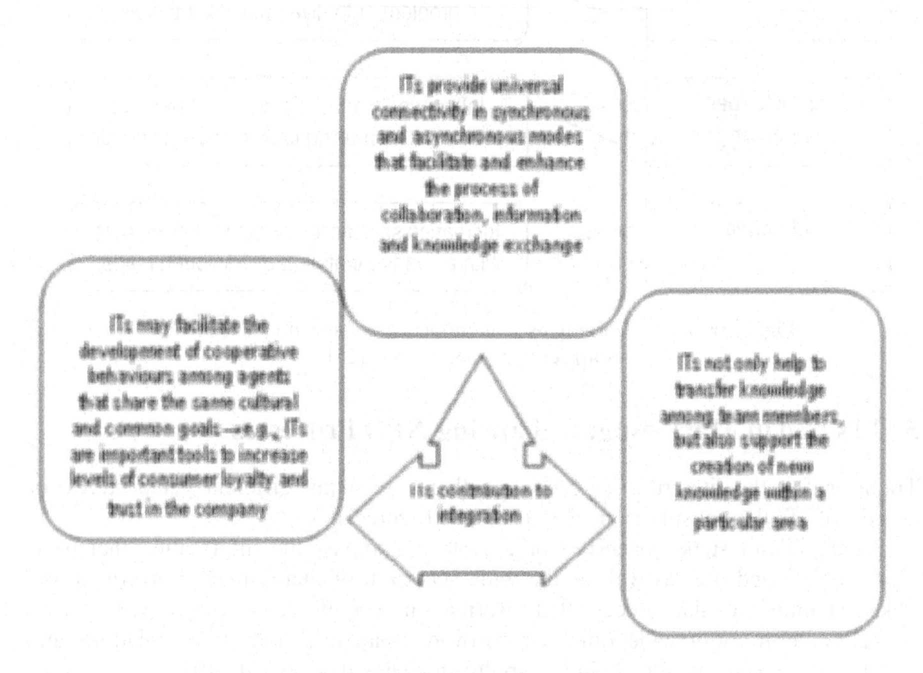

Fig. 3. ITs´ contribution to integration
Source: Own elaboration

consumer-related information that simplifies marketing decision processes. On the other hand, ITs makes possible the marketer with extraordinary capability to target specific groups of individuals precisely and enable them to practice mass-customization and one-to-one marketing strategies, by adapting communications and others elements of the marketing mix to consumer segments [20]. In fact, ITs not only affect how individual activities are performed, but through new information flows, greatly enhance a company's ability to exploit linkages between activities, both within and outside the company [19]. Thus, ITs can create new, strong linkages between internal activities, and even coordinate these actions more closely with their consumers and suppliers to facilitate integration within the company [14]. As previously noted, a greater degree of integration is a key factor for internal relationships, through involving consumers and other functional agents in teamwork to develop new products [13]. ITs increase integration in different ways (see Figure 3).

Even if partners do not a have a common location, culture, history or future, ITs can enhance collaboration and knowledge transfer and use [26]. With the widespread use of ITs global or virtual teams have became a reality. [21], analysing the ability and willingness to cooperate, suggests that ITs increase teamwork integration in two ways, firstly facilitating and speeding knowledge transfer, both tacit and explicit, and second, reinforcing the levels of trust and confidence that normally develop in face-to-face meetings.

6 Conclusions

Many studies have noted the importance of client cooperation, the availability of market intelligence and the intensive use of ITs in product innovation. However, there exists relatively little development regarding the relationship between these factors, and the way in which the intensive use of ITs in the development of marketing processes promotes the creation of synergies and increases the chances of success for innovation and the launch of new products. The integration of the client in NPD offers the company important benefits which, from the marketing perspective, such close relationships become understood as a strategic asset to be exploited. These benefits include an increase in customer confidence and commitment, which in turn increases the degree of involvement with the company, leading to a more active and cooperative role in the development of certain processes. In the same way, the establishment of close relationships with the client increases the quantity and quality of the information that is supplied to the company, and directly impacts in the development of overall market intelligence. As a strategic resource, information constitutes a key element in all business processes. Within it rests the base of organizational knowledge and part of the organizational ability to respond quickly and efficiently to market changes. In the NPD process, the importance of market information is even greater, such that the product innovation process constitutes in itself a learning process, the result of which should ideally be a manifestation of the preferences and requirements of the market.

The intensive use of ITs supports the development of both elements and promotes the creation of synergies, increasing the chances of success for both the innovation process in general and the launch of new products to market. Many works highlight the importance of ITs as a key element in integrating marketing processes within NPD, leading to an increase in company innovation through the development of new products that are adapted to market needs and reduce technological, strategic and marketing risk. In the creation of market intelligence, ITs are an important source of data acquisition and generation. Argyres [2] explains how ITs constitute one of the most appropriate means for getting close to the environment and developing extensive knowledge about the different agents operating in the field. They therefore allow a company to access a huge quantity of important and up-to-date information in a simple, quick and economic way. Furthermore, ITs are a vital element in the generation, transmission, diffusion and use of knowledge within the organisation. They not only give companies the necessary means to treat, manage, analyse and store information [27], but they also make it easier to transmit and diffuse that information throughout the company, for its later use in the strategic decision-making process.

On the other hand, various authors have also indicated the importance of ITs as an element that improves consumer integration in the innovation process, enabling cooperation and communication. As a source of information, and prior to the establishment of a relationship, ITs help to identify and determine the degree of attractiveness of possible strategic partners. Also, as a communication channel, ITs are the means by which the company sets up fluid communication with its cooperation partners, through the transmission –or even creation– of knowledge, and the breaking-down of barriers of time, space and economy that limit the effectiveness and efficacy of the process. Finally, as a combination of all of these, ITs act as a socialising factor that enables the foundation of relationships by permitting continuous and intense communication among members, providing the basis for agreement and consensus, and the development of a climate of trust and commitment grounded on social and affective values.

References

Achrol, R.S., Kotler, P.: Marketing in the Network Economy. Journal of Marketing 63(Special issue), 146–163 (1999)

Argyres, N.S.: The Impact of information Technology on Coordination: Evidence from the B-"Stealth Bomber". Organization Science 10(2), 162–180 (1999)

Appleyard, M.M.: The Influence of Knowledge Accumulation on Buyer-Supplier Codevelopment Projects. Journal of product Innovation Management 20, 356–373 (2003)

Atuahene-Gima, K.: Differential Potency of Factors Affecting Innovation Performance in Manufacturing and service Firms in Australia. Journal of Product Innovation Management 13, 35–50 (1996)

Bhattacharya, C.B., Sen, S.: Consumer-Company Identification: A Framework for Understanding Consumers' Relationship with Companies. Journal of Marketing 67, 76–88 (2003)

Brady, M., Saren, M., Tzokas, N.: Integrating Information Technology into Marketing Practice –The IT Realize of Contemporary Marketing Practice. Journal of Marketing Management 18, 555–577 (2002)

Deeds, D.L., Rothaermel, F.T.: Honeymoons and Liabilities: The Relationship between Age and Performance in Research and Development Alliances. Journal of Product Innovation Management 20(6), 468–485 (2003)

Griffin, A., Hauser, J.R.: Patterns of Communication among Marketing, Engineering and Manufacturing –A Comparison between Two Product Teams. Management Science 38(3), 360–373 (1992)

Grönroos, C.: Relationship Marketing: Interaction, Dialogue and Value. Revista Europea de Dirección y Economía de la Empresa 9(3), 13–24 (2000)

Hamel, G., Prahalad, C.K.: Competing for the Future. arvard Business School Press, Boston (1994)
Jaworski, B.J., Kohli, A.K.: Market Orientation: Antecedents and Consequences. Journal of Marketing 57, 53–70 (1993)
Johnson, J.L., Sohi, R.S., Grewal, R.: The Role of Relational Knowledge Stores in Interfirm Partnering. Journal of Marketing 68, 21–36 (2004)
Kahn, K.B.: Market Orientation, Interdepartmental Integration, and Product Development Performance. Journal of Product Innovation Management 18, 314–323 (2001)
Leenders, M.A.A.M., Wierenga, B.: The Effectiveness of Different Mechanisms for Integrating Marketing and R&D. Journal of Product Innovation Management 19(4), 305–317 (2002)
Maltz, E., Kohli, A.K.: Intelligence Dissemination across Functional Boundaries. Journal of Marketing Research 15, 47–61 (1996)
Michael, S.C., Palandjian, T.P.: Organizational Learning and New Product Introductions. Journal of Product Innovation Management 21, 268–276 (2004)
Nonaka, I.: The Knowledge-Creating Company. Harvard Business Review 69(6), 96–104 (1991)
Pitta, D.A., Franzak, F.: Boundary Spanning Product Development in Consumer Markets: Learning Organization Insights. Journal of Product & Brand Management 6(4), 235–249 (1997)
Porter, M.E., Millar, V.E.: How Information Gives You Competitive Advantage. Harvard Business Review 63(4), 149–174 (1985)
Prasad, V.K., Ramamurthy, K., Naidu, G.: The Influence of Internet-Marketing Integration on Marketing Competencies and Export Performance. Journal of International Marketing 9(4), 82–110 (2001)
Roberts, J.: From Know-how to Show-how? Questioning the Role of Information and Communication Technologies in Knowledge Transfer. Technology Analysis & Strategic Management 12(4), 429–443 (2000)
Santoro, M.D.: Success Breeds Success: The Linkage between Relationship Intensitive and Tangible Outcomes in Industry-University Collaborative Ventures. Journal of High Technology Management Research 11(2), 255–273 (2000)
Sinkula, J.: Market Information Processing and Organizational Learning. Journal of Marketing 58, 35–45 (1994)
Slater, S.F., Narver, J.C.: Does Competitive Environment Moderate the Market Orientation-Performance Relationship? Journal of Marketing 58, 46–55 (1994)
Slater, S.F., Narver, J.C.: Market Orientation and the Learning Organization. Journal of Marketing 59, 63–74 (1995)
Smith, P.G., Blanck, E.L.: Leading Dispersed Teams. Journal of Product Innovation Management 19, 294–304 (2002)
Swan, J., Scarbrough, H., Hislop, D.: Knowledge Management and Innovation: Networks and Networking. Journal of Knowledge Management 3(3), 262–275 (1999)
Tatikonda, M.V., Stock, G.N.: Product Technology Transfer in the Upstream Supply Chain. Journal of Product Innovation Management 20, 444–467 (2003)
Thomke, S., Von Hippel, E.: Customers as Innovators: A New Way to Create Value. Harvard Business Review 80(4), 74–81 (2002)
Tuominen, M., Möller, K., Rajala, A.: Marketing Capability: A Nexus of Learning-Based Resources and Prerequisite for Market Orientation. In: 26th EMAC Conference, vol. III, pp. 1220–1240. Warwick, UK (1997)
Tzokas, N., Saren, M.: Building Relationship Platforms in Consumer Markets: A Value Chain Approach. Journal of Strategic Marketing 5, 105–120 (1997)

Vilaseca, J., Torrent, J.: ICT and Transformations in Catalan Business. Research Report II. Universitat Oberta de Catalunya (2003), http://www.uoc.edu/in3/pic/eng/pdf/PIC_empresa_abs_eng.pdf

Von Hippel, E.: Lead User Analysis for the Development of New Industrial Products. Management Science 34(5), 569–582 (1988)

Vorhies, D.W., Harker, M., Rao, C.P.: The Capabilities and Performance Advantages of Market-Driven Firms. European Journal of Marketing 33(11/12), 1171–1202 (1999)

Webster Jr., F.E.: The Changing Role of Marketing in the Corporation. Journal of Marketing 56, 1–17 (1992)

Weerawardena, J.: The Role of Marketing Capability in Innovation-Based Competitive Strategy. Journal of Strategic Marketing 11, 15–35 (2003)

Wei, Y.S., Morgan, N.A.: Supportiveness of Organizational Climate, Market Orientation and New Product Performance in Chinese Firms. Journal of Product Innovation Management 21, 375–388 (2004)

Management Guidelines for Database Developers' Teams in Software Development Projects

Lazar Rusu[1], Yifeng Lin[2], and Georg Hodosi[1]

[1]Stockholm University/Royal Institute of Technology, Stockholm, Sweden
[2] Bwin Games AB
lrusu@dsv.su.se, yifeng.lin@bwin.org, hodosi@dsv.su.se

Abstract. Worldwide job market for database developers (DBDs) is continually increasing in last several years. In some companies, DBDs are organized as a special team (DBDs team) to support other projects and roles. As a new role, the DBDs team is facing a major problem that there are not any management guidelines for them. The team manager does not know which kinds of tasks should be assigned to this team and what practices should be used during DBDs work. Therefore in this paper we have developed a set of management guidelines, which includes 8 fundamental tasks and 17 practices from software development process, by using two methodologies Capability Maturity Model (CMM) and agile software development in particular Scrum in order to improve the DBDs team work. Moreover the management guidelines developed here has been complemented with practices from authors' experience in this area and has been evaluated in the case of a software company. The management guidelines for DBD teams presented in this paper could be very usefully for other companies too that are using a DBDs team and could contribute towards an increase of the efficiency of these teams in their work on software development projects.

Keywords: database developers' teams, agile software development, capability maturity model, management guidelines, software development project.

1 Introduction

Recent studies shows that the SQL developer job market has increased with 33.6% in the first three quarters of 2008 corresponding with the same period in 2007 [18]. In opinion of Colleen Graham [21] this aspects could be explained because "as the popularity of the data-intensive initiative continues to grow, the relational database management system is receiving ongoing attention". On the other hand in opinion of Mike Perks - solution architect at IBM most projects fail therefore for the improvement of the success of software project development he is recommending a list of best practices [20] that a project team should follow. Furthermore teams have become the strategy of choice when organizations are confronted with complex and difficult tasks [24] because teams have the potential to offer greater adaptability, productivity, and creativity than any one individual can offer [14][25]. As a new role in a company, there are no instructions for the database developer managers to administrate efficiently a database

M.D. Lytras et al. (Eds.): WSKS 2009, CCIS 49, pp. 377–386, 2009.

developers (DBDs) team. In the review of DBDs tasks we have found in [15] mentioned some of the typical tasks that a DBD should do. However, these studies are limited because they do not have clear boundaries for the tasks between DBDs and other roles. In fact according to Paulk [19] the fundamental problem about software productivity and quality is the inability to manage the software process therefore Capability Maturity Model (CMM) is a framework for an effective software process. On the other hand we have not found published any research about how to execute CMM's practices in DBDs teams. There are also studies like for example those done by Fritzsche and Keil [13] that have verified that some CMM process areas can be covered by agile software development methods. Moreover agile methodologies showed slightly better delivery performance than rigorous methodologies in terms of business performance, customer satisfaction, and quality [7]. Furthermore according to Wang and Vidgen [27], a process needs continuous adjustment and adaptation to specific situations to avoid process rigidity and deterioration. In this way by using the agile improvement practices, the CMM process is shaped and re-shaped through a series of interactions between change agents and adopters [4] and more appropriate practices can be found.

2 Research Methodology

To respond to the problems described in the previous section our research questions has looked to investigate the followings issues: "*What are the tasks and boundaries for DBDs teams?*" and "*What practices from CMM and agile software development can be used in managing more efficiently the DBDs teams?*"

In this perspective the main goal of our research paper has looked to improve the software development process for DBDs teams by providing them with a set of management guidelines which includes a list of tasks, boundaries and practices that can be usefully for a company/manger that wants to setup a DBDs team. The case study methodology [28] has been used and the research has been conducted through one of the projects in which the authors have been involved in a DBDs team. For collecting the data we have conducted semi-structured interviews with developers and different team managers in a software company in Stockholm, Sweden. The research data has been collected between February to April 2008 and from August 2008 till January 2009 a desk analysis of these data has been performed by the authors. For the collection of the data and before the project has started three domains has been analyzed: (1) previous work about DBDs tasks in a software development project; (2) CMM stand and practices; and (3) agile software development methodologies. Moreover during the work on a software project, direct observation has been used as a data gathering tool by one of the authors for collecting the data directly and frequently. The interviews have been conducted with the developers and managers and were recorded and transcribed and later on the principal ideas from these interviews were submitted back to the interviewers for review and confirmation. In this study we have used the analysis technique as a research approach by collecting the data from a case study and by comparing with the existing theories [12].

3 Research Background

This section presents the research background for our research. Firstly, we will present what DBD tasks should be in a software development project, an overview of the Capability Maturity Model (CMM) and last we will discuss about the selection of agile software development methodologies for DBDs teams.

3.1 DBDs Tasks in a Software Development Project

According to Info-tech research group [15]: "the DBDs role is to strategically design and implement database across the organization and ensure a high level of data availability". In our opinion and also as [11][15] the tasks for DBDs should include the followings phases: (1) database installation; (2) schema design and implement; (3) data migration and verify; (4) ensure data entry, which we will mention as "data monitor"; (5) Extract Transfer Load (ETL); (6) program optimization and schema; (7) written descriptions about the program, which we will mention as "design document"; and (8) identify data entry which we will mention in our paper as "reporting data support". In the next section all these 8 tasks described before will be detailed and explained in the case of a software development project.

3.2 An Overview of Capability Maturity Model

The dominant perspective on software development is rooted in the rationalistic paradigm, which promotes a product-line approach to software development using a standardized, controllable, and predictable software engineering process [10]. The Capability Maturity Model (CMM) is one of this kind process models that could help people to understand an organization's process maturity level. The CMM has 5 levels of maturity that are from low to high and are the followings: Initial, Repeatable, Defined, Managed, and Optimizing [5]. For each maturity level except for level 1, there are several Key Process Areas (KPAs) that can be considered as characteristics of this maturity level. Moreover each KPA is described in terms of key practices that are describing the activities and infrastructure that contribute to the implementation and institutionalization of the KPAs [19]. Although a higher level will bring better processes for a company and this because most of the processes at level 4 and 5 are unattainable without sacrificing some agile bedrock [13]. Therefore in our research, we have used only the practices from level 2 and 3.

3.3 Comparison of Agile Software Development Methodologies

In 1990s, agile software development has appeared as a contrary idea to the traditional process [1]. It is a style of software development characterized by emphasis on people, communication, working software, and responding to change [8].

In the Agile Manifesto from 2001 [2] a lot of agile methods are introduced like are eXtreme Programming (XP) [1], Scrum [26], Crystal Clear [6] and Lean Software Development [22]. A comparison and analyses of all these methodologies has been done by Coffin and Lane [9] regarding their strengths and weaknesses. The results of their analyses shows: that XP and Crystal methodology cannot work for distributed teams. Therefore we cannot choose them in our case because the DBDs team needs to be shared by several projects and has to be distributed. On the other hand both Scrum and Lean works well with small teams, highly volatile, distributed teams and multiple customers/stakeholders. According to Bern et al. [3] the customer is perhaps the most important component of the business environment and has a lot of influence on the entire process of software development. Moreover the customer participation and steering is one of the strengths of Scrum, which is appropriate for DBDs teams because the team's customers are in the same company and can be easy invited to participate in the projects. Based upon these arguments in our research we have used Scrum as a typical method of agile software development.

4 Case Study: Development of Management Guidelines for Database Developers Teams in a Software Project

In this section a software development project has been used as a case study within a software company in Stockholm, Sweden looking to provide a set of management guidelines for DBDs teams for improving their efficiency in software development projects.

4.1 Software Project Background

The company is a software one doing business in software games area. The project used in this case study is described below. The company managers have decided to start a "bonus system" to attract more players to join games. The basic idea of this bonus system was to send out some promo tickets, and while the players meet the request, the bonus will be released and the player will get some beneficence. The tasks for the project team are listed as the followings:

- To setup the database server for this project.
- Develop programs that can record and fetch tickets, bonus, beneficence, and player information.
- Provide aggregated data for BI/DW team to analyze the effects of this system.

4.2 DBD Tasks in the "Bonus System" Software Project

For working as DBD in a software development project there are 8 tasks to be done (as we have seen in the previous section) and these are described below in the case of "bonus system" project.

Database installation
The project team needs to decide the database version, resource, and schedule to install database. These decisions should be done in project planning phase, which is a KPA for CMM level 2. It involves developing estimates, establishing necessary

commitments and defining the plan to perform the work [19]. Moreover project planning meeting is also one practice in Scrum.

In "bonus system" project, the customer has decided that the DB server should be MySQL, and the DBDs team has decided that the version should be 5.1. In order to provide an environment for programmers start to work as soon as possible, the DBDs team has setup the development environment and later on the company's DBAs setup up the test and production environments.

Schema design and implement

Schema design phase should not only to implement customer's requirement, but also to interact with the customer to get a reliable requirement. Requirements management is a KPA for CMM Level 2 and during this process, we have used Entity-Relationship (ER) model to help the customer and programmers have a common understanding, and it is easy to translate into a data model. While the ER model was being established, we have found that the DBDs team can also give suggestions to find out some potential attributes for non-functional requirement (i.e. space or speed). For these attributes, the customer figures out the priority, and then the DBDs team work for database design. Another thing need to be decided is the life cycle of the data: where is the data from (i.e. load data from an old system) and how long time the data will exist (after one period they will be archived or removed). In this project, one programmer found that some modifications cause problems due to less communication and mistakes, so peer review was suggested that is a KPA for CMM level 3 and focus on finding out the software defects for removing them in the early phase.

Data migration and verify

In this project, the DBDs team developed scripts and programs, and they are parts of the project's product. Configuration management (CM) is a KPA for CMM level 2 that looks to establish and maintain the integrity of the products of the software project throughout its life cycle [19]. It includes a set of activities performed to identify and organize software items and control their modification [16]. In this project, the source codes include database schema script, logic code script (i.e. store procedure) and data migration program.

Data monitor

Project tracking and oversight is a KPA for CMM Level 2 and it is used to check the project process. For the DBDs team we convert this task to track and oversight the database server, which includes checking the database server to find out the applications' status and taking actions to correct them when they're out of expectation. There are three different frequency monitoring: online, daily and monthly, and they are shared by DBA, DBD, BI/DW teams. In this project, the DBDs team works for daily check. A program is developed to check the data's validity, and all these checks are against the replication servers. The monitoring runs automatically, and if something goes wrong it will submit an email to correlative person.

Extract Transfer Load (ETL)

In Extract Transfer Load (ETL) process, data from different projects need to be collected to support BI/DW team's request. This needs intergroup coordination and intergroup coordination is a KPA for CMM Level 3, which is to set up a channel to improve communication between different project teams. This is also one of the important purposes to establish a DBD team. The DBD team is distributed, but at least the team members can be brought together at the start, end or other pivotal points during the projects [17]. Therefore the "Scrum meeting" practice is brought into the DBDs team for the purpose to share DBDs' knowledge. In Scrum's rules, one typical Scrum meeting takes 10-15 minutes every morning and each team member should answer 3 questions about the project [26]. In DBDs team, this kind Scrum meeting is borrowed and all DBDs are supposed to attend, but the meetings are set up only twice per week and the questions are different.

Program optimization and schema

From a developer's point of view, the software optimization can be done by enhancing developers' skills ability. The training program is a KPA for CMM Level 3 that is looking to develop the skills and knowledge of individuals so they can perform their roles effectively and efficiently [19]. In this perspective the different aspects for DBD's training can include:

- Database new production training: When a new version database is released, both DBDs and DBAs will attend the new production training.
- Process management training: When some management processes (i.e. Scrum) are used in the company, all the related personnel need to be trained.
- Program language and tool training: the DBD team works with the programmers together and he/she must know the program languages and tools that are used.

There are also internal trainings in the software company and the DBDs team has been also requested to organize this kind of training. In this project, a training meeting has been hosted to explain the old system's tables.

Design document

Document is a very import part in CMM and now there is a widely discussion in Agile methodology field about how many documents are necessary. The design documents that a DBD needs to write in this project includes database schema design, application-database schema relationship, and detail design. Moreover some key figures to explain why the table is designed like that are also written and these key figures will be saved in another place. The detail design is for the program that the DBDs team develop (i.e. data migration program) and includes how to organize the data, the time limitation for extracting these data, and how to run these applications.

Reporting data support

Since the DBDs have the access to fetch production system's data, a lot of persons come to ask for the information. These persons include the project manager, BI/DW team, customer service or some customers. But because we don't know whether these data contain secret information this task will be done by the BI/DW team. In this way all questions should come only from BI/DW team, and we are only providing answers to the DBD team's questions.

5 Results and Discussions

From the interviews we have conducted with the project team and managers in the software company part of our case study we have come out with the following results. In case of the project team their feedback to our findings includes:

- The DBDs team prepares the develop environment and schema design faster than programmers.
- Code reviewing gets really effective. On the other hand DBD team reviewing requires more effort but it's more important.
- Data verification is very important and it should be done from the end user's perspective.
- Training and the experienced employees sharing their knowledge are very usefully for the project team.
- Lack of design document could be a problem for new coming persons in one project. Therefore database related document is more necessary because the schema design is more abstract than programs.

In case of the managers from the software company their feedback to our findings includes:

- Group meetings with DBDs team are not necessary to have it every day. Probably twice a week is enough.
- DBA team should be responsible for the online data verification. DBDs team should be responsible to provide data to BI/DW team that will analyzed this data for business purpose.
- DBDs team should own the schema of the database and all changes on schema must be addressed to the DBDs team. But the data should be owned by BI/DW team and the analysis of data should be the BI/DW team's task, and only they will have to request data from DBDs team.
- DBDs should provide suggestions for optimization to the programmers and the programmers will decide whether to take these suggestions or not.

As the final result of our research approach we have come out with the following set of tasks, boundaries and suggested practices that are described in Table 1, and which are forming the management guidelines that a DBDs team should use for improving his work efficiency.

As we have mentioned before in this paper the management guidelines for a DBDs team that are presented in table 1, work well for smaller teams usually no more than 10 team members since Scrum methodology is used in our case. So in case the DBD team size is larger than 10 members as Rising and Janoff [23] points out, the large team can be divided into collections of smaller sub-teams and this will work if the sub-teams are independent and the interfaces well defined. But when the overlap is considerable and the interfaces poorly understood, the benefits are not so great.

Table 1. The list of tasks, boundaries and practices for DBDs team

Tasks for DBDs	Boundaries for the tasks	Suggested practices for DBDs team
DB installation	Only respond for develop environment	1. The database version, the person responsibilities and resources should be defined. 2. A higher priority for developing the environment setup.
Schema design and implement	Only owns schema design, the BI/DW team owns the data	3. Required documents should include data's model, lifecycle and capability. 4. A clear responsibility between data and schema. 5. Higher the priority and coverage for database relative review.
Data migration and verification	Result should be approved by data owner	6. Separate the program code with database scripts. 7. Separate the clean-up script and setup script.
ETL	Extracting data not the analysis	8. Have weekly DBDs team meetings, across projects, to share knowledge between the projects' work. 9. Use a common place to put tips/solutions.
Optimize program and schema	Only provide suggestions	10. The training for DBDs team to be on database, process or program training. 11. Expert teams for training in the same company.
Data monitor	Basic logic check	12. Clarify the responsibility for different levels of database track. 13. Document the measurement data and submit warnings automatically if it's possible.
Reporting data support	Only answer to the questions from BI/DW team	14. Clarify responsibility for schema and data and separate the privilege for them.
Design document	Only for database schema and extracting program	15. Separate common design with key figure. 16. Create common repository for database relative tips/solutions. 17. Higher the priority for document.

6 Conclusions

In this research paper we had firstly done an analysis for the new need for managing efficiently a DBDs team, by specifying the tasks for this new position, and secondly we have developed a set of management guidelines that are including tasks, boundary, and practices to help the managers in improving the work efficiency with this new team. The results of our research have come out from analyzing different software process development methods (CMM and Scrum) including the practices coming from the authors experiences in a DBD role. Moreover, the collected data have been also complemented from the direct observations of the authors within a software development project and the interviews performed with a DBDs team and other managers from a software company. In summary the management guidelines proposed in this paper are:

- Eight tasks and their boundaries.
- Eight KPAs from CMM that are reachable for DBDs tasks and 17 suggested practices for the DBDs team to reduce bugs, reduce developing time, and enhance project team's stability by adding process management practices (from CMM) into each iteration (from Scrum).

On the other hand, these results are having as limitation the fact that are based on one verified software development project. However, these management guidelines (tasks, boundaries for the tasks and practices) developed in our research have been used before in other software development projects with different requirements and the feedback received from the project teams and managers confirmed our results. In conclusion we believe that the management guidelines described in this paper will work properly too in the case of other software development projects/companies that are using a small DBDs team (up to 10 members) and following them will improve their work efficiency.

References

1. Beck, K., Andres, C.: Extreme Programming Explained: Embrace Change, 2nd edn., p. 19. Addison-Wesley Professional, Reading (2004)
2. Beck, K., Beedle, M., Bennekum, A., Cockburn, A., Cunningham, W., Fowler, M., Grenning, J., Highsmith, J., Hunt, A., Jeffries, R., Kern, J., Marick, B., Martin, R.C., Mellor, S., Schwaber, K., Sutherland, J., Thomas, D.: Manifesto for agile software development (2001), http://www.agilemanifesto.org/ (accessed on October 21, 2008)
3. Bern, A., Pasi, A., Nikula, U., Smolander, K.: Contextual Factors Affecting the Software Development Process - An Initial View. In: The Second AIS SIGS and European Symposium on Systems Analysis and Design, Gdansk, Poland, June 5 (2007)
4. Börjesson, A., Martinsson, F., Timmerås, M.: Agile improvement practices in software organizations. European Journal of Information Systems 15, 169–182 (2006)
5. Carnegie Mellon Software Engineering Institute. Capability Maturity Model Integration (CMMI^{SM}), Version 1.1, pp. 11–14 (2002), http://www.sei.cmu.edu/pub/documents/02.reports/pdf/02tr012.pdf (accessed on October 2008)
6. Cockburn, A.: Crystal Clear: A Human-Powered Methodology for Small Teams. Addison-Wesley Professional, Reading (2004)
7. Cockburn, A., Highsmith, J.: Agile software development: The people factor. IEEE Computer 34, 131–133 (2001)
8. Coffin, R., Lane, D.: A Practical Guide to Seven Agile Methodologies, Part 1 (2006a), http://www.devx.com/architect/Article/32761 (accessed on October 21, 2008)
9. Coffin, R., Lane, D.: A Practical Guide to Seven Agile Methodologies, Part 2 (2006b), http://www.devx.com/architect/Article/32836/0/page/1 (accessed on October 21, 2008)

10. Dyba, T.: Improvisation in small software organizations. IEEE Software 17, 82–87 (2000)
11. EI Group: Database developer (2008), http://www.schoolsintheusa.com/careerprofiles_details.cfm?carid=466 (accessed on October 21, 2008)
12. Eriksson, L.T., Wiedersheim-Paul, F.: Att utreda, forska och rapportera, 6th edn. Malmö, Liber Ekonomi (1999)
13. Fritzsche, M., Keil, P.: Agile Methods and CMMI: Compatibility or Conflict? e-Informatica Software Engineering Journal 1(1) (2007), http://www.e-informatyka.pl/e-Informatica/attach/Issue1/Vol1Iss1Art1eInformatica.pdf (accessed on October 27, 2008)
14. Gladstein, D.L.: Groups in context: A model of task group effectiveness. Administrative Science Quarterly 29, 499–517 (1984)
15. Info-tech research group: Database Developer Info-Tech Advisor: Job Description (2009), http://www.infotech.com/ITA/Research%20Centers/Data%20,-a-,%20Digital%20Assets/General/Database%20Developer.aspx (accessed on October 27, 2008)
16. Jalote, P.: CMM in Practice: Processes for Executing Software Projects at Infosys, p. 195. Addison-Wesley Professional, Reading (1999)
17. Miller, A.: Distributed Agile Development at Microsoft Patterns & Practices. Microsoft (2008), http://download.microsoft.com/download/4/4/a/44a2cebd-63fb-4379-898d-9cf24822c6cc/distributed_agile_development_at_microsoft_patterns_and_practices.pdf (accessed on April 9, 2009)
18. OdinJobs.com, SQL developer Job Market Overview (2008), http://www.odinjobs.com/Sql-developer_job_market_overview.html (access on March 20, 2009)
19. Paulk, M.C., Weber, C.V., Curtis, B., Chrissis, M.B.: The Capability Maturity Model: Guidelines for Improving the Software Process, pp. 3–4, 32, 39, 133, 180, 213. Addison-Wesley Professional, Reading (1994)
20. Perks, M.: Best practices for software development projects (2006), http://www.ibm.com/developerworks/websphere/library/techarticles/0306_perks/perks2.html (access on March 20, 2009)
21. Petty, C.: Gartner Says Worldwide Relational Database Market Increased 14 Percent in 2006, Gartner Inc. (2007), http://www.gartner.com/it/page.jsp?id=507466 (accessed on October 27, 2008)
22. Poppendieck, M., Poppendieck, T.: Lean Software Development: An Agile Toolkit. Addison-Wesley, Reading (2003)
23. Rising, L., Janoff, N.S.: The Scrum software development process for small teams. IEEE Software 17, 26–32 (2000)
24. Salas, E., Cooke, N.J., Rosen, M.A.: On Teams, Teamwork, and Team Performance: Discoveries and Developments. Human Factors: The Journal of the Human Factors and Ergonomics Society 50(3), 540–547 (2008)
25. Salas, E., Sims, D.E., Burke, C.S.: Is there a "big five" in teamwork? Small Group Research 36(5), 555–599 (2005)
26. Schwaber, K.: Agile Project Management with Scrum, p. 138. Microsoft Press (2004)
27. Wang, X., Vidgen, R.: Order and chaos in software development: a comparison of two software development teams in a major IT company. In: European Conference on Information Systems 2007, pp. 816–817 (2007)
28. Yin, R.K.: Case Study Research: Design and Methods, 3rd edn. Sage Publications, Thousand Oaks (2003)

ERP and Four Dimensions of Absorptive Capacity: Lessons from a Developing Country

María José Álvarez Gil, Dilan Aksoy, and Borbala Kulcsar

Department of Business Administration, Universidad Carlos III de Madrid, C/Madrid, 126 28903 Getafe (Madrid) España
maria.alvarez@uc3m.es, daksoy@emp.uc3m.es, bkulcsar@emp.uc3m.es

Abstract. Enterprise resource planning systems can grant crucial strategic, operational and information-based benefits to adopting firms when implemented successfully. However, a failed implementation can often result in financial losses rather than profits. Until now, the research on the failures and successes were focused on implementations in large manufacturing and service organizations firms located in western countries, particularly in USA. Nevertheless, IT has gained intense diffusion to developing countries through declining hardware costs and increasing benefits that merits attention as much as developed countries. The aim of this study is to examine the implications of knowledge transfer in a developing country, Turkey, as a paradigm in the knowledge society with a focus on the implementation activities that foster successful installations. We suggest that absorptive capacity is an important characteristic of a firm that explains the success level of such a knowledge transfer.

Keywords: Knowledge Transfer, Absorptive Capacity, ERP.

1 Introduction

In the last few decades, global competition has forced business organizations increase in intensity and complexity. The competing companies seek for crucial capabilities such as fast product development, customized manufacturing and quicker distribution in order to satisfy customer needs and desires.

Information systems, which are the means by which organizations and people, utilizing information technologies, gather, process, store, use and distribute information in business processes, are utilized by most firms to succeed in challenges such as obtaining and sustaining such capabilities.

Organizations provide capability to perform precise, punctual, and efficient operations that lead to effective management and competitive advantage by the contribution of information systems. Those organizations that make use of information systems to enhance how they do business in both domestic and global markets are more likely to obtain significant advantages against their competitors. However, the transfer of such technologies between national cultures might present additional problems of learning and adaptation. From a view point of a developing country, SWOT analysis could be useful as a way to evaluate the advantages and disadvantages of IT applications.

M.D. Lytras et al. (Eds.): WSKS 2009, CCIS 49, pp. 387–394, 2009.

Strengths: Turkey, as a developing country, has certain advantages for creating a robust IT sector and transforming its economy into an information-based economy. The following advantages, some of which are common to other developing countries, can be listed. (1) Turkey has some advantages as a latecomer to the IT scene. No heavy investments have been made in preliminary technologies. IT departments in the country enjoy the advantage of developing more reliable software in a shorter time on more reliable hardware platforms implementing the new technologies transferred. (2) The hardware costs have decreased steeply during the last few decades. Almost all major international computer vendors operate in the country and therefore the computer market is highly competitive and advantageous for the buyer. (3) In the 1980s Turkey invested heavily in its communications infrastructure which makes it possible to build a potential information-based economy where dissemination of information is of prime importance. (4) Turkey has a large young population with the potential of becoming the key resource of a robust IT sector.

Weaknesses: One major issue that the developing countries face with is the operational needs (e.g. developing reliable software). Cultural differences from Western societies also infer difficulties in transferring knowledge to a developing country. The dominant western mentality in this area forces the developing countries to design and produce software reflecting this dominant western mentality. Therefore, the developing countries have to either adopt this mentality or to create their own alternatives in IT applications. Another important problem of the developing countries is the insufficient qualitative and quantitative formation given to the personnel. This issue is gaining importance since the sophistication of IT is increasing sharply.

Opportunities: Turkey is an attractive location which has an important positioning for foreign companies to enter the emerging central Asia through engaging in joint ventures with Turkish partners. Most of the organizations in Turkey which have implemented ERP are the local subsidiaries of multinationals. This is not surprising since they tend to follow their parent companies' global IT strategies which commonly incorporate ERP. They train their personnel appropriately about the importance of IT and introduce them about the advanced IT applications in the other subsidiaries and headquarters. The required support is provided from the sister organizations or headquarters, they are able to attract and retain high-qualified personnel due to their financial strength, and generally they have the managerial and technical infrastructure needed for the ERP.

Threats: A survey conducted in Norway by Karlsen and Gottschalk confirms that IT project success is significantly related to knowledge transfer [6], [13]. If sufficient training is not provided, the project staff learns on the job, and project duration increases. Unfortunately, numerous studies show that training time is often underestimated and training budgets are often set at very low levels [8], [28]. Robey *et. al.* conclude that firms that managed to overcome knowledge barriers and successfully assimilate new processes invested approximately 15–20 percent of their project budgets in formal team training. However, the companies that were less successful spent a maximum of 10 percent of their Project budgets on training [24].

We propose that IT applications and implementations, such as ERP, can be seen as knowledge transfer integrated with practical experience. Consequently, absorptive capacity framework could enhance the development of knowledge society by explaining the success and failure of this knowledge transfer performed through ERP implementations. Depending upon prior studies and related theoretical framework we suggest that due to cultural and organizational constraints in transferring technology to a developing country that is lack of the western mentality, extensive training and absorptive capacity have crucial effects on the success of IT projects.

The rest of the paper proceeds as follows. First of all, we focus on those aspects of absorptive capacity that critically determine the success of an ERP implementation that brings high profits. Secondly, a brief description on ERP is given and its benefits for companies are stated. Afterwards, three companies in Turkey that have implemented ERP software successfully were examined to understand the key points leading to profitable installations. Finally, we integrate the relevant theoretical background with the cases and conclude with the contributions that absorptive capacity brings to successful IT projects.

2 Literature Review

Background on absorptive capacity: Levinthal and Cohen define a firm's absorptive capacity as the ability of a firm to recognize the value of new external information, assimilate it, and apply it to commercial ends. Absorptive capacity varies across firms since it is generated in a variety of ways [5]. For instance, Levinthal and Cohen suggest that prior related knowledge effects the ability to evaluate and utilize outside knowledge to a great extend. Moreover, they argue that absorptive capacity might be built by sending firms' personnel for advanced technical training.

In a recent article, Zahra and George elaborate the absorptive capacity model [31]. According to Zahra and George, absorptive capacity has four dimensions - acquisition, assimilation, transformation, and exploitation - where the first two dimensions form potential absorptive capacity; the latter two dimensions form realized absorptive capacity. They argue that more attention should be given to studying the realized absorptive capacity which emphasizes the firm's capacity to leverage the knowledge that has been previously absorbed. As put forward by Zahra and George, firms can acquire and assimilate knowledge but might not have the capability to transform and exploit the knowledge for profit generation [31].

We follow the idea of Zahra and George and emphasize the four dimensions of absorptive capacity in ERP projects. Acquiring and assimilating capabilities, in the case of an ERP project, might be analogous to buying it from the vendor and training the personnel with the aim of assimilating this new knowledge. The transformation capability could be parallel to the customization or development processes of the ERP package taking into account firms' specific needs. On the other hand, exploitation capability might increase due to management support and employee belief in the project. Based upon this four-dimensional absorptive capacity framework, we suggest that success of an ERP project varies to a great extend with the transformation and exploitation capability of a firm which is also called the realized absorptive capacity.

An overview of ERP system: Enterprise Resource Planning (ERP) system is defined as "A software solution that carries out all the functions of an enterprise to succeed in organizational goals as a collection of integrated subsystems." ERP provides organizations with efficient and consistent data gathering. It also establishes a common data infrastructure for the use of various functions and activities.

Companies implementing the standard ERP Package need the adaptation of the ERP package to suit their requirements. In order to achieve this, technical aspects of the ERP package as much as the functional aspects of the business have to be well understood. The flexibility of ERP Package allows project team to meet the company process requirements with the capabilities of the standard package in the implementation process. ERP vendors don't design the standard packages with the expectations of fulfilling the companies' own business processes. Henceforth, company processes are generally changed to obey package functionalities.

One important issue in ERP implementation is determining the extent to which organizational processes need to be changed to fit into the enterprise system framework and the extent to which the enterprise system must be customized to address existing routines in the organization.

Various aspects of a firm's operations and performance are affected by ERP implementation. Therefore, commitment of top management and employees to the project is highly desirable. During the last decade, numerous failed ERP implementation cases indicated that ERP implementation transforms a company's organizational structure, business process, and employee involvement extensively [1], [29]. The literature on ERP implementation has suggested various critical practices during the implementation of ERP system. For instance, experiencing the inconsistency between the ERP package and firms' current production process and organizational structure would be inevitable. Whether to revise the ERP package or to adapt their current business process to meet the requirement of IT package implementation would be an important decision to be made by the managers. Employees' behavioral dynamics in ERP implementation should not be underestimated near to that of managers and staff members. Top management commitment, employee involvement, and functional area support are suggested to be the foundations to successful ERP implementations by prior studies [3].

3 Case Studies in the Implementation of ERP Systems

Many researchers have been suggesting the case study research methodology for developing conceptual and descriptive understanding of complex phenomena [9], [19], [30]. ERP implementation process, which extends across the whole organization, might be considered as a complex issue taking into consideration the expensive and extensive activities undertaken including planning, justifying, installing and commissioning of the installed system [22].

There are many benefits of the case study method including the ability to observe causality and harvest evidence and logic to build, develop or support theory [18]. When compared to survey research methods, it enables researchers for more valuable follow-up questions to be asked and answered and can result in more extensive findings and insights that are valid, generalizable and rigorous [20].

In this study, we use a longitudinal case study methodology analyzing three diverse companies with the aim of describing the steps in the process and to investigate the complex relationships between these steps. The companies included in this study are manufacturing companies competing in different industries.

3.1 Company A

Since 1999, Company A's single business focus was on consulting services that it rendered relating to drilling. In 2002, due to changes taking place in the business, they decided to diversify and started to export natural stone. Currently, Company A exports natural stone to U.S., Italy, France, Netherlands, U.K. and Canada. They maintain their own warehouses in the U.S. and Italy. They work with about 80 producers in Turkey; they do not have production operations of their own and engage in trading only.

In 2005, the company came to the conclusion that the order processing they were performing using Microsoft Excel and accounting software was proving to be too problematic, and they decided to search for a solution that would fit their organization. So, they decided to purchase a software program. The mistake they made at the time was not to assign a dedicated person to be the responsible contact point for this software in the organization.

For about a year, they tried in vain to get the program to conform to the way the company do business. For everyone, it was much easier to continue to use Excel than to transition to the use of the new software. Then in January 2007, an industrial engineer was hired to help in the planning activities of the company. This new person was assigned to the task of running that project, which, looking at where the company is now, has proven to be a very accurate decision.

The manager of the company states that the flexibility of the ERP system, delivered with accompanying source code, they have been using is of great importance. He also points out that their business processes are very distinct, and accordingly, a number of customizations have been developed that are unique to the company. The packaging process required special attention. Development was performed that made it possible to show the corresponding sales activity for each purchase order. This provided them with the ability for traceability within the system. One of the major benefits for the company has been the dramatic drop in error rates.

3.2 Company B

The company has completed 42 years in the automotive supplier industry, together with 2 manufacturing factories, 10 district retailers, 199 authorized service points and 700 client network, has been serving all over Turkey. The company has become a known and sought brand in more than 30 countries with its foreign dealers and customers, having been the provider for the main automotive manufacturers in Turkey and abroad, increasing the success and development in the international market as much as in Turkey.

52% of the production is exported. Expanding accelerates increasingly. Being the leader in the sector, the company has to increase the efficiency continuously and has to pursuit and plan the production closely. The need of continuous development, made ERP inevitable for the company. The company was ready in all sides before starting the ERP project. The necessary personnel had been provided. The management believed in the project and made the employees believe in it.

The managers emphasize that there are several other package programs in the market that limit the firms in several ways. Henceforth, using such commercial packages might turn into a disadvantage for firms in the process of employing ERP. They also suggest that users should give up with those packages that include narrow templates and habits in order to achieve successful ERP implementations.

3.3 Company C

The company was built as a partnership between a Turkish firm and a French firm in 1994. Essentially, it emerged as a continuation of the Turkish firm that was found in 1984 and was producing plastic and metal components for the automotive supply industry. In 1994 through the direction of the need of the sector, the projects in Europe started to be transferred to Turkey synchronously. Accordingly, with the need of low project investment costs, foreign partnerships gained importance among the main automotive manufacturers. For instance, the first Renault Megane project run in Turkey is such a project which came out in France and was in serial production in Turkey 7-8 month afterwards.

The company started to work with a Turkish ERP vendor in 2005. The company states that an important reason to work with them was that their Turkish partnership was working with them and a serious synergy was generated consequently. They transferred the implementation experience from the Turkish partner to the company. Both of the two companies work in the same business line, producing metal and plastic for the automotive supply industry. Since the business processes were similar, the company was able to integrate the system more easily by providing them with sufficient raining. Another important characteristic of the software preferred was its open-source code that made the company able to develop it according to their needs.

The company believes that the greatest lack would be the support of the company management to the ERP project. If the company management doesn't support or supports the projects with question marks, then it is inevitable for the employees to get de-motivated and difficulties arise in the project.

4 Conclusions

ERP is getting to become widely used day by day for companies in developing countries and they see that if they implement ERP software successfully, everything will be in order. However, low implementation success rates force the companies develop alternatives of implementation. While some companies develop their own ERP systems, some of them prefer standard ERP packages and some of them prefer to have a hybrid system with an open-source code so that they can develop the system according to their specific needs.

In line with our proposition, firms that achieve successful ERP projects in this study have concentrated to a great deal on planning, justification, training, installation and development practices that are the key issues in successful knowledge transfer through an increased absorptive capacity.

Future studies can focus on determining whether the theory aforementioned can be used as a framework for the success of other integrated technology adoptions, or if unlike technologies have some specific success characteristics.

References

1. Akkermans, H., Bogerd, P., Yucesan, E., van Wassenhove, L.N.: The impact of ERP on a supply chain management: Exploratory findings from a European Delphi study. European Journal of Operational Research 146, 284–301 (2003)

2. Barney, J.B.: Firm Resources and Sustained Competitive Advantage. Journal of Management 17, 99–120 (1991)
3. Berinato, S.: A day in the life of Celanese's big ERP Rollup. CIO Magazine, 54–63 (January 15, 2003)
4. Birdoğan, B., Dereli, T., Baykasoğlu, A.: An investigation on the readiness of Turkish companies for enterprise resource management. Journal of Manufacturing Technology Management 15, 50–56 (2004)
5. Cohen, W.M., Levinthal, D.A.: Absorptive Capacity: A new Perspective on Learning and Innovation. Administrative Science Quarterly 35(1), 128–152 (1990)
6. Dixon, N.M.: Common Knowledge. Harvard Business School Press, Boston (2000)
7. Escribano, A., Fosfuri, A., Tribo, J.A.: Managing External Knowledge Flows: The Moderating Role of Absorptive Capacity. Research Policy 38, 96–105 (2009)
8. Fletcher, P.T., Bretschneider, S.I., Marchand, D.A.: Managing information technology: Transforming country governments in the 1990s. Syracuse University School of Information Studies, Syracuse, NY (1992)
9. Flynn, B.B., Sakakibara, S., Schroeder, R.G., Bates, K.A.: Empirical research methods in operations management. Journal of Operations Management 9(2), 250–284 (1990)
10. Hitt, M.A., Dacin, M.T., Levitas, E., Arregle, J., Borza, A.: Partner Selection in Emerging and Developed Market Contexts: Resource-based and Organizational Learning Perspectives. Academy of Management Journal 43(3), 449–467 (2000)
11. Jacobs, F.R., Bendoly, E.: Enterprise Resource Planning: Developments and Directions for Operations Management Research. European Journal of Operational Research 146, 233–240 (2003)
12. Jansen, J.J., Van Den Bosch, F.A.J., Volberda, H.W.: Management Potential and Realized Absorptive Capacity: How do Organizational Antecedents Matters? Academy of Management Journal 48(6), 999–1015 (2005)
13. Karlsen, J.T., Gottschalk, P.: An empirical evaluation of knowledge transfer mechanisms for IT projects. Journal of Computer Information Systems 44(1), 112–119 (2003)
14. Kirlidog, M.: Information Technology Transfer to a Developing Country: Executive Information Systems in Turkey. OCLC Systems & Services 13, 102–123 (1996)
15. Lane, P.J., Lubatkin, M.: Relative Absorptive Capacity and Interorganizational Learning. Strategic Management Journal 19(5), 461–477 (1998)
16. Lane, P.J., Salk, J.E., Lyles, M.A.: Absorptive Capacity, Learning, and Performance in International Joint Ventures. Strategic Management Journal 22(12), 1139–1161 (2001)
17. Li, L., Marcowski, C., Xu, L., Markowski, E.: TQM—A predecessor of ERP implementation. International Journal of Production Economics 115, 569–580 (2008)
18. Maffei, M.J., Meredith, J.R.: Infrastructure and flexible manufacturing technology: theory development. Journal of Operations Management 13(4), 273–298 (1995)
19. McCutcheon, D.M., Meredith, J.R.: Conducting case study research in operations management. Journal of Operations Management 11(3), 239–256 (1993)
20. Meredith, J.R.: Building operations management theory through case and field research. Journal of Operations Management 16, 441–454 (1998)
21. Minbaeva, D., Pedersen, T., Bjorkman, I., Fey, C.F., Park, H.J.: MNC Knowledge Transfer, Subsidiary Absorptive Capacity, and HRM. Journal of International Business Studies 34(6), 586–599 (2003)
22. Muscatello, J.R., Small, M.H., Chen, I.J.: Implementing enterprise resource planning (ERP) systems in small and midsize manufacturing firms. International Journal of Operations & Production Management 23, 850–871 (2003)

23. Plaza, M., Rohlf, K.: Learning and Performance in ERP Implementation Projects: A learning-curve model for analyzing and managing consulting costs. Int. J. Production Economics 115, 72–85 (2008)
24. Robey, D., Boudreau, M.C., Ross, J.: Learning to implement enterprise systems: An exploratory study of the dialectic of change. In: CISR Working Papers. MIT, Cambridge (2000)
25. Sarkis, J., Gunasekaran, A.: Enterprise resource planning: Modeling and Analysis. European Journal of Operational Research 146, 229–232 (2003)
26. Selekler-Gökşen, N., Uysal-Tezölmez, H.: Control and Performance in International Joint ventures in Turkey. European Management Journal 25, 384–394 (2007)
27. Vandaie, R.: The role of organizational knowledge management in successful ERP implementation projects. Knowledge-Based Systems 21, 920–926 (2008)
28. Wu, L., Rocheleau, B.: Formal versus informal end user training in public and private sector organizations. Public Performance and Management Review 24(4), 312-321 (2001)
29. Xu, L.: Editorial. Enterprise Information Systems 1(1), 1–2 (2007)
30. Yin, R.K.: Case Study Research: Design and Methods, 2nd edn. Sage Publications, Newbury Park (1994)
31. Zahra, S.A., George, G.: Absorptive Capacity: A review, Reconceptualization, and Extension. Academy of Management Journal 27(2), 185–203 (2002)

Approaches of E-Records Management in E-State Transformation Process in Turkey

Fahrettin Özdemirci and Özlem (Gökkurt) Bayram

Ankara University, Faculty of Letters,
Department of Information and Records Management,
Atatürk Bulvarı, 06100 Sihhiye, Ankara, Turkey
odemirci@humanity.ankara.edu.tr, ozlembayr@gmail.com

Abstract. In Turkey, e-state transformation has gained speed since 2002 with the reforms of public administration. In this frame, an important part of the targets determined on the activity plan of information society strategy has been realized. However, there are difficulties in structuring of electronic records management system, which is within the e-state components. Considering the OECD indicators, deficiencies in applications which will provide information and service sharing between citizen and institution and inter institutions attract attention. It can be possible that the institutions open their records and services to public on a national portal in the frame of the standards and the laws forming the background of e-record management. Furthermore, there is a need for a national policy which should watch how the adaptation to the mentioned standards and laws realizes and should guide the institutions in this way. This study aims at exposing the principles of national records management and directing the forming of a policy that will give speed and function to the integrated state structuring in the public.

Keywords: e-state management, e-state transformation, national e-records management, Turkey.

1 Introduction

Like the other countries, Turkey has passed various stages on the way to become e-state, as well. In this process dealt with targets of transformation to information society, communication with interaction based on information and records on every field has gradually increased and gained more importance. For this reason, in the first section of the study, introducing the stage in which Turkey is in e-state transformation briefly, e-records management as an e-state component is taken into consideration. Evaluating the existing situation, determining national policy and strategy principles, which will provide working together in e-records management programs for institutions, is considered.

2 E-Transformation Process and E-Transformation Components in Turkey

Turkey is a big country which has a population of 72 millions whose 30 % live in the country. In spite of the difficulty in organization of services and units for increasing

M.D. Lytras et al. (Eds.): WSKS 2009, CCIS 49, pp. 395–403, 2009.

life quality of citizens by modernizing public administration in the structure of information society due to its bigness, Turkey has realized important progresses in e-transformation in a short time.

Development process of e-state in Turkey has followed different stages. In 1970s and 1980s, studies intensified on the automation of back office of some functions such as taxes and population procedures. In 1990s, studies on defining information society and economy based on information were given importance. In 2000-2002, discussions on application of e-state gained importance. The existing stage of e-state application started with beginning work of a reformist government with an agenda having reform of public administration, political and economic stability in November 2002 [14].

E-transformation in the level of the country is provided by Turkey Executive Board of E-Transformation in the presidency of deputy president [16]. An important success level is reached by applying national based projects in basic public services such as health, judgment, education, tax, social security and bid.

In 2003, "E-Transformation of Turkey Project Short Period Action Plan" [9] and then "2005 Action Plan" [7] were put into practice and concluded for the purpose of bringing analyzing approaches and supporting public institutions, private sector and the citizens. Studies on legal and technical substructure were carried out with 2003-2004 Action Plan projects, studies on spreading reflections of activities on daily life were carried out with 2005 Action Plan, and Strategy of Information Society and Action Plan-2006-2010 targeting 2010 has been prepared [5], [6], [16].

"The Strategy of Information Society 2006-2010" [1] forming the road map Turkey will follow in transforming to information society in the Period till 2010 and "The Strategy of Information Society Action Plan 2006-2010" [2] consisting 111 actions under 7 titles were put into practice in 2006 [24]. Actions were mostly started in 2006, finishing dates of the projects were spread. 2006-2007 was planned as substructure preparation and fast acquiring period; 2007-2008 as basic and comprehensive actions providing transformation to information society period; 2009-2010 as application stage and receiving the results period. It is estimated that necessary budget for the projects in the 2006-2007 Action Plan will be much more than the whole public investments [2], [1].

There are targets among the basic targets of The Strategy of Information Society as following:

- Providing modernization in public administration and functioning by considering work processes in the public,
- Presenting services of the public for citizens and business world as more effective, faster, easy available and productive,
- Making the citizens take advantage of the opportunities of information society in maximum level, reducing the numerical gap, increasing the employment and productivity,
- Making the business enterprises use information and communication technologies widely and effectively,
- Establishing competitive milieu which will provide wide, high quality services with appropriate fees, and making the sector of information and communication technologies grow and locating as a global competitive sector.

Actions in the fields of "Citizen Focused Service Transformation" and Social Transformation" started to be realized as prior [2]. These targets determine realization strategies of not all of them but an important part of them. We can summarize the components of e-state as making communication in electronic realm with the citizens and the institutions by using informatics technologies, founding network systems, reducing cost price and sharing information [13]. Taking into consideration the e-transformation indicators until today, it can be said that a part of the determined targets has been realized with various projects; in another part of them projects and substructure works have still been continuing. Due to firstly giving importance to using information and communication technologies, important cost disposition on public services has been aimed. Considering the OECD indicators, it is understood that important dispositions on health spending, financial functioning and control, tax revenue and banking procedures have been gained [14].

Such progresses on structuring of e-state bring organizational change in public administration and accelerate the democratizing processes. Development of e-services is given priority by intensifying on internet as major communication mechanism especially between the public and citizens and business world [8], [1], [14]. Turkey has a technology having a wide band network structures in point of internet substructure. The universities (ULAKNET-National Academic Network) [21] and National Judgment Network Project (UYAP) [23] have their own wide band networks. E-Health [10] enterprise started founding its own network for data receiving and giving [14].

From the point of applications on providing communication between the public, private sector and the citizen, which we give within the components of e-state, having a computer and internet opportunities is a preliminary condition for the citizen but it is not enough in servicing for this purpose. Considering the indicators about the theme in point of using information and communication technologies, the availability level of the citizen to information and technology services is on the 52nd line among 100 countries according to the indicators of 2006-2007 [12]. Workings carried out in the frame of EU also give ideas on using electronic services. EU has determined 20 basic public services for contrasting the developments in using electronic service of the member countries, and made an evaluation effectively measuring the rate of 12 services consisting of sending application form and paying on internet for citizens and 8 services for business world. OECD makes same measurement for Turkey and gives its comparison with EU countries. The rate of online services Turkey presents to the business world is higher than it presents to the citizens. In fact, rate of full-interactive e-service presented to the business world is close to the average of AB-28 (including Norway, Iceland and Switzerland) and higher than AB-10 countries [14].

Table 1. Rates of interactive service presented on internet.

The rate of 20 EU basic public services presented as interactive (%)	Turkey	EU-18	EU-28	EU-10
Services for citizens	25	37	36	33
Services for business world	63	74	67	55

Source: Turkey OECD data and Cap Gemini "Online Availability of Public Services: How is Europe Progressing?"(June 2006).

E-State Gate [4], which was designed for presenting necessary integrated public services for citizens and business world from one gate, was opened in 2008. It is planned that it will present a wider service in future. In e-state gate, online presentations of 22 public services are made in addition to informing the citizens and the business world about various public services. Today, working on widening of interactive services for the public are encouraged, yet the level of the citizen can take advantage of it is lower than the level of the business world.

A certain success has been gained in applications of e-state projects of Turkey that we can regard as national scaled. Among them, there are Ministry of Justice UYAP (National Judgment Network Project), Ministry of Internal Affairs-MERNİS Project, Deed and Cadastral Information System (TAKBİS), Security General Directorate POLNET, Ministry of Finance Tax Offices Automation Project (VEDOP), Government Material Office E-Sale, Ministry of Labor E-Declaration, Undersecretary of Customs Modernization of Customs Administration (GÜMOP), Ministry of Finance Say2000i Project, Prime Ministry Laws Information System [14], [2].

Public institutions also support the e-transformation process with their own projects. One of the important repulsive powers on this theme is encouraging ICT (information and communication technologies) projects in public institutions as an e-state transformation device. Considering the 2002-2008 Investments of Public Information and Communication Technologies, it is determined that 271 ICT projects consisting different services in government institutions have been supported. In the content of this project, there are innovations such as software, hardware, information process network, founding information system, transferring public information accumulation into e-milieu, digitizing, and forming e-archive [20].

It can be said that deficiency in technologic substructure of public has entered in a solution process. However, there are still deficiencies about e-transformation decisions of ICT. Although e-records workings are taken among e-state components and e-records management applications are regarded as a part of e-transformation, deficiencies in integration of the control of information and records on business processes in the public with e-state forming policy attract attention.

3 E-Records Management as an e-Transformation Component in Turkey

E-records applications present service to e-transformation devices in point of giving opportunity to watch business processes and records in office and out of office, increasing total service quality of the institution, reducing serious costs such as paper, toner, work force and time costs, and physical depot costs of archives. For this reason, e-records management is among e-state actions as an important component. E-records management applications make e-state more accessible, more responsible and faster.

The evidence of activities and services of institutions are records. From this perspective, one of the most important components of e-transformation is e-records management. There is a need for e-records management system in institutions in order to provide access to all past and update records for production and usage of e-records. Planning the process of registering and using of current and semi-current records and archive records, organizing and executing should be taken as a managerial activity field as the other administrative works of the institution and e-records management should be seen as a part of e-transformation. In addition to managing e-records for legal, managerial and cultural reasons in the institutions, they should be preserved for archive purpose, as well. In e-state structure, there is a need for electronic records management system that will provide managing all processes from the production of e-records. For this reason, in transition period to e-state, e-records procedures necessitate more important strategic approaches and applications than the records produced in traditional media need. Today, transferring of information accumulations and memories of nations to the next generations is depended on management of e-records.

In institutions, keeping and using every kind of information and record as evidence of works in electronic media are as necessary as executing work processes in electronic media. Realistic approaches on producing, storing and sharing of institutional information and records in electronic media safely in the e-transformation process, are needed. Appropriate systems for the needs of the institutions need to be developed in order to follow and preserve information and records belonging to current and past work processes in the same system. E-records management is an application that consists of all records and information flow, life period of each record and e-archiving processes in an institution.

Following production processes of records produced for inner and outer activities of institutions is both legally worthy for institutional productivity and also for preserving interests and prestige of the institution. In this point, the most important element reflecting the indicator of legal worth is using e-signature. The connection between e-state and e-signature is provided with e-records management. E-signature is a legal process which provides executing of e-archive applications safely and making them gain legal validity. E-signature is the most important thing for solving legal dimension of e-records management applications. E-signature also has an important role for archive program and strategies consisting of national approaches. Existence of a valid, reliable, institutional e-archive cannot be thought without an effective e-signature application which is integrated with e-records management structure.

In Turkey, with Electronic Signature Law [11] numbered 5070, safe electronic signature (e-signature) usage having same legal effect with wet signature according to the concerned secondary law was placed in a legal basis in 2004. Then, with Prime Ministry Circular numbered 2004/21 [17], it was decided that high quality electronic certificate needs of public institutions and foundations will be meted by Public Certification Center founded in body of TÜBİTAK-UEKAE (National Electronic and Cryptology Search Institute) [22]. Moreover, institutions, where private foundations and persons apply for e-certificate, were determined [8].

UYAP executed by the Ministry of Justice in Turkey founded an outstanding information network system in e-signature application. Furthermore, institutions such as Banking Regulation and Supervision Agency (BDDK), Ministry of Labor and Social Security, Turkish State Railways (TCDD) and Turkish Statistical Institute (TÜİK) have made e-certificate applications to Public Certification Center for using in their projects that are in preparation and test stages. This application constitutes the substructure of policies that will provide inter institutions integration.

Considering studies on e-state applications, not enough realizing sharing of information and records in institutions and between institutions in electronic media are among the frequent problems. In e-state transformation, e-records integration means presenting public activities and services in electronic media in a way of finding solutions and working together.

Public institutions and foundations can present service in different settlement places and consist of different departments and units in a hierarchical structure. Institutions also have different working flow systems, methods of records treatments and arrangements according to their works' types and features.

For possibility of information share between institutions, exposing various information, the institutions have and need, in an open and clear way is necessary. After defining the available sources, the institutions should determine a methodology on who can access which information under what conditions.

So, harmony with standard and laws will be real if institutions know that which ways, methods and rules they should follow in information and records share both inner and outer communications and also in work processes. However, technical guides and application procedures prepared by the institutions themselves may prevent developing of standard applications in working together, secrecy, security and data share.

Developing e-records management standards in back office applications for working together has also great importance in fulfilling of determined criteria in the action plan.

In 2005 e-Transformation Action Plan [7], determining minimum standards about working in a harmony of electronic record systems of institutions and effective management of them. In the result of the studies of General Directorate of State Archives which is responsible for the mentioned action, Criteria of Electronic Records Management Systems were prepared and they were published as TSE standard [3] in 2007. With Prime Ministry Circular [18] dated 2008 it was asked the institutions for making their e-records management systems compatible with this standard in two years and obeying this standard in newly found systems was made an obligation.

1,0 demand of Guide of Working Together Basis, which had been put into application in 2005, was nullified with the circular dated 28 February 2009 and numbered 2009/4 [19], and 2,0 demand was put into practice in 2009, and the basis and standards in the guide were made obligations for institutions to obey [8]. The standards which have to be obeyed in e-records management systems for working together are given in 2009 Guide.[1]

Turkey should continue to develop data standards and technical standards with the aim of encouraging data change and working together. However, cooperation with private sector has importance for providing harmony between data standards of public sector and the data standards developed in private sector [14].

Although publishing these standards is a positive step, any mechanism that supervises application of them in institutions has not been formed yet. For this reason, developing national policies and strategies on e-records management including necessary standards and legal arrangements of institutions with a high level structuring should be in public administration of the state.

---

[1] ISO 15489/1-2:2001 Information and documentation -- Records management -- Part 1: General and Part 2: Guidelines was translated into Turkish and published by TSE in 2007. TSE 13298: 2007 Information and Documentation–Electronic records management. ISO 15836:2003 Information and documentation--The Dublin Core metadata element set. ISO 23950:1998 Information and documentation—Information retrieval (Z39.50)--Application service definition and protocol specification. ISO/IEC 11179 Information technology--Metadata registries (MDR)--Part 5: Naming and identification principles.

4 Policy of National Records Management

In developed countries, e-state applications are considered as an crucial component of knowledge society strategy. "*A knowledge society strategy will ensure that all business operators and the public sector have sufficient skills needed in a rapidly developing information society*" [15]. An important part of public institutions has technologic structure that can operate information in electronic media. However, they have not completed the studies that must be done on management of information processed in electronic media.

National policy and strategies should be defined for success of e-records management in e-state applications. A national action plan should be prepared with these definitions. Three basic needs should be met with means of the action plan. First, each institutions adaptation to the given standards should be followed and reported, and the deficiencies on this subject should be met. Beginning of making projects in the institutions which has not records management substructure should be planned. Second, a supervising and guiding mechanism for controlling whether the institutions, which have records management systems, have "criteria of working together" should be formed. Third, new standards and laws needed in meeting the needs that will occur in this process should be determined.

The duty of executing and continuing the action plan consisting of these steps should be given to National Archive. There is a need a national archive law which will strengthen the institutional structure of National Archive in order to execute this duty effectively.

Taking the prior countries in e-transformation into consideration, it is seen that e-records management policy and strategies are determined by national archives of that country and they direct the public institutions. National archives of these countries announce their rules as a guide in order to facilitate supervising of records produced in public institutions in an integrated structure. Information needed to be underline in such a guide as following:

- There should be standards and laws, which provide public institutions with forming institutional records management systems and information share between institutions.
- There should be directing information on production for inner procedures providing supervising of applications on the existing records management and archive systems of institutions.
- The necessity of planning educational activities on forming institutional culture for the determined procedure should be stressed.
- The necessity of defining and employing qualifications of human power for e-records management of institutions.
- The necessity of organizing of e-records management and archive program and policies as a part of institutional records management should be mentioned.

5 Conclusion and Suggestions

Turkey has taken important steps in widening ICT projects in all sectors especially in public institutions since studies on e-transformation process began. The reports of OECD shows that ICT services and productions has more interactive usage fields in business world than citizens and the developments are in an unbalanced distribution in the existing situation. For this reason, Turkey has given importance to improving services for citizens and increasing the usage.

In Turkey, a national portal, which provides access to public information and services in one gate, was opened in order to overtake the difficulty of presenting information and services to citizens from many different web sites. This development is a turning point in transition to "integrated state" in which institutions share business and information processes in e-transformation process. Such an enterprise also will provide opportunity for functioning of structure providing information and records share within and between institutions and between institutions and citizens. For Turkey, completing e-state maturing stage will be realized by forming information management systems including e-records management approaches.

The biggest gap in electronic records management applications in our country is some institutions' lack of information and conscious in forming records management systems that are compatible with laws and standards. In this point, adopting institutional culture should be taken into consideration as prior step. It is very important to make senior administrators adopt that the matter has a process on common policy and applications needing scientific and administrative approaches beyond a technological necessity. Therefore, widening of e-records management applications in national level can be provided. It will be possible to integrate records management with e-state applications with changes in laws and standards in the frame of these policies.

It is a fact that important problems will occur in records share and access between institutions if there is no standards, technical guides and legal arrangements that provide integration. For this reason, central management should prevent institutions from developing separate and repeated e-records management applications by preparing standards, guides and necessary laws for cooperation in using common database and services and developing electronic service. Central management should give authority and duty to National Archive as a pioneer and authoritative institution for developing and practicing e-records management solutions in a common way. National Archive should take a role supervising and directing the compatibility to the principles of working together and data share in projects of transferring institutional records into e-media. In the frame of the concerned laws and standards, e-records management program should be formed, it should be structured within managerial functions, and it should be defined as a part of national records management policy.

The purpose of national records management is to provide working together between all institutions, firstly public institutions, presenting service to public in electronic media and to provide executing a guiding structure by supervising the adaptation to technical standards formed with authority, responsibility, method and criteria of institutions in this frame.

An e-state transformation integrated with national e-records management policy can provide forming a public administrative structure having an effective, transparent and simplified business processes, and present a higher quality, and faster public service to business world.

References

1. Bilgi Toplumu Stratejisi 2006-2010. Devlet Planlama Teşkilatı Müsteşarlığı, Ankara (2006)
2. Bilgi Toplumu Stratejisi Eylem Planı 2006-2010. Devlet Planlama Teşkilatı Müsteşarlığı, Ankara (2006)

3. Bilgi ve Dokümantasyon – Elektronik Belge Yönetimi Standardı (TSE 13298). Türk Standartlar Enstitüsü, Ankara (2007)
4. E-Devlet Kapısı (Republic of Turkey Government Gate), https://www.turkiye.gov.tr
5. E-Dönüşüm Türkiye Projesi 2003-2004 KDEP Uygulama Sonuçları ve 2005 Eylem Planı. Devlet Planlama Teşkilatı Müsteşarlığı, Ankara (2005)
6. E-Dönüşüm Türkiye Projesi 2005 Eylem Planı Sonuç Raporu. Devlet Planlama Teşkilatı Müsteşarlığı, Ankara (2006)
7. E-Dönüşüm Türkiye Projesi 2005 Eylem Planı. Devlet Planlama Teşkilatı Müsteşarlığı, Bilgi Toplumu Dairesi Başkanlığı, Ankara (2005), http://212.175.33.22/2005EP/2005EylemPlani.pdf
8. E-Dönüşüm Türkiye Projesi Birlikte Çalışabilirlik Esasları Rehberi: Sürüm 2.0. Devlet Planlama Teşkilatı Müsteşarlığı, Bilgi Toplumu Dairesi Başkanlığı, Ankara (2009), http://www.bilgitoplumu.gov.tr/yayin/eDTrBirlikteCalisabilirlikv2.pdf
9. E-Dönüşüm Türkiye Projesi Kısa Dönem Eylem Planı 2003 – 2004. Devlet Planlama Teşkilatı Müsteşarlığı, Bilgi Toplumu Dairesi Başkanlığı, Ankara (2003), http://www.bilgitoplumu.gov.tr/kdep.asp
10. E-Sağlık Çalışmaları, http://sbu.saglik.gov.tr/esaglik
11. Elektronik İmza Kanunu: K.No: 5070. T.C. Resmi Gazete (Republic of Turkey Official Gazette, 25355 (23.01.2004), http://rega.basbakanlik.gov.tr
12. Dutta, S., Mia, I. (eds.): The Global Information Technology Report 2007-2008. Palgrave Macmillan, Hampshire (2008)
13. Kuran, H.: Türkiye İçin E-devlet Modeli Analiz ve Model Önerisi. Bilgi Üniversitesi, İstanbul (2005)
14. OECD E-devlet Çalışmaları Türkiye. Devlet Planlama Teşkilatı Müsteşarlığı, Ankara (2007)
15. Sammour, G., Schreurs, G., Zoubi, A.Y., Vanhoof, K.: Knowledge Management and E-Learning in Professional Development. In: Lytras, M.D., et al. (eds.) Proceedings of WSKS 2008. CCIS, vol. 19, pp. 178–183 (2008)
16. Başbakanlık, T.C.: 2003/48 Sayılı Genelge. T.C. Resmi Gazete (Republic of Turkey Official Gazette, 25306 (4/12/2003), http://rega.basbakanlik.gov.tr
17. Başbakanlık, T.C.: 2004/21 Sayılı Genelge. T.C. Resmi Gazete (Republic of Turkey Official Gazette, 25575 (06/09/2004), http://rega.basbakanlik.gov.tr
18. Başbakanlık, T.C.: 2008/16 Sayılı Genelge. T.C. Resmi Gazete (Republic of Turkey Official Gazette, 26938 (16/07/2008), http://rega.basbakanlik.gov.tr
19. Başbakanlık, T.C.: 2009/4 Sayılı Genelge. T.C. Resmi Gazete (Republic of Turkey Official Gazette, 27155 (28/02/ 2009)., http://rega.basbakanlik.gov.tr
20. 2008 Yılı Kamu Bilgi ve İletişim Teknolojileri Yatırımları. Devlet Planlama Teşkilatı Müsteşarlığı, Bilgi Toplumu Dairesi Başkanlığı (2008), http://www.bilgitoplumu.gov.tr/yatirim/2008KamuBITYatirimlari.pdf
21. Ulusal Akademik Ağ (ULAKNET), http://www.ulakbim.gov.tr/ulaknet
22. Ulusal Elektronik ve Kriptoloji Araştırma Enstitüsü (UEKAE), http://www.uekae.tubitak.gov.tr/
23. Ulusal Yargı Ağı (UYAP), http://www.uyap.gov.tr
24. Yüksek Planlama Kurulu 2006/38 sayılı Kararı. T.C. Resmi Gazete (Republic of Turkey Official Gazette, 26242 (28/07/2006), http://rega.basbakanlik.gov.tr

Supplier Networks and the Importance of Information Technology: Outlook on the European Automotive Industry

Maria José Alvarez Gil, Borbala Kulcsar, and Dilan Aksoy

Department of Business Administration, Universidad Carlos III de Madrid, C/Madrid, 126 28903 Getafe (Madrid) España

maria.alvarez@uc3m.es, bkulcsar@emp.uc3m.es, daksoy@emp.uc3m.es

Abstract. The trends in the automotive industry changed radically from the beginning of the 80s. Increasing competition, new systems and developments compelled the companies to re-evaluate and re-design their investments and processes, by extending their networks to other parts of the world in order to gain more market. This trend could be observed first in the Western-European countries and later in Eastern-Europe. With entering new areas the companies had to face with several difficulties coming inter alia from the decisions of supplier network and information system implementation. In our study we analyze the strategic decisions of major carmaker companies entering the Eastern-European market. Our research includes two case studies of the Hungarian automotive sector.

Keywords: supply chain management, automotive industry, FDI, case study.

1 Introduction

With the downfall of the socialist era in Eastern-Europe, the countries faced a complex situation of multiple possibilities and threats. The transition process helped effectively the adaptation to developed countries, although it required radical changes, fast learning skills and adequate strategic and organizational decisions. As knowledge transfer played an important role in emerging countries, it highly contributes to the view of the knowledge society.

This transition period was studied by many scholars before, as well as the effect of foreign direct investment (FDI)[1] in the emerging countries and their beneficent/ noxious impact on the economies. An interesting detail was the focus on multinational companies (MNE) as the major source of foreign capital inflow and the adapting government decisions.

In our study we will analyze the appearance of multinational carmaker companies to Eastern Europe, especially focusing on the situation of Hungary and the restructuring of the automotive industry in the country. We take into account in such a

[1] FDI: Foreign Direct Investment.

M.D. Lytras et al. (Eds.): WSKS 2009, CCIS 49, pp. 404–410, 2009.

context the R&D investment decisions of the country, as strategic actions with the combination of institutional, financial, and organizational restructuring decisions.

The number of R&D collaborations and networks show big differences by sectors, regions or countries. Hungary became a relatively large share of FDI in the CEE[2] region. With this push of the economic activity, the innovation system was modernized.

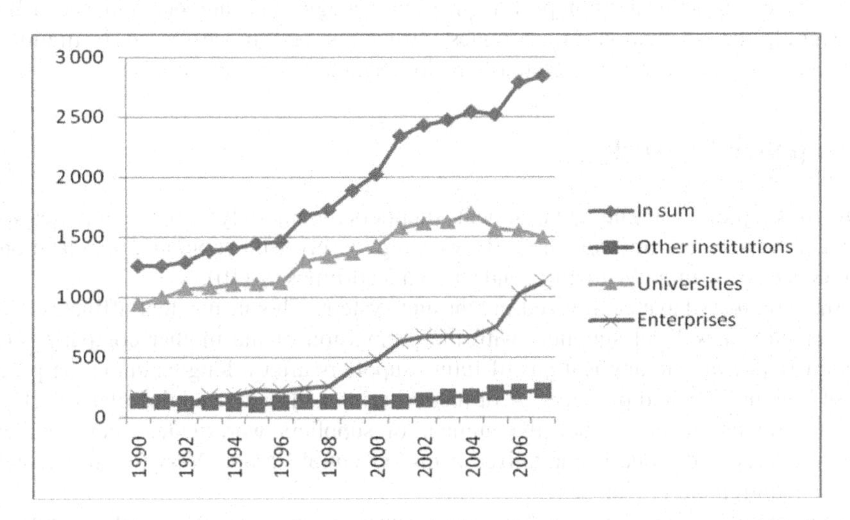

Fig. 1. Number of R&D locations in Hungary (data obtained from the Hungarian National Statistical Institution)

2 Theoretical Framework

As suggested by (11) the transition economies have contribution to main theories such as, the transaction cost theory (TCT), agency theory (AT) resource based theory (RBT) and institutional theory (IT). We argue that in our context RBT and IT are the main theoretical cornerstones.

RBT places the main focus on idiosyncratic resources and capabilities as key drivers of firm performance. Resources constitute a basis for competitive advantage in a transition context. In a highly idiosyncratic environment, context-specific resources such as business networks and process-related capabilities such as strategic flexibility may be important. The context thus influences the way firms manage their resources. Partner selection and organizational learning are also highly important factors. The main challenge lies in identifying the resources and change processes that create value in the specific context. The complementarities of resources are believed to be crucial for the success of different strategically partnerships, such as alliances or joint ventures (JVs).

With the creation of alliances and JVs, learning, knowledge transfer or technological transfer is usually the first objectives to achieve of the host companies. The concepts of absorptive capacity and relational embeddedness may enhance our understanding of

[2] CEE: Central and Eastern Europe.

what resources enable organizations to receive, adopt, and apply external knowledge (11). This contributes to the importance of knowledge and learning in the local companies, which is in turn an important aspect for the knowledge society.

The role of institutions (IT) is particularly important in acquisitions. In CEE countries, the institutions surrounding privatization set the context for foreign acquisitions, which have a direct bearing on post-acquisition strategies (11) and performance (17).

Overall, the adaptation of strategies, structures and processes to institutional idiosyncrasies has been recognized as a major challenge for managers (11).

3 Supplier Networks

In today's rapidly changing environment, suppliers are not only a source of materials; they can be a source of innovation. By working closely with suppliers, organizations can reduce costs, improve quality, and shorten lead-times (9)(10).

Big carmakers have centralized purchasing system. This means that affiliates may only employ new local suppliers with the permission of the mother company. The headquarters evaluate applications of future suppliers after a long examination process of both product and producer. A further criterion is the size and the related lack of capital. Big carmakers reduce the number of suppliers worldwide. They contract therefore large and properly capitalized firms in Central Europe. Very few firms qualify this requirement in Hungary.

The conditions under which local parts suppliers become integrated (or not) in cross European production networks largely depends upon the internationalization strategies and the related intentions for local outsourcing.

As we know Hungary, Slovenia and Slovakia, are 'the second tier countries'. They have become integrated in networks of component supply. The assembly operations they contain generally share a lower local content, whereas their trade orientation is much more export oriented. The country that has been most successful in these terms is Hungary regarding to build up own car assembly capacity from scratch by attracting a non-European investors. Therefore, the Hungarian example deserves further analysis.

Suzuki has the widest local supplier network in the country. General Motors however it follows a different strategy, has a high impact on the Hungarian automotive sector. We will further analyze the position of these two companies in the country.

4 The Role of Information Technology – ERP

Today, information and communication technologies play a key role in Supply Chain Management, which aims to integrate all key business processes throughout the entire supply chain (12). Some of these processes can be supported by ERP (enterprise resource planning) software.

Software vendors like i2, Oracle, or SAP are the major developers of supply chain management software. Considering the high costs involved in the implementation of Supply Chain Management Software is important to analyze and evaluate the usage of such software solutions (2).

The business world's embrace of enterprise systems may in fact be the most important development in the corporate use of information technology in the 1990s. When used appropriately, ERP software integrates information used by the accounting, manufacturing, distribution, and human resources departments into a seamless computing system. A successful ERP can be the backbone of business intelligence for an organization, giving management a unified view of its processes. Companies implement such systems to obtain a common IT platform.

A survey study about the usage of ERP system in the automotive industry was conducted by (2) where the authors asked 1000 car manufacturers, suppliers and distributors about ERP. They found that 48% of the contestants has implemented or plan to implement ERP solutions. From these contestants the majority used a program developed by SAP. The results of this survey indicated that information technology and concerning systems are becoming more important. However, there are still some serious barriers to overcome in their diffusion.

Still, regarding the ERP and the implementation of information systems we need to analyze further aspects of this issue. We argue that such technologies will be key drivers to success in the long run, by implementing it on a wider range of affiliates in car companies. We would like to study how companies in the car industry can implement successfully the ERP system.

5 Research Questions and Methodology

RQ1: Is the supplier network of the major car companies in the country efficient? What kind of strategies (relationships) can be observed by the companies with local suppliers?

As mentioned before, different companies follow different 'best practice' strategies with suppliers. Also to build up a local supplier chain is a burdensome process for companies. We will analyze and compare in our case studies the two major car investors in Hungary and their supplier strategies.

RQ2: *Does information technology contribute to the performance of the companies?*

As we mentioned before the importance of information technologies such as ERP will be a key factor on the long run when we talk about cooperation between different plants, departments and affiliates in Europe. Therefore we suggest studying further this issue regarding the automotive sector.

In the methodological part we will analyze two case studies, by reviewing the two major contributors to the Hungarian automotive industry.

Hungarian Suzuki

Magyar Suzuki, a Japanese-Hungarian joint venture located in Esztergom, commenced commercial production of compact cars in October 1992. As the company aimed to increase its market share in Europe it introduced 10 new models until 2007. In Hungary, to add these new cars to the current product lines, some $100 million had been invested at the Esztergom plant (6)(7).

As Hungary only joined the EU in May 2004, Hungarian Suzuki had to reach 60 per cent EU content in order to export its cars to EU markets. Moreover, it bought certain parts from its local suppliers – initially it only produced 30,000-40,000 cars a year – but followed a single-sourcing strategy. Therefore, it had very strong incentives to build a local supply base in the beginning. With Hungary's accession to the EU, however, it has fundamentally changed, and accordingly Hungarian Suzuki's supplier strategy has been revised (6). Together with its Japanese suppliers, it had conducted a thorough technological and financial audit, covering literally every single aspect of doing business from purchasing inputs through production methods and machinery, to accounting, sales and management. Then joint efforts had also been made to improve the selected supplier's technical level and economic performance.

Hungarian Suzuki has provided its Hungarian and other Central European suppliers with various sorts of technological and managerial knowledge (know-how) on purpose, as it did need a strong local supply base to reach the required EU content. To achieve this goal, close co-operation with the selected suppliers was required, previously accustomed to the standards and norms of the planned economy, in order to 'drive' them into a different system, namely market economy.

Once the EU content had been achieved, Hungarian Suzuki didn't had strong incentives to continue this supplier strategy. Since then, it has made far less significant efforts to develop its local supply base. The current assistance, however, is still important.

GM/Opel

Opel Hungary Vehicle Manufacturing Ltd. opened the Hungarian car assembly plant and an engine factory in a customs-free zone at Szentgotthárd, close to the Austrian border, in 1992. GM Opel's investment Hungary had totaled 600 million Euros by 2003 (18).

Opel Hungary, as opposed to Hungarian Suzuki, has never been 'forced' to reach a certain level of local content, as the cars assembled in Szentgotthárd were sold in the domestic market, and thus EU rules controlling access to the EU markets did not matter. Most of its local suppliers were Hungarian subsidiaries of its long-established Western European partners (5)(18). This second distinctive characteristics of Opel Hungary, compared to Hungarian Suzuki is, that it prefers working with its well-known suppliers, and so to set up their operations in Hungary, either investing in green-field[3] investments or by takeover of domestic firms, and transferring their technologies as well as managerial techniques to upgrade their skills and competences.

Tacit knowledge, gained at Opel, is transferred to other companies in various ways. One form is when employees leave to suppliers for a higher position. Firstly, it is a loss from the point of view of Opel Hungary; but the resources used to train these employees are not wasted. As finding new suppliers has become an important task recently, it is obviously easier to work with suppliers where former Opel employees are in high positions.

These sources and tactics of the foreign investors are worth to mention and to study in a deeper context as we can observe how environmental circumstances and strategic decisions contribute to the evolution of the domestic market and organizational structures.

---

[3] Green-field investment: to invest in an area or country that has not such activities before.

6 Summary

Foreign firms' strategies in emerging countries and dynamics of national innovation systems are with great importance in understanding the transition to capital market structure. In this period the focus is on attracting foreign investors and foreign capital to the area. For this reason it is more fruitful to create an attractive, favorable environment for R&D and innovation by maintaining a sound, well-performing higher education and research system, providing the necessary physical and institutional infrastructure, facilitating industry-academy co-operation and other forms of networking.

Automotive investment activities across borders have significantly intensified in the last twenty years in an attempt to cut costs via re-location of production, and to get closer to the ultimate customers in emerging markets. These intensified investment activities have had crucial bearings on the Hungarian automotive industry.

Supply chain management is a highly complex task in companies. In major car producer companies it is a crucial aspect to success. Moreover in multinational, worldwide known firms the decision of expanding the supplier network is with critical importance. Regarding the external factors, the cultural differences and ability to production, firms may face difficulties. To obtain a local supply network in newfound environment is a challenging task for investors. Strategic decisions on the creation of such a network differ among companies due to their existing practices and supplier chains. Therefore it is important to study this question and place a major importance on existing trends and on future possibilities as well.

References

1. Bartlett, D., Seleny, A.: The Political Enforcement of Liberalism: Bargaining, Institutions, and Auto Multinationals in Hungary. International Studies Quarterly 42(2), 319–338 (1998)
2. Buxmann, P., von Ahsen, A., Díaz, L.M., Wolf, K.: Usage and evaluation of Supply Chain Management Software – results of an empirical study in the European automotive industry. Info Systems J. 14, 295–309 (2004)
3. Demeter, K., Gelei, A., Jenei, I.: The effect of strategy on supply chain configuration and management practices on the basis of two supply chains in the Hungarian automotive industry. International Journal of Production Economics 104, 555–570 (2006)
4. Gil, M.J.A., de la Fé, P.G.: Strategic alliances, organizational learning and new product development: the cases of Rover and Seat. R&D Management 29, 4 (1999)
5. Havas, A.: Local, regional and global production networks: re-integration of the Hungarian automotive industry. In: von Hirschhausen, C., Bitzer, J. (eds.) The Globalization of Industry and Innovation in Eastern Europe - From Post-socialist Restructuring to International Competitiveness, pp. 95–127. Edward Elgar, Cheltenham (2000a)
6. Havas, A.: Changing Patterns of Inter- and Intra-Regional Division of Labour: Central Europe's long and winding road. In: Humphrey, J., Lecler, Y., Sergio Salerno, M. (eds.) Global Strategies and Local Realities: The Auto Industry in Emerging Markets, pp. 234–262. Macmillan, Basingstoke (2000b)
7. Havas, A.: The Interplay between Innovation and Production Systems at Various Levels: The case of the Hungarian automotive industry, Working paper (2003)

8. Iwasaki, I.: Corporate Governance in Transition Economies Part II: The Case of Hungary. IER Discussion Paper Series (B) (2005)
9. Laseter, T.: Balanced Sourcing: Cooperation and Competition in Supplier Relationships. Jossey-Bass Publishers, San Francisco (1998)
10. Lewis, J.D.: The Connected Corporation. The Free Press, New York (1995)
11. Meyer, K.E.: International business research on transition economies. In: Rugman, A., Brewer, T. (eds.) The Oxford Handbook of International Business, pp. 716–759. Oxford University Press, Oxford (2001)
12. Oliver, R.K., Webber, M.D.: Supply-chain management: logistics catches up with strategy, Outlook, Booz, Allen and Hamilton Inc. In: Christopher, M. (ed.) reprinted in Logistics: The Strategic Issues, pp. 63–75. Chapman Hall, London (1982)
13. Worrall, D., Donnelly, T., Morris, D.: Industrial Restructuring: The Role of FDI, Joint Ventures. Acquisitions and Technology Transfer in Central Europe's Automotive Industry (2003)
14. van Tulder, R., Ruigrok, W.: European Cross-National Production Networks in the Auto Industry: Eastern Europe as the Low End of European Car Complex. Erasmus University of St. Gallen, Berkeley Roundtable on the International Economy (University of California, Berkeley) (1998)
15. Lung, Y.: The Changing Geography of the European Automobile System. Synthesis in CoCKEAS workpackage # 4 (2002)
16. Hudson, R.: Changing industrial production systems and regional development in the New Europe. Transactions of the Institute of British Geographers 27(3), 262–281 (2002)
17. Galgóczi, B., Tóth, A.: The development of the automobile industry in Hungary and the changing patterns of employee interest representation. SEER SouthEast Europe Review for Labour and Social Affairs (2002)
18. Somai, M.: The Hungarian automotive industry, Working paper, Hungarian National Bank (2004)
19. Motwani, J., Subramanian, R., Gopalakrishna, P.: Critical factors for successful ERP implementation: Exploratory findings from four case studies. Computers in Industry 56, 529–544 (2005)
20. Uhlenbruck, K., De Castro, J.: Foreign acquisitions in Central and Eastern Europe: Outcomes of privatization in transitional economies. Academy of Management Journal 43, 381–402 (2000)

Answer or Publish – Energizing Online Democracy

Miklós Antal[1] and Dániel Mikecz[2]

[1] Budapest University of Technology and Economics, Department of Environmental Economics, Sztoczek u. 2, Building St, Room 420, 1111 Budapest, Hungary
[2] Budapest University of Technology and Economics, BME-UNESCO Information Society Research Institute (ITTK), Sztoczek u. 2, Building St, Room 108, 1111 Budapest, Hungary
antalmi@gmail.com, mikecz.daniel@ittk.hu

Abstract. Enhanced communication between citizens and decision makers furthering participation in public decision making is essential to ease today's democratic deficit. However, it is difficult to sort out the most important public inputs from a large number of comments and questions. We propose an online solution to the selection problem by utilizing the general publicity of the internet. In the envisioned practice, decision makers are obliged either to answer citizens' questions or initiatives or to publish the letter received on a publicly accessible web page. The list of unaddressed questions would mean a motivation to consider public inputs without putting unnecessary burdens on decision makers – due to the reliance on the public, their workload would converge to the societal optimum. The proposed method is analyzed in the course of the existing Hungarian e-practices. The idea is found valuable as a restriction for representatives and a relief for some other officials.

Keywords: electronic democracy, public deliberation, inclusive governance, e-transparency, communal decision making.

1 Introduction

Contemporary democracies face enormous challenges from several aspects [1], [2]. Basic democratic expectations like inclusion, deliberation, transparency of decision making, accountability, and legitimacy are hardly met in a variety of countries amid massive socio-economic changes. Symptoms of a significant democratic deficit can be observed at different levels of the emerging multi level decision making systems [3].

First, it is never easy to appropriately include people in decision making processes in large groups. Direct forms of democracy are virtually unviable in large political units [4]. To preserve democracy, the emergence of some sort of representation seems to be inevitable. In order to guarantee that representatives stand for the people's will, communication between voters and representatives is necessary between elections. Though, it is usually impossible for individuals to affect public decisions through direct communication with their representatives, since personal views are typically ignored. The incapability of individuals to take meaningful parts in public discussions and the present decline of civil associations [5] resulting in a deeply individualized society render political self-expression difficult. In the lack of platforms and tools to

M.D. Lytras et al. (Eds.): WSKS 2009, CCIS 49, pp. 411–419, 2009.

influence decisions, mass societies may take shape, where people are exposed to manipulation and societal outcomes can diverge from the public interest [6]. Consequently, each step favoring inclusion supports a stronger democratic system.

A second persistent question in representative democracies is how public deliberation can be facilitated to move towards an optimal tradeoff between consensus decision making and representative democracy [7]. The question is especially urgent today, when knowledge and information are absolutely critical resources: production and application of knowledge and access to relevant information are structuring forces in the emerging knowledge societies. It is argued by many, that the process of deliberation and reliance on diverse sources of information can boost governmental performance [8]. The loss of reliable first hand information is increasingly inadmissible in the age of indirect information flows. Elections, where poorly informed masses vote to elect representatives can no longer substitute participation in deliberative processes, where people can contribute to much simpler decisions and make direct use of their personal experiences.

However, difficulties to organize, evaluate and use results from deliberations hinder wide-spread applications of such solutions. It is hard to set convenient rules for the deliberation process to facilitate practical reasoning [9]. Thus, further efforts are needed to assist knowledge based rational reasoning to reach morally correct decisions. Another cumbersome and controversial task is to select the relevant information during and after the deliberation process. This selection largely determines outcomes, so it is decisive to elaborate methods fitting in the democratic conception.

A third concern is to maintain or create transparency in decision making systems and ensure accountability of decision makers. Globalization poses significant problems to traditional democratic mechanisms. It is harder and harder to identify actors behind decisions affecting the public. The importance of corporations, lobby organizations, think tanks and other elusive players in the political battlefield is growing. The rise of transnational companies, the increasing interconnectedness of financial systems and in broader terms the loss of governments' economic and political sovereignty as a consequence transform the role of decision making bodies. Hierarchical methods are inefficient in today's increasingly complex and dynamic environments. Besides command and control governance, potentials for deregulation are also limited due to market failure [10]. Governance is more and more only possible within policy networks, where mutual interdependencies shift the administrators' emphasis towards coordination between actors. Traditional representative systems together with the traditional sources of transparency and accountability are thus declining. People feel alienated from politics, trust in different organizations including all significant actors in public decision making (representatives, parties, companies, international organizations etc.) falls [11]. Without the slightest intention to decide whether network governance and democracy were compatible [12], we can state that these changes necessitate the development of novel forms of publicity to uphold accountability and transparency. New methods of communication between decision makers and citizens are required.

Fourth, legitimacy of decisions is crucial to avoid the accumulation of tensions in society. Affected by all the previous factors (inclusion and participation, deliberation, transparency, and accountability) plus the social, economic, and political outcomes and various other properties of the given democratic system [13], the exact origins of legitimacy can hardly be revealed. Although, there is one undeniable fact: there are serious legitimacy problems in our restructuring societies [1], [14]. No matter in which direction we are trying to find the way out of the current crisis, we will have to rely on a broad consensus to reclaim public support for societal decisions. Voting in itself in today's complex systems is not sufficient any more to rebuild trust and ensure the enforcement of public will.

Not surprisingly, a revival of regional politics, non-politicized participation in decision making, and increased reliance on electronic communication systems (a regime change in information systems [15]) in these public activities are deemed to be crucial to move towards a more responsive and deliberative democracy enjoying enhanced legitimacy [16].

In the next section we demonstrate a new, widely applicable internet-based method that could expand the political action space of citizens and facilitate better access to relevant information about public issues. In Section 3, we study the potential role of the envisaged method in the course of Hungarian e-practices to see its contribution to knowledge society through positive effects on the quality of public discussions. Section 4 investigates practical aspects through concrete examples and concluding remarks are given in Section 5.

2 The Vision in Brief

The idea outlined in the present paper addresses a major problem in the communication of decision makers and engaged citizens. Decision makers usually receive scores of emails every day, many of which – with or without a reason – are not considered important enough to be answered. Consequently, a number of potentially beneficial civil initiatives and insightful questions are dismissed. At the same time, representatives receive numerous emails that are not to be answered according to the general public opinion. Due to a shortage of time, we can not expect decision makers to consider all public inputs. Thus, the question arises, whether it is possible to separate important and less important letters. The present article strives to answer this question by utilizing the freely available general publicity of the internet, opening the door for an efficient mechanism offering better chances for public deliberations.

The essence of the idea is very simple: *decision makers should be obliged either to answer citizens' questions or initiatives, or they should publish the letter received on a publicly accessible web page.*

Advantages of this solution are clear. Important questions will not likely be forgotten, because anyone can browse the list of unanswered letters and see which problems a given decision maker did not address. Political opponents would be inspired to search for socially important topics in the list of unaddressed issues. At a minimum, we can expect better outcomes than today when there are essentially no incentives to reply to citizens. On the other hand, the workload of decision makers would not unnecessarily grow – actually, it would converge to the societal optimum.

Fortunately, this solution does not include hidden problems: in the case of inadequate answers it is possible to ask again, while senseless, pugnacious questions can be sent to the web page instead of bothering with answers. Pre-edited letters sent by members of pressure groups can be answered by a single relevant answer, though; one answer would presumably be needed.

Moreover, there are no technical problems of the implementation: storage capacities are virtually infinite in terms of such documents. Obviously, perfect functioning of this system is hindered by uneven access to computer technology. Though, uneven distribution of resources and powers means no motivation to limit potentially beneficial behavior of masses of people. Nevertheless, access to modern communication technology should be improved to promote equal chances.

A further advantage of the idea is that it can be extended to numerous other fields, where sorting important and unimportant information causes troubles. Certain authorities struggle, because they are obliged to investigate each and every reported case. For them, it is a great opportunity to alleviate bureaucratic burdens and create time for important cases. Presumably, instead of political opponents, professionals and civil actors would check unaddressed cases of different authorities. Traditional methods can also be combined with the present idea (stipulating conditions when traditional solutions should be used), giving flexibility to formerly rigid social structures.

All in all, when serving the public, whom else should we rely on, if not the public?

3 The Vision in the Course of Hungarian E-Practices

After the concise description of the idea, let us see how the proposed solution could contribute to the success of electronic democracy in today's knowledge-based societies. In line with the problems outlined in the introductory part, we focus on the key aspects of inclusion, participation, deliberation and knowledge transfer, transparency, accountability, and legitimacy. Obviously, individual e-practices deliver only partial solutions to the problems of democratic functions; their role is to alleviate the democratic deficit, to complement and bolster up offline institutions. Thus, we study how the proposed idea fits into the course of Hungarian e-practices.

Here we distinguish two approaches regarding the usage of online tools in the democratic process. E-government is a management-oriented approach with practical goals. Economic efficiency, cost reduction, simplification of bureaucratic processes, and a smooth functioning of the governance system are at the center of the idea. E-democracy is much more about basic democratic principles with a special emphasis on equal chances. The permanent and diverse interaction between citizens makes grass-roots e-democratic applications the breeding ground for new ideas. Hence these opportunities of horizontal exchange of information are catalysts of the knowledge society.

The majority of state-run practices in Hungary belong to the domain of e-governance, and only a few, mostly self-organizing solutions have tighter links to the 'harder' e-democratic functions. European standards are of profound importance on Hungarian administrations giving the reason for this significant bias towards e-governance as opposed to e-democracy. As we shall see in the followings, the proposed idea could help to mitigate this bias.

Apart from the strictly administrative electronic mechanisms (e.g. online taxation) we can distinguish three types of governmental or municipal e-government applications. The first type is purely about informing citizens. According to the Hungarian law (2005.XC), all public institutions including the parliament, government agencies, municipalities, courts etc. are required to publish information about their organization's structure, management, budget and activities. Furthermore, politicians, high-ranking officials and other politically exposed persons also have to publish data about their financial situation. These legislations have undoubtedly positive effects on transparency and accountability. However, if there are ambiguities about the published data or questions arise concerning the published information people do not really have the chance to contact the respective parties and to demand more information. The potential application of our proposal can assist involvement in such cases. Not only does it foster participation in the democratic process, but also facilitates a more effective fight against the abuse of power or authority, and thus steers processes closer to public interests. Arguably, electronic informative systems backed up by our visionary idea could partly restructure top-down communication. By supporting more effective vertical knowledge transfer, it would enhance transparency and accountability, contributing to legitimacy as well.

The second electronic governance tool already present in Hungary requires active participation from citizens and representatives: online consultations with elected officials are deemed increasingly important to establish direct contacts between voters and representatives with the primary aim to broaden the horizon of local decision making processes. To date, online consultations are far from general in Hungary, but some officials (especially mayors) discuss different topics with their voters online. Though, consultations are accidental and weakly formalized, there are no constrains, consequences are not traceable. In these regards we expect improvement from the envisioned practice. In the lack of obligations to respond, publicity could serve as the only legitimate tool to evoke answers from decision makers. Politicians' reputation – one of their main assets – could be threatened if they do not consider reasonable public inputs. Besides transparency, inclusion would also be assisted by the idea: aware of the growing importance of their opinions, more people would be inspired to participate in discussions. Moreover, if the list of unanswered questions is browsable by topics or geographical regions, comparison between cities or regions becomes easier, too. People struggling to get through with similar messages could find and help each other: this kind of horizontal knowledge transfer would contribute to the success of knowledge societies by dismantling the information barriers in problems of collective actions.

Online public debate of legislative drafts is the third e-governance practice already applied in Hungary. Civil and professional organizations, recently even individuals can post electronic comments to express their opinions during the codification process. However, participation for non-professionals is difficult, and it is a rather reactive way of inclusion – legislative proposals are always made by governmental or municipal bodies, citizens can only comment on them. The practice we suggest could ease these problems from two aspects. On the one hand, the public concern formulated in laypeople's letters could motivate capable contributors to translate these inputs into professional language when making their contributions to public debates – these horizontal information flows could assist the effective usage of knowledge in society. On the other hand, the governmental commitment to better consider public inputs could enhance the significance of public proposals vis-à-vis governmental initiatives adding to the diversity of information sources.

As for practices from the field of e-democracy, online forums are the most notable means to foster discussions about collective problems in Hungary. The internet serves as the open public space for free deliberation: presumably some of the unanswered letters would soon become forum topics, again contributing to horizontal knowledge transfer. These self-organized spontaneous discussions and citizen activities are deemed crucial by proponents of deliberative democracy. However, further prerequisites of a well functioning deliberative system [17] are missing: people do not have equal access to the rational debate and there are no incentives to put consensus opinions – the outcomes of the deliberations – into practice. While the access to the debate can be assisted by further improvements in internet access and digital literacy, we argue that the envisioned method could help to select substantial contributions from the mixed quality debates and thus could serve as an effective filter which connects deliberations and decision (or policy) making. Accordingly, e-democratic practices offering inclusion for masses could be elevated to a higher level of governance without jeopardizing its efficiency goals by inadequate inputs.

4 Technical Aspects

We would like to emphasize that the present article gives a rather theoretical description of our idea; its implementation should always be suited to the local circumstances, preferably fitted to existing practices.

As for Hungary, citizens can ask written questions from public administration offices that are required to answer (2004.CXL). Clients are allowed to publish answers; however, there are no forums where the publication of these questions and answers is regular. If a web portal was launched for this purpose, it could be complemented by the pages where unanswered letters addressed to politicians, mayors, or other representatives are published. Online consultations could also be organized through these pages. These pages together could evoke vivid public discussions and increase the openness of knowledge societies.

The portal would be maintained officially and would be directly linked to the page of the decision making bodies under consideration. Presumably it would not cause serious problems for municipalities, government agencies or other public bodies to provide storage capacities and maintenance to run the system.

Another case, where the appropriate knowledge management is essential, is the functioning of environmental agencies. Nowadays in Hungary, these government agencies have to split their time between duties where they serve as official authorities and issues reported by citizens. While the former may include very complicated environmental impact assessments (e.g. licensing procedures of power plants); the latter may be very petty issues (e.g. the reproduction of protected but very common grasshoppers reported by an aquarist). As these agencies are notoriously underfunded and understaffed, they operate very slowly and sometimes on a poor scientific level. However, no matter how important it would be, for example, to devote more time to conduct a complicated assessment and gain knowledge about fundamental industrial processes, the agency has to investigate and document each and every case reported by citizens. According to time constraints, this is a very serious limitation on the appropriate functioning of a knowledge-based system. We claim that the envisioned practice of this paper could mitigate these burdens and thus could aid the emergence of the knowledge society.

However, as it may have become obvious from the previous parts of the paper, the primary focus of the application would be on local elected officials, because the role of direct communication with citizens is the most remarkable in their case. Though, higher levels of decision making and non-elected officials could also be involved. In fact, organizations that are either not obliged to answer at all, or those that are obliged to answer to each and every input could potentially be considered; essentially there are no restrictions.

It is important to note that a self-amplifying process could lead to the widespread usage of the system. As more and more people browse the pages of unanswered letters, the motivation to participate in decision making grows due to the increasing interest of decision makers to reply and the enhanced publicity in the case of ignorance. Although, it also means that in the beginning the opportunity has to be advertised to jumpstart usage.

To help orientation, letters should be labeled by topics, decision makers, and geographical locations. Having a clear overview of the topics, political rivals, lobbyists, and committed citizens will have better chances to push some letters or ideas to the top by giving further publicity to them.

5 Conclusions

In this paper we envisioned a new method in the electronic communication between citizens and decision makers and studied how the proposed rules of communication could contribute to the efficiency and legitimacy of democratic decision making. We argue that the obligation of decision makers to either answer or publish letters received from people could have very positive effects from several aspects.

Applied to politicians at any level of the decision making system, the regulation would amplify the increasingly necessary communication between representatives and the represented. On the one hand, people would be better motivated to participate in discussions, because they would know their representatives are politically interested to answer rational questions. People would know that they either receive an answer or draw potential attention to the topic raised in the letter. On the other hand, representatives would be better informed without entering into lengthy and irrational debates. These deliberations would further inclusion up to the point where the public opinion sees the ideal tradeoff between consensus decision making and effective representation, between meaningful rational deliberation and senseless dispute.

Potentially combined with other online practices, the envisioned solution could be conducive to enhance transparency and accountability, contributing to the re-emergence of trust, too. The implemented solution could link deliberations to political actions; or, in the case when consensual outcomes are ignored, at least this ignorance would be publicly known. Positive sides of the traditional representative and the increasingly demanded deliberative solutions are combined in the idea. The appropriate levels of inclusion, deliberation, and transparency are tuned by the public.

The same holds, if we apply the method in the opposite direction: instead of imposing new regulations on decision makers, general obligations to respond could be lifted in the case of some civil servants to make time free for the most important tasks.

The conception fits into the optimistic discourse about the social impact of information communication technologies. Individual actions in the extended political action space would probably lead to stronger ties between decision making bodies and civilians. This kind of inclusion in decision making would not risk but strengthen deliberation and so it would add to the deliberative character of democracy but preserve the basic representative function. The bias towards e-governance as opposed to e-democracy could be alleviated without risking the efficiency and functionality goals of governance.

The proposal is in line with the recommendations of the European Council's ad hoc committee on electronic democracy [18]. As required by the document, it is "additional to traditional democratic processes", "widens the choices available to the public", "supports the democratic roles of intermediaries between citizens and the state", and "promotes opportunities for meaningful and effective public deliberation and participation in all stages of the democratic process, responsive to people's needs and priorities". New communication tools are thus applied to back up basic functions of the envisioned knowledge-based democracy.

Obviously, the presented idea is not a cure-all, only a slight step towards a more effective and legitimate democratic system supported by modern technology. However, as it seems to be easily doable, we deem its implementation worthwhile.

References

1. Beck, U.: What is Globalization? Polity Press, Cambridge (2000)
2. Goodhart, M.: Democracy, Globalization, and the Problem of the State. Polity 33, 527–546 (2001)
3. DeBardeleben, J., Hurrelmann, A. (eds.): Democratic Dilemmas of Multilevel Governance, Legitimacy, Representation and Accountability in the European Union. Palgrave Macmillan, Basingstoke (2007)
4. Sartori, G.: The Theory of Democracy Revisited. Chatham House Publishers, Chatham (1987)
5. Putnam, R.: Bowling Alone: America's Declining Social Capital. J. Democr. 6, 65–78 (1995)
6. Kornhauser, W.: The Politics of Mass Society. Ill. Free Press, Glencoe (1959)
7. Habermas, J.: Between Facts and Norms. Contributions to a Discourse Theory of Law and Democracy. Polity Press, Cambridge (1996)
8. Bessette, J.: The Mild Voice of Reason: Deliberative Democracy & American National Government. University of Chicago Press, Chicago (1994)
9. Blattberg, C.: Patriotic, Not Deliberative, Democracy. CRISPP 6, 155–174 (2003)
10. Börzel, T.A.: Organizing Babylon – on the Different Conceptions of Policy Networks. Public Admin. 76, 253–274 (1998)
11. GlobeScan/World Economic Forum: Trust in governments, corporations, and global institutions continues to decline. Global Public Opinion Survey (2005), `http://www.globescan.com/news_archives/WEF_trust2005.html`
12. Klijn, E.-H., Skelcher, C.: Democracy and governance networks: Compatible or not? Public Admin. 85, 587–608 (2007)
13. Pitkin, H.: The Concept of Representation. University of California Press, Berkeley (1967)

14. Zürn, M.: Global governance and legitimacy problems. Gov. Oppos. 39, 260–287 (2004)
15. Christiaanse, E.: The Human Web and the Domestication of the Networked Computer. In: Lytras, M.D., Carrol, J.M., Damiani, E., Tennyson, R.D., Avison, D., Vossen, G., De Pablos, P.O. (eds.) The Open Knowledge Society. A Computer Science and Information Systems Manifesto. CCIS, vol. 19, pp. 117–124. Springer, Heidelberg (2008)
16. Castells, M.: The Power of Identity, 2nd edn. The Information Age: Economy, Society and Culture, vol. 2. Blackwell Publishing, Oxford (2004)
17. Gabardi, W.: Contemporary Models of Democracy. Polity 33, 547–568 (2001)
18. CAHDE Ad hoc Committee on E-Democracy: Recommendation Rec(2009)X of the Committee of Ministers to member states on electronic democracy (e-democracy). Directorate General of Democracy and Political Affairs, Directorate of Democratic Institutions, Good Governance in the Information Society Project (2008)

CICERON-e: Interactive Tourism for SMEs*

Luis de-Marcos[1], Roberto Barchino[1], José-María Gutiérrez[1], José-Javier Martínez[1], Salvador Otón[1], Fernando Giner[2], and Ricardo Buendía[3]

[1] Computer Science Department, University of Alcalá,
Ctra Barcelona km 33.6, 28871, Alcalá de Henares, Madrid, Spain
{luis.demarcos,roberto.barchino,josem.gutierrez,
josej.martinez,salvador.oton}@uah.es
[2] Area de Organización de Empresas, University of Alcalá,
Plaza de la Victoria 4, 28801, Alcalá de Henares, Madrid, Spain
f.giner@uah.es
[3] Anova IT Consulting, S.L.
Avda.Punto Mobi, 4. Alcalá de Henares, Madrid, Spain
ricardo.buendia@anovagroup.es

Abstract. This project presents a new way of sightseeing, by adapting itself to new times and new technologies. It has been created a system able to guide the tourist at all times during a visit to a city by means of a PDA with GPS technology; it shows the most interesting points (both in text and multimedia contents), the most important routes, a street directory, a calendar events, etc. As additional support, the system has an internal web page that helps to manage in an easy way the touristic contents. Moreover, it is used a Wi-Fi wireless technology to update the PDA contents automatically. The system has a server that allows to use the web page and update the PDA, hiding the access to the database to other kind of services not allowed by the system. Therefore, the most noteworthy aspects of this project are the use of a Wi-Fi wireless technology and a GPS positioning technology.

Keywords: GPS, wireless technologies, Wi-Fi, e-tourism, PDA.

1 Introduction

Tourism in Spain is an important part of the national GDP; it provides a great income to the country, year after year [1-4]. Furthermore, with the advent of the new technologies the habits of tourists have changed; from the selection of destination to the behavior while they are enjoying their holidays. The idea of developing this project came up from these thoughts. This project involves a tourist guide integrated in a PDA which locates the tourist by GPS at all times, and shows and explains the sights of a city or the oddities of a monument, among many other things that will be explained below.

---

* This Project has been co-funded by the Spanish Ministry of Industry, Tourism and Commerce under contract *TSI-030100-2008-189* in the *Avanza PYME* program.

M.D. Lytras et al. (Eds.): WSKS 2009, CCIS 49, pp. 420–429, 2009.

Most of the current multimedia tourist guides still require GPRS connectivity, a feature that many users dislike due to the cost of connection [5]. Another example of multimedia tourist guides is given by the audio guides for MP3 players [6], but they do not provide information in real time to the tourist, since they cannot locate him/her physically on a map. CICERON-e represents the solution for all these limitations.

This project has created a PDA software, with all advantages it entails: visualization of multimedia content (images, video) and GPS location. It is precisely this technology which allows to locate the tourist on a map, indicating at all times where he/she is and all sights around (museums, monuments, hotels, restaurants, etc.). The system also includes a BackOffice module that will be the vehicle used for the remote management of the guide contents. This module is in turn made up of a web page and a server, and it will allow to manage the cities, sights, events, etc. in an easy and fast way. In addition, the server will enable the PDA to update all current contents of the system just by selecting the option update, after entering a password. This is an automatic update, using Wi-Fi.

Therefore, the most important technologies used in this project are:

- Wi-Fi: it is used to update the PDA contents; deleting all the data that have been erased, entering new data from the system and modifying those that have been modified since the last update.
- GPS: it is used for the purpose of locating the tourist on the map, indicating where he/she is at all times.

To sum up, this system is made up of the following modules:

- **Mobile device:** because of its easy and intuitive use no previous learning is required. It allows to guide the user at all times; to visualize contents about each sight, street and events in the city; and to follow those routes predefined or created by the user. Moreover, the user will be able to consult the most visited points by other tourists.
- **Web page:** for the internal management of content, it enables add and delete users, searches of sights, routes, events, etc. In addition, it will provide the possibility of visualizing statistics about the most visited and consulted places, together with the most visited routes.
- **Server:** it allows to update the PDA with the current contents of the system, as well as access of the web page to the database in order to fulfill its commitment.

2 Main Aims

The main aim of this project has been to develop a system able to guide a tourist by GPS in an unknown city, showing him/her multimedia contents about the sights, predefined routes; and letting he/she even create new ones. The application does not require an extensive technological knowledge, since it has been developed with an intuitive and easy-to-use interface.

Specific aims are the data collection for further statistics, the searches for streets, the information about events and the content update.

The system is made up of the parts explained in the paragraph above, which are:

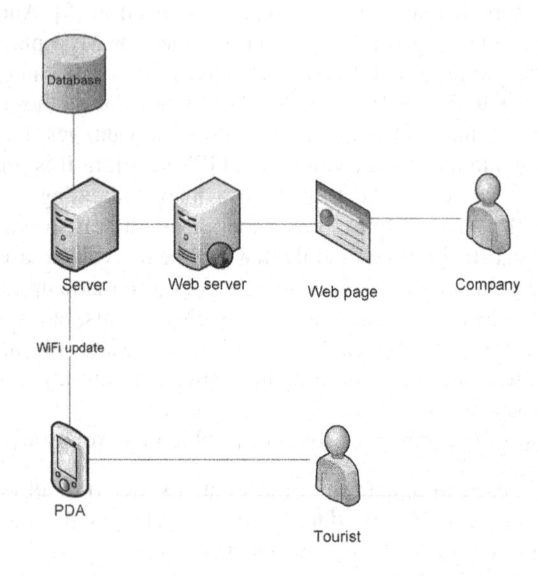

Fig. 1. CICERON-e System Architecture

On the one hand there is a web page which will manage all contents, sending orders to the web server to access to the database, always through a server. In this manner the data will be protected from unauthorized access, since it won not allow direct access to the database. Moreover, the web page is internal to the company, so nobody else should have access through it.

On the other hand there are a set of PDAs (the diagram only shows one, but it is simplified, in fact there will be several PDAs) that will be used by the tourists contracting the service in each city. These PDAs will communicate with the server by Wi-Fi only to perform the content update. An important characteristic of this project is that there is no cost in communications, since the PDA only communicates with the server when updating, and it is free because it is performed via Wi-Fi: the rest of the time no communication is required, since the contents are integrated.

Another interesting feature of the system is that it has been designed for the different modules of the system to be installed in a distributed way, ie each one in a different component: on the one hand the database, server on the other, and web server on the other hand. That is, everything can be integrated in the same server or, on the contrary, each part can be distributed in a different one.

The system, in more detail, is made up of the following elements:

Database: where all information related to any aspect of the city is stored, ie: servers and its contracted PDAs, sights, types of points, calendar events, routes, introductory videos to the city and the routes, etc. It includes a Database Management System in PostgreSQL 8.2. version.

Server: it is the intermediary between the database and the other two systems (PDA and web page). It provides, to the web page, access to the database; and to the PDAs the possibility of updating contents in an automatic and intelligent way: it detects information that has been added, deleted or modified since the last update and sends anything needed.

Web page: it provides to the company the possibility to manage contents in an easy-to-use way, through an intuitive interface which allows add, delete and modify contents in each of the cities and distributors, and visualize the statistics. It also allows to create a file for a particular distributor, that will be subsequently unzip in a PDA, providing (to the PDA) the information about all contents of the city.

Web server: it is required in order to serve the web page. It could be integrated in the contents server or in another different computer.

PDA: each of the PDAs has the software of its corresponding city, and it does not allow access to any other application, neither the telephone, nor the Operating System itself. It shows just those contents of the city for what the application was designed.

CICERON-e meets the following basic functions that initially were proposed as requirements:

- To provide information in **several languages**, not only in Spanish.
- **To locate by GPS** the tourist at all times.
- **To show multimedia contents** (images and videos) about the city contents.
- **To show the predefined routes** for the city and **to allow creating new ones** by the tourist.
- To collect relevant data and use it **to obtain statistics** about the most visited points and routes, together with the most visualized points through the application.
- **To search streets** within the city.
- **To visualize the most important city events.**
- Automatic and easy **update contents.**

3 Project Description

Here is described CICERON-e, specifying its scope, technological environment and main users.

3.1 Determining the Scope of the System

This project stems from the need of develop an innovative multimedia system within the sector. As stated above, nowadays systems of this kind usually involve an associated cost due to the continuous connection they require to communicate with a server. Furthermore, there are other touristic guides working just with audio (audio-guides), but it prevents the location in real time of the tourist on a map.

As mentioned above, the whole system is made up of different modules:

- A database containing all the information related to the different points, routes, events, etc. of each of the cities registered in the system.
- A server as an intermediary between the web page (in fact, the web server)

and the database. Besides, it provides an automatic PDAs update, sending all required information to do so.
- A web server in charge of providing service to the web page, by communicating with the server.
- A web page that receives the operations to be performed through the company staff, and indicates them to the web server.
- Several PDAs, which are mobile devices where the application with the contents of a particular city will be integrated. It will be the device carried by the tourist once he/she has contracted the service, in order to be located and to be able to look at the city information.

Thanks to the GPS device it will be possible to locate the tourist and, at the same time, to store data about the points he/she has visited, in order to (afterwards) obtain statistics about the most visited points.

Furthermore, the system controls that the tourist cannot access to certain application sections, as a safety measure. The main characteristics of the system are detailed below:

- Support for multiple languages.
- Real-time GPS location of tourists on a map.
- Showing multimedia contents such as images and videos.
- Visualization of predefined routes and creation of routes with points determined by the tourist him/herself.
- Search for streets of the city.
- Information about city events.
- Automatic update (upon administrator's request) by Wi-Fi in order to keep the city contents updated.

3.2 Identification of the Technological Environment

In the following subsections it is carried out a definition of the technological environment that has been required to meet the needs of the project, specifying the possible constraints and restrictions. To do so the technological environment has been taken into account.

3.2.1 Area of Action

The area of action will be focused on those places where tourist usually stay at or look up data about a city; such as hotels, rural cottages, information/tourist offices, etc. These establishments can be found anywhere within the Spanish geography, or even abroad, thanks to the availability of the application in several languages.

3.2.2 Server

The system comprises a server that, as it was mentioned above, acts as the intermediary between the information contained in the database, the web page and the PDA. This structure was chosen to protect the database from unwanted accesses.

This server can be integrated together with the database and the web server, or can be implemented in a different device, since the system has been designed to operate in a distributed way, as long as the configuration has been correctly done.

For the implementation of this application the following technologies have been used:

- The programming language has been Java, from Sun Microsystems.
- To access the database they have been used both JDBC and Hibernate.
- The communication is made through sockets, both with the web server and the PDAs when updating. In the case of the PDAs, the communication technology is by Wi-Fi.
- For the data representation XML files have been used.

3.2.3 PDAs

This application has been fundamentally developed to be used in a PDA by the tourists. The final system has a great number of PDAs, since every distributor (hotel, rural cottage, etc.) usually has more than one.

It is maybe the most important part of the system, as it deals directly with the final user (the tourist). For that reason, it has an easy-to-use interface by people unfamiliarized with the NTIC. Moreover, it also has several languages available that can be selected at the beginning of the execution (or later, from the menu).

The application of this device has been designed to block the access to the rest of the PDA functions, so the tourist can only access to this program.

The technologies used for the software development of this device have been:

- As language programming C# .Net, and the technology LinQ for searches similar to the consults carried out in the database access, thus allowing more speed of the application.
- As standard format of data representation, XML files have been managed.
- For the communication with the server and the update, Wi-Fi technology has been used, through sockets.
- In regard to the location of the tourist on a map, it has been used a GPS positioning technology. The coordinate system has been the geodetic coordinate system (decimal degrees), to avoid problems when calibrating in cities or towns located between two UTM zones.

3.2.4 Database

The system has a database PostgreSQL 8.2 containing the following data:

- Cities: all cities where the service is contracted.
- Distributors: those establishments where the PDAs will be distributed to the tourists.
- PDAs: those PDAs that have been ever used the update service, since they are automatically registered.
- Events: interesting activities taking place in the city.
- Sights: most interesting points to be visited by a tourist, such as monuments, museums, restaurants, etc.
- Routes: they contain more than one point, and they are usually made up of points related to a specific topic (eg "tapas route", monuments route, etc.).
- Types of points: categories in which each point can be classified.

- Languages: all languages for which the application is designed.
- Points visits: record of the points visited by each tourist (including date and hour).
- Number of times each point was visited: number of times (with date and hour) a tourist has consulted each point using the application.
- Routes visits: record of the routes visited by each tourist (including date and hour).

3.2.5 Web Server

The web server provides the web page with those operations offered by the server who has access to the database. It was chosen Tomcat in the implementation of the system, since it is one of the most popular and used web servers, and it works both in Windows and Linux.

3.2.6 Web Page

The web page available for the company staff has been created using a large variety of technologies: Java, Javascript, Struts, Servlets and JSP. It allows to carry out multiple and varied operations:

- Add, delete and consult of the city data.
- Consult of every city included in the system.
- Modification of the introductory video of a city.
- Add, delete, modification and consult of the data from a distributor.
- Consult of every distributor in a city.
- Consult of the PDAs of a distributor.
- Automatic production of the contents of a city in a zip, for a specific distributor.
- Add, add in another language, delete, consult and modification of the data from a particular point.
- Consult of all points of a city.
- Add, add in another language, delete, modification and consult of the data from a particular route.
- Consult of all routes of a city.
- Add, add in another language, delete, consult and modification of the data from a particular event.
- Consult of all city events.
- Add, add in another language, delete and modification of the types of points.
- Visualization of statistics:
    - Most visited points.
    - Most visited routes.
    - Points of which its information has been the most consulted in the PDA.
- Massive load of points in a city, through a XML file with a particular format.

All performed operations will affect the database (except for the case of the consults; they do not modify anything), always through the server.

3.3 Identification of System Users

In the following paragraphs they will be identified the users who carry out each part of the system.

3.3.1 Tourist

This user is really the final user; the last one using the developed software for the PDA. From its use they will be obtained data to subsequently obtain statistics to show in the web page. This user can perform the following tasks with the PDA:

- To visualize the introductory video of the city, if there is one.
- To visualize the map of the city, as well as being located on it.
- To search city streets.
- To consult the top10 of the most visited points.
- To choose a predefined route to be shown in the map.
- To create a route with the points he/she wants to visit to be shown in the map.
- To search interesting points according to the type or direction and, once found, to visualize the information (both text and multimedia content).
- To consult city events.
- To change the language of the application.
- To consult the help and credits.

The system is also required to follow some standard usability guidelines for PDA developments [8].As an example, below they are shown some of the software captures executed in the PDA:

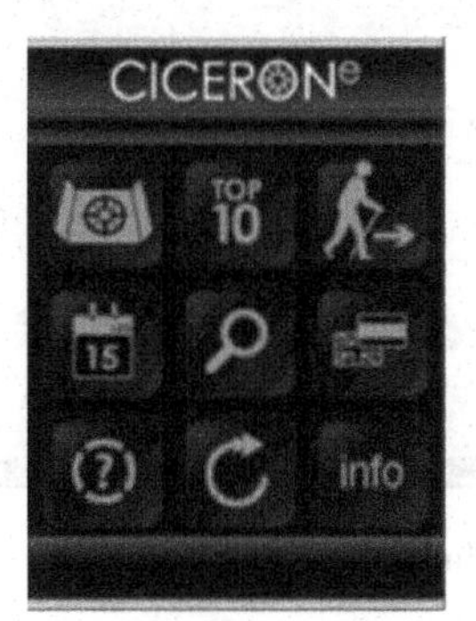

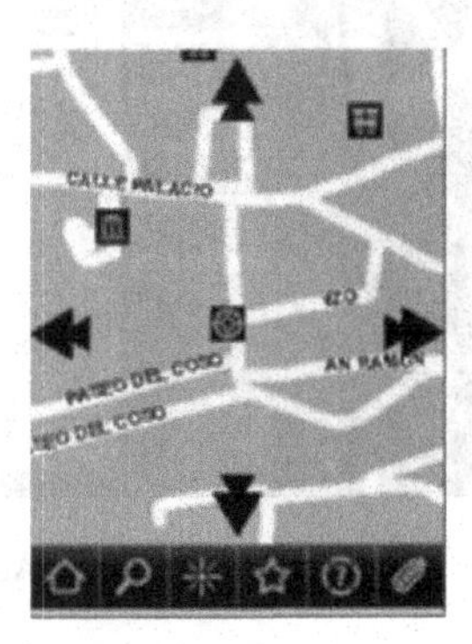

Fig. 2. CICERON-e PDA Application

The image above left shows the main menu of the application; while the image above right represents an example of the map screen, when the tourist is being located by GPS in real time.

The image below left shows the information of a sight; and the image below right shows a picture (multimedia content) of the sight, after clicking on the small image of the image below left.

Fig. 3. CICERON-e Tourist Point Information & Multimedia

3.3.2 Administrator

This user is the responsible for introducing and modifying the system contents. Administrator will be a people from the company that will exploit the product commercially. The tasks that can carry out are detailed in paragraph 3.2.6, because they are basically the operations that the web page allows to perform. Next it is shown a capture of the web page:

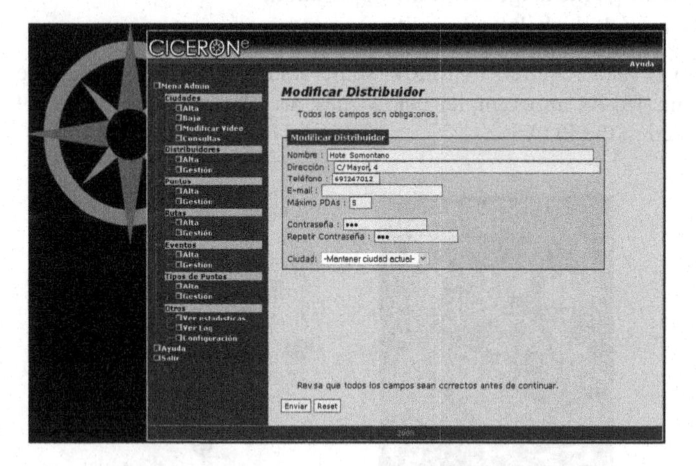

Fig. 4. CICERON-e Administration Application

3.3.3 Distributor

This user will be at the place where the PDAs will be distributed to the tourists, such as hotels, rural cottages, tourist and information offices, etc. The distributor is in charge of manage the PDAs, ie, to offer them to the tourists. In addition, he/she has to update the software periodically, in order to obtain the most updated contents of the city. Therefore, to sum up, his/her include the distribution of PDAs among tourists and the updating of the PDA software periodically.

4 Conclusions

There were two main aims for this project; to fulfill functionality and that the product had the desired aspect. This last characteristic has not been developed for this part directly (as it was designed by graphic designers), but the functionality did. A detailed list of the achieved functionality consists of the tasks that can be performed with the PDA software, together with the tasks that can be performed in the web page.

In addition, there was a particular period of time to do it, since the project had, in advance, a committed date to be implemented in a city.

Once finished, it is important to value the following aspects about the carried out work:

- **Scope:** the project has been initially developed to be exploited in Spain. However it is possible to implement it in another countries, thanks to its support in several languages (nowadays there are only two, but they can be added a total of six). Therefore, it can be stated that there is an important scope.
- **Possible improvements:** despite the fact that this project meets all functionalities initially agreed, it is possible to improve the system. Perhaps the most important improvement would be to make the routes in some way taking into account the shapes of the streets, since the current implementation only guides the tourist in a straight line; it does not consider the crossroads.
- **Acquired knowledge:** the execution of this project has provided some knowledge about technologies not very used so far. Also it has allowed to go into the knowledge of other ones (more known) in depth.

For more information about Ciceron-e, including licensing, please visit http://www.anovagroup.es and http://www.ciceron-e.es/.

References

1. Rivas García, J.: Problemas de estructura y economía del mercado turístico. Septem, Madrid, España (2007)
2. Ramírez Cavassa, C.: Marketing turístico. Trillas, Madrid, España (2006)
3. Rey Graña, C.: Economía del turismo estructura de mercado e impacto sobre el desarrollo. Asociación Hispalink-Galicia, Galicia, España (1998)
4. Lodeiro Hermida, M.: Indicadores estadísticos del sector turístico. Asociación Hispalink-Galicia, Galicia, España (1998)
5. Ortega Jiménez, S.: Aplicacion turística para Pocket PC mediante navegación GPS. Trabajo Final de Carrera, Universidad de Alcalá, España (2007)
6. Castro Gil, M.: Diseño y desarrollo multimedia Sistemas, imagen, sonido y vídeo. Ra-ma, Madrid, España (2002)
7. GPS - NMEA sentence information, `http://aprs.gids.nl/nmea/` (accessed on 23-03-2009)
8. Dumas, J.S.: A practical guide to usability testing. Intellect, Westport, CT, USA (1999)

Developing e-Business Capabilities to Bridge the Knowledge Divide in Mediterranean Countries

Giustina Secundo, Gianluca Elia, Alessandro Margherita, and Giuseppina Passiante

e-Business Management Section, Scuola Superiore ISUFI – University of Salento, Via per Monteroni s.n. 73100 Lecce Italy
{giusy.secundo,gianluca.elia,alessandro.margherita, giuseppina.passiante}@ebms.unile.it

Abstract. This paper presents the results achieved in terms of e-business capabilities developed in an International Master framed within an Euro-Mediterranean cooperation. In particular, an e-Business Design Laboratory is here described which has been set up for designing and implementing innovative solutions to bring digital and organizational innovation in traditional and new industries. The most significant highlights of the last two editions of the Master are also reported in terms of the human, social and structural capital being generated.

Keywords: e-Business Capabilities, Digital Divide, Learning in Action, Learning Laboratory, Mediterranean Countries.

1 Introduction

In the current knowledge society, the productivity and competitiveness of countries depends upon their capacity to generate, share and apply efficiently knowledge in the effort of innovating development models, processes and dynamics. Moreover, the development of technological and managerial competencies, the capability to enter in global networks and adopt effectively Information and Communication Technologies (ICTs) into daily working processes, become the pillars for ensuring a sustainable development of private and public organizations.

In this scenario, several Southern Mediterranean Countries – and especially the Middle Eastern and North African (MENA) regions – are asked to overcome their isolation from global knowledge flows and learning networks, by leapfrogging in the knowledge-based economy and fulfilling their "knowledge divide".

In this view, it is necessary to invest in human capital (capabilities, skills and experiences), structural capital (technologies, processes, practices, demonstrators, patents and brands) and social capital (the web of relationships and cooperation) [1]. This represents a challenge for universities that are required to invent new ways of re-organizing their knowledge base, new ways of producing knowledge and new learning processes, as well as to adopt entrepreneurial behaviors.

Starting from these premises, some fundamental questions arise: *How can be e-business capabilities created to bridge the knowledge divide? And how this could be done in practice through a higher education program?*

M.D. Lytras et al. (Eds.): WSKS 2009, CCIS 49, pp. 430–439, 2009.

In this framework, the papers presents the empirical results obtained from the International Master Program in e-Business Management (IMeBM) promoted by the "Mediterranean School of e-Business Management" (hereafter Med School – www.ebms.it) and devoted to talented people coming from several Mediterranean countries, with a particular focus on the MENA Region. The Med School is an initiative launched by the e-Business Management Section (eBMS) of Scuola Superiore ISUFI – University of Salento (Italy) that has initiated the program in 2005, and other local stakeholders and partners in Morocco, Tunisia and Jordan. The Med School aims at proposing a systemic approach to the "Digital and Knowledge Divide" in the Mediterranean Area, through the creation of a network of Competence Centres for developing Human, Structural and Social Capital. Among the initiatives of the Med School, since 2006 three editions of the IMeBM Program have been launched to develop e-Business capabilities. The Master represents the main focus of this paper.

Next section introduces the issues related to digital and knowledge divide (section 2). The Med School Initiative and the International Masters' are then presented (sections 3 and 4). A discussion of key highlights is then reported (section 5) and some conclusions are drawn in the last section (section 6).

2 Digital Divide and ICT Capabilities in MENA Countries

Digital Divide is the expression of an asymmetry in knowledge access and in the capability to use it in order to radically renew models, processes, and development dynamics. More and more it is needed to overcome the unequal capabilities to seize the benefits coming from the diffusion and the effective use of ICTs and leap forward in the age of globalization, digital networks and knowledge. Nevertheless, a recent report by the World Information Society [2] indicates that, nowadays, the digital divide is narrowing in terms of internet usage and evolving from inequalities in basic access to ICTs and their availability, to differences in the quality of the user experience. Several Mediterranean countries in the Middle East and North Africa (MENA) region are trying to encounter and overcome digital divide, by facilitating the diffusion and sharing of knowledge and experience between these countries, and sometimes within the societies themselves [3]. At country and industry level, the penetration of networks and digital innovation is captured by indicators such as the NRI - Networked Readiness Index 2003-2004 and the KEI - Knowledge Economy Index 2004. These measures highlight sensible differences among southern and northern Mediterranean countries in terms of innovation, education and ICT adoption. For example, values of NRI and KEI for France are namely 4,6 and 8,08 whereas Algeria's values are 2,75 and 2,39.

From the other hand, it must be noticed that countries' success in the Knowledge Economy increasingly relies on highly-skilled and qualified people, which in turn requires rapid, effective, and less expensive education and training. In a few words, it is required to invest in Human Capital, Structural Capital and Social Capital to leapfrog into the Knowledge Economy, and so, to narrow the Digital Divide. In this sense, the World Bank mentions the need of a leapfrog strategy to sustain a quantum leap in the development of countries and their economic and institutional framework. Governments, universities and researchers are challenged to understand how Internet and digital networks can be used to facilitate knowledge sharing and collaborative work, and which public and private efforts should be enacted to streamline digital and organizational change. This is a challenge for Universities that are required to invent new ways of re-organizing their knowledge base, new ways of producing knowledge and new learning processes, as well as to adopt entrepreneurial behaviors. Moreover, it is necessary to note that knowledge production is no more an exclusive domain of Universities, rather it is a social phenomenon that originates in the complex interactions between universities, enterprises, public institutions and the society as a whole. It's in this debate that the experiment of the Mediterranean School of e-Business Management is framed. The following section provides more details about the School, along with the strategic and operational perspective to develop the e-Business capabilities.

3 The Mediterranean School of e-Business Management: A Challenge to Reduce the Digital and Knowledge Divide

The MeD School is an initiative launched by the eBMS of Scuola Superiore ISUFI – University of Salento, Italy that has initiated the program in 2005 with other local private and public stakeholders and partners: the Al Akhawayn University in Morocco, The University of Jordan in Jordan, the Technopole Elgazala in Tunisia. The value proposition of Med School is to facilitate and accompany the diffusion of Digital and Organizational Innovation in local firms and public institutions, with particular focus on SMEs and regional authorities, using human development programs, joint research activities, e-Business and e-Government training programs, pilot projects and prototypes. The main objective of the Med School organizational model is to integrate learning, research and entrepreneurial processes (aspects which are often isolated) and leverage their virtuous complementarities. The ultimate purpose is to develop a network to enhance the conditions, at both context and organizational level, which can facilitate the introduction of e-business in SMEs located in developing countries, as well as Euro-Mediterranean business cooperation and partnerships. A constant focus on integrating business and technology aspects was addressed in the design and development of three key components which build up the organizational model: people (Human Capital), purpose (Social Capital) and process (Structural Capital) (see fig. 1).

The people component (human capital) is founded on an integrated capacity building strategy which aims to prepare multi-cultural leaders able to manage technological change, disseminate innovative practices, and develop the scientific and managerial competencies necessary to modernize Mediterranean SMEs. In this perspective, a key element is represented by "competency and skills development" actions, which include higher education programs, International Masters', and Ph.D. programs with a mixed academic-industry faculty, coming from leading institutions. Another element is the set of "experiences" and demonstrative actions, showing successful cases of ICT adoption in public and private organizations. Other two elements are the "leadership attitudes" and entrepreneurial support actions, to facilitate the launch of start-ups and spin-offs based on innovative ICT applications, and the "awareness development" actions, in the form of executive programs for public and corporate managers aimed to spread knowledge related to e-business benefits and to develop core skills for managing innovation projects.

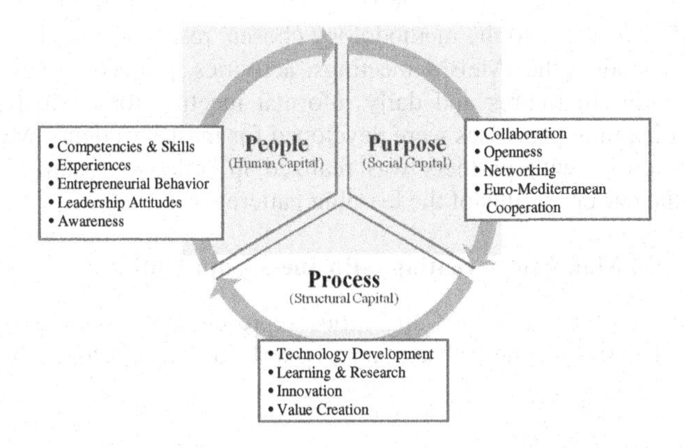

Fig. 1. The Organizational Model of the Mediterranean School

In terms of processes (structural capital), the model is based on investments in learning and joint research and innovation programs focused on strategic issues like knowledge management and e-business, and with the ultimate objective to create business and social value through setting up research laboratories, technological infrastructures, brands, and patents. Finally, the purpose component (social capital) is developed through the creation of cooperative open networks among universities, private companies and public institutions. Networks are the way to creating and disseminating strategic know-how, resulting in enhanced relationships among SMEs in different sectors. Collaboration is promoted both vertically, with European countries, and horizontally between different Mediterranean countries, by promoting entrepreneurial behaviours and supporting Euro-Mediterranean cooperation. The categories of actors involved in the organizational model are represented by Academic, Public and Industry Actors, including local Firm's associations.

4 Developing e-Business Capabilities in the International Master's

In the frame of the Med School, the case study of the International Master in e-Business Management (IMeBM) – Graduate Diploma in Innovation and Change Engineering is here presented. The mission of the IMeBM is to prepare outstanding people in leading Business Innovations based on the use of ICTs. The aim is to create business Engineers and Change Managers capable to identify and exploit the distinctive potential of the new ICTs for reconfiguring traditional business contexts, which we refer as e-Business capabilities [4]. The Program is organized in mobility between Italy and the Southern Mediterranean Countries and is offered on annual basis, full–time, for a total of 1940 hours. It's devoted to people graduated in Business, Engineering or Computer Science coming from Italy, Morocco, Tunisia, Jordan, Egypt and other Southern Mediterranean Countries.

4.1 Research Methodology

A participative observation [5] has been applied to the study of the IMeBM community, from January 2006 till March 2008. During the observation time, more than 40 students coming from Mediterranean Countries have been involved in the 2 cycles of the IMeBM. According to the methodology chosen, researchers and executive staff actively participated in the IMeBM meetings, activities, projects and learning processes. Weekly, questionnaires and daily informal meeting for evaluating students' satisfaction on learning processes were developed for master students. Monthly meeting and focus group with partners was realized in order to focalize the learning strategy with the current results of the learning patterns.

4.2 The IMeBM Model for Creating e-Business Capabilities

The development of e-Business Capabilities represent an evolution of traditional management education in the 21st century [4], and can be developed by integrating

Table 1. The Change Manager and Business Engineer profile

IMeBM BUILDING BLOCKS	KNOWLEDGE
Internet worked global Business	a - understand **global macroeconomics** paradigms and globalisation scenario
	b - apply the **global business strategy planning** as a critical process for organizations operating in a dynamic and unpredictable environment.
	c- compare the different methodologies and tools for **competitive macroeconomics**
Internet Business Management	a – apply an holistic framework to represent the Business
	b- imagine and structure new initiatives of business by developing a marketing and **busines s plan.**
	c - recognize and apply the right instruments and metrics for the measurement of processes performance and whole **business performance.**
	d - explain how **operations management** impacts the success of a business and how can be used for optimising the business organisation
	e - understand how the organization's strategy drives the need for different types of **management accounting** information
Organizational Learning, Innovation and Leadership	a - apply the **global business strategy planning** as a critical process for organizations operating in a dynamic and unpredictable environment.
	b – evaluate the key role of **innovation management** in the creation of a sustainable competitive advantage and in the achieving of a market leadership.
	c - define the architecture of a **knowledge management** System and the processes involved with its implementation and evaluation
e-Business Design and Implementation	a – design and Evaluate an **e-business** architectures and e-Business model.
	b - identify the phases of an innovation process, the critical internal and external forces and limits and apply the tools of **Project Management.**
	c – configure new ***e-business models*** in particular industries

SKILLS	ATTITUDES
Managing diverse culture	Adaptability and responsiveness
Dealing with ambiguity, uncertainty and paradox	Intuition
Decision making	Demanding excellence
Project Management	Perseverance and tenacity
Ability to make complex simple	Curiosity and creativity
Presentation skills	Self Awareness
Listening and observation	Self Confidence to involve others
Networking and collaboration	Energy to motivate and energize
Teambuilding and team-working	Capacity to learn

management and technological-based curricula. The basic principles of the IMeBM program are the following:

IMeBM profile. The overall aim of IMeBM is to create outstanding people defined as Change managers and Business engineers. This is a multidisciplinary profile pursued through the creation of inter-disciplinary competences profiles able to identify the distinctive potential of emerging technologies integrating both technological and business knowledge [6]. Table 1 shows the profile in terms of knowledge, skills and attitude.

IMeBM content. An inter-disciplinary approach characterizes the IMeBM program. Knowledge coevolves with the context of application, dynamically and beyond the contribution of any single discipline. The focus will be an integrated offering of various disciplinary fields, through integrative modular program design, rather than offering semester length courses in particular specialist topics such as marketing, finance, accounting etc. [7]. Learning patterns are structured in terms of knowledge, skills and attitudes in an integrated learning architecture, focused on the building blocks illustrated in Table 2, integrating Business Management with Technology Management.

Table 2. e-Business Knowledge developed in the IMeBM

IMEBM PROGRAM BUILDING BLOCKS	KNOWLEDGE DEVELOPED IN PARTICIPANTS
Internet Business Management	Holistic approach to business management, business performance, business planning, operation management, accounting management HW&SW management, data management, network management, technology platforms, Internet marketing, e-business, knowledge management.
Internet worked Global Business	Globalization, Network and Digital Innovation, Digital innovation.
Strategic Management and Entrepreneurship	Strategy and models for managing company in global Environment, Strategic Management, Entrepreneurship and New Business Creation.
Organisational Learning, Innovation and Leadership	Change management & leadership, innovation management, project management, Intellectual Property Management, Digital Business Innovation.
e-Business Design and Implementation	Network and web technologies for e-Business; Project Management, e-Business Architectures, Innovative Management of e-Tourism, Innovative Management of e-Agribusiness, Innovative Management of e-Government, Innovative Management of new product development in Aerospace.

IMeBM learning approach. 'Learning in action' strategy lies at the center of MeD School business leaders' creation process: it refers to learning processes triggered by experiential laboratories in which participants think, work and learn together how to invent new ICT-based business solutions. Applications and experiences are not means to learn, but ends from which learning originates, in a process that relies on a series of trials/experiences, observation, reflection, and systematization [6]. Lectures and seminars are held by faculty coming from different Universities and Business Schools located in USA, Europe ad Euro-Mediterranean countries and from an International Networking of Academic and Industrial Partners. This network is set up to expand knowledge horizons because in rapidly changing contexts much of the critical resource base that will determine the success of an organization lies outside its boundaries and escape its direct control [8].

4.3 The Laboratory in Practice: Developing e-Business Capabilities

Within the IMeBM, activities take place beyond classroom, in the research and development laboratory. Laboratory represents the "learning environment" of all the student activities that bring to the development of innovative e-Business models and solutions in some business contexts, like Agri food, Tourism, New Product Development and e-Government. The dominant philosophy is that students learn best not only by receiving knowledge but also by experimenting it, interpreting it, and learning. Especially during the Laboratory phase the e-Business capabilities are developed. The laboratories activities of the IMeBM 2006 and 2007 editions have been organized into three main phases.

Phase 1: e-Business Design (common to all the participants and organized at eBMS – Lecce. Duration: 7 weeks). The activities of this phase aim to provide participants with a set of competencies related to the benefits deriving from the application of ICTs to business as well as to the methodologies supporting business and organizational change. Moreover, students will be provided with a methodology for the implementation of organizational change through the ICT-driven innovation of business processes. The competencies to acquire in this phase are preliminary to the activities for the experimentation of the following step, and consisting of ICT-based applications for the innovative management of traditional business domains. The competences to acquire during this phase are related to: e-Business models, Network and Technologies for e-Business, e-Business Architecture, e-Business strategy and Project Management.

Phase 2: e-Business Implementation (participants divided in groups, and organized at eBMS – Lecce. Duration: 7 weeks). The activities of this phase are focused on the application of methodologies and tools to verify in "action" the benefits deriving from the e-Business applications in some traditional industries. This phase is characterized by the adoption of a "Project based" learning strategy. Participants are involved in different applicative learning activities related to the Innovative management of Agri-Food Value chain, Tourist destinations, New Product development in aerospace industry and e-Government. The projects developed within the experiential laboratories serve as the "central core" of the learning program, by which the knowledge and skills learned are put in practice.

Phase 3: e-Business in practices (in mobility in Morocco, Tunisia and Jordan – Duration: 10 weeks).This phase is characterized by the analysis and contextualization of e-business models and solutions developed during the phase 2 to firms operating in the Mediterranean area. Getting involved into complex projects and real issues is the ultimate learning experience. Participants may contextualize the e-Business capabilities with a company that corresponds to a field of interest or specialization area pre-existing within phase 2 and represents a variety of industry including Agri-food, Tourism, software companies and other types.

The three phases of the laboratory have been supported by the use of a web learning platform allowing the experimentation of a blended learning experience for the development of e-Business capabilities. The laboratory's experience was structured according to the delivery of some Web learning courseware, mixed to both face-to-face meetings with mentors and e-Business experts. The technological platform adopted for the Web learning activities was the "Virtual eBMS" an integrated Web learning and Knowledge Management (KM) system, completely web-based, developed by combining different market products and some components developed ad hoc; at the base of KM and Web learning systems integration process there is the use of open source technology [9].With respect to the 2 Master's editions, the 2006 edition was much more focused of the use of CD, DVD, some web learning courses and especially face to face seminars. The 2007 edition was more focused on the use of the "Virtual eBMS" system and face to face seminars delivered to a target of 14 young talents coming from Morocco, Tunisia and Jordan. Furthermore, the availability of different ways of knowledge delivery (use of multimedia presentations, Web resources, knowledge objects, SCORM objects), jointly with collaboration tools such as chat, forum and shared notes provide the means for learners to decide how to personalize their experience according the learning style, in order to maximize the effectiveness of their learning [9]. During the laboratory an efficient tutorship activity was conducted to provide learners with online support, both from a technological point of view – for technical issues related to the use of the Virtual eBMS Web Learning platform, and from the content point of view – to explain or clarify some doubts related to the knowledge domain to analyze.

5 Discussion of Findings

In the two IMeBM editions, the e-Business Design Laboratory allowed developing human capital (competences developed in the e-Business context), Social Capital (networks and cooperation with local institutions and companies) and Structural Capital (innovation projects). In order to highlight the distinctive features of each laboratory phase, the results will be described as follow:

- the Human Capital developed will be presented mainly referring to phase 1 (e-Business design), since phase 1 is the more theoretical one, with the aim to develop basically the human capital dimension (basic capabilities on e-Business);
- the Structural Capital will be described mainly referring to phase 2 (e-Business Implementation), since phase 2 is conceived more practically, oriented to develop and configure the structural capital dimension (research and innovation projects);
- the Social Capital will be described mainly referring to phase 3 (e-Business in practices), since phase 3 aims basically to expand the Social Capital dimension (cooperation and relationship with Mediterranean companies and institutions).

Concerning the Human Capital dimension, in Phase 1: e-Business Design, almost 20 students have been involved (35% coming from Morocco, 48% coming from Jordan and 17% coming from Tunisia). More than 25 mentors and speakers mostly coming from Europe, North-America, and testimonials have been involved in face to face seminars. During this phase, master students learning activities were monitored and assessed; periodically, a set of reports containing the main information related to learner activities were generated by the master's coordinator and mentors. The students' final evaluation made by the mentor was based on the following elements:

- Results of learners' intermediate test structured according to a set of single choice, multiple choice, matching questions, as well as true/false questions.
- Results of learners' final test structured as the intermediate test.
- Assessment of the deliverable related to the project assigned.

These three elements have had a different weight on the final evaluation, according to the following percentages: 20% for the intermediate test, 50% for the final test, and 30% for the project deliverable. The mentor assigned a specific score, from 0 (poor) to 5 (excellent) to both the two tests and the deliverable. The final average score for the initiative is 3.78 on a maximum of 5 points; therefore we can consider the final results as fulfilling in respect to the objective of the laboratory experience.

The e-Business capabilities created during Phase 1 have also contributed to develop the Structural capital – what is usually defined as projects, software and databases during Phase 2: e-Business Implementation. These act at the same time as enabling factors (since structural capital is generated for matching projects' requirements) and as catalysts of the learning-in-action environment of the IMeBM. The projects' development, then, reinforces the dynamics of interactive learning which constitute the core mission of the Med School. In particular, specific projects were developed by Masters' participants with the objective to design and implement e-Business solutions in the agrifood industry (e.g. innovative e-business models, suites for business processes automation, digital marketplaces), tourism industry (destination management systems, B2C, B2B and B2G architectures), e-government, and new product development process (integration of different technological platforms, distributed design groups, design simulation).

Finally, the "Phase 3: e-Business in action" contributed to create the Social capital in terms of relationship and partnerships with local institutions and companies located in Morocco, Jordan and Tunisia where the students spent this phase contextualizing the e-Business models and solutions developed during the previous phases. We refer at first to the agreement with 15 companies and institutions (35% in Morocco, 48% in Jordan and 17% in Tunisia) in the field of Tourism, ICT, Agri food, Government and Aerospace in which the Master's students spent the phase 3 of the Laboratory. In particular, in Morocco the 20% of companies belong to agri food, 20% to aerospace, 30% to tourism and 30% to governmental institutions; in Jordan the 50 % of companies belong to software and ICT industry, 20% to governmental institutions, 15% to tourism and 15% to telecommunication industry. In Tunisia, the 20% of companies belonged to Agri food, 30% to aerospace, 20% to tourism and 30% to governmental institutions.

6 Conclusions

In this paper we have presented a case of Euro-Mediterranean cooperation aimed to leverage human capital as a strategic foundation of development in the context of the knowledge economy. In particular, it is here presented the case of the International Master in e-Business Management, a higher education program finalized to the creation of e-Business capabilities in different industries such as agrifood, tourism, public administration, and aeronautics. The core phase of the program is represented by the "e-Business Design Laboratory", an integrated environment where the learning strategy becomes more experiential-oriented and based on simulations and action-learning projects. The laboratory has been organized into three main phases related to the Design, Implementation and practicing the e-business capabilities acquired. All the three complementary phases were supported by technologies and tools of web learning. In the editions 2006 and 2007 of the Master's, the three phases of the laboratory significantly contributed to develop human, social and structural capital, and to create a first nucleus of people trained to activate the diffusion of organizational and digital innovation to modernize traditional and emerging industries in Mediterranean Countries. Future developments will be aimed to activate similar initiatives, by involving all the managerial levels to promote and diffuse at large scale the e-Business capabilities and practices that represent the real strategic and sustainable lever to leapfrog Mediterranean Countries in the Knowledge Economy.

References

1. Seeman, P., et al.: Building Intangible Assets: A Strategic Framework for Investing in Intellectual Capital. In: Morey, D., Maybury, M., Thuraisingham, B. (eds.) Knowledge Management: Classic and Contemporary Works. The MIT Press, Massachusetts (2000)
2. World Information Society Report (WISR), International Telecommunication Union (ITU), Geneva, Switzerland (2007), http://www.itu.int/osg/spu/publications/worldinformationsociety/2007/index.html
3. Assaf, W., Elia, G., Fayyoumi, A., Taurino, C.: Prospect of e-Learning: the case of Jordan'. In: Proceedings of e-Society 2007 - IADIS Multi Conference on Computer Science and Information Systems, Lisbon, Portugal (2007)
4. Romano, A., Elia, V., Passiante, G. (eds.): Creating Business Innovation Leadership. An Ongoing Experiment, The e-Business Management School at ISUFI, Edizioni Scientifiche Italiane, Naples (2001)
5. Yin, R.K.: Case Study Research. Design and Methods. Sage, California (1994)
6. Secundo, G., Passiante, G.: An innovative approach to creating business leaders: evidence from a case study. Int. J. Management in Education 1(3), 214–230 (2007)
7. Lorange, P.: Strategy means choice: also for today's business school! Journal of Management Development 24(9), 783–790 (2005)
8. Gulledge, T.R., Sherwin, J.: Technology transfer: combining educational resources and industrial experience. Int. J. of Innovation and Learning 2(4), 335–356 (2005)
9. Elia, G., Secundo, G., Taurino, T.: Towards unstructured and just-in-time learning: the "Virtual eBMS" e-learning system. In: Proceedings of m-ICTE2006 - Current Developments in Technology-Assisted Education, Seville, Spain, vol. 2, pp. 1067–1072 (2006)

Securization of e-Cognocracy

José María Moreno-Jiménez[1], Joan Josep Piles[2], José Ruiz[2], José Luis Salazar[2], and Alberto Turón[1]

[1] Grupo Decisión Multicriterio Zaragoza, University of Zaragoza, Gran Vía nº 2, 50005 Zaragoza, Spain
[2] Grupo de Tecnología de las Comunicaciones, University of Zaragoza, María de Luna nº 1, 50018 Zaragoza, Spain
{moreno,jpiles,jruiz,jsalazar,turon}@unizar.es

Abstract. E-cognocracy [1], [2], [3] is a cognitive democracy oriented to the extraction and sharing of knowledte related with the scientific resolution of public decision making problems related with the governance of society. This model of democracy takes advantage of the potential of the knowledge society by means of the incorporation of the knowledge and the preferences of the actors involved in the decision making process. This is carried out by using different rounds to incorporate the preferences and a collaborative tool for the discussion stage. The technological security of the voting process has been studied in previous papers of our research group [8], [9], [10], [11]. Now we cope with the security of the discussion stage, taking into account either registered and non registered individuals.

Keywords: E-cognocracy, e-democracy, security, voting, discussion.

1 Introduction

The hypothetical loss of legitimacy that surrounds the citizens' –political– representatives conduct in the last years is spuring on more and more calls for a greater involvement of citizens in the decision making related to the governance of society [1], [2], [3].

It is not enough, as erroneously arises in many public forums, to ask the represented people for the participation in the political parties blogs in order to raise suggestions and proposals to be considered by the political representatives. Such participation, restricted solely to the debate of ideas, has a decisional character basically unidirectional (unidecisional participation).

The citizens provide the representatives with ideas, and these finally make the decisions they want, taking into account or not the suggestions made by represented people. This unidecisional interpretation of citizen participation is causing the lack of interest and the "unhooking" of the individuals from one of the most important patrimony of the human being, at least in the Western culture: democracy.

In order to favor the involvement of the citizenry in the resolution of the complex problems that usually arise in the field of the public decision making and, in passing, improve the transparency of the democratic system and its control by its true owners –the

M.D. Lytras et al. (Eds.): WSKS 2009, CCIS 49, pp. 440–446, 2009.

citizens–, Moreno-Jiménez [1], [2], [3] suggested the use of the cognitive democracy model known as e-cognocracy, which apart from improving the transparency, control and participation, pursues the creation of a better society by means of training the citizenry in the scientific resolution of high complexity problems associated with the governance of society.

This cognitive approach to the traditional democracy (e-cognocracy) is based on the evolution of living beings [4] and responds to the new challenges and requirements arisen in the context of the Knowledge Society.

It can be understood as a space oriented to the talent, intelligence and imagination of the human being, the true figure of the new society. Taking advantage of the catalytic potential of democracy in knowledge creation, and that of ICT in its extraction and spreading, e-cognocracy, besides informing the citizen expects training him in the scientific decision making.

When talking in this context about scientific decision making, it has not to be understood –as happens with the traditional approach to Science– as the exclusive consideration of the objective and rational. The new scientific method [4], according to the new Knowledge Society, handles the objectivity, verifiability and rationality of classical science as well as the emotional, intangible and subjective aspects associated with the human factor. To that end it is enough to provide a rational treatment to the emotional or, if preferred, an objective treatment to the subjective [5].

So the new model of citizenry representation (e-cognocracy) favors the evolution of specie since it contributes to alleviating the three cognitive limitations of the human being [6]: ignorance, stupidity and passion. In this sense [5] ignorance refers to the lack of information, stupidity to the lack of training and passion to the lack of "reason", i.e. the control of emotion over reason.

In order the e-cognocracy to be effective when it comes to democratizing knowledge it is necessary to achieve the participation, or rather the implication, of the citizenry in the public decision making. To that end the technological tools used to put the process followed in this cognitive democracy into practice need to be reliable [7].

It is specifically required to warrant the technological properties usually demanded to this kind of electronic participation [8], [9], [10], [11].

In previous work the authors have proposed a specific methodology, e-cognising [10] to guarantee the technological security of the e-voting process in e-cognocracy. A voting with its own characteristics, since several rounds are allowed (usually two), what forces to link the ballots, and since not only the options preferences but also the distribution of intensities among them are incorporated.

In what follows the authors will present the design of their own methodology to warrant the technological security of the other key stage in e-cognocracy: the e-discussion carried out by means of a collaborative tool (forums, wikis…).

2 E-Cognocracy

E-cognocracy seeks to convince citizens and not to defeat them (e-democracy), by aggregating the results obtained from political parties (representative democracy) and citizens (participative democracy) by assigning different weights (w_1 and w_2)

depending on the context of the problem (local, regional, national or supranational) and the objectives of the system. Its characteristics can be seen in [12], [7].

The key idea of e-cognocracy is to educate people (intelligence and learning), promote relations with others (communication and coexistence), improve society (quality of life and cohesion) and construct the future (evolution) in a world of increasing complexity [12].

The e-cognocracy process follows these stages [10]:

1. *Problem Establishment.* In this stage, and by means of the web, the relevant aspects of the problem are identified: context, actors and factors (mission, criteria, subcriteria, attributes and alternatives), as well as their interdependencies.
2. *Problem Resolution.* This stage provides the priorities for the alternatives being compared. To this end we use Analytic Hierarchy Process [13]. The steps of this methodology are: (i) Modelling, where a hierarchical representation of the problem (which should include all the relevant aspects of the decision problem) is established; (ii) Valuation, in which the actors incorporate their judgements through pairwise comparisons between the elements considered in the problem. A positive, reciprocal pairwise matrix is provided for each node of the hierarchy; (iii) Prioritization, where the local and global priorities are obtained by using any of the existing prioritization procedures and the Hierarchical Composition Principle, and (iv) Synthesis, in which the individual total priorities are derived by applying additive or multiplicative procedures. One of the main characteristics of AHP is the existence of a measure to evaluate the inconsistency of the decision maker when eliciting their judgements. Finally, the social priorities of the alternatives are obtained by aggregating the individual preferences or judgments [14].
3. *Model Exploitation.* This stage of the resolution process derives the patterns of behaviour of the actors involved in the resolution process. This is done by considering the incorporation of uncertainty through interval judgments and the identification of the opinion groups and the critical points of the problem by using analytical and graphical visualization tools.
4. *Discussion.* Using any media or collaborative tool (forum in our case), the citizens' representatives (through their respective political parties) and the citizens themselves (through the network) give their motives and justify the decisions. From these comments and messages, the arguments that support the alternatives, as well the attributes that are more relevant in the resolution process are identified. This new knowledge about the problem, explicit knowledge, is put into the network. In this way, the actors involved in the resolution process would learn about the problem and its resolution before beginning a new round in preference elicitation.

   In order to achieve the proposed objectives concerning the knowledge creation and extraction these comments need to be linked to the preferences of the alternatives set up by its author, but not at the expense of the security warrants. Moreover, the messages published can either be anonymous or identified by its author's name, in which case his identity must be verified.

5. *Second round in problem resolution.* After updating the individual preferences with the explicit knowledge derived in the previous resolution stage, the participants are provided with fresh information that may induce them to change their preferences. In a second voting round the priorities of the alternatives are obtained. This process (discussion followed by a later voting round) can be repeated several times.
6. *Knowledge Extraction and Democratisation.* Using the information on the preferences of the two rounds and the quantitative information included in the comments elicited in the discussion stage, the comments that support each pattern of behaviour and each preference structure are identified. From these comments, and by using text-mining techniques, we will determine the motives of preference changes between the two rounds. In this context, and after measuring the changes in preferences, we also identify the social leaders, that is to say, the actors whose opinions provoke these preference changes and whose arguments are followed by the majority [15].

The procedure concludes with the social democratisation of the knowledge through the web. This final step is the main objective of the cognitive democracy we propose to deal with public decision making; e-cognocracy can be considered as one of the most promising formulas for citizens' participation in e-government, which is one the current social demands [16], [17], [18].

3 Security of e-Discussion

In previous work the authors have proposed a specific methodology, e-cognising [10] to guarantee the technological security of the e-voting process in e-cognocracy. A voting with its own characteristics, since several rounds are allowed, what forces to link the ballots, and since not only the options preferences but also the distribution of intensities among them are incorporated. Our system of e-voting warrants the following technological properties:

- **Authentication.** Only voters in the census shall be able to vote.
- **Democracy.** Each voter shall be able to vote only once in each round. If one voter cast two opinions in the same round, he should use the same linking-tag and would be detected.
- **Anonymity.** A voter shall not be linked to its vote. Anonymity is guaranteed by the kind of signature used, and the probability of guessing the actual voter is 1/N, where N is the number of voters in a given group.
- **No coercion**. A voter shall not be able to prove its vote. In exchange for his voting, the voter receives only a signed linking-tag and time-stamp and does not bear a relation to the content of the vote. As the voter is requested to sign a random message, he is prevented from selling his vote (otherwise, the vote could be easily sold by providing both the message and its signature to a third party, who could then proceed to vote).
- **Accuracy**. It shall not be possible to remove a valid vote from the final counting. Each voter signs as a member of a valid group, therefore, he must know the private key of a member, which is impossible to fake provided it has a suitable length.

- **Reliability**. It shall not be possible to include a non-valid vote in the final counting. Each voter has a signature of the linking-tag that he or she sent to the RA and a list of these linking-tags will be published before the recount; therefore, even if RA is compromised, the ballots cannot be deleted, since this action will be reported by the affected voters who will present their signed linking-tags to support their objection.
- **Veracity**. Only each voter can cast its vote. A ballot cannot be sent to RA (even if RA is compromised), because it would be necessary to obtain a valid signature, and that is not possible without the private key of the voter.
- **Verifiability**. Voters shall be able to verify that their vote has been correctly accounted. For each vote received, the RA gives back to the voter a signed linking-tag. Later, when beginning the recount, RA publishes a list of the linking-tags of the ballot. If a voter has a ballot that is not included in the list, he could report it to the RA so that it undertakes the appropriate action.
- **Neutrality**. For each round the vote should be secret until the recount phase. RA decrypts the votes once each round is finished.
- **Linkability**. Two votes from the same voter in different rounds of the voting shall be linked together, but not to the voter who cast them. Every voter uses the same linking-tag in every round. We link all the ballots with the same linking-tag as coming from the same person.

The e-discussion carried out by means of collaborative tools (forums, wikis etc.) is included in the voting process as an ordinary stage in which the openly expressed opinions are linked with the preferences indicated in the vote. Therefore, as just another round in the process, the e-discussion guarantees the same technological properties as the other stages, with the exception of those that, due to the nature of the debate, do not need to be guaranteed. Table 1 shows the relationships of the properties that are guaranteed in this stage of the debate.

Up to now only registered persons have been allowed to express their opinions. At the time present we are analyzing the way of adding the opinion of not registered people and link this opinion with the changes in the preferences of the participants in the problem resolution (registered people).

Table 1. Properties of e-discussion

Authentication	Warranted for all voters in the census
Democracy	Not necessary; it does not need to be warranted since each voter may cast as many votes as he/she wants
Anonymity	Voluntary; it can be warranted by the trust placed on the Recount Authority
No coercion	Not necessary
Accuracy	Warranted
Reliability	Warranted
Veracity	Warranted
Verifiability	Warranted
Neutrality	Not necessary; it does not need to be warranted since the comments are published at the moment they are expressed
Linkability	Warranted

4 Conclusions and Future Work

Following with the analysis of the security from the technological point of view of the process followed in e-cognocracy, this paper focuses on the properties that the e-discussion process proposed as intermediate stage of this cognitive democracy must verify in order to extract the arguments that support the decisions made by the actors involved in the problem resolution.

Considering the discussion as a new round of the e-voting process, Section 3 has specified the desired properties; at present we are analyzing the way of warrant these properties allowing the participation in the discussion of not registered persons. In this sense, the participation in the debate process of people who show interest and may contribute with information and knowledge about the resolution of the problem is advisable from the point of view of the knowledge society, aimed at the joint creation of a better society.

Within this context, the approach we have introduced in this article can be applied to most of the large set of public decision making situations that arise into the administration, such as some real experiences of e-participatory budget allocation, budgetary planning or eliciting the opinion and preferences of the citizenship regarding the location of public services, carried out by our research group *(GDMZ)* for the City Council of Zaragoza or the Regional Government of Aragon (Spain).

References

1. Moreno-Jiménez, J.M.: Las Nuevas Tecnologías y la Representación Democrática del Inmigrante. IV Jornadas Jurídicas de Albarracín. Consejo General del Poder Judicial (2003)
2. Moreno-Jiménez, J.M.: E-cognocracia y Representación Democrática del Inmigrante. Anales de Economía Aplicada (2004)
3. Moreno-Jiménez, J.M.: E-cognocracia: Nueva Sociedad, Nueva Democracia. Estudios de Economía Aplicada 24(1-2), 559–581 (2006)
4. Moreno-Jiménez, J.M., Polasek, W.: E-democracy and Knowledge. A Multicriteria Framework for the New Democratic Era. Journal Multicriteria Decision Analysis 12, 163–176 (2003)
5. Moreno-Jiménez, J.M.: El Proceso Analítico Jerárquico. Fundamentos. Metodología y Aplicaciones. In: Caballero, R., Fernández, G.M. (eds.) Toma de decisiones con criterios múltiples. RECT@ Revista Electrónica de Comunicaciones y Trabajos de ASEPUMA, Serie Monografías, vol. 1, pp. 21–53 (2002)
6. Kaufmann, B.E.: Emotional arousal as a source of bounded rationality. Journal of Economics Behaviour & Organization 38, 135–144 (1999)
7. Moreno-Jiménez, J.M., Piles, J., Ruiz, J., Salazar, J.L., Sanz, A.: Some Notes on e-voting and e-cognocracy. In: Proceedings E-Government Interoperability Conference 2007, Paris, France (2007)
8. Piles, J.J., Salazar, J.L., Ruiz, J., Moreno-Jiménez, J.M.: Security Considerations in e-cognocracy. In: Levi, A., Savaş, E., Yenigün, H., Balcısoy, S., Saygın, Y. (eds.) ISCIS 2006. LNCS, vol. 4263, pp. 735–744. Springer, Heidelberg (2006)
9. Piles, J.J., Salazar, J.L., Ruiz, J., Moreno-Jiménez, J.M.: Enhancing the Trust and perceived security in e-cognocracy. In: Alkassar, A., Volkamer, M. (eds.) VOTE-ID 2007. LNCS, vol. 4896, pp. 125–136. Springer, Heidelberg (2007)

10. Moreno-Jiménez, J.M., Piles, J., Ruiz, J., Salazar, J.L.: E-cognising: the e-voting tool for e-cognocracy. Rio's Int. Jour. on Sciences of Industrial and Systems Engineering and Management 2(2), 25–40 (2008)
11. Salazar, J.L., Piles, J., Ruiz, J., Moreno-Jiménez, J.M.: E-cognocracy and its voting process. Computer Standards and Interfaces (30/3), 124–131 (2008)
12. Moreno-Jiménez, J.M., Polasek, W.: E-cognocracy and the participation of immigrants in e-governance. In: Böhlen, et al. (eds.) TED Conference on e-government 2005. Electronic democracy: The challenge ahead. Schriftenreihe Informatik, vol. 13, pp. 18–26. University Rudolf Trauner-Verlag (2005)
13. Saaty, T.: The Analytic Hierarchy Process. McGraw-Hill, New York (1980)
14. Escobar, M.T., Moreno-Jiménez, J.M.: Aggregation of Individual Preference Structures. Group Decision and Negotiation 16(4), 287–301 (2007)
15. Moreno-Jiménez, J.M., Cardeñosa, J., Gallardo, C.: Arguments that support decisions in e-cognocracy: A qualitative approach based on text mining techniques. In: World Summit on the Knowledge Society, WSKS 2009 (2009)
16. Grant, G., Chau, D.: Developing a Generic Framework for E-Government. Journal of Global Information Management 13(1), 1–30 (2005)
17. Tan, C.-W., Pan, S.L., Lim, E.T.K.: Managing Stakeholder Interests in e-Government Implementation: Lessons Learned from a Singapore e-Government Project. Journal of Global Information Management 13(1), 31–53 (2005)
18. Chen, Y.N., Chen, H.M., Huang, W., Ching, R.K.H.: E-Government Strategies in Developed and Developing Countries: An Implementation Framework and Case Study. Journal of Global Information Management 14(1), 23–46 (2006)

Business and IT Alignment: An Evaluation of Strategic Alignment Models

Mohamed El Mekawy, Lazar Rusu, and Nabeel Ahmed

Department of Computer and Systems Science (DSV),
Stockholm University/Royal Institute of Technology (KTH), Sweden
{moam,lrusu}@dsv.su.se, nahaseeb@kth.se

Abstract. The process by which business and IT are brought inline with each other to enhance the performance of business and to achieve business goals is called strategic alignment. Some models have been developed for the assessment of strategic alignment but an evaluation and comparison of capabilities of the models has not been studied due to the absence of a structured evaluation criteria.. The purpose of this paper is to evaluate the strategic alignment models, collected through a comprehensive survey. For this purpose, we develop a strategic alignment evaluation framework which is a composition of a number of criteria. We further use this framework for the evaluation of six main alignment models. These models present concrete ways to evaluate strategic business IT-alignment.

1 Introduction

Business and IT are said to be aligned if the functions of IT are developed inline with the business to achieve business objectives [1, 2, 3]. IT is employed to raise the performance of business to achieve business goals [3, 4]. Alignment is not a static process, instead it is an ongoing process. Therefore, for each strategic change in business it is required to re-access the appropriateness of business-IT alignment.

Emergence of internet has given businesses a new trend and as a result IT has emerged as an established tool for business optimization [5, 6]. However, for an efficient use of IT, it is required to align it with business and its use has been a major concern for CIO's to achieve business objectives [7]. Misalignment may lead to poor performance in business [8], increasing inefficiencies [9] and hindrance of smooth running [10] of a business. Therefore, it is important to study the evaluation of alignment in-order to assess its impact on business. Efforts have been made to measure business-IT alignment, as a consequence alignments models are developed to be used for evaluating business and IT alignment.

Over the years, business and IT alignment has been studied and some models [3, 11, 12] have been developed. However, a comprehensive approach for measuring the capabilities of models has never been studied. Although some studies are available like [2, 11] but a structured comparison of business-IT alignment models is missing. This is due to the absence of established criteria that can be used for evaluating the models. Criteria are available in literature, but they are spread across many sources.

M.D. Lytras et al. (Eds.): WSKS 2009, CCIS 49, pp. 447–455, 2009.

For example, different studies [2, 11, 13, 14] have emphasized different criteria. The purpose of this paper is to study and compare the capabilities of alignment models collected through a comprehensive survey. But a comprehensive method for evaluating the models is not available.

Therefore, we have developed an alignment model's evaluation framework. This framework consists of a number of criteria sorted into four groups for better understanding and for simplicity of presenting. Most of the criteria are collected from literature and some important criteria have also been developed. Furthermore, this structured framework is used to evaluate the capabilities of alignment models.

The rest of the paper is organized as follows: in section 2 we present a brief introduction of business-IT alignment models under investigation. Section 3 contains the methodology used in the paper. Section 4 contains a framework (remember it consists of criteria) for the evaluation of alignment models. Section 5 contains the experiments done to compare the models and section 6 contains the results of the models' evaluation. Finally, we present our conclusions in section 7.

2 Research Methodology and Importance of Alignment

The purpose of this paper is to evaluate and compare capabilities of business-IT alignment models.

For that, we did a comprehensive survey to collect business-IT alignment models. These are the models under investigation in this paper. An established framework for evaluation of capabilities of business-IT alignment is missing. Although, a number of criteria are available but they are spread across various sources, and different researchers have emphasized different aspects. However, criteria are not presented in a structured form. Therefore, in this paper we have developed a framework called alignment model's evaluation framework. The framework consists of 23 criteria collected from different sources and presented in a structured way. The structure consists of four groups.

In this paper present the results of the strategic alignment models' evaluation and a comparison of business-IT alignment.

Importance of Alignment. The importance of alignment can be seen not only by the documented research stating its pros but also by scrutinizing the impact of misalignment and its cons to an enterprise [21]. Research also advocates alignment playing a key role in better understanding and improvement of business-IT relationship. Enterprises today, by seeing alignment significance, do not only focus on their business objectives but also try to lean their infrastructure with IT innovation [1] In the hunt of alignment significance, researchers found both the efficiency and effectiveness of business-IT alignment. Effectiveness entails doing the things right on a right time while efficiency entails doing things right, former being long-term and latter being short term [17].

Why alignment is important does not remain the question, in fact how it can be achieved and matured is the real concern of business executives nowadays. From business strategy to IT-Governance, all entails a dynamic multi-dimensional superset of alignment components, such as organization infrastructure and IT-infrastructure and their processes, changing the status of alignment, from a serene to a continuous multidiscipline developing process [18].

3 Alignment Models

In this section we present a brief introduction of the business-IT alignment models under investigation. In order to collect these models we searched through business related digital libraries (by Emerald, Elsevier, IEEE, ACM, Inderscience, Taylor & Francis) and several other journals indexed by ISI. For that we used several keywords like, alignment models, business IT alignment, etc.

On searching we found 19 models and finally selected 6 models for evaluation and comparisons because the rest of 13 authors don't present a concrete model for evaluating business-IT alignment. Out of the six models, four are the core alignment models whereas 2 models are extensions of existing models.

1. ***Strategic Alignment Model – SAM.* (1993).** Henderson and Venkatraman proposed a strategic alignment model in 1993 which became till now the most famous and discussed one among researchers [11]. SAM is based on building blocks known as strategic fit and functional integration. It represents a distinction between the internal focus of IT (i.e. IT processes and infrastructure) and external perspective of IT (i.e. IT Strategy [15]). It also distinguishes in the same way between the internal focus of business and the external perspective of business. Each of the division further subdivides into different alignment perspectives; former splits in to strategy executer and technology transformer, and latter splits up to competitive potential and service level [2].
2. ***Integrated Architecture Framework – IAF. (2000).*** [3] As the name portrays, that it corresponds mainly to supporting the integrated architectural design of business and IT. The authors of IAF advocate architectural design as a catalyst in aligning business and IT. They also enhanced the SAM model by introducing the architecture of the information / communication / knowledge infrastructure by splitting up a single internal domain into structural and operational levels. In fact, by representing IAF, they divide alignment into structural and operational alignment which relates to variable like architecture and capabilities, and processes and skills respectively. The model follows a top down approach for aligning between main architecture areas.
3. ***Luftman's Alignment Model – LAM. (2000).*** [12] One of the most elaborated models in the information systems domain by J. Luftman, .The model presents strategic alignment as a complete holistic process which encompasses not only establishing alignment but also its maturity by maximizing alignment enablers and minimizing inhibitors. LAM follows a bottom-up approach by setting goals, understanding the linkage between Business and IT, analyzing and prioritizing gaps, specifying the project management actions, choosing and evaluating success criteria and consequently sustaining alignment by all these processes.
4. ***Reich & Benbasat Model – RBM. (2000).*** [16] One of the most discussed models in the strategic information systems domain is the Reich & Benbasat Model (RBM). The authors of RBM lay down factors related to the social dimension that can potentially apprehend alignment between business and IT objectives. The four factors included; shared domain knowledge, IT implementation success, communication between business and IT executives, and connections between business and IT planning. The Model follows a

top- down approach by focusing on the antecedents along with the current practices that directly influence alignment. No doubt, that there are two dimensions of strategy creation, namely intellectual and social, but the authors of RBM have selected the social domain as it would more scrutinize people involved in the creation and maturity of alignment.

5. ***Sabherwal and Chan Alignment Model – SCAM. (2001)*** [17] This model has an emphasize on strategic content rather than processes, realized rather than intended strategies, and IS strategy rather than IT and information management (IM) strategies. They use business strategy types from Miles and Snow's [22] typology of defenders, analyzers and prospectors. They have defined business strategy and IT strategy as the two main domains for alignment. Further, each one of them is divided in attributes and types.
6. ***Hu Huang Alignment Model – HHAM. (2006).*** [18] Hu and Huang added relationship management as an antecedent along with using a Balanced Scorecard as a tool and expanded the RBM for achieving, managing and sustaining alignment. The authors advocate a top-down approach to create an effective alignment system. Moreover, they have used the balanced score-card not as a ready-made performance indicator but rather as a management system (or IT-Management Tool) by using four different perspectives of an enterprise's business, named as innovation and learning perspective, internal process perspective, customer perspective and financial perspective [19, 20].

4 Strategic Alignment Models' Evaluation Framework

In this section we present the alignment model's evaluation framework which is further used in section 5 for the evaluation of business-IT alignment models. The framework consists of twenty three criteria collected from literature. For simplicity in understanding we have grouped them into four groups. The criteria are collected through a combination of quantitative and qualitative approaches called triangulation.

Selection of the criteria is based on a) a series of interviews with key business and IT persons from 37 organizations, b) published empirical studies, c) classified into criteria related to design, organizational, analysis and utilization aspects of model. After many considerations of different studies and interviews, we have identified twenty three (23) criteria that are used as a benchmark for evaluating the capabilities of the business-IT alignment framework. Definitions of these criteria are as follows:

Design-Related Group: Includes all criteria that deal with the design aspects of the model and how it was originally built. It also draws attention on different domains of a model and their relation with organizations' business and IT domains.

- Conceptualizing and directing the emerging area of strategic management of IT.
- Does the model consider architecture of business and IT units?
- Consideration of business objective and goals. Does the model centralize the business goals?
- Modularity; does the model support phase wise analysis which provide a separate picture of every module?
- Method of development; how the model is developed, is it theoretical, survey or a case study?

Organizational-Related Group: Includes all criteria that are related to the governance of organisations, and how IT and business work are organised. The criteria also refer to the size of organisation and the business domain where the organisation acts in its marketplace.

- Organizational size. Large, medium, and small
- Does the model leverage IT capabilities to shape and support their business strategies?
- Futurity. Does the model understand the potential of IT for tomorrow's organizations?
- Domain specific. Is the model useful for a specific set of business?

Analysis-Related Group: Includes all criteria that show how the model goes deeply to analyse different aspects of business and IT domains in organisations.

- Assessment Purpose, possible values are IT investment, perspective (strategic role), productivity or business value, factors that impact the alignment.
  - Level of IT investment. Does the model consider the effect of IT investment in the organization?
- Level of correspondence. The level of which the model can be used (Strategic, tactical, and operational)
- Information strategy content. Which level of information strategy is considered in the model? The possibilities are IS, IT, IM.
- Level of planning. At which level business and IT strategic planning is integrated? (functional, business process, intra organizational, and inter organizational)
- Level of business aspects. To which level does the model apply on business aspects?
- Measurability. Can the alignment be quantified by the model?
  - If yes, can the measurement be verified?

Utility-Related Group: Includes all criteria that deal with how the model can be used and applied on organisations. It also considers different levels of integration, dividing tasks and taking responsibility.

- Flexibility. Is the model capable to measure alignment in case of changes in business or IT?
- Risk aversion (Level of risk). Does the model give any indication about expected risk areas in an organization?
- Strategic Fit. The interrelationship between external and internal components.
- Strategic integration. The integration between business and IT aspects.
- Complexity of using and applying. Who can use the model and apply it? (Top business/IT management people, and senior managers)?
- Effectiveness and efficiency. Short-term and long term reliance.
- Sharing responsibility and risk. Does the model help to assess the level at which risks and responsibilities can be shared between business and IT?
- Error Accumulation. Does the model's components (dependent/independent), will effect the overall quality of the model?

Table 1. Evaluation of Strategic Alignment Models

		SAM 1993	IAF 2000	LAM 2000	RBM 2000	SCAM 2000	HHAM 2006
Design Related	1. Conceptualizing the IT	+/+	+/+	+/+	+/-	+/-	+/+
	2. Architecture consideration	+/+	+/+	+/+	+/-	-/-	-/-
	3. Consideration of Buss. Goals	+/-	+/+	+/+	+/-	+/-	+/-
	4. Modularity	+/+	+/+	+/+	+/-	+/+	+/-
	5. Method of Development	Th.	Th/CS	Survey	Survey	Survey	Th/CS
Organizational Related	6. Organizational Size	L/M/S	L/M/S	L/M/S	M/S	M/S	L/M/S
	7. Leverage IT capabilities	+/+	+/+	+/+	+/-	+/-	+/-
	8. Futurity	+/+	+/+	+/+	+/+	+/-	+/+
	9. Domain Specific	+/+	+/+	+/+	+/-	+/-	+/+
Analysis Related	10. Assessment Purpose	Per	Per/Fact	Per/Fact	Fact	Per/Fact	Per/Fact
	11. Level of Correspondence	Str/Tac	Str/Tac/Opr	Str/Tac/Opr	Str/Tac	St/Tac	Str/Tac/Opr
	12. Information Strategy Content	IT/IS	IT/IS/IM	IT/IS/IM	IT	IS	IT/IM
	13. Level of Planning	Bus. Pros.	Bus. Pros.	Func.	Intra.	Intra	Bus. Pros.
	14. Level of Business Aspects	+/-	+/+	+/-	+/-	+/-	+/+
	15. Measurability of Alignment	+/+	+/+	+/+	+/-	-/-	+/+
Utility Related	16. Flexibility	+/+	+/+	+/+	-/-	-/-	+/-
	17. Risk Aversion	+/-	+/+	+/+	-/-	-/-	+/-
	18. Strategic Fit	+/+	+/+	+/+	+/-	+/-	+/+
	19. Strategic Integration	+/+	+/+	+/+	+/+	+/+	+/+
	20. Complexity of Use and Apply	Med	Com	Com	Sim	Sim	Com
	21. Effectiveness & Efficiency	+/-	+/+	+/+	+/+	+/-	+/+
	22. Sharing Responsibility & Risk	+/-	+/+	+/+	-/-	+/-	+/+
	23. Error Accumulation	+/+	+/+	+/-	+/-	+/-	+/-

Th = Theoretical, CS = Case Study. L = Large, M = Medium, S = Small. Per = Perspective, Fact = Factors. Str = Strategic, Tac = Tactical, Opr = Operational. IT = Information Technology, IS = Information System, IM = Information Management. Bus.Proc = business process, Fun = functional, Intra = Intra-organizational, inter = inter-organizational. Med = Medium, Com = Complex, Sim = Simple.

5 Results of Evaluation

In this section we present the evaluation results the strategic alignment models under investigation. For the evaluation of every criterion, it is studied that a) whether discussion about the criterion is given in the paper or not, b) whether the criterion is included in the model or not. The evaluation results are presented in z/x format, where the value of z is + if discussion about the criterion is available, and - if there is no discussion about the criterion. Similarly, the value x is + if the criterion is included in the model and - if the criterion is not included. The detailed results are presented in table 1.

There are some exceptional criteria whose results cannot be presented in z/x format. Therefore, we use some abbreviations to present results of the evaluation. The details of abbreviations are given under the table. Here, we briefly discuss some evaluation' results of models.

Two models (IAF and LAM) have not only considered business goals but also business goals are part of the alignment model. In contrast, the remaining models (SAM, RBM, SCAM, HHAM) include some discussion about goals, however, the goals do not become a part of the alignment model.

Most of the models (SAM, IAF, LAM, HHAM) are applicable for all sizes (small, medium, and large) of organizations. Contrarily, RBM and SCAM are applicable only to medium and small size organizations.

In half of the models (SAM, IAF, HHAM), business and IT strategic planning integration is at low level (also called business process level). While in LAM, it is at functional level. Contrarily, in RBM and SCAM, it is at a high level called intra organizational level.

The complexity of IAF, LAM and HHAM is high therefore business people cannot apply these models directly. However, RBM and SCAM are relatively simple to apply on an organization. Therefore both business and IT people can apply these models for the alignment of business and IT. In contrast to the above, the complexity of applicability of SAM is medium.

6 Discussion and Conclusions

Strategic alignment is an on going process that aligns business and IT to effectively achieve objectives of an enterprise. In the past, models for alignment of business and IT are presented, however an evaluation and comparison of these models is missing. This is due to the absence of structured criteria that can be used for evaluating the strategic alignment. Therefore, we develop a framework based on criteria that is used for the evaluation of strategic alignment collected through a comprehensive survey. In table 2, we conclude our evaluation about business-IT alignment models.

There are also some limitations of this study which we plan to overcome in an extended version of the paper. In section 4, the criteria are described very briefly and motivation of including each criterion is not presented due to space limitations. The motivation for selection of each criteria and its related reference will be presented in the extended version of the paper.

Table 2. Evaluation of Strategic Alignment Models

	Evaluation
SAM 1993	It is a comprehensively designed model because it differentiates domains and sub domains of all parts of both business and IT. However, it does not go to the operational level. Although the model can be applied to measure the alignment but it is does not highlight the risks in the company.
IAF 2000	As it is an extension of SAM, it deals with the drawback of SAM. The improved model can go deeper to the operational level for the analysis. Business and risk aspects also can be measured. However, these added components increase the complexity of using and applying the model.
LAM 2000	One of the well established and comprehensive models as it follows bottom up approach starting from the factors that affect the alignment towards business and IT domains at the top level. One of the few drawbacks is the complexity of applying and using the model, and the need for expert having understanding of both business and IT.
RBM 2000	Although it has added social factor as a new added aspect to the alignment which has not been considered by other models. However, it lacks the deep analysis of business domain within an organization.
SCAM 2000	The model is not complete as it focuses on general aspects of business and IT domains without going deeply for e.g. IT and business architecture, as well as it works only at strategic and tactical level, without considering business and IT processes.
HHAM 2006	It is an extension of RBM. It has added balanced score card as a very useful tool for enhancing business measurability and how it can be integrated with IT through all business processes. Therefore, it also added analysis at operational level. However, the model lacks organizational and architectural aspects as well as modularity. The model remains also complex with the balance score card implementation.

References

[1] Luftman, J.: Assessing Business Alignment Maturity. Communications of AIS 4, Art. 14 (2000)

[2] Strassman, P.A.: The Squandered Computer, Evaluating the Business Alignment of Information Technologies. Strassmann, Inc. (1997)

[3] Maes, R., Rijsenbrij, D., Truijens, O., Goedvolk, H.: Redifining Business – IT alignment through a unified framework. PrimaVera Working Paper 2000-19, Universiteit van Amsterdam (2000)

[4] Love, P.E.D., Holt, G.D.: Construction business performance measurement: the SPM alternative. Business Process Management Journal 6(5), 408–416 (2000)

[5] Kehoe, D., Boughton, N.: Internet based supply chain management: A classification of approaches to manufacturing planning and control. International Journal of Operations & Production Management 21(4), 516–525 (2001)

[6] Vollmann, E.T., Berry, W.L., Whybark, D.C., Jacobs, F.R.: Manufacturing Planning and Contol systems for supply chain management. McGraw-Hill Pub., New York (2005)

[7] Hu, Q., Huang, C.D.: Aligning IT with Firm Business Strategies Using the Balance Scorecard System. In: Proceedings of the 38th Annual Hawaii International Conference on System Sciences, HICSS (2005)
[8] Pongatichat, P., Johnston, R.: Exploring strategy-misaligned performance measurement. International Journal of Productivity and Performance Management 57(3), 207–222 (2008)
[9] Piplani, R., Fu, Y.: A coordination framework for supply chain inventory alignment. Journal of Manufacturing Technology Management 16(6), 598–614 (2005)
[10] Mitchell, R.K., Morse, E.A., Sharma, P.: The transacting cognitions of nonfamily employees in the family businesses setting. Journal of Business Venturing 18(4), 533–551 (2003)
[11] Avison, D., Jones, J., Powell, P., Wilson, D.: Using and Validating the Strategic Alignment Model. Journal of Strategic Information Systems 13(3), 223–246 (2004)
[12] Luftman, J., Papp, R., Brier, T.: Enablers and Inhibitors of Business-IT Alignment. Communications of the Association for Information Systems 1(11) (1999)
[13] Earl, M.J.: Integrating IS and the organization: A framework of organizational fit, In Information management: The Organizational Dimension. Oxford University Press, Oxford (1996)
[14] Tallon, P.P., Kraemer, K.L.: Investigating the Relationship between Strategic Alignment and IT Business Value: The Discovery of a Paradox. In: Shin, N. (ed.) Creating business value with information technology: challenges and solutions. IGI Global (2003)
[15] Henderson, J.C., Venkatraman, N.: Startegic Alignment: Leveraging Information Technology for transforming Organizations. IBM Systems Journal 38(2&3) (1999)
[16] Reich, B.H., Benbasat, I.: Factors That Influence The Social Dimension of Alignment Between Business And Information Technology Objectives. MIS Quarterly 24(1), 81–113 (2000)
[17] Sabherwal, R., Chan, Y.E.: Alignment Between Business and IS Strategies: A Study of Prospectors, Analyzers, and Defenders. Information Systems Research 12(1), 11–33 (2001)
[18] Hu, Q., Huang, C.D.: Using the Balanced Scorecard to Achieve Sustained IT-Business Alignment: A Case Study. Communications of the Association for Information Systems 17, 181–204 (2006)
[19] Kaplan, R.S., Norton, D.P.: The Balanced Scorecard: Measures That Drive Performance. Harvard Business Review 70(1), 71–79 (1992)
[20] Otley, D.: Performance Management: A Framework for Management Control Systems Research. Management Accounting Research 10(10), 363–382 (1999)
[21] Luftman, J., Papp, R., Brier, T.: The Strategic Alignment Model: Assessment and Validation. In: Proceedings of the 13th Annual International Conference on Information Technology Management Group of the Association of Management (AoM), pp. 57–66 (1995)
[22] Miles, R.E., Snow, C.C.: Organizational Strategy, Structure and Process. McGraw-Hill, New York (1978)

A Study for the Organization of the Greek Publishing Enterprises at the Beginning of the 21st Century

Petros Kostagiolas[1], Banou Christina[1], and Stelios Zimeras[2]

[1] Department of Archive and Library Science,
Ionian University, Ioannou Theotokis 72, Corfu 49100, Greece
[2] Department of Statistics and Actuarial-Financial Mathematics
University of the Aegean, 83200, Karlovassi, Samos, Greece
pkostagiolas@ionio.gr, cbanou@ionio.gr, zimste@aegean.gr

Abstract. Book is indeed a unique commodity. The book publishing industry internationally is deeply influenced by the changes resulting from the increasing competition and the rapid introduction of novel information technologies and the Internet. In that framework, new departments, activities, roles and responsibilities are indeed emerging. This paper discusses issues for the organization of book publishing companies and provides an overview of the Greek publishing industry. Thereafter, the results of a survey concerning the structural organization of the publishing companies in Greece are presented. The study has been conducted during the second quarter of 2007 through a specially designed semi-structured questionnaire. In the study participated 123 representatives of Greek publishing houses, of all sizes and categories, from a total of 239 publishing houses that took part at the Pan-Hellenic Book Exhibition organized by the Athens Book Publishers Association.

Keywords: book publishing industry, publishing companies, organization, management, empirical research, Greece.

1 Introduction

The publishing industry since the invention of printing and the "Printing Revolution" [1] has been in the core of information evolution [2]. Over the last two decades, significant changes have taken place in the publishing industry worldwide [3, 4, 5]. Some of the features of today's publishing industry are the predominance of large publishing groups and the acquisitions and mergers between large and small publishing houses [3]. Nowadays, the current "New Information Revolution", being the fourth one, after the invention of writing, of the written book and of the printed book [6] has redefined and still develops the publishing policy, the organizational structure and the management of publishing companies.

The printed book is a unique product [7] due to its dual nature as content and as object. Due to this dual nature, the life cycle of the book as a commodity exceeds the biological limits of human life. The book through its commercial and cultural dimension serves different uses, depending on the cultural, scientific and social life [8, 9]. The impact of the new information technologies and the knowledge society is

M.D. Lytras et al. (Eds.): WSKS 2009, CCIS 49, pp. 456–465, 2009.

undoubtedly great in the current, changing period and in the hybrid environment where the printed publications coexist with the electronic/digital ones [10]. It must also be pointed out that the printed book nowadays flourishes [2]; in that framework, book publishing companies are called upon to satisfy the special needs of the reading audience, to exploit the opportunities of technology and to meet the demand for innovation. The purpose of the paper is to investigate the way in which current changes in the publishing industry and the information technology influence the management and organization of book publishing companies. The profile of the Greek book publishing companies is outlined and an empirical study is presented. The empirical study has been conducted during the National Book Exhibition from the 3[d] to the 14th of May 2007 in Athens which was organized by the Athens Book Publishers Association.

2 Issues for Management and Organization of Publishing Companies

Since the first decades after the invention of printing, the division of labour in the printing-publishing companies of the time was substantial, especially in large companies, such as the one of Christoph Plantin in Antwerp [11]. During the 19th century, the Industrial Revolution offered new opportunities to the publishing market as to the production, distribution and promotion of books [8, 12]. Therefore, the need for further segmentation and systematization of the publishing work emerged.

The course of the text from the author to the reader presupposes the appropriate organization of the publishing activity. Nowadays, this activity is more complex than ever and is based on the knowledge intensity and the expertise of all parties involved in the publishing activity [13, 14]. It is noteworthy that the author can proceed to the self publishing of the work on the Internet [10]. The Open Access movement and self publishing form an additional reason for the redefining of the role and the policy of the publishing company [15].

Traditional organization models of the publishing companies are based on the central role played by the publisher, who is called to take all the important decisions concerning the selection, design and publication of the titles (list building), as well as the development of the marketing and promotion strategies. Diachronically, the publisher has been the protagonist, deciding, concentrating almost all the responsibilities and supervising all the phases of the publishing activity. He created the publishing policy and the profile of the publishing house, depending on his/her ideology, aesthetic perceptions, taste, values and objectives [2].

The "printing revolution" was the first one to reduce the time and the cost of spreading information and knowledge [1, 8]. New jobs related to the new means of information and to the development of printing techniques brought upon significant changes in the production, distribution and consumption of books [8]. In that context, the publisher-printer was collaborating with scholars, editors, correctors, translators, bookbinders, artists for the design, production and distribution of his publications [8, 12]. These collaborations expressed the aims, the values and the policy of the publisher-printer. Thereafter, the Industrial Revolution, due to technological, economic and social changes, further systemized the production, distribution and promotion of printed books, while the French Revolution had pointed out that information and the printed book was a primary good [12]. The enlargement of the reading audience altered the publishing policy and the organisational structure of publishing companies.

Nowadays, the organizational structure of publishing houses has evolved compared to the past, due to the new needs and expectations created by globalization and competition and due to the demand for specialized publishing activity according to the kind of text produced [10]. According to the Association of American Publishers [16], the departments that may form a publishing house are the following: administrative, advertising, audio, art and design, editorial, finance, human resources, information technology, internet development, legal contracts, managing editorial, marketing, production, publicity, publisher's office, purchasing, sales, subsidiary rights and permissions. Naturally, the size of the publishing enterprise plays an important role to the (partial or not) development of a formal organizational structure, thus to all the departments in a publishing house. In smaller publishing enterprises, for instance, there are only some of the above mentioned departments, whereas the responsibilities and work of some of them are combined.

Table 1. The annual production of titles and the number of publishing companies for the last six years (source for years 2001, 2003, 2004: National Book Centre, 2007; source for years 2002, 2005, 2006, 2007: National Book Centre, 2008)

Publishing enterprises categories	**Number of publishing companies and percentages of book production**						
	2001	**2002**	**2003**	**2004**	**2005**	**2006**	**2007**
Large	14	16	20	18	20	19	21
Percentage of annual book production	33.8%	32.5%	38.3%	35.4%	36.5%	36.9%	38.0%
Medium	148	151	138	148	161	166	163
Percentage of annual book production	47.6%	47.5%	44.5%	44.1%	43.5%	43.2%	43.0%
Small	476	552	460	563	618	636	614
Percentage of annual book production	18.1%	19.4%	16.8%	19.6%	19.0%	18.7%	17.9%
Total	638	719	618	729	799	821	798
Percentage of annual Production	100%	100%	100%	100%	100%	100%	100%

3 An Overview of the Publishing Market in Greece

The publishing industry in Greece, even though not isolated from global developments, presents traditional unique characteristics, mainly due to the language. Therefore, foreign conglomerates and large publishing groups do not conduct any significant activities in Greece till now; consequently, traditional, "family" structures survive [2]. Thus, in Greece the publishing houses are usually family companies, where the administration is hereditary and the publisher is also the owner and the manager [2]. The Greek publishing industry constitutes a market rather representative of the small publishing markets [2]. A categorization of the publishing companies in Greece is based on the size of a publishing house, i.e. Small Publishing Companies (annual production up to nine titles), Medium sized Publishing Companies (annual production from ten to eighty titles), Large Publishing Companies (annual production exceeding eighty titles). Additionally, the publishing houses are distinguished, depending on the type of the published text on which they focus [10] to: General, Scientific-Technical-Medical, educational publishers, reference books publishing, children's books publishing, university/academic presses. The classification suggested by the National Book Centre of Greece is equivalent, with the main thematic sections being formed as follows [17,18]: A. humanities and social sciences, B. literature, C. children's books, D. pure and applied arts, E. general, practical and self-help books, F. art books and G. auxiliary school books and ELT.

Table 1 provides the number of publishing enterprises belonging to each of the three size categories (large, medium and small) with the corresponding percentage of annual new title book production. Over the years, the increase of overall new book title production is quite substantial. In 2007, the number of new publications reached the 9,566 whereas in 2002 the new titles were 7,893, whereas a decade ago were about 3,500 [17, 18]. For example, in 2006 the large publishers were nineteen, the medium size publishers were 166 and the small ones reached the 636 (Table 1). For the same year the large publishers produced around 36.9% of new titles book production, the medium size publishing houses produced the 43.2% and the small publishers, although significantly larger in numbers produced only the 18.7% (Table 1). In Greece the medium size publishers continue to form a substantial part of the publishing activity.

Table 2. The annual production of titles per thematic category of publishing companies during the last six years (source: National Book Centre, 2008)

Categories depending on the subject	Number of titles and percentage per year					
	2002	2003	2004	2005	2006	2007
Humanities & social sciences	2,037 (25,8%)	2,077 (26.0%)	2,223 (26.0%)	2,290 (26.6%)	2,422 (25.1%)	2,421 (25,3%)
Pure and applied arts	593 (7.5%)	676 (8.5%)	632 (7.4%)	619 (7.2%)	622 (6.4%)	587 (6,1%)
Literature	1,720 (21.8%)	1,766 (22.1%)	1,782 (20.9%)	1,925 (22.4%)	2,014 (20.9%)	2,084 (21,8%)
Children's books	1,479 (18.7%)	1,481 (18.6%)	1,722 (20.2%)	1,822 (21.2%)	2,115 (21.9%)	1,781 (18,6%)
Auxiliary school books & ELT	907 (11.5%)	627 (7.9%)	611 (7.2%)	474 (5.5%)	803 (8.3%)	775 (8,1%)
Art books	420 (5.3%)	451 (5.7 %)	571 (6.7%)	537 (6.2%)	686 (7.1%)	726 (7,6%)
General, practical and self-help books	737 (9.3%)	902 (11.3%)	998 (11.7%)	943 (11.0%)	985 (10.2%)	1,192 (12,5%)
Total	7,893 (100%)	7,980 (100%)	8,539 (100%)	8,611 (100%)	9,648 (100%)	9,566 (100%)

Table 2 provides evidence for the annual new book titles production per thematic category between the years 2002 to 2007. The category "humanities and social sciences" holds the leading part in book production, with a percentage of 25.3% in 2007. "Literature" has also a high percentage (21.8%) and the category "General, practical and self-help books" follows with 12.5% of the total printed book production in 2007. "Auxiliary school books & ELT" (8.1%), "pure and applied arts" (6.1%), "art books" (7.6%) are categories aiming at a special reading audience. Compared to previous years (Table 2), it is observed that the number of produced titles per year in children's books was gradually increasing till 2006 (in 2002 the percentage amounted to 18.7%, whereas in 2006 it reached 21.9%), and in 2007 there is a decrease (18.6%). Concerning Art Books and General, practical and self-help books no fluctuation is observed in their annual publication. Auxiliary school books & ELT, although presented a fall since 2003, are now gradually rising and reach 8.1% of the book production per thematic category. Finally, a decrease is observed in books of "Pure and Applied Arts" (6.1%), while "Literature" (21.8% in 2007) is rather steady. The above synopsize the demand and the profile of the reading audience.

4 An Empirical Study for the Publishing Enterprises in Greece

The purpose of the study is to exhibit the organizational structure of the Greek publishing enterprises. For the empirical study, which was carried out, a structured questionnaire was developed, comprising twenty (20) closed-ended questions and one open-ended question. A pilot study was first conducted in order to improve the reliability of the study questions as to clarity and relativeness to the objectives of the study. The pilot study was conducted by experts from both the academic area and the publishing industry. In its final form, the structured questionnaire comprised three sections with twenty (20) closed-ended questions and an open-ended research question. In Section A. included eight questions regarding basic information for the publishing house and the responded (name, years of operation, annual output of new titles, number of employees, thematic category/categories of titles, annual turnover, existence of printing infrastructure and existence of responsible for title selection). Section B. included questions regarding the existence or non existence of specific departments (administrative, advertising, audio, art and design, editorial, finance, human resources, information technology, internet development, legal contracts, managing editorial, marketing, production, publicity, publisher's office, purchasing, sales, subsidiary rights and permissions). Finally, Section C. included an open-ended question providing the opportunity for making further comments. The questionnaire was distributed in May 2007 during the National Pan-Hellenic Book Exhibition of the Athens Publishers-Booksellers Association. In the survey representatives from 239 publishing houses were asked to participate, while the 123 of them agreed to participate, i.e. a response rate of around 51.5%. However, it is significant to note that 19 out of the 21 large publishing companies did agree to participate in this study.

4.1 Study Results

The majority of the publishing house representatives (52.03%) stated that the publishing house operate continuously for at least twenty years; while in regard to the annual output of titles, the 41.46% (51 publishing companies) of the participants stated that they publish up to 9 titles per year (small publishers), 40.65% (50 publishing houses) stated that they produce ten to eighty titles (medium size), whereas 15.45% (19 publishing companies) that their annual production exceeds eighty titles (large publishing houses). According to the responses concerning to the number of the employees working in the publishing companies which participated in the study, the 63.41% (78 publishing companies) stated that they employ one to nine people, the 19.51% (24 publishing companies) employ ten to forty nine people, whereas the 8.13% (10 publishing companies) employ up to two hundred and fifty people.

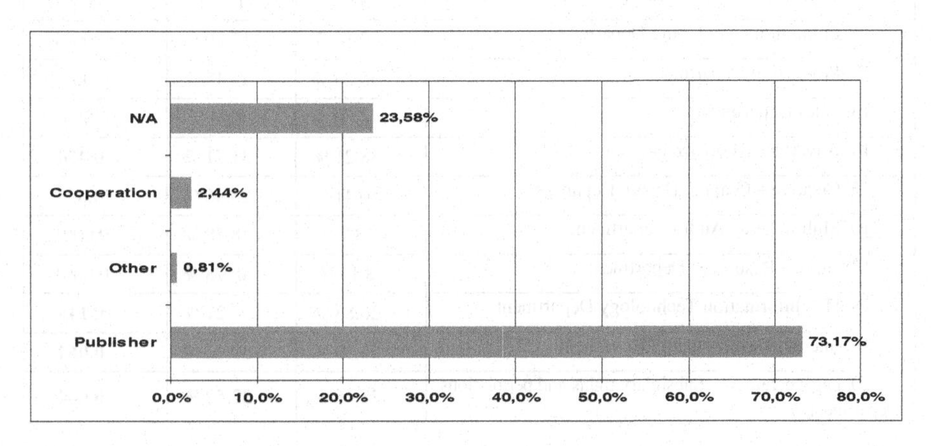

Fig. 1. Results on the main responsibility for the selection of titles under publication (Section A –Question 8)

In the question relative to thematic categories, 35.0% of the participants stated that they focus on general books (literature etc.), 19.0% on children's books, 18.0% on scientific-technical-medical books, 2.0% on auxiliary school texts and ELT, and 12.0% on other categories of titles. In regard to the annual turnover, 47.15% of the participants (58 publishing companies) stated that it does not exceed two MEUROS, 7.32% (9 publishing companies) that it ranges between 2 and 10 MEUROS, whereas the other participants did not answer the question. In the question (Section A –Question 7) relative to the existence or non existence of printing infrastructure 15.45% (19 publishing companies) they stated that there are printing facilities in the publishing house and 76.42% (94 publishing companies) that there are not.

The last question of the first section of the questionnaire (Section A –Question 8) concerns the main responsibility for the selection of titles to be published, whether it depends entirely on the publisher, or whether it is conducted in collaboration. It is interesting that 73.17% of the participants (90 publishing companies) answered that the publisher has exclusively the responsibility and decides on the selection of the titles to be published (Fig. 1). Only a small percentage 2.44% (3 publishing companies) answered that the publishers decide in collaboration with the heads of the several publishing house's departments, whereas only 0.81% (1 publishing company) answered that somebody else decides in the publishing house. Therefore, the main responsibility for the compilation of the publishing house's catalogue (backlist and frontlist) and the development of the publishing policy belongs to the publisher.

Table 3. Existence or non existence of departments (Section B–Questions 9 to 20) in the organization of publishing companies, according to the answers of the participants

Section B Publishing companies' organization structure	Specific departments existence or non existence		
	YES	**NO**	**p<0.05**
9. Editorial Department	75.61 %	13.01 %	0.0001
10. Finance – Accounting Department	69.92 %	18.7 %	0.0001
11. Promotion - Marketing Department	60.16 %	17.89 %	0.0001
12. Production Department	55.28 %	21.14 %	0.0001
13. Sales Department	74.80 %	8.94 %	0.0001
14. Advertising Department	38.21 %	31.71 %	**0.450**
15. Creative – Graphic Design Department	50.41 %	28.46 %	0.008
16. Multimedia – Audio Department	18.7 %	48.78 %	0.0001
17. Human Resources Department	8.13 %	48.78 %	0.0001
18. IT – Information Technology Department	26.83 %	38.21 %	**0.146**
19. Internet Development Department	24.39 %	39.02 %	**0.064**
20. Legal contracts – Subsidiary rights and permissions Department	28.46 %	32.52 %	**0.644**

The results of the empirical study for the structure of the Greek publishing houses are presented in Table 3. In the first column of Table 3 the specific departments are presented; whereas in the subsequent columns the actual percentages showing the existence or non existence of the respective departments are exhibited. In the last column the p-values larger than 0.05 was statistically significant for the basic categorical variables compared among the groups by x^2-test or Fisher exact test as appropriate. As the study shows, more than the two thirds of the representatives (75.61%) stated that the publishing house has an editorial department, whereas only 13.01% answered negatively to its existence and 1.63% did not answer the question. With almost the same percentage (74.80%) the participants identified the sales department; while the majority also stated that there is a finance/accounting and promotion/marketing department with 69.92% and 60.16%, respectively. Moreover, substantial are the percentages on the existence of a Production (55,28%) and a Creative/Graphic Design department (50.41%). In the question on the existence of an Advertising department, the 38.21% gave a positive answer, which possibly means that the Promotion/Marketing department undertakes the advertising as well. Furthermore, the 28.46% stated that there is a Legal Department; whereas only a small number of representatives stated that there are Human Resources (8.13%) and a Multimedia/Audio (18.7%) department. The departments of IT and Internet Development were identified in 26.83% and 24.39% of the publishing houses included in this study, respectively. Although these are rather small percentages, they exhibit the gradual introduction of a new organizational structure due to the advent of the information technologies and the internet. However, around 70% of the publishing companies that participated in this study have an editorial, a finance-accounting and sales department, whereas more than half have a promotion-marketing department, a production department and a Creative/Graphic Design department. Considering the answers concerning the existence or no of specific departments, it was found, after the application of the binomial non-parametric test, that significance differences were indicated for all the departments, except the Departments of Advertising (p=0.45), Information Technology (p=0.146), the Internet (p=0.064) and Legal/Copyrights (p=0.644) (Table 3). The analysis of the open-ended question reveals that most of the publishing companies outsource a significant part of the book printing production collaborating with printers, bookbinders, correctors, artists, scientists etc. Indeed, the 76.42% stated that they did not have any printing facilities. Representatives of the larger publishing enterprises stated that in the near future the organizational structure of the publishing houses will further develop in order to accommodate production practices based on novel information technologies, the Internet and electronic publications.

4.2 Discussion and Further Analysis

In many experimental situations we wish to compare a set of observed frequencies with a set of theoretical ones. The observed frequencies are those obtained empirically by direct observations or experiments, while the theoretical frequencies are usually called the expected frequencies. The x^2 may be employed as a descriptive measure of the magnitude of the discrepancies between the observed and expected frequencies. The larger these discrepancies, the larger x^2 tend to be. If no discrepancies exist, the observed and expected ones are the same, so x^2 will be close to zero. It must be noted that negative values cannot occur. Table 4 illustrates the resulting p-values applying x^2-test (Fisher exact test) considering the investigated factors.

Table 4. p-value using x^2-test (Fisher exact test)

	years	respons ibility	Edito rial	Financ e	Promot ion	Produc tion	Sales	Adver tising	Creativ e	Multim edia	Human	IT	Intern et	Lega l
years														
responsibility	0.378													
Editorial	0.899	0.344												
Finance	0.389	0.507	**0.001**											
Promotion	0.682	0.339	**0.001**	**0.001**										
Production	0.352	0.361	**0.001**	**0.001**	**0.001**									
Sales	0.996	0.582	**0.001**	**0.011**	**0.001**	**0.001**								
Advertising	0.944	0.579	**0.001**	**0.001**	**0.001**	**0.001**	**0.001**							
Creative	0.202	0.262	**0.001**	**0.001**	**0.001**	**0.001**	**0.001**	**0.001**						
Multimedia	0.528	0.638	0.156	**0.001**	0.225	**0.017**	0.158	**0.043**	**0.001**					
Human	0.184	0.745	0.348	0.402	**0.021**	0.280	0.178	0.303	0.78	**0.007**				
IT	0.358	0.659	**0.026**	0.145	**0.035**	**0.005**	**0.029**	0.117	**0.007**	**0.001**	**0.002**			
Internet	0.667	0.695	**0.006**	**0.003**	**0.037**	**0.001**	**0.026**	**0.014**	**0.001**	**0.001**	0.066	**0.001**		
Legal	0.529	0.664	**0.006**	0.09	**0.001**	**0.001**	0.068	**0.001**	0.09	**0.027**	**0.048**	**0.016**	**0.018**	

Considering Table 4, it is clear that the years of operation of the publishing companies as well as the person whom has the main responsibility for the selection of titles under publication have not any association with the Publishing companies' departments (p-values was found no significant with significant level a=0.005). Also it can be seen that that there is not strong association between Multimedia departments and Human Resources Departments with the remaining Publishing companies' departments, meaning that most of the companies have not these specific departments into there companies (48.78% negative response for Multimedia and 48.78% negative response for Human Resources). Finally, important departments are Editorial, Sales,

Financial, Promotion, Production and Creative, where these entire departments are associated with each other (grey values).

Applying discrete principal components analysis (PCA) was found that Editorial, Promotion, Sales, Creative and IT Departments are related to each other with communalities given by Table 5. Thus, instead having five variables, based on the multivariate analysis we proposed two groups (factors F_i), which explain the 78.82% of the initial information; these factors with the appropriate explanations are: 1st factor F_1: Editorial, Promotion, Sales Departments; 2^{d} factor F_2: Creative and IT Departments.

Table 5. Calculation of the communalities for each variable using PCA

Variables	Editorial	Promotion	Sales	Creative	IT
Communalities	0.725	0.926	0.993	0.904	0.899

5 Conclusions

The structural organization demonstrates the allocation of roles and responsibilities in the publishing houses for the design, production, distribution and marketing of the printed book. Furthermore, it enlightens the extent to which the cultural and social changes of the knowledge society influence the publishing activity. Although this work provides empirical evidence for the Greek publishing enterprises, the proposed framework can be employed for studying the organizational structure of publishing houses internationally. The Greek publishing companies remain family-traditional companies, which, however, strive to exploit new technological changes and opportunities in a competitive environment. Statistical analysis (using Fisher exact test) indicates that the most important department for the publishing companies were found to be the Editorial, Sales, Financial, Promotion, Production and Creative. These results introduce a strong association between these departments. The study confirmed the important role of the publisher who decides on the creation of the publishing policy and the selection of the titles to be published. Therefore, the publisher's personality, ideology, taste, aims, values and perceptions play an important role. Publishing activity and information services are dominant in a sensitive and unique book production environment, which integrates both tangible and intangible characteristics.

Acknowledgements

The empirical research conducted has been carried out in the thesis dissertation by Miss Leonora Triantafyllopoulou, MSc, Department of Archive and Library Science, Ionian University. The authors would like to thank all those contributed to the research and especially Miss Leonora Triantafyllopoulou.

References

1. Eisenstein, E.: The Printing Revolution in Early Modern Europe. Cambridge University Press, Cambridge (2000)
2. Banou, C.: Inferences for the Greek Publishing Industry through the Analysis of Market Data. I.J. of the Book 5(2), 141–147 (2008)

3. Schiffrin, A.: The Business of Books. How International Conglomerates Took Over Publishing and Changed the Way We Read. Verso, London – New York (2001)
4. Sparks, S.: From brand extension to migration: the business publishing industry today. Business Information Review 19(4), 16–21 (2002)
5. Thompson, J.B.: Books in the Digital Age. The Transformation of Academic Publishing in Britain and the United States. Polity Press, Cambridge (2005)
6. Drucker, P.F.: Management challenges for the 21st century. Elsevier, London (2007)
7. Baverstock, A.: Are Books Different? Marketing in the Book Trade. Kogan Press, London (1993)
8. Febvre, L., Martin, H.J.: The Coming of the Book. The Impact of Printing, pp. 1450–1800. Verso, London – New York (1990)
9. Johns, A.: The Nature of the Book. Print and Knowledge in the Making. The University of Chicago Press, Chicago (1998)
10. Clark, G., Phillips, A.: Inside Book Publishing, 4th edn. Routledge, London (2008)
11. Vervliet, H.D.L.: Concentration vs. Specialization in the /technical Printing Processes from the Fifteenth to the Nineteenth Centuries. In: Cavacciocchi, S. (ed.) Produzione e commercio della carta e del libro secc. XII-XVIII. Atti della Ventitreesima Settimana di Studi (15-20/4/1991), pp. 511–519. Leo Olschki, Firenze (1992)
12. Barbier, F.: Histoire du livre. Armand Colin, Paris (2001)
13. Epstein, J.: Book Business. Past, Present and Future. W. W. Norton & Company, New York – London (2001)
14. Keh, H.T.: Evolution of the Book Publishing Industry. Structural changes and strategic implications. J. of Management History 4(2), 104–123 (1998)
15. Banou, C., Kostagiolas, P.: Assessing and managing quality for the book publishing industry. In: Proceeedings. Third International Conference on Typography and Visual Communication "From Verbal to Graphic", University of Macedonia, Thessaloniki, June 2007, under publication (2007)
16. Association of American Publishers: About Publishing: Major / Department Guide (2007), http://www.bookjobs.com/page.php?prmID=12
17. National Book Centre of Greece: The Book market in Greece 2007. National Book, Athens (2008), http://www.ekebi.gr [in Greek]
18. National Book Centre of Greece, The Book market in Greece. Athens: National Book Centre (2007)

A Middleware Framework to Create and Manage Data Structures for Visual Analytics

Juan Garcia, Diego A. Gomez, Antonio Gonzalez,
Francisco J. Garcia, and Roberto Theron

Departamento de Informática y Automática, Universidad de Salamanca,
Plaza de la Merced s/n. 37008, Salamanca, Spain
{ganajuan,dialgoag,agtorres,fgarcia,theron}@usal.es

Abstract. Visual Analytics require the use of large datasets and most of the time visualizations use their own data structures definition or make use of simple standards such as XML, CVS, or text files. Our proposal presents a middleware framework that focuses on providing a way to manage those data structures, allowing to access heterogeneous data sources. The framework allows defining data structures for different visualizations using the same dataset. It dynamically generates a visual interface that allows users to define data structures according to the specific visualization requirements. It also provides a repository to store data structures to reuse them in other visualizations or by means of inheritance, composition and associations. We have implemented a first prototype, which is discussed in the results presented on this paper.

Keywords: Middleware framework, data structure modeling, visual analytics.

1 Introduction

Visual Analytics facilitates analytical reasoning by means of automated analysis techniques and interactive visualizations to synthesize information and derive insight from massive, dynamic, ambiguous, and often conflicting data; detect the expected and discover the unexpected; provide timely, defensible, and understandable assessments; and communicate assessment effectively for action [1]. This is critical in our current knowledge society, which is overwhelmed by the huge availability of data. In this scenario, data transformation into knowledge is challenging. Visual Analytics supports this task and allows the active participation of users through the use of interaction techniques. Moreover, several papers that address the overwhelming nature of our current knowledge society have been published during the last years [2] [3].

Visual Analytics frequently takes into consideration data extraction, preparation and mining. However, the process of producing the input data for visual representations is very complex. Furthermore, this complexity increases when data sources are heterogeneous and one of the visualization design goals is to produce visualizations independent from data source types.

M.D. Lytras et al. (Eds.): WSKS 2009, CCIS 49, pp. 466–473, 2009.

The aim of this paper is to propose a framework to support visualizations that are independent of data source types and structure. Following, we analyze which file formats and data structures are used by most common visual libraries.

Most visual libraries, such as Prefuse [4] and Processing [5], use their own data structures (frequently defined in XML format). Consequently, users working with visual libraries must adequate their data to already defined data structures, which is an exhausting and difficult process. In addition, it implies that using the same dataset with other visualization tools requires data transformation to the appropriate data structure or format.

Graphviz [6] is an open source Graph Visualization Software that represents structural information as diagrams of abstract graphs and networks. Graphviz layout programs and take descriptions of graphs in a simple text language, GXL (Graph eXchange Language) [7] or XML. Grappa is a Java graph drawing package used by Graphviz to display and manipulate graphs. This package provides methods to build, manipulate, traverse and display graph nodes, edges and subgraphs. Graphs can be created using the class methods provided by the library or interactively with the mouse. Graphviz offers the possibility to create visualizations that define their own data structures, but it also allows creating its data structures dynamically by reading an XML schema, text file, etc. JGraph [8] is a diverse set of visualizations that also allows defining data structures from XML and GXL files.

ManyEyes [9] is another set of visualization tools that provides some datasets and offers the possibility to upload new datasets. The format is very simple and consists of tables where the values in each row are separated by tabs. The first row of each table should be the "headers" row, which values describe the columns. Actually, this is a very simple format that can be easily exported from databases or spreadsheets. Nevertheless this kind of data representation cannot be considered as a well-defined data structure, but just a data organization.

GGobi [10] is an open source visualization program for exploring high-dimensional data. It provides highly dynamic and interactive graphics such as scatter plot matrixes, parallel coordinates, time series plots and bar charts. The data format used by GGobi is XML and CSV (comma-separated variables).

GVF (Graph Visualization Framework) [11] uses GraphXML [12], a textual descriptive language to define graphs in a human readable form. The same authors developed "Royere", based on the GVF project, which reads GML [13] file formats.

Mondrian [14] is a visualization framework that focuses on the visualization of statistical datasets. This framework supports standard tab-delimited ASCII files either databases. As previously described GXL, GraphXML and GML have been defined to be used exclusively with graphs. More complex data structures cannot be defined using these languages.

Other frameworks such as ORCA [15], have been implemented as an approach to deal with data structures, nevertheless they have not been completely successful. OpenVisuals [16] is another interesting visualization project that uses CSV as the input format for the datasets. This project aims to provide the collaboration among different users by submitting either datasets or visualizations.

1.1 Contributions and Structure of This Paper

As explained above, most visualization tools use their own datasets or some standards that are subsets of XML language. We consider that by providing the data with a standardized structure, the information to be represented can be easily managed; resulting in portability and the possibility to define more complex data structures.

Consequently, we propose a framework to support visualizations that are independent of data source types. Furthermore, the framework helps in the process of defining and storing new visualization data structures in a repository where they can be publicly accessed. Our proposal represents a middleware or intermediate tier between the source (where the data are stored) and the final visualization. This intermediate tier allows users to make an easier process of modeling and managing data.

The framework is intended to extract information from heterogeneous data sources and to provide a visual interface that allows defining data structures that will be used by visualizations. Overall, the framework is able to access most common data sources (Databases, XML, Excel or text files) and dynamically produce a visual interface from which the visualization designer choose fields to create a data structure for the visualization.

The goal of the framework is to offer a platform that allows creating user defined data structures to feed visualizations and manage heterogeneous data sources. An advantage of the proposed Framework is that having a dataset, and having a specific structure, the data can be mapped into the data structure to be displayed by the visualization.

The definition of data structures allows the easily movement of data between visualizations. The definition of the data structures, as proposed in this paper, is object oriented compliant and supports the creation of specialized structures through inheritance, composition and association. These concepts strengthen the possibility to define new and more powerful structures.

Finally, the framework produces a set of Java classes that generate the data structures, and map original dataset values into the generated data structures. In the next sections we discuss how to define data structures and their components, the results of this research and the Conclusions and future work

2 Providing a Structure to the Information

Our proposal is intended to create new data structures that can be used by different visualizations. The objective is to provide users with a framework where data structures are defined to share datasets among visualizations. The idea relies on the fact that defining abstract data structures, would simplify datasets treatment. The user is provided with a tool to define a specific data structure without taking care of the source where the data is read. Once this data structure has been defined, the framework generates the Java classes that will be used to create the data structure and map data into it.

Portability is one of the advantages of offering the possibility to generate data structures independent from datasets and visualizations. For instance having a hierarchical dataset, it can be viewed using different visualizations such as treemap, radial or hyperbolic graphs, etc. Another advantage is the possibility to define new abstract data structures by reusing existing ones by means of adding new attributes and relations, applying inheritance, composition or creating associations.

The framework is intended to have a repository to manage and share the structures created previously. Data structures can be labeled to indicate with which visualizations have been successfully tested.

2.1 Mapping Data to Data Structures

The process starts by defining a data structure and specifying which data column is associated to each value in the structure. Once a data structure has been defined, the structure is populated with real values. This process is transparent to the user. Each value (represented by a column) in the data source is mapped to one column of the data structure, so the final result is a set of data items to be represented by the visualization. This set of items represents a new data structure dataset that can be used by any visualization that recognizes it. Basically, our proposal aims to map a dataset from any data source type into a visualization data structure.

2.2 Storing and Managing Data Structures

The data structures are stored in a shared repository where all users can search for an existing final data structure or from which a new data structure can inherits. The idea is that this repository contains all new data structures as well as the Java classes that were generated by the framework. The framework also offers the possibility to export data structures, with the mapped data, into different formats such as XML, or Java serialized classes that can be shared among visualizations.

2.3 Defining Structural Components

Defining the components of a data structure is done by defining a general abstract data structure from which all the new data structures are created as sub-structures: it means a more specialized structure that takes as it base the most general one. This new structure is filled out by creating the internal components that define the main functionality. Two sections form the abstract data structure. The first section corresponds to the data structure information. The fields of this section are listed in table 1. The second section is related to the components of the structure. All attributes and relationships that belong to the structure should be listed. Internal attributes are defined by the properties listed in table 2.

The relationships between two attributes are basically defined as a mathematical function. It means that a dependency between two attributes is a relationship where an element of a given set (the domain) is associated with an element of another set (the range) after the function evaluates their values. The definition of functions empowers

Table 1. This table shows the attributes used to describe the structure

Field	Description
Name	An identifier field that names this data structure.
Description	It is used to describe the purpose of this structure and information such as its application field.
Author	Author of the structure.
Date	The date when the structure was created.
Comments	Comments about how to use the data structure.
Type	A field to indicate if this structure represents a hierarchical, timeline, graph or any other dataset type.

the process of data treatment and is helpful for the creation of more complex data structures. For instance, it could be helpful to define a new attribute referring to the background color, which depends on the value of another attribute in a specific range. This type of relationships are defined by the elements listed in table 3.

Table 2. This table shows the elements for each attribute in the structure

Element	Description
Identifier	An identifier for the attribute.
Type	The identifier should have an assigned data type, i.e. Numeric, Boolean, Alphanumeric, Structure, Color and Date.
Value	The value assigned to the attribute according to the type. This value is assigned during the filling out process the data structure.
Description	It describes the reason why this attribute was created, and what its value represents.
Quantitative	This Boolean field indicates whether or not this attribute represents a quantitative value. In the value of the field is false, it stores a quantitative value.
Constraints	It indicates the constraints applied to values in the field, i.e. a numeric attribute contains only positive values. It allows defining regular expressions for alphanumeric values. There is no limit to the number of constraints for this attribute.

Table 3. This table shows the elements belonging to each declared relationship

Element	Description
Identifier	An identifier for the relationship.
Description	It describes the relationship, what it does represent or why it has been declared.
DomainAttribute	Represents the attribute for the values in the domain dataset. It should belong to the structure itself and comply with the properties and constrains defined above.
RangeAttribute	This field represents the attribute for the values in the dataset range. It also belongs to the structure itself and should comply with the properties and constrains defined above.
Function	This element is the most important one in a dependency relationship. It is a mathematical operation over the attribute value in the domain and it will give the correspondent value for the attribute in the range. It is defined by a Java source code class. It is intended that the defined function accords to the dataset in order to correctly associate it.

3 Results

The framework allows the definition of basic data structures, from which more complex data structures are derived. Basic data structures support similar relationships to the ones defined by the basic object oriented theory. In addition, data structures fields could be derived from the calculation of other field values. The framework is focused on the creation of Java components and the tool was implemented using the same programming language. The framework proposal is depicted on figure 1. As the first part of the implementation, we have designed and programmed the basic and most general abstract data structures. We have used these data structures as the base for the definition of new data structures that have been tested in our visualization tools. Currently, our implementation just covers a part of the whole framework. This first approach has been used to test the functionality with the creation of the structures.

Our tests included the creation of new specific data structures. These structures were used with different visualization such as the spiral timeline, vimoodle, treemap

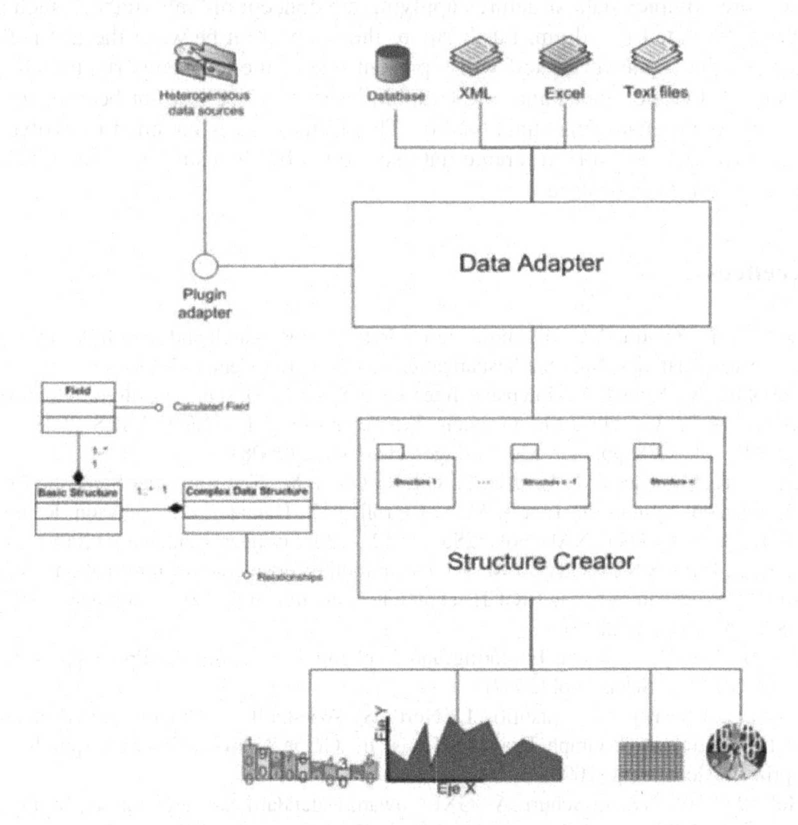

Fig. 1. The framework represents a middleware tier between the source containing the data and the visualization techniques. The framework includes a section to create structures, a repository to store them and another section to get the data directly from the source and to fill the structures with the values.

and table lens. We have been able to test the connection between the encapsulated dataset and the visualization techniques. The structures were filled with values by using a non-graphical method because this feature has not yet been implemented. The data sources that have been used were taken from diverse sources such as XML files, Java serialized files and text files. The process to adequate the data source formatting to the created structures was done by hard-coding, which is not desirable. This process will be implemented to be executed automatically as part of the original proposal.

4 Conclusions and Future Work

This first prototype was programmed to create and test new data structures and validate different aspects. The main goal to reach was to generate the same data structures that we had been using in our visualizations previously. This first goal was successfully reached even when we had to hardcode the connection between the data sources and the framework. There is a lot of work to be done starting with testing the creation of new more complex data structures applying the concept of "inheritance" such as in the Object Oriented Paradigm. Furthermore the connection between the diverse data sources and the framework itself will represent one of the most important challenges to be solved. On the other hand, the Graphic User Interface has not been developed yet as well as the managing functionality. This framework is intended to evolve as a general purpose framework to create data structures but focusing on visual analytics and information visualization.

References

1. James, J.T., Kristin, A.C.: Illuminating the Path: The Research and Development Agenda for Visual Analytics. National Visualization and Analytics Center (2005)
2. Valtolina, S.: Knowledge-Intensive Interactive Systems Design in Cultural Context. In: Lytras, M.D., Carroll, J.M., Damiani, E., Tennyson, R.D. (eds.) WSKS 2008. LNCS (LNAI), vol. 5288, pp. 523–530. Springer, Heidelberg (2008)
3. Garcia, R., Gimeno, J.M., Perdrix, F., Gil, R., Oliva, M.: The Rhizomer Semantic Content Management System. In: Lytras, M.D., Carroll, J.M., Damiani, E., Tennyson, R.D. (eds.) WSKS 2008. LNCS (LNAI), vol. 5288, pp. 523–530. Springer, Heidelberg (2008)
4. Heer, J., Card, S.K., Landay, J.A.: Prefuse: a toolkit for interactive information visualization. In: Proceedings of the SIGCHI conference on Human factors, in computing systems, USA, pp. 421–430 (2005)
5. Fry, B.: Visualizing Data: Exploring and Explaining Data with the Processing Environment. O'Reilly, Sebastopol (2007)
6. Ellson, J., Gansner, E., Koutsofios, L., North, S., Woodhull, G.: Graphviz and Dynagraph - Static and Dynamic Graph Drawing Tools. In: Graph Drawing Software, pp. 127–148. Springer, Heidelberg (2003)
7. Holt, R.C., Winter, A., Schurr, A.: GXL: toward a standard exchange format. In: Proceedings Seventh Working Conference on Reverse Engineering, pp. 162–171 (2000)
8. Bagga, J., Heinz, A.: JGraph— A Java Based System for Drawing Graphs and Running Graph Algorithms. Springer, Heidelberg (2002)

9. Viégas, F.B., Wattenberg, M., van Ham, F., Kriss, J., McKeon, M.: Many Eyes: A Site for Visualization at Internet Scale. IEEE transactions on visualization and computer graphics (2007)
10. Swayne, D.F., Lang, D.T., Buja, A., Cook, D.: GGobi: evolving from XGobi into an extensible framework for interactive data visualization, computational statistics and data analysis, pp. 423–444 (2003)
11. Marshall, M.S., Herman, I., Melançon, G.: An Object-Oriented Design for Graph Visualization
12. Herman, I., Marshall, M.S.: GraphXML, An XML-based graph description format. In: Marks, J. (ed.) GD 2000. LNCS, vol. 1984, pp. 52–62. Springer, Heidelberg (2001)
13. Himsolt, M.: GML - Graph Modelling Language (1997), `http://infosun.fmi.uni-passau.de/Graphlet/GML/`
14. Theus, M.: Interactive Data Visualization using Mondrian. Journal of Statistical Software (2002)
15. Sutherland, P., Rossini, A., Lumley, T., Lewin-Koh, N., Cook, D., Cox, Z.: ORCA: A Visualization Toolkit for High-Dimensional Data, NRCSE-TRS No. 046 (2000)
16. OpenVisuals framework, `http://www.openvisuals.org/`

Factors Influencing e-Business Adoption in the Greek Hotel Sector

Ir. Samanta and P. Kyriazopoulos

Technological Educational Institute of Piraeus
P. Ralli & Thivon 250 12244
Aigaleo Athens
isamanta@teipir.gr, pkyriaz@teipir.gr

Abstract. The purpose of this research is to identify the impact of business process improvement in the area of e-marketing in the hotel industry. The research identifies the barriers which block organizational change effort. A sample of thirty hotels in the city of Athens was used.This paper presents a SWOT analysis of the hotel sector, identifying the strengths, weaknesses, opportunities and threats that firms faced in the process of change. The results show that the majority of firms use, to a small extent, the e-marketing concept to improve their communication strategy and reach market segments.

Keywords: hotel sector, e-commerce, e-marketing, e-business.

1 Introduction

Fundamental changes tend to influence business environments in several aspects. These changes can be observed both in the internal and external environment of an organization. Technological and socioeconomic sectors of a business environment tend to undergo major changes to a great extent.

Under such circumstances, and with the emergence of globalization, which could be argued to be a heavily responsible factor for these changes, many industries need to review previous strategies and systems. New pattern developments that will provide appropriate guidance may become essential.

In the hospitality business environment, these changes are even more evident than in any other type of organization. The main factors affecting hotel success are globalization, technological developments, changes in customer preferences, differences in competition amongst hotels, horizontal/vertical integration and vertical applications. Hotels must cope with economic, political, and market conditions, which are much more complex than before.

The service sector seems to adjust much quicker to the reality of new technological achievements [1] [2]. E-services hold almost two-thirds of the world's economic activity and trade in services is continuously growing due to internationalization of domestic services and transmission technologies and the opening up and regulatory reforms of state monopolies in transport and communications.

In the hotel sector in 2008, approximately 40% of all hotel bookings were be generated from the Internet (one-third in 2007, 29% in 2006). At least another third of all

M.D. Lytras et al. (Eds.): WSKS 2009, CCIS 49, pp. 474–483, 2009.

hotel bookings will be influenced by the Internet, but done offline (call centre, walk-ins, and group bookings). By the end of 2010, over 45% of all hotel bookings will be completed online [3]. The hotel industry in Greece constitutes the major power of economic development, contributing more added-value than manufacturing. In the last decade, the development of the hotel industry has focused on high-class categories, and on the renovation and modernization of the remaining categories. The impact of the 2004 Olympic Games on the Greek hotel industry was a determinant key of the industry's success. Hotel firms used the Games as a springboard for long-term growth of the sector. In 2006, new investments via the European Union (EU), for re-engineering and upgrading, materialized in the hotel industry for.

The purpose of the present research is to identify the effect of the e-business concept on hotels in Greece.

2 Literature Review

2.1 The Development of e-Business on the Tourism Sector

Some authors define e-business as a way of selling a company's products over the Internet [3]. A broader, more inclusive definition is that the sharing of business information, maintaining business relationships and conducting business transactions by means of telecommunications networks are all part of the phenomenon of electronic business [2].

Table 1. Online sales: B2C and B2B

Year	B2C sales Actual and estimated in $ billions	B2B sales (including electronic data interchange E.D.I.) Actual and estimated in $ billions
2007	240	6800
2006	190	5300
2005	150	4100
2004	130	2800
2003	100	1600
2002	80	900
2001	70	730
2000	50	600
1999	25	550
1998	10	520
1997	5	490
1996	Less than 1	460
Adapted from reports by CyberAtlas (http://cyberatlas.internet.com/); eMarketer (http://www.emarketer.com); Forester research (http://www.forester.com); and the Statistical Abstract of the United States, 2006, Washington: U.S. Census Bureau.		

E-business influences all sectors and markets and is related to the total strategy and operations of enterprise. It incorporates the total operations that relate to the term e-commerce, by connecting these operations with networks that exceed the limits of enterprise [2]. Table 1 presents the actual and estimated online sales in business-to-consumer (B2C) and business-to-business (B2B) categories.

Holloway and Taylor [4] and Holloway and Taylor [5] recognized the need of the emergence of e–business in the tourism sector as a result of a fast-forwarding society. Potential tourists are now able to find all information needed available on the web, compare prices, tailor their own package tour, make reservations and pay for them *"without leaving their armchair"*. Buhalis [6] explained that information and communication technologies (ICTs), including the Internet, information systems and telecommunications, are the tools that brought about the evolution of tourism demand and developed a business model of tourism organizations, in collaboration with e–business and e–management, including e–planning and e–strategy.

The term e–tourism emerged from the digitalization of the procedures and the entire value chain in tourism including travel, transport, leisure, hospitality and catering industries. Buhalis [6] also sets out the perspective that e–commerce's application in e–tourism maximizes *"the efficiency and the effectiveness of the tourism organization"*.

Braun and Hollick [7] support the view that e–learning for a destination is still in "*its infancy*" and the whole industry is now "*learning to grow its service technology*" [7]. In Greece, presently, only up to 5% of hotel reservations are made through the Internet [8].

2.2 The Influence of e-Business on Marketing

According to Lemmergaard [9], the areas of industry practice facing the greatest changes due to e-business and the technologies supporting it are the traditional marketing practices, customer service and sales. In addition, Papamichali et al [10] and Pickton and Broderick [11] state that the field of marketing is at the centre of changes caused due to the entrance of the industries to the digital environment [12]. Tourism marketers should acknowledge the four characteristics of services: the intangibility, inseparability, variability and perishability for the development of sustainable overall tourism products [11]. Service marketers should adopt a promotional material that would be adequate to provide their message clearly to potential customers, enough to create expectations and a sense of travelling.

Pickton and Broderick [11] conclude that the marketing communications mix embraces sales promotion, public relations, advertising, sponsorship, direct marketing communications, packaging and personal selling activities. The combination of the

Table 2. The e – marketing mix in Tourism

Product	Services & tangibles products
Price	Discounting, value for money, special web offers
Place	E - medias (internet)
Promotion	On - line direct sales through internet, destination web sites

Developed from: Swarbrooke and Horner [13].

use of these elements will contribute to an integrated marketing effort (Table 2) only if the messages can be transmitted to the target groups. According to Procter and Ackroyd [12], the five main components of the overall tourism product includes the destination's access, attractions and facilities; the costing of travelling; and the total of the ideas and beliefs that has been created for a tourism destination.

2.3 Travel Website Content and Structure

More analytically, Smith [14] proposed that a well-developed website should include a number of key quality factors in order that customer requirements be met and satisfied. The *"clarity of purpose"* is the first factor that a business on which a business should focus. This includes all the elements that comprise the purpose of the website's existence. The website should clearly state what it is offering, such as information or information and transactions, orders or exchange platforms (links, feedback, fill- in forms), so that the user can benefit from the site. All of these elements should be presented in a way so as to reflect the firm's image by matching the company's profile with the virtual reality of the web. A well–structured website should have *"richness"* of information, including language selections, contact details, hotel, room and food and beverage facilities, photo albums, pricelist, online availability, special offers, distances, and currency exchange in order that the site be competitive to bring potential customers to the hotel. Sinha [15 and Chuang and Chong [16] added that these kinds of *"comprehensive"* sites are those that add value to all available travel products, as they help potential travellers to have an estimation of what they are going to experience.

3 Research Methodology

In order to identify the organizational background of e-business in the hotel sector, the sample is defined by selecting firms operating in Greece. The sample used consisted of 100 hotel units in three main categories: luxury hotel, four-star and three–star hotels. This specific categorization took place in order to illustrate how the hotel industry has adopted and implemented e-commerce practices. In the final analysis, we used data from 30 businesses at a response rate of 35%. Of these, 46.4% were luxury hotels, 35.7% were four-star hotels and 17.9% were three-star hotels. The questionnaire developed consisted of 128 questions,. and rather 47 of the questions were evaluated using a five-item, five-point Likert scale, ranging from 1 to 5 (mostly: 1='not at all', 5='very much'), 76 nominal variables (Yes–No type) and four scale variables. Management provides the researcher with opinions and perceptions of e-business implementation, therefore. The main questions asked from the participants included issues concerning operations, technology, marketing, pricing, customer value and staff. The study was undertaken in the Attica in Greece through private interviews. The research took place from June to September 2008.

The data processing included a set of interactive procedures, and, in order to explore and extract the data and acquire knowledge concerning e-business, two different approaches were applied. Typical statistical analyses, such as descriptive frequency analysis were used in order to picture the main characteristics of the sample. The second approach concerned factor analysis, and the orthogonal factor model was used with the principal component analysis. The correlation matrix of variables was analysed and the Varimax method of factor rotation used to interpret the factors.

3.1 Results and Discussion

According to the findings, 58.3% of the participating hotel units are taking part in international sales systems. The majority of their customers are, mainly, from Germany (91.7%), Italy (87.5%), UK (79.2%), Switzerland (62.5%) and Austria (62.5%). Information material provided by the hotels is mostly in English (91.7%), Greek (91.7%) and French (33.3%). It is quite remarkable that, although a great number of the clients are German, the information brochures do not include this language. The percentage of forthcoming customers from one year to another is 40%. The basic target groups of these hotels are 25% business travellers and 58.3% people who are travelling for both business and pleasure. Among the participants, 62.5% take part in exhibitions whereas 37.5% are also being indirectly promoted through group or regional administration stands. All participants used e-mail, and their Internet sites are mostly in at least four languages: English, German, Russian, and Swedish, with only 79.2% in Greek. The hotel unit sites are mostly used for essential information about the hotels, while traditional communication means are used for further contact (fax and telephone), up to 83.3%. Moreover, the sites can be used to order an information brochure online (87.5%), answer more specialized questions (91.7%) and to book online (95.8%), but with an offline settlement. A third of the site visitors can check room availability and other facilities for the season as well as prices, while 58.3% can also be informed about transportation possibilities to and from the hotel. Finally, 8.3% of the site visitors used the Internet to pay fully or partially online for the chosen package offer. The updating of sites, as well as the response to customers, is performed by the hotel personnel (up to 83.3%). It is notable that only 8.3% of the hotels use other specialized service suppliers for such matters. A third (33.4%) of the customers visiting the Internet sites of the participants were Europeans and Americans, while 12.5% of the customers who booked online were European residents. The participants believed that the Internet, as a marketing communication tool, contributed to promoting the hotel improve the firm's profile as well as to increasing sales and developing relationships with the customer, as shown in Table 3.

Table 3. Internet as a marketing tool

Internet is useful for promotion	100%
Internet is useful for increasing sales	95.8%
Internet is useful for reducing advertising costs	83.3%
Internet helps to develop relationships	91.7%

58.4% of the participants believed that, in order to have access to the Internet, the company must have the resources to invest in employees' training, and expensive equipment, while 85.5% thought that they would not be able to write off the invested capital. 87.5% thought that they could increase the turnover of the hotel unit.

Concerning those who have not yet participated in an e-market, the findings are very interesting: 42.9% of the sample believed that participating in e-business activities is not important, 57.1% believed that it is expensive, while 57.2% claimed that the reason for not using it is the lack of technological infrastructure in the company. Moreover 57.2% suggested that their business partners were also not taking part in it. Around three quarters (71.5%) of the participants estimated that they will be on the Internet shortly, whereas 85.7% thought that this will happen in the long term. Finally, 42.9% believed that the reason they are not using the Internet is the fact that there are no adequate e-markets that correspond to their needs. As far as the advantages deriving from their entry to the Internet are concerned, they believed that saving neither time nor saving money are important motives. Expansion of customers and of partner relations were also not considered of importance, while possible benefits such as broadening of knowledge or information, creation of advantages in competition and upgrading of products and services were not believed to be important.

3.2 Factors Influencing e-Business Adaptation in the Greek Hotel Sector

3.2.1 Factor Analysis

In Factor analysis (Table4), the orthogonal factor model with the principal component method was used. The correlation matrix was analysed and the Varimax rotation method used for better interpretation of the factors. The variables that were used for the growth of the factor model were all of the categorical–ordinal type. We calculated that KMO =0.478. It is judged not satisfactory as it is below 0.5. This is probably because the data size is too small. Significance =0.000, which means that the variables are correlated. The diagonal elements constitute the Measure of Sampling Adequacy (MSA) for the corresponding variable. The smallest MSA is 0.208 and the greatest 0.637. Moreover, many from the diagonal elements of the matrix are not near zero (of – diagonal elements are the partial autocorrelations of variables but with the opposite sign).

In the above factor analysis, factor 1 has eigenvalue 3.093 and interprets 30.932% of the total variance. Factor 2, meanwhile, has eigenvalue 2.050 and interprets 20.592% of the total variance. Together with factor 1 they interpret 51.434% of the total variance. Finally, factor 3 has eigenvalue 1.664 and interprets 16.640% of the total variance. With factors 1 and 2, it interprets 68.074% of the total variance. From examining the above factors, it appears that the majority of those asked believed that a developed information system can increase the turnover of the companies that use it as a basic means of increasing the companies' incomes. Moreover, respondents seem to agree on the fact that the Internet is, nowadays, an essential element for efficient administration. At the same time, it improves the image of the company and can contribute to improving human relations.

It is, however, necessary for all managers to know how to work with e-business, in order for it to be effectively used by everyone. Moreover, it is a tool that requires expensive investments, and it would appear that it is not easy to write off the invested money. It is, on the other hand, extremely useful in making tasks easier and faster in activities such as finding new suppliers and customers, information on the activities of the partners, and also in other dealings, such as price listing and other services.

Table 4. Factors influence e-business adaptation in the Greek hotel sector

Factors	Associated variables	Loadings
1. Opportunities of e-business operation to e-marketing and promotion activities through the web	The e-business through e-marketing improves the public image of the company	0.902
	The company cannot be efficiently administered without the use of an operating site	0.840
	E-business can increase incomes	0.723
2. Strength of Internet use in customer value and marketing and weaknesses of Internet use in operation of technology	The e-business can help to unify environmental moves	0.858
	The e-business helps to increase customer value of customers and making relationships	0.824
	The company has to invest on expensive equipment to have access to the e-business	–0.630
3. Threats in e-business operation and lack of innovation knowledge from staff	The managers are obliged to know how to operate the e-business	0.819
	The time needed is related to the flow of information	0.693
	The e-business does not write off the invested money	–0.638

3.3 SWOT Analysis of the Hotel Sector in Greece

The three factors emerged from the above factor analysis assist and allow hotel firms to have an understanding of the parameters that take place in the internal and external environment of the hotel sector. Firms taking into consideration these factors and analysing them can compose a SWOT analysis, which comprises the basis for their e-marketing plans. SWOT analysis is an important tool for auditing the overall strategic position of a business or business sector and its environment. It involves sizing-up firms' internal strengths and weaknesses and external opportunities and threats. From the SWOT analysis, we can see some of the major strengths, weaknesses, threats and opportunities of the hotel sector emerging from the findings of factor analysis, as seen in Table 5. It can help the hotel sector to identify how they may be different when compared with offline operations. What new information is can be offered by developing a website or adopting Internet technologies? How business operations may be improved with the use of Internet technologies and, finally, what special issues affect the online platform?

From the strengths, we can see the advantages derived from e-business implementation. From the weaknesses, we can see what could be improved, what the hotel

Table 5. SWOT Analysis in hotel sector in Greece

	Strengths	***Weaknesses***
Internal Environment	Flexible customization Availability of services regardless time. Geographic availability of services Speed of service delivery E-business helps with the procedure of making relationships E-business can help to unify environmental moves	Inadequate international marketing programmes for the promotion of e-services Unfriendly interface that is as a result the misjudgement of the service provision. The hotel firm has to invest on expensive equipment to have access to the e-business
	Opportunities	***Threats***
External Environment	Appearance of new partners New customer segments Collaboration and communication opportunities E-business through e-marketing improves the public image of the company E-business can increase incomes Decrease direct procurement costs through e-procurement. High information input the Internet offers a growing range of new possibilities to increase Seek better supplier deals as e-business can help to unify environmental moves	The lack of innovation culture towards e-business managers, are obliged to know how to operate the e-business Low investment in Information technology infrastructure The e-business does not write off the invested money

sector does poorly and what problems could be avoided. From the opportunities, we can see good opportunities and interesting trends. Lastly, from the threats, we can see the obstacles that the hotel sector faces in improving performance and culture. The gains of the transition from business provision to e-business provision were many, but the threats and weaknesses continue to exist.

4 Conclusions

The Internet in the hotel sector in Greece offers another means of communication, with the basic function of exchanging information and customer services. It is rather characteristic that, although there is contact via e-mail, the traditional ways of communication are mostly used. It is those hotel units that operate a site themselves that do not pay attention to improving them, since the sites are supported either by the company personnel or outsourcing. As a result, it can be used neither as a sales nor as an advertising tool. However, it is still considered an expensive means of promotion and publicity. On the other hand, there is still an important percentage that believes that the Internet is not a vital tool. They do not see any benefits, such as saving time and money, in participating in any e-market, nor do they see it as an innovative way of finding new customers. On the contrary, they see more barriers than chances, since they think that they will have no relative advantage in competition, but only high costs for training.

It is clear that, in order to make further use of e-business, a change of organizational culture is necessary to be shared by members of hotel firms that operate unconsciously, and that define, in a basic "taken-for-granted" fashion, an organization's view of itself and its environment". Changing the hotel sector' culture is not an easy or quick process, as cultural attitudes evolve slowly, and take time to change. There need to be changes in infrastructure, such as better training of personnel by specialists, the creation of upgraded sites, but also the strategic planning of use and development of the e-business.

Consequently, in order for the hotel sector's delivery through the Internet and, generally, through electronic channels to be valuable and successful, it must have a well-organized process control and well described objectives. The result is the continuous improvement of e-business and the creation of innovative e-services. In a rapidly changing environment, the Greek hotel sector, in order to be developed, has to design its future investment of its competitiveness using contemporary innovative tools of marketing.

4.1 Knowledge Management Implications

The results of this research conclude that the hotel unit, through its business experience as well as the increasing tendency derived from Internet usage, should take advantage of the new way of functioning in the tourist market and use that knowledge in a new management philosophy, separating its concept from marketing orientation to centre-custom orientation.

In order for the new knowledge management to come true, it is necessary for the hotel units to invest in technology so as not only to have a presence on the Web but also to provide complete e-commerce services by using new strategies from the traditional marketing to e-marketing. The use of technology offers initiatives to customers in order to work in a peer-to-peer environment by choosing all the services they wish from the space they themselves choose according to their personal needs. Tourists can be characterized as"wired travellers" who seek information through the Internet and ask for travel services/products and information of high quality and monetary value. Tourism consumers have become more critical and they ask for higher performance for a lower price.

References

1. Vladlena, B.: Perceptions of trust and experience: potential barriers to web 2.0-based learning. International Journal of Knowledge and Learning 4(5), 427–437 (2008)
2. Kalakota, R., Robinson, M.: From E-Business to E-Services: Why and Why Now? Addison-Wesley, New York (2003)

3. Lagrosen, S.: Effects of the internet on the marketing communication of service companies. Journal of Services Marketing 19(2), 63–69 (2005)
4. Herbig, P., Hale, B.: Internet: the marketing challenge of the twentieth century. Internet Research Electronic: Networking Application and Policy 7(2), 95–100 (1997)
5. Holloway, J.C., Taylor, N.: The Business of Tourism, 7th edn. Pearson Education, Prentice Hall, London (2006)
6. Buhalis, D.: eTourism – Information Technology for Strategic Tourism Management, 1st edn. Pearson Education, Prentice Hall, London (2003)
7. Braun, P., Hollick, M.: Tourism skills delivery: sharing tourism knowledge online. Education and Training 48(8/9), 693–703 (2006)
8. Fisk, R.: Wiring and growing the technology of international services marketing. Journal of Services Marketing 13(4/ 5), 311–318 (1999)
9. Jeanette, L.: Interfirm knowledge management through a web-based bench learning system. International Journal of Knowledge and Learning 4(4), 317–328 (2008)
10. Papamichail Nadia, K.: A decision-analytic tool for assessing decision making: a life-cycle management approach. International Journal of Knowledge and Learning 5(1), 37–49 (2009)
11. Pickton, D., Broderick, A.: Integrated Marketing Communications, 2nd edn. Prentice Hall, England (2005)
12. Stephen, P., Stephen, A.: Strategies for flexibility technology-centered and labour-centred flexibility in UK manufacturing. International Journal of Manufacturing Technology and Management 1(4/5), 366–380 (2000)
13. Swarbrooke, J., Horner, S.: Consumer Behaviour in Tourism, 2nd edn. Butterworth-Heinemann (Elsevier), Oxford (2007)
14. Smith, A.: Information exchanges associated with Internet travel marketplaces. Online Information Review 28(4), 292–300 (2004)
15. Sinha, I.: Cost transparency: the Net's real threat to process and brands, pp. 43–52. Harvard Business Review, Boston (2000)
16. Chuang, T., Chong, P.P.: Searching Advertising Placement in Cyberspace. Industrial Managements and Data Systems 104(2), 144–145 (2004)

The Impact of e-Customer Relationship Marketing in Hotel Industry

Irene Samanta

Technological Educational Institute of Piraeus
P. Ralli & Thivon 250 12244Aigaleo Athens
isamanta@teipir.gr

Abstract. The present research investigates the extent to which Greek hotels had developed the electronic customer relationship marketing (E-CRM). The study verifies the practices that frequently appear in relationship marketing process within online operations or whether their Internet presence mainly depends on the basic actions of "supplying information" and "reservations". Also, it investigates the effects of e-CRM system on customer loyalty and satisfaction as well as the impact of relationship marketing practices to customer retention and acquisition. They have understood the importance of using electronic channels instead of traditional ones to implement their marketing strategies. Thus, e-crm system has assisted hotel business to manage more effectively their reservations and serve their customers as fast and as effective as possible. They did not seem to apply many of the relationship marketing strategies to emphasize customer retention and continual satisfaction because of difficulties in staff training.

Keywords: e-CRM – hotel sector-e-marketing.

1 Introduction

The introduction of Internet technology to international business has led to its wide-scale application in the hotel industry [16]. Hotel internet marketing helps firms establish interactive relationships with their customers and address individual needs and preferences of customers [28]. As competition increases, the customer is becoming more demanding and hospitality business need to develop new, better targeted products; 'satisfying the customer' constitutes a significant factor that creates strong competitive advantage and provides crucial motives for further business advancement. According to Zineldin [28] 'Hotels have the possibility of receiving feedback from customers'. Customer complaints, costumer reports, customer research, and customer interviews are only some examples. In these terms, the implementation of an e-CRM system may seem as a natural development in the particular research path.

Empirical studies show that e-customer relationship marketing (e-CRM) can have a positive impact on customer satisfaction in the case of Internet-based services [17] identifying consumers' different preferences. Berry and Parasuraman [3] state that a firm which applies a relationship marketing strategy can monitor customer behaviour

M.D. Lytras et al. (Eds.): WSKS 2009, CCIS 49, pp. 484–494, 2009.

by directly managing its customer database. E-marketing utilises the electronic medium to perform marketing activities with the purpose of finding, attracting, winning and retaining customers. Taking a relationship marketing perspective the hotel industry have to use the full potential and challenges of the Web as a strategic tool to facilitate the development of relationships with customers [12].

On the other hand, different opinions are raised by academics and practitioners about the beneficial side of e-crm. Most of these arguments tend to focus on the inflexibility of use of the system, and the rigid database that constitutes severe challenge for front office manages since they will need to keep the system constantly updated. Other arguments focus rather on the costs of the extra training hours of the staff in order to be able to use it.

This research focuses on hotel bookings as a major sector of the travel industry. Web specialists estimate that in 2010, at least one-third of all hotel bookings will be completed online [26], and the role of the e-CRM system. Issues, implications, opportunities and threats of the particular system will be examined and analyzed always in close connection to the nature of the accommodation sector. Furthermore, the aim of this research is to examine whether the use of e-CRM practices followed by first category hotels of Attica build mutual beneficial relationships with their guests and take advantages from the opportunities provided by e-CRM.

2 Literature Review

2.1 Optimizing Marketing and Customer Relationship Marketingon the Internet

E-marketing gives businesses the opportunity to have a global presence and enables them to establish their image, sell a wide range of products and services and enlarge relationships between firms and customers. E-marketing through the use of electronic communications technology such as Internet, e-mail, databases and mobile devices improves customer loyalty [24]. Then the company can deliver better targeted communications and online services that match its customer individual needs. The steps followed in the marketing mix process are:

Scope: A company has to determine the strategic role of the web activities. A web site could be transactional or promotional. Transactional sites include basic functionality such as the ability to view recent and past transactions. Promotional sites publish content with information about hotel services and do not allow any transactions.

Site: Web experience. Customer oriented content. Law and Huang found that travellers often use multiple web sites to compare prices before they book [19], [18].

System: provides an outline of technical factors concerning the secure, safe, cost-efficient, user friendly operation, market intelligence reports and performance analysis. There are projects in measuring, evaluating, and understanding system performance.

CRM strategy is a response to the changing dynamics of the marketplace and a systematic effort to enable companies to build long lasting relationships aiming at profit and maximizing the customer's lifetime value to the firms [6]. According to Gordon, relationship marketing involves the creation of new and mutual value between a

provider and individual consumer [14]. Reichheld and Sasser [22] claimed that a 5% improvement in customer retention can result in an increase in profits of between 25 and 85 percent depending on the industry [7]. Relationship marketing takes over after the sale for existing customers by stimulating the intention to buy and sometimes after the sale through post-purchase reinforcement.
Gilbert and Powell-Perry [12] proposed the following *five steps* of how firms can become e-customer-centric:

Identify more about the customer through database analysis. The role of the hotel Web site could measure and track guests' reactions to different offer being posted by a hotel on a Web site which helps to create their profile [23] [29].

Improve and make the service more attractive. The compilation of online questionnaires, targeted to frequent guests, tracks their attitudes to services.

Inform to build guests' knowledge of the company. Key information, e-newsletters, lists and locations of key agents, special offers could all be posted on the hotel Web site [30].

Tempt guests to purchase more regularly, try different kind of services. The Web could force this objective by direct e-mailing to frequent hotel guests promotion leaflets with special offers for existing or new services.

Retain the customer by developing different forms of loyalty schemes. Entry into the Web site could be restricted through the use of identification numbers and passwords for the exclusive use of hotel loyal guests.

The attitudes and intentions affect behavior in order to understand the complexity of customer loyalty [21]. Gustafsson, [15] focus on: a) customer satisfaction, b) affective commitment c) calculative commitment and d) normative commitment.

Customer satisfaction: Barnes, [2] and Bolton and Ruth [5] state that the interpersonal element of customer service in service organizations (such as hotels) influences customers' perceived service quality and satisfaction influence the consumer behaviors such as positive word of mouth and repurchase [5].

Affective commitment: originates from identification, common values and similarity [1] can be defined as one part desire to maintain a relationship with a provider because they enjoy working with it, and because both experience a sense of loyalty and trust [9].

Calculative commitment: can be defined as the cold calculation of costs and profits, involving also investments and available alternatives to replace or build-up for foregone investments [31].

Normative commitment: is reflected in the moral obligation to stay in a relationship. Geyskens, [10] states that firms are increasingly looking to have fewer, yet more intense relationships with their partners and "*the use of global commitment measures - which measure intention to continue a relationship without consideration of the underlying motivation- could confound or mask different, and possibly even opposite effects*".

Trust is defined as a belief that the one will act in the best interests of the other relationship partner [27] and has been referred to as the key element of successful relationship development [13], [20]. Trust is more than ever a central concern for hospitality industry as by nature, the service cannot be tried before purchase.

3 Research Methodology

In this research the questionnaires have been sent by e-mail and by fax to the companies. For a small part of the respondents it was arranged an appointment in order to have also personal conduct. The questionnaire refers to the evaluation of e-marketing strategies which are used by (five star) hotels to specifically examine the impact of electronic relationship marketing practices on customer retention and acquisition. A five-point scale used in the questionnaire (strongly agree, agree, neutral, disagree, strongly disagree). As it is commonly known, deluxe category hotels (belonging to large enterprises) appreciate more electronic procedures and potential of strategies. That was also the reason why they were selected for this specific research. The total numbers of luxury category hotels that are in Attica are 29. However 25 hotels were selected because 4 belong to the same owner. The questionnaire is divided in four sections each one reflects to the research objectives. The respondents were mainly managers from marketing or sales departments or general managers and this increase the validity of the received data.

3.1 Managers' Opinion from the Use of e-Marketing

The 80% of the respondents (60% agree and 20% strongly agree) answered that are able to offer more customized tourism services by using the Web and explained orally that this operates as a key motivator for encouraging tourists to purchase online. However, one fifth of the managers (20%) neither agree nor disagree that the web can facilitate product customization.

The 68% of the respondents argue (52% agree and 16% strongly agree) that the use of internet reduce significantly the time of information distribution than any other traditional method. Almost one third of deluxe hotels in Attica (28%) consider that time saving is not exclusively the main point for the use of e-marketing than any other traditional marketing method.

Hotel enterprises use a plethora of different distribution channels to enable service purchase. E- Newsletter is one of those channels in which hotels can contact their customers directly and intermediaries like travel agents are bypassed. The largest proportion of respondents (52% agree and 36% strongly agree) believe that e-newsletters reduce cost and time of information distribution. On the other hand, the above statement (12%) of the respondents finds this opinion neutral. Concerning the implementation of an e-CRM system to the improvement of company's profit and sales the notions were positive: an amount of 72% of respondents agree (44% strongly agree and 28% agree) and one fifth was neutral.

In a relatively high proportion 72% (40% agree and 32% strongly agree) of respondents assume that customers find e-methods of communication are flexible and more useful than traditional ones. However, the other seven out of 25 respondents have a neutral opinion.

In order to identify the e-marketing strategies that are used from deluxe category hotels of Attica it is useful to analyze the relation between the mean (M) and the standard deviation (S.D.) of each of the following strategies.

Safe e-transactions is one of the tools used by hotels more frequently as the mean (M) is 4, 56 and the standard deviation (S.V.) is small 0, 65. Managers pay also attention to service promotion as the (M) is 4, 44 and the (S.V.) is 0, 51. Information update is appeared in the (M) of 4, 36 even though the (S.V.) is a little bit higher ((S.V.=0,76) showing that the data points are not exclusively clustered closely around the (M). Advertising like banners and sponsored links appear with mean= 4, 24 while (S.V.=0,93) is neither close to zero nor that high and that shows that firms use this process in their e-marketing strategy. Motivation of customers to buy new services is one more tool that deluxe hotels of Attica include in their e-marketing strategy ((M) =4, 16 and (S.V.) =0,85) but it does not seems to be a priority.

According to the companies survey and personal experience (Table 1), customers prefer primarily more customized services to his/her needs (M =4, 36, S.V.=0,64). In a relative lower (M) 4,32 but also lower (S.V.) 0,48 customers are searching for an attractive site. With a mean 4,20 and (S.V.) 0,82 is obvious that customers pay attention to information quality in order to visit a company again. Receiving an extensive feedback motivates guests to visit a company again ((M) =4, 12, (S.V.) =0,73). Guests of luxury hotels of Attica find that e-methods of communication are flexible and more useful than traditional ones in a (M) of 4, 04 and (S.V.) in 0, 79.

Table 1. Customer's expectations and preferences

According to company's survey and personal experience, customer	Mean	Std. Deviation
Prefer more customized services to his/her needs	4,36	0,64
Searching for an attractive site	4,32	0,48
evaluate information's quality in order to visit company again	4,20	0,82
Receive an instant and extensive feedback which motivate him/her to visit company again(physically or electronically)	4,12	0,73
Assume that e-methods of communication are flexible and more useful than traditional ones	4,04	0,79

The most important feature of the e-CRM tool used by companies is the safety. Managers assume that it is proved that e-CRM lead to money and time savings. One more characteristic of e-CRM system is that it is designed in order to produce accurate data/performance (M) =4, 20, (S.D.) =0, 87). E-CRM that uses accurate data to enhance every guest interaction demonstrate clearly that managers value the guest and want to build a strong and lasting relationship. However, in a low (M) of 3,80 and (S.D.) in 1,00 managers believe that e-CRM is not the power tool assisting in future customer forecasting and communication link between company's departments

(M=3,72 S.D.=0,89). On the other hand, e-CRM it is not exclusively adjusted to each guest preferences and needs (M=3, 68, S.D. =1, 07).

3.2 Achieving Customer Retention and Acquisition by e-Methods practices

In order to retain its customers (Table 2), the companies find most important the data gathering and editing of results of customer's profile. Moreover, hotels offer services that response to customer preferences. Hotels try to provide information about new services to the guests and roughly with the same frequency motivate customer to buy new services through e-newsletter. The method that hotels are using less in order to maintain guests is the suggestion of special offers depending on frequency in hotel site.

Table 2. Customer retention

In order to retain customers, company:	Mean	Std. Deviation
Gathering data and edit customer's profile	4,68	0,56
Offering services that response to customer preferences	4,32	0,90
Informing customers about new services	4,16	0,75
Motivating customer to buy new services through e-newsletter	4,16	0,80
Suggesting special offers depending on frequency in hotel site	3,76	1,05

For the customer acquisition hotels priority is the privacy of customers data (M =4,28, (S.V.) 0,98) and the speed of correspondence seems to be equally in sequence (M=4,28, (S.V.) =0,89). Furthermore, classification of data base in the (M) of 4, 12 and (S.V.) in 0,88 is one more process that hoteliers pay attention to acquiring new customers. The dialogue from both directions (hotel-guest and guest-hotel) is one more function assist hotel in customer acquisition (M=4,04, (S.V.) = 0,73). The experience during the ages and self-confidence in the shifting external environment and customer preferences are the basic weapons for managers to be proactive among rivals (M=4,00, (S.V.) =0,96). From the less valuable tools for customer acquisition managers use recording and monitoring of customer behaviour (M= 3, 92 (S.V.) =0,81) and pricing strategy (M= 3,88 (S.V.) =0,88).

3.3 The Implementation of e-CRM Strategies and Its Contribution to Hotels

E-customer relationship marketing (e-CRM) help business deals to be perfect and precise. Most enterprises (56% of the respondents agree and 36% strongly agree) argue that e- CRM used by their firm is designed in order to produce accurate data/performance and 80% of the sample argue that it is proved that it saves time and money for the company. However, the remaining one fifth (20%) of the sample give a neutral opinion of how obvious it is that the system is used to save time and money and gave an oral explanation that guests sometimes are not familiar with electronic procedures and it is preferable to develop relationships with them traditionally.

The most important feature of e-CRM system is that it is highly customizable. Thus, 64% (44% agree and 20% strongly agree) of managers argue that e-CRM system that they used is adjusted to each customer needs. With the help of extended application support, it can be integrated easily with any application. Thus, an existing e-CRM application can use the required features of e-CRM with minimal changes and optimized costs. The results are showing that 82% (36% agree and 52% strongly agree) of the respondents argue that e-CRM system in their company is safe and the same percentage (52% agree and 36% strongly agree) consider e-CRM as the key tool in managing loyal and satisfied customers.

It is commonly accepted that e-CRM is doubtless a tool that tracks customer preferences and other personal data but it is not generally used as a predicting tool. However, it is a tool that is used as a communication link between company's departments. Total of 68% of the sample agree (52% agree and 16% strongly agree) that e-CRM facilitates communication through different departments while just 20% have a neutral opinion and 12% disagree that it could be a combination link.

Hotel managers have explained orally that 'e-CRM provides customer information data sharing and allows a company to better target its customers and offer them the products and services they are most likely to purchase based on history and demographics. When someone visits the web site it doesn't mean that they are an exclusively potential guest. It can be family travellers, group leaders or vacation planners. In order to succeed it is necessary for hoteliers to know the customer'.

According to the present research hotels implement e-CRM strategies in order to create the feeling that quality is their main aim an amount of 88% (56% agree and 32% strongly agree) of respondents and 72% targeting to enhance long term loyal customer relationship.

Almost one quarter (24%) of the managers argue that long term customer relationship is not their main target to implement e-CRM strategies. On the other hand, 92% (36% agree and 56% strongly agree) of them focus on customer acquisition and a bit less (84%) on customer retention.

Customer service is primarily a reactive function supporting transaction, while e-CRM as a whole is a proactive long term strategy. E-CRM is not just a tool to achieve and enhance customer satisfaction. 88% of managers use e-CRM in order to offer high quality services and 72% focus on achieving high standard services. A satisfied customer and high quality service leads to customer loyalty.

Moreover, e-CRM is used by managers in order to achieve long term objectives of the companies: 88% of the respondents (44% strongly agree and 44% agree) while 76% (48% agree and 28% strongly agree) try to improve company's procedures.

3.4 Factors Affecting Customer Retention and Aquisition

(94%) managers consider that collecting data and editing customer's profile is the key of customer retention and 92% argue that in order to maintain customers, a company has to offer services that meet customer preferences.

Information update about new services seems to be one more significant point for customer retention. 88% of the respondents think that providing information about new services gives customers the opportunity to have more interest on the specific firm as it is not a routine product that they purchase but a total refreshing one.

76% of managers (36% agree and 40% strongly agree) motivate their customer to buy new services through e-newsletter as a strategy to maintain an ongoing and lasting relationship with them. E-newsletter inform subscribers of new reasons to visit company's website and reward customer's referrals by coordinating newsletter content with sales promotions department.

Hoteliers create regular opportunities to reach out to the market, improve customer retention rate and increase regular sales or targeted promotional sales. About one quarter (24%) neither agree nor disagree that motivation for purchasing via e-newsletter is leading to customer loyalty.

56% of the respondents (32% strongly agree and 24% agree) argue that they pursue customer retention by offering special offers depending on the frequency of visits in the Web site. On the other hand 32% neither agree nor disagree that suggesting a discount or an offer influence customer to have a loyal behaviour and 12% disagree.

Managers should periodically review sales figures as a regular process of marketing strategy. They should know who bought and what service and, of course, at what price. For customers who visit the site frequently and purchase regularly, managers can approach them with special offers. Special offers can include a coupon sent via e-mail or generally a form of discount.

In order to acquire customers, managers find e-CRM more useful (52% strongly agree and 32% agree) to keep the privacy of customer data . Information requested from customers ought to be used exclusively to provide professional services.

Speed of correspondence seems to be the second important method that managers (52% strongly agree and 28 % agree) pay attention to during implementation of e-marketing strategy. E-CRM system operates in a collaborative environment through unique features including Application Sharing and Text Chat. This leads guests to being closer to the transaction and using it in an easy, safe and collaborative way.

Bidirectional dialogue and classification of data base are used (76% of deluxe hotels of Attica) as two main points in their online presence. Bidirectional dialogue is the first step for a successful communication. However, almost one quarter (24%) of the respondents neither agrees nor disagrees on that.

Guest can easily compare different prices and get information about services that offered as the price becomes most of times an important decision criterion. [6] state: "*hotel groups are not forced to offer very low prices, as long as the customer is receiving perceived value for money.*" One fifth of managers have a neutral opinion on the implementation of pricing strategy even though they know that a shift on price is able to change dramatically both gross margins and sales volume. A very small percentage (8%) of them disagree that pricing is a tool that facilitates a company acquiring new customers.

4 Conclusions

According to the results Luxury category hotels of Attica are interested to some extent in practicing e-marketing with respect to e-CRM strategies. The highest percentage focus on high standard services and are willing to offer services that are personalized to each customer needs. Additionally, managers utilize e-CRM and e- marketing tools both to maintain and acquire new customers. They have understood the importance of using electronic channels instead of traditional ones to implement their marketing strategies. The above findings highlight that managers have perceived the importance of well-designed e-CRM processes and the perspectives of customers in relationship building. The results imply that managers are aware of the different effects of each stage of e-CRM. Therefore, they have planned appropriate strategies for the knowledge of customer and generate and reinforce satisfaction and commitment.

On the other hand, managers of the remaining of the deluxe category hotels of Attica express neutral opinion on the majority of the questions concerning the e-marketing and e-CRM process that they follow. However they give emphasis to employees of the organization to ensure that they are communicating effectively among the departments. They did not seem to apply many of the relationship marketing strategies to emphasize customer retention and continual satisfaction.

These results imply that managers have not been aware for the new technological improvements and they don't take into consideration that the customer knowledge is essential for their businesses to know where they can improve and what they should expand on and personalized marketing messages with information that is suitable for its customer thus result in more premium-driven loyalty for the hotel services [25].

It is important therefore, hospitality sector in Greece obtain high position worldwide as it has the full advantages by nature it needs. The managers should benefit from the technological advantages and offer high standard services that are closed to customer preferences in order to maintain them and acquire new ones. Personalizing the customer experience in online marketing is considered as a powerful communication and retention tool. Customizing interaction with the most valuable guests provide significant long-term rewards.

References

1. Bansal, H.S., Irving, P.G., Taylor, S.F.: A three-component model of customer commitment to service providers. Journal of the Academy of Marketing Science 32(3), 234–250 (2004)
2. Barnes, J.G.: The Role of Internal Marketing: If the Staff Won't Buy It, Why Should the Customer? Irish Marketing Review 4(2), 11–21 (1989)
3. Berry, L.L., Parasuraman, A.: Marketing Services. Competing through Quality. Free Press/Lexington Books, Lexington (1991)
4. Bolton, R.N.: A Dynamic Model of the Duration of the Cus-tomer's Relationship with a Continuous Service Provider: The Role of Satisfaction. Marketing Science 17, 45–65 (Winter 1998)
5. Boulding, W., Kalra, A., Staelin, R., Zeithaml, V.A.: A Dynamic Process Model of Service Quality: From Expectations to Behavioral Intentions. Journal of Marketing Research 30, 7–27 (1993)
6. Cataldo, G.: A new way to understand marketing – the CRM. Szczecin University Publisher (2006)
7. Christopher, M., Payne, A., Ballantyne, D.: Relationship Marketing. Heinemann, London (1991)

8. Evans, J.R., Laskin, R.L.: The relationship marketing process: a conceptualisation and application. Industrial Marketing Management (1994)
9. Fullerton, G.: The service quality-loyalty relationship in retail services: Does commitment matter? Journal of Retailing and Consumer Services 12(2), 83–97 (2005)
10. Geyskens, I., Steenkamp, J.-B.E.M., Scheer, L.K., Kumar, N.: The effects of trust and interdependence on relationship commitment: A transatlantic study. International Journal of Research in Marketing 13(4), 303–317 (1996)
11. Geyskens, I., Steenkamp, J.-B.E.M., Kumar, N.: Generalizations about trust in marketing channel relationships using meta-analysis. International Journal of Research in Marketing 15(3), 223–248 (1998)
12. Gilbert, D.C., Powell-Perry, J.: A current overview of Web based marketing within the hotel industry. In: Proceedings of Tourism in Southeast Asia and Indo-China Forth International Conference, Changmai, Thailand (2000)
13. Goodman, L.E., Dion, P.A.: The determinants of commitment in the distributor-manufacturer relationship. Industrial Marketing Management 30(3), 287–300 (2001)
14. Gordon, I.: Relationship Marketing: New Strategies, Techniques and Technologies to Win the Customers You Want and Keep Them Forever. John Wiley and Sons Publishers, Chichester (1999)
15. Gustafsson, A., Johansson, M.D., Roos, I.: The Effects of Customer Satisfaction, Relationship Commitment Dimensions, and Triggers on Customer Retention. Journal of Marketing 69 (2005)
16. Cathy, H.: Journal Importance of Hotel Website Dimensions and Attributes: Perceptions of Online Browsers and Online Purchasers of Hospitality & Tourism Research 30(3), 295–312 (2006)
17. Karmarkar, U.: Will you survive the services revolution? Harvard Business Review 82(6), 100–108 (2004)
18. Kerner, S.M.: Online is destination for travel booking? (2005), `http://www.clickz.com/showPage.html?page=3528131`
19. Law, R., Huang, T.: How do travelers find their travel and hotel websites? Asia Pacific Journal of Tourism Research 11(3), 239–246 (2006)
20. Moorman, C., Zaltman, G., Deshpandé, R.: Relationships between providers and users of market research: The dynamics of trust within and between organizations. Journal of Marketing Research 29(3), 314–329 (1992)
21. Oliver, R.L.: Whence Consumer Loyalty? Journal of Marketing 63(Special Issue), 33–44 (1999)
22. Reichheld, F., Sasser, W.: Zero defects: quality comes to services, pp. 105–111. Harvard Business Review, Boston (1990)
23. Schall, M.: Best practices in the assessment of hotel-guest attitudes. Cornell Hotel and Restaurant Administration Quarterly 44(2), 51–65 (2003)
24. Smith, P.R., Chaffey, D.: E-Marketing eXcellence: at the heart of eBusiness. Butterworth Heinemann, Oxford (2001)
25. Starkov, M., Price, J.: Building an eCRM Strategy in Hospitality: How to Establish Mutually Beneficial Interactive Relationships with Your Customers (2007)
26. Starkov, M., Price, J.: Hotelier's online marketing resolutions. Hospitality eBusiness Strategies, HeBS (2007)
27. Wilson, D.T.: An integrated model of buyer-seller relationships. Journal of the Academy of Marketing Science 23(4), 335–345 (1995)
28. Zineldin, M.: Beyond Relationship Marketing: Technologicalship Marketing. Marketing Intelligence & Planning 18(1), 9–23 (2000)

29. Vladlena, B.: Perceptions of trust and experience: potential barriers to web 2.0-based learning. International Journal of Knowledge and Learning 4(5), 427–437 (2008)
30. Jeanette, L.: Interfirm knowledge management through a web-based bench learning system. International Journal of Knowledge and Learning 4(4), 317–328 (2008)
31. Stephen, P., Stephen, A.: Strategies for flexibility: technology-centered and labour-centred flexibility in UK manufacturing. International Journal of Manufacturing Technology and Management 1(4/5), 366–380 (2000)

An Ontology for Musical Phonographic Records: Contributing with a Representation Model

Marcelo de Oliveira Albuquerque[1], Sean Wolfgand M. Siqueira[1], Rosana de Saldanha da G. Lanzelotte[1], and Maria Helena L.B. Braz[2]

[1] Department of Applied Informatics (DIA/CCET), Federal University of the State of Rio de Janeiro (UNIRIO): Av. Pasteur, 458, Urca, Rio de Janeiro, Brazil, 22290-240
{marcelo.albuquerque,sean}@uniriotec.br, rosana@lanzelotte.com
[2] DECivil/ICIST, IST, Technical University of Lisbon: Av. Rovisco Pais, Lisbon, Portugal, 1049-001
mhb@civil.ist.utl.pt

Abstract. Music is a complex domain with some interesting specificities that makes it difficult to be modeled. If different types of music are considered, then the difficulties are even bigger. This paper presents some of the characteristics that makes music such a hard domain to model and proposes an ontology for representing musical phonographic records. This ontology will provide a global representation that can be used to support systems interoperability and data integration, which provides disseminating music worldwide, contributing to culture in the knowledge society.

Keywords: Music, Model Representation, Concept, Ontology, Musical Phonographic Records, Data integration.

1 Introduction

With the increase in the number of musical information sources available on the Web, the problems related to representing, querying and sharing this kind of data became more evident [1]. This happens because music domain has some specificities that are not perceived by cataloguers and information systems developers that usually deal with the domain of traditional bibliographic records.

Records, cassettes, CD's, and other music recordings come under a general category, in this work called phonographic records. There is no specific metadata standard for cataloging musical phonographic records. So, what happens is that each institution having a music data source adopts an existing metadata standard or builds its own model based on existing ones. However, different sets of metadata describe different things [2] and the use of different sets/standards contributes to the structural and semantic heterogeneity in data sources. Therefore, the formal definition of concepts related to the field of music is essential to represent information in music catalogs that can be understood by users and by machines.

In order to formalize the representation of knowledge in Computer Science, theories and techniques of Information Science begin to gain importance. Among them there is the Concept Theory of Dahlberg [3], in which a concept is a unit of knowledge used to describe a category or class of entities, events or relations.

M.D. Lytras et al. (Eds.): WSKS 2009, CCIS 49, pp. 495–502, 2009.

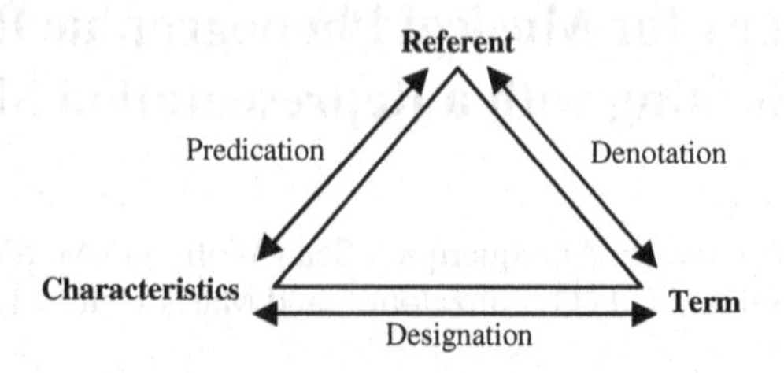

Fig. 1. Dahlberg Conceptual Triangle [3]

The conceptual triangle of Dahlberg (Fig. 1) shows that concepts are formed by the characteristics (properties) allocated by a community to the referent (formal object), that is represented by a term. To have the definition of a concept all the conditions must be achieved. A concept is a mental construct, so it is something far beyond the term or linguistic symbols that denote it. The concept has the characteristics necessary to define a referent of a kind and is the product of a community, so their perception is the same regardless of language.

As the music area has ambiguity in different concepts [4], the use of ontology is being proposed with the aim of formalizing the definition of relevant concepts to represent information in the music domain, ensuring their semantics.

The representation model for musical phonographic records aims at improving the interoperability of such resources and related systems, thus disseminating songs and culture while also contributing to improve music knowledge. It is also important to notice that music can contribute in other scenarios such as learning and health, while the ontology approach can enable machine understanding of the resources' semantics.

The rest of this paper is organized as follows: section 2 presents related work, section 3 proposes an ontology for musical phonographic records, and section 4 presents some final considerations.

2 Related Works

Riley et al. [5] considers the diversity (of styles, cultures, formats and uses) in the music domain as a factor that hampers the representation of information. Examples of this diversity are: the issue of ambiguity of the word "work" and its granularity; representation of musical instruments; the specification of a formation for performing a musical piece; and the musical genre classification. Moreover, another factor that increases the heterogeneity between data sources is that each institution and user has special needs and, therefore, the most suitable metadata sets are used over interoperability restrictions.

Some solutions were proposed to deal with information representation in the music field, but they do not really solve the problems. They are isolated, often ad hoc, and largely do not address the subject of conceptual definition, making difficult their use by other peers in the community.

Examples of these proposals are: Music Ontology; the definition of a metadata dictionary [6]; a thesaurus of instrumental/vocal formation [7]; and the musical genre classification [8].

The definition of musical work and its granularity is a difficult issue to be solved. A musical work refers to something that is not static, and therefore may have changes or different interpretations. In music, this fact is clear. Just consider that there is a different version of the same work and it can be interpreted as a new work or a musical piece composed by parts as a symphony where each part can be considered a work by itself.

Another issue is the conceptualization of musical work. It may be referencing an entire career of a musician or even a single song. So, removing the ambiguity in the definition of work is important for the representation of information related to music domain.

The issue of classification by musical genre in computer systems is also an open problem [9]. In [8], the few number of researches in this area is questioned as well as whether they should be abandoned. According to the authors it is a very promising field, but that still needs to develop. Hampering this development some aspects were pointed out: problems of definition (ambiguity) and in the use of classifications, the importance of the genres definition and how the work has progressed on theses issues.

The music genre classification tries to organize the works in groups according to some aspects. Each group is divided considering the analyses of music domain specialists. In [8], some rules for definition of genres are outlined, such as formal and technical aspects, a semiotic approach and social or ideological habits.

The representation of musical instruments in systems for cataloging and information retrieval is a relevant aspect that is being addressed by some authors as [7, 10]. These initiatives are important as they will contribute to facilitate the cataloguers work and assist users when searching for music information.

Music Ontology (MO) is the first work in the direction of having a generic way for representing information related to the field of music using an ontology. According to Raimond et al. [10], the MO aims to "provide a set of identifiers and structured representation for the field of music". This ontology is based on Functional Requirements for Bibliographic Records (FRBR) [11] and its concepts and relationships were extracted from the metadata defined by MusicBrainz [12]. The MO is formed by a combination of some ontologies as Timeline Ontology, the FOAF Ontology (Friend-of-a-friend Ontology), the ABC Ontology and the Event Ontology.

The MO was defined based on FRBR, which were developed for common bibliographic records. The FRBR entities consider three groups: products, people and subject. The MO uses only the product level that is subdivided into **work**, **expression**, **manifestation** and **item**. The problem is on the use of these four entities to represent music, because, before applying the model, it is necessary to define the boundaries between these entities. As the MO concepts belonging to the group product (first description level of FRBR) are not clearly defined, their use is subjective and depends on the user's understanding about the terms.

The development of the MO is an attempt to model music domain through the definition of classes and relationships. But, as shown, the work presents some open issues that may be considered conceptual and structural problems.

3 The Proposed Ontology

The literature review showed the lack of a model for structural and semantic representation of concepts about music phonographic records while proposals such as MO do not satisfy various aspects of the domain. Therefore, a model to represent music information about phonographic records is proposed. To formalize this proposal an ontology was built.

The concepts of the ontology have been identified through research in the music bibliography. For this, we used the web service "Oxford Music Online" that references the digital "Grove Encyclopedia", the "Oxford Dictionary of Music" and "The Oxford Companion to Music" [13]. Among these references, Grove Encyclopedia is the most comprehensive. It is an indispensable reference about music and musicians [14], containing more than 60,000 items including biographies, articles and concept definitions.

The "Oxford Music Online" allows the simultaneous consultation on the mentioned references. Among search options, the most useful for this research allows searching for concepts, and the results obtain were used as basis for the definition of concepts in the construction of the proposed ontology.

However, even with the great number of definitions on the music domain that were available through the web service, some concepts were extremely difficult to define. An example is the definition of the musical work. According to Smiraglia [15] the semiotic analysis of the term suggests a variety of social and cultural roles when considering artwork, and especially for music.

The IFLA (International Federation of Library Associations and Institutions) presented in FRBR, in 1998, a definition [16], which considers that the term **work** designate a particular intellectual or artistic creation, limiting the understanding of musical work to a simple composition. However in the Grove Encyclopedia, the term musical work was used to describe the sum of all the works composed or interpreted by a musician throughout his career.

The concepts chosen by this research were also defined based on the "Oxford Music Online" that proposed metadata for bibliographic description in music domain.

Below are listed all the concepts used in the ontology[1] construction according to the concept definition by Dahlberg:

- Concept: Album.
  - Term: Album_musical.
    - Definition: Set of songs, released together commercially in an audio format, recorded in some kind of support.
    - Properties: codigo_ISRC (International Standard Recording Code – ISRC code), data_da_gravacao (recording date), idioma (language), lado_do_album (side), local_de_gravacao (recording place), notação_na_colecao (collection notation), numero_na_serie (series number), numero_de_faixas (track number), tempo_de_duracao (time), tipo_de_suporte (media type), titulo_alternativo (alternative title), titulo_uniforme (uniform title), titulo_da_serie (series title).

[1] This ontology was built in Portuguese, so the terms, properties and relations are defined in this idiom.

    - Relations: Album x Collection (e_contido_por = is_contained_by, contem = contains); Album x Music (e_parte_de = is_part_of, possui = has).
- Concept: Collection.
  - Term: Colecao.
    - Definition: Set of items, with no relationship, randomly accumulated.
    - Properties: instituicao_de_guarda (custody institution), quantidade_de_itens (quantity of items).
    - Relations: Collection x Album (e_contido_por = is_contained_by, contem = contains).
- Concept: Gender.
  - Term: Genero_musical.
    - Definition: Categorical and typological construct that identifies musical pieces as belonging to a particular category and type of music that can be distinguished from other types of music.
    - Properties: descricao (description), termo_de_referencia (reference term).
    - Relations: Gender x Music (e_englobado_por = is_included_by, engloba = includes)
- Concept: Musical instrument.
  - Term: Instrumento_musical.
    - Definition: Device to produce sound and that is used to perform a musical composition.
    - Properties: descricao (description), termo_de_referencia (reference term).
    - Relations: Musical Instrument x Instrumentalist (toca = plays, e_tocado_por = is_played_by).
  - Term: Instrumento_de_corda (sub-class String instrument).
    - Definition: Musical instrument that produces sound through the vibration of strings.
  - Term: Instrumento_de_metal (sub-class Brass instrument).
    - Definition: Musical instrument made of metal, which produces sound by air flow.
  - Term: Instrumento_de_percussao (sub-class Percussion instrument).
    - Definition: Musical instrument that produces sound through impact, scraping or agitation, with or without the aid of stick or drumstick.
  - Term: Instrumento_de_sopro (sub-class Wind instrument).
    - Definition: Musical instrument not made of metal, which produces sound by air flow.
  - Term: Instrumento_eletronico (sub-class Electronic instrument).
    - Definition: Musical instrument that produces or modifies sounds electronically.
- Concept: Music.
  - Term: Musica.

- Definition: Grouping rhythmic, melodic or harmonic sound, composed specifically as a form of auditory communication.
- Properties: data_da_composicao (composition date), data_da_gravacao (recording date), idioma (language), local_de_composicao (composition place), local_de_gravacao (recording place), numero_da_faixa (track number), opus, tempo_de_duracao (time), titulo_alternativo (alternative title), titulo_uniforme (uniform title).
- Relations: Music x Composer (e_composta_por = is_composed_by, compoe = composes), Music x Interpreter (executa = executes, e_executada_por = is_executed_by), Music x Cultural Period and Gender (e_englobado_por = is_included_by, engloba = includes), Music x Album (e_parte_de = is_part_of, possui = has).

- Concept: Musician.
  - Term: Musico.
    - Definition: Person who interprets and/or composes music.
    - Properties: data (date), nacionalidade (nationality) nome_uniforme (uniform name), nome_alternativo (alternative name).
    - Relations: Musician x Musical Work (possui_uma = has_a, pertence_a = belongs_to), Musician x Cultural Period and Gender (e_englobado_por = is_included_by, engloba = includes).
  - Term: Compositor (sub-class Composer).
    - Definition: Musician who performs the activity or the process of music composition.
    - Relations: Composer x Music (e_composta_por = is_composed_by, compoe = composes).
  - Term: Intérprete (sub-class Performer).
    - Definition: Musician who brings to reality a composition by performing musical instrument and/or voice.
    - Relations: Performer x Music (executa = executes, e_executada_por = is_executed_by).
  - Term: Grupo_musical (sub-class Musical group).
    - Definition: Performers group bringing to reality a composition by performing musical instrument and / or voice.
    - Relations: Musical group x Solo performer (integra = integrates, e_integrado_por = is_integrated_by).
  - Term: Solista (sub-class Solo performer).
    - Definition: Interpreter who brings a composition to reality through the use of musical instrument and / or voice.
    - Relations: Solo performer x Musical group (integra = integrates, e_integrado_por = is_integrated_by)
  - Term: Instrumentista (sub-class Instrumentalist).
    - Definition: Interpreter who performs a musical composition using a musical instrument.
    - Relations: Instrumentalist x Musical Instrument (toca = plays,

e_tocado_por = is_played_by).

- Term: Regente (sub-class Conductor).
  - Definition: Interpreter responsible for aspects of the performance of a musical group, showing through gestures or patterns how music should be interpreted.
- Term: Vocalista (sub-class Singer).
  - Definition: Interpreter that brings to reality a composition by voice.

- Concept: Musical Work.
  - Term: Obra_musical.
    - Definition: All the artistic achievements of a particular musician.
    - Relations: Musical work x Musician: (pertence_a = belongs_to, possui_uma = has_a).
- Concept: Cultural Period.
  - Term: Periodo_cultural.
    - Definition: Time period that encompasses a unique set of artistic styles and performance practices.
    - Properties: data_do_periodo (period date), descricao (description), termo_de_referencia (reference term).
    - Relations: Cultural Period x Music and Musician (e_englobado_por = is_included_by, engloba = includes).

4 Conclusions

This paper presented a conceptualization of musical phonographic records by analysis of concept definitions in the music domain found in specialized literature. The problems of lack of specific standards for cataloging music, the use of Dahlberg Theory of Concepts and some related works were discussed.

To solve the lack of a model to represent concepts, relations and properties in the music domain an ontology was proposed. It was built based on the most used and recommend items to catalogue musical phonographic records. It was developed through research considering various projects in this domain and its content has been specified using well known bibliographic sources from the area of musicology. As future work it is planned to develop an application based on the proposed ontology.

The authors consider that having a model defined by an ontology will contribute to solve the problem of heterogeneity of data sources, and increase interoperability between institutions that preserve musical collections increasing access and visibility.

Preservation and access to cultural heritage such as music data sources is an important aspect that can not be forgotten in the construction of a knowledge society. As future works, the proposed ontology could be combined with other ontologies for richer experiences such as the one proposed in [17] for describing emotions. In addition, a new system could be developed for cultural context, for instance, considering the approach proposed in [18].

Acknowledgments. Special thanks to FCT Portugal, through the ICIST funding, for the financial support, making possible the accomplishment of this work.

References

1. Lai, C., Fujinaga, I., Descheneau, D., Frishkopf, M., Riley, J., Hafner, J., McMillan, B.: Metadata infrastructure for sound recordings. In: International Conference on Music Information Retrieval, Vienna (2007)
2. Corthaut, N., Govaerts, S., Verbert, K., Duval, E.: Connecting the dots: music metadata generation, schemas and applications. In: 9th International Conference on Music Information Retrieval, Philadelphia (2008)
3. Dahlberg, I.: Concept and Definition Theory. Classification Theory in the Computer Age, pp. 12–24 (1988)
4. Lai, C., Fujinaga, I.: Metadata data dictionary for analog sound recordings. In: Joint Conference on Digital Libraries, Chapel Hill (2006)
5. Riley, J., Dalmau, M.: The IN Harmony Project: Developing a Flexible Metadata Model for the Description and Discovery of Sheet Music. The Electronic Library (2007)
6. Lai, C.: Metadata for phonograph records: Facilitating new forms of use and access. PhD Thesis – McGill University, Schulich School of Music. Montreal (2007)
7. Mannis, J.A.: Adequacy of the thesaurus structure to represent instrumental/vocal formation in method of cataloging of musical documentation in Marc. ANPPOM Congress, 14. Porto Alegre (2003) (in Portuguese)
8. McKay, C., Fujinaga, I.: Musical genre classification: Is it worth pursuing and how can it be improved? In: Int. Conf. on Music Information Retrieval, Victoria (2006)
9. Ferrara, A., Ludovico, L.A., Montanelli, S., Castano, S., Haus, G.: Semantic Web ontology for context-based classification and retrieval of music resources. ACM Trans. on Multimedia Computing, Communications, and Applications (2006)
10. Raimond, Y., Abdallah, S., Sandler, M., Giasson, F.: The Music Ontology. In: International Conference on Music Information Retrieval. Vienna (2007)
11. FRBR - Functional requirements for bibliographic records: final report. IFLA Study Group on the Functional Requirements for Bibliographic Records. International Federation of Library Associations and Institutions. IFLA Universal Bibliographic Control and International. Munich (1998)
12. MusicBrainz, `http://musicbrainz.org`
13. The Oxford Music Online. Oxford University Press, Oxford (2008), `http://www.oxfordmusiconline.com`
14. Fairtile, L.B.: The New Grove Dictionary of Music Online. Journal of the American Musicological Society (2003)
15. Smiraglia, R.P.: Musical Works as Information Retrieval Entities: Epistemological Perspectives. In: Annual International Symposium on Music Information Retrieval. Indiana (2001)
16. IFLA – The International Federation of Library Associations and Institutions, `http://www.ifla.org`
17. López, J.M., Gil, R., García, R., Cearreta, I., Garay, N.: Towards an Ontology for Describing Emotions. In: Lytras, M.D., Carroll, J.M., Damiani, E., Tennyson, R.D. (eds.) WSKS 2008. LNCS (LNAI), vol. 5288, pp. 96–104. Springer, Heidelberg (2008)
18. Valtolina, S.: Knowledge-Intensive Interactive Systems Design in Cultural Context. In: Lytras, M.D., Carroll, J.M., Damiani, E., Tennyson, R.D. (eds.) WSKS 2008. LNCS (LNAI), vol. 5288, pp. 523–530. Springer, Heidelberg (2008)

A Semantic Support for a Multi-platform eGovernment Solution Centered on the Needs of Citizens

Luis Álvarez Sabucedo and Luis Anido Rifón

Telematics Engineering Department, Universidade de Vigo, Spain
{Luis.Sabucedo,Luis.Anido}@det.uvigo.es

Abstract. Despite of all the efforts devoted to eGovernment solutions, a number of flaws can be identified in current approaches. Projects in the domain usually fail at addressing the citizen in the actual channels the citizen, the final user, is actually operating. This feature is required to engage the citizen in eGovernment services. The goal of this paper is to provide a solution that can delivery operations/interactions with the citizen in his/her preferred channels. This involves an effort on modeling a solid back office support and on developing proper interfaces for the number of possible environments. The paper addresses this situation and suggests both an modeling tool and a design for the supporting architecture.

1 Introduction

The provision of services in the domain of eGovernment involves currently an increasing number of areas and functionalities. Despite of all the efforts and improvements undertaken in the area, there is still an aspect of these solutions not yet fully covered: the multi-channel support for services. Nevertheless, some solutions have already been presented and are currently operative. A review of them is presented on Section 2.

The provision of a system capable of providing a multichannel support requires a long-term bet on Public Administrations (PAs hereafter) and the design of entire platforms suitable for that purpose. The process towards such solutions requires, as a first step, the identification of the business model (see Section 3). As this model is expected to be used under a number of different solutions, it is clear the need of an abstract definition of contents. The chosen technique to make that possible is the use of semantic descriptions on the identified business model. Therefore, a brief review on this technology is provided (check Section 4) and its application to the domain is proposed (check Section 5). On the top of this model, a common set of supporting services, regardless of the user environment, is described (check Section 6).

A subset of possible devices and environments are proposed for this study. In particular, we set our goals on light-weighted clients, in particular Web-based solutions, WAP-enabled interfaces, and MHP systems. Even this set of options

M.D. Lytras et al. (Eds.): WSKS 2009, CCIS 49, pp. 503–512, 2009.

could be not exhaustive, it covers the must common solutions and environments for services. Besides of the provision of services and its description, some general grounds about usability and user interface design are presented as general guidelines (check Section 7). Making use of those premises, and the previous results, a design of the system is presented. In this phase we must consider not just changes on user interfaces, but also changes on functionalities as not all devices are suitable for the same services. Therefore, it is presented a software architecture to derive the entire system according to these ideas (see Section 8). Finally, some conclusions are presented to the reader (see section 9).

2 Review

In order to classify the status of maturity in the domain of eGovernment we can take advantage of several already defined indexes. In particular, we would like to mention the classification of services provided by the UN[1]. We can speak about different categories:

1. Stage I - Emerging. Information is presented only and documents are available only for download.
2. Stage II - Enhanced. The citizen can search for documents and perform more advanced operations; nevertheless, he/she can submit a very small amount of information to PAs.
3. Stage III - Interactive. Interactive services are available and government officials can be contacted via email, fax and telephone.
4. Stage IV - Transactional. Two-way interaction is supported and complex services (such as taxes and fees payment and postal services) are available.
5. Stage V - Connected. It conforms the final level that integrates all services under ICTs platforms and supports a two-way open dialog between citizens and PAs.

Regretfully, citizens dealing with public administrations are normally forced to use web-based interfaces to access services. Informally, we can classify the available services on web portals according to the following categorization[2], from simpler to more complex:

- Presence. Just information available on the web such as downloadable forms, web maps, etcetera.
- Information about services. It is possible to access periodically updated information about specific topics related to the PA: minutes from meetings in a council hall, useful telephone numbers, call for public works, etcetera.
- Interaction. Support for assistance from the PA in a digital manner: email support for information and some operations.
- Complex operations. Real operations supported in a holistic manner: support for personal folder, tax payments, security features, etcetera.
- Support for political management. Services related to eDemocracy are available: support for polls, interaction with the agenda of meetings, etcetera.

Limitations and drawbacks can be identified in the domain and these are even more patent on those solutions that make attempts to develop solutions for other environments. These problems are usually related to poor user interfaces in terms of usability, troubles locating the wished services, drawbacks for authentication mechanisms, etc. A complete study of this situation can be found in the already finished project McEgov[3]. For more concrete projects offering particular solutions we can focus on some particular projects such as: the National Health Service (NHS) (United Kingdom)[1], paying taxes in Spain[2], Railway timetable and electronic ticketing (Germany)[3], ...

3 Business Model

From the review of already deployed solutions we can conclude that interactions between the citizen and PAs are driven by the exercise of a right or the fulfilling of an obligation. Therefore, the definition of services provided by the administration in these terms is suggested. Actually, it is possible to focus on what the citizen is requesting and not on the PA itself as it happens many times. Within this proposal, we define *LifeEvent* (LEs hereafter) as any particular situation a citizen must deal with and requires assistance, support or license from a PA. We can consider as LE situations such as getting certifications, paying a fine, getting married, moving, ...

The definition of a LE includes: task, a title for the considered operation; description, a high level description of the desired operation expressed in natural terms from the point of view of the citizen; input documents, all operations carried out by the administration require some input document; output documents, it will vary its content from the expected document (i.e., a certification, a license, ...) to information about potential failures in getting the expected document; scope, the scope of the operation (local, national, international, ...) in which we want it to be recognized; security conditions, the security mechanism involved in the whole process, such as identification of both parties, cipher methods, etc; cost, the amount of money to be paid, the time it will take, other penalizations involved, ...; version, LE can be modified and changed from one version to another; time to respond, the maximum life span to complete the service; and others.

The proposed algorithm goes as follows:

1. Identify the particular problem we are dealing with in terms of features and PAs involved.
2. Decompose the main problem into several different ones that may be completed in a single step, i.e., each step must produce a document meaningful for the citizen. These new problems may become in new LE as well.
3. For each procedure identified in the previous step, look for the input documents, scope and cost. These must be expressed in terms of the ontology of the system. If required, the ontology will be expanded.

[1] Check www.nhsdirect.nhs.uk
[2] Check http://www.aeat.es
[3] Check http://www.bahn.de

4. Identify internal partial steps that the citizen could be interested in. These steps usually involve internal documents that may carry no meaning for the citizen but are relevant for the administration. Nevertheless, the citizen will be informed about future and past steps of his request.
5. Identify all possible documents created as potential final steps of the operation.
6. Update all services and agents that may be aware of the new service.

4 Overview on Semantics

According to Sir Tim Berners-Lee, the semantic web is[4]

> an extension of the current web in which information is given well-defined meaning, better enabling computers and people to work in co-operation

Therefore, semantics is compelled to provide information, not just data. In other words, semantics introduces meaning into the data in order to allow computers to deal with this information in a more interoperable manner. The aim for this discipline is the provision of information understandable by machines. To achieve this ambitious goal, ontologies are a key element. In the literature, several definitions or approximations to the concept of ontology are provided. A quite suitable definition for ontology may be[5]:

> An ontology is a formal, explicit specification of a shared conceptualization of a domain of interest.

By means of this definition we are addressing an ontology as a support to put in a concrete way abstract information about a certain domain by means of a machine-understandable data format.

To express an ontology in a formal manner different languages[6] are at our disposal. OWL (Ontology Web Language)[7] is the W3C Recommendation that covers most of DAML+OIL and it is intended to provide a fully functional way to express ontologies. To make possible different levels of complexity, OWL provides different sublanguages with increasing expressivity: OWL Lite, OWL DL and OWL Full. By using OWL, we are addressing a standard, solid and interoperable platform for the provision of this solution. Besides, this option, OWL Full, allow us to include metadata expressed by means of RDF Schemas.

5 Semantics Applied

Therefore, LEs are expressed in a formal way by means of a semantic support. In particular, we take advantage of OWL to express the information relevant for the system. Nevertheless, we must keep in mind that OWL is just a tool to express knowledge with all its potential and limitations.

As developing an ontology is a common task, standard methods have been defined. Among all of them, Methontology[8], a process recommended by the

Foundation for Intelligent Physical Agents (FIPA)[9] to develop ontologies, has been chosen. This methodology imposes several stages and phases to construct an ontology in an organized manner. The aim of this ontology is to provide support for LEs and, hence, LE is the main class in this ontology. The main slots are name, version, description, cost, etc. as shown in the previous section. Additionally, some classes are included:

- Citizen. It supports the characterization of the citizen itself. This class is tagged with metadata and provides properties that explain its functionalities.
- PA. This class models the behavior of Public Administrations. Every PA involved in the system must be characterized according to some properties to make it possible interaction such as name, Public PKI Key, scope, etc.
- Document. This class defines legal proofs and includes properties such as owner, issuing entity, etc. Besides, for each sort of document supported, a new sub-class is defined; thus, each new document is modeled as a new instance in the proper subclass.

In this ontology, we have reused previously proposed metadata. For example, in the task of defining the citizen, one of the main classes in the system, FOAF (Friend of a Friend)[10] has been reused. Also, to mark documents in the system, metadata from European standards has been reused, in particular, CWA 14860[11] from CEN[12]. This is part of a general philosophy leading toward the maximum possible agreement and reusability both for the ontology and the software based on it.

Several properties have also been identified regarding to LEs. These ones allow the definition, to a certain extension, of the business logic in the system as they provide with mechanisms to decide on operation in the system such as: discover which LE can be invoked, how is in charge for a certain LE, how long a LE may last,... For example, by means of the property *generates*, it is possible to discover which document can be achieved as the result of the completion of a certain LE.

6 Supporting Services

So far, a complete characterization of services for eGovernment has been provided. This support must be capitalized by means of useful tools that increase the quality of service for citizen. In the frame of this proposal, several types of operations have been identified in separated blocks to allow a distributed management of services.

6.1 Searching Services

Advanced mechanisms to deal with LEs, i.e., desired services, become available with this proposal. Provided LEs from several PAs are uploaded to a central server that makes possible to:

- locate all LEs that fulfill a certain task.
- search all LEs that can be executed by a citizen. Note that the condition for its availability depends on the existence of input documents and the user profile among other criteria.
- discover which LE are required to get a certain document. As the orchestration of services can be considered as the search for the chain of LE to get in position to invoke the desired one, this feature is the foundation for simpler mechanisms for orchestration.
- Discover which PA may be involved in a certain service.

These operations are possible by means of SPARQL[13] queries. These ones can obtain the information making good use of the properties mentioned above. Besides, as LEs are described semantically, the client agent can particularize the information according to the environment of the user. For example, if the agent is running on a WAP-enabled device in order to save space on the screen, only the information whose meaning is relevant, is displayed on the screen. Another case of adaptation is related to data customization. In case the user has stated that only local PAs are suitable, the agent can take into account only those LEs supported by PAs of that scope.

6.2 Invoking Services

In order to invoke a service, only the LE description and the required documents are needed. Those LEs depending on documents that are in the client agent can be invoked from any user interface. Therefore, it is possible to identify a set of LEs suitable for mobile environments according to simple requirements about input documents. It makes little sense to support taxes paying on a mobile device but, on the contrary, it can be quite useful to be able to pay a fine (a service that may require only the digital ID) with a mobile phone. To accomplish this, the only requirement is to modify the user interface due to the fact that the mechanism to invoke LEs is completely device independent.

6.3 Tracking Services

Once the operation is invoked, it is considered wise to have the user engaged in the system. And the best strategy is to keep him updated about the evolution of the operation. To accomplish that goal, it is used the result of the previous operation..

The information about the steps performed is presented in a RSS[14] file. This file contains all data and it can be used in the more convenient manner by the client agent, depending on the environment. Therefore, in the same way that RSS readers can be used to read news on digital newspapers, citizen can be subscribed to their own invocations of LEs in order to track its execution. Besides, this way, it is possible to provide simple user interfaces where citizen can check the current status of operations on nearly any device such a WAP-enabled phone, a PDA, ...

7 Usability

The next step is the provision of user interfaces for the client agents considered previously. In order to face the development of those prototypes, some decisions have to be made at different levels. We have to take into account the different circumstances for the user interface itself and also for the kind of services available on them. These differences are due to several circumstances: different input devices (remote control for TV, limited keyboard for mobiles, normal keyboard for web environments), different sizes and qualities on screens, different amount of available data (personal files on a PC and just typed information on a MHP-set), different capacity to process data, etc.

In order to make the proper definition of the users interfaces from the point of view of the graphical design, we can take into account recommendations from Gestalt[15]. The Gestalt psychology or gestaltism is a theory for the interpretation of the information on the human mind. According to this theory, the human brain processes all the information in a holistic manner according to some general principles. These ones can be used to properly design graphical user interfaces. In particular, the following laws from this cognitive theory were specially taken into account:

- Law of similarity. The human mind groups similar elements if they have a similar appearance.
- Law of proximity. Those elements that are close tend to considered as related, even they are not.
- Law of Symmetry. The human brain links those object with some sort of symmetry.
- Law of Common Fate. Elements that show the same pattern are perceived as linked somehow.

Therefore, when designing interfaces under strict graphics limitation, we must take advantage of them to facilitate the identification of information and interface elements.

Also it is required to identify the most suitable services for each platform. Regardless of considerations about the technical viability of different implementations for the access to services, it seems compulsory to identify the most demanded services from each platform. Therefore, we can accept as likely that services to check the current status of an operation will be more demanded than invocation of services on mobile solutions. Also the set of information to be displayed on MHP-based solutions should be larger than on those one based on WAP devices. In the same way, the more complex solutions will be more demanded on Web-based environment, where the user will more likely be able to access their own files with information to fed the platform.

8 Designing the System

With these ideas on mind, the entire system is sketched. From Figure 1 it is clear the different layers in the system. From the bottom to the top, we can identify the following parts in the system:

- Data Layer. The information in the system is stored according to the semantic support described previously. Therefore, records of data are actually information from the point of view of semantic contents achieving, this way, a high level of interoperability in case of interaction with external systems.
- Semantic Layer. Semantics does not just define which is the information in the system but also establishes the rules for its access, providing the business logic of the system, and facilitates the information managing by means of semantic engines that arbitrate how data is accessed and updated.
- Presentation Layer. At this point, the differentiation for different sort of agents starts. So for each kind of device, a different interface is provided. This interfaces varies not only the final display of information but also the kind of services available. Thus, for computer-based environments services for invoking LEs will be a highlight and, on the other side, for WAP-based developments, checking the status of an invocation is the preferred operation.
- Middleware support. For the interchange of data, the use of Web Services[16] was chosen. The main reason for this is due to its features regarding to openness, scalability and security using the support of standardized extensions.
- Presentation Layer on the client side. Of course it is required to provide the agent to be executed on the client side. These agents designed to be light-weighted must invoke this own version of the system on the other side using the already presented interfaces.

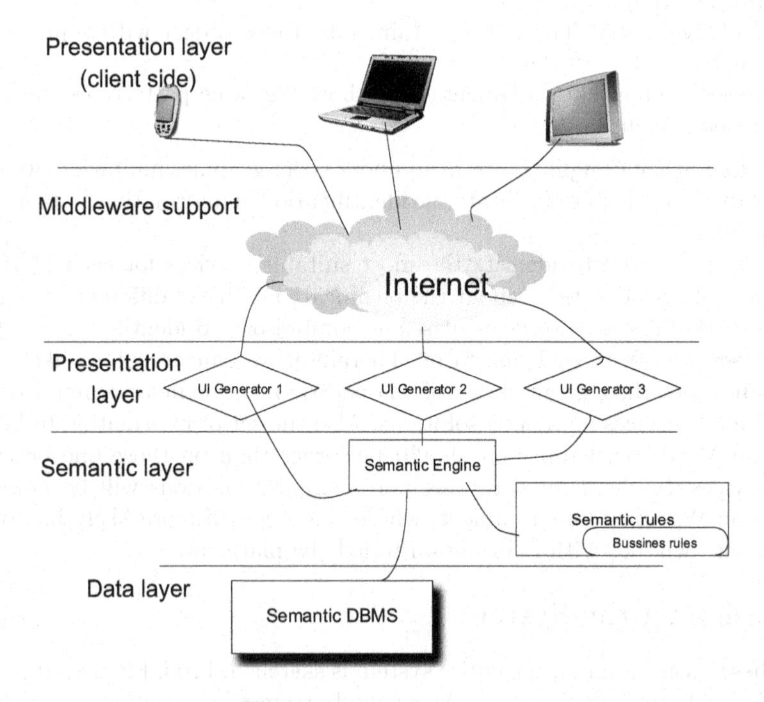

Fig. 1. Design of the system

9 Conclusion

The next step in the provision of solutions for the domain of eGovernment is tackling the multi-channel support for solutions. In order de achieve this goal it is required a long term bet that focus its efforts in two different areas:

- Providing a back-office support for services. Services must be defined in abstract and interoperable terms. So, they can be used by a number of platforms and clients.
- Adapting the interfaces for the service. This involves the review of the user interface itself and also the election of those services more suitable for the given scenario.

In this process, several technologies turn out to be quite useful. This way, we can outline the use of semantic support to describe contents and information; and also the use of Web Services as a reliable and simple-to-develop middleware technology for the interchange of information. Besides, the use of both technologies combined is quite feasible in terms of development cost and maintenance.

On the drawbacks of the proposal, it must be mentioned the cost of developing and maintaining the different versions for each client. Even, a large part of the code is common, some parts of the system must be developed *ad-hoc* for each environment and maintained. Also, the use of semantic technologies may lead to some burden as the software platforms are not yet completely evolved.

As a final remark, we can state that in close future all solutions in the domain will have to take into account the multi-channel feature for their systems as it will be a highly demanded property from citizens. Actually, it also can be considered as the natural evolution for solutions in the domain that must take full advantage of the state-of-the-art technologies to fulfill its mission.

Acknowledgment

This work has been funded by the Ministerio de Educación y Ciencia through the project "Servicios adaptativos para e-learning basados en estándares" (TIN2007-68125-C02-02) and by the Xunta de Galicia, Consellería de Innovación e Industria "SEGREL: Semántica para un eGov Reutilizable en Entornos Locais" (08SIN006322PR).

References

1. United Nations: Un e-government survey 2008: From e-government to connected governance (2008), http://unpan1.un.org/intradoc/groups/public/documents/UN/UNPAN028607.pdf
2. Esteves, J.: Análisis del Desarrollo del Gobierno Electrónico Municipal en España (2008), http://www.sumaq.org/egov/img/publicaciones/3.pdf
3. Manzoni, M., et al.: MC-eGov (2009), http://www.mcegov.eu/

4. Berners-Lee, T., Hendler, J., Lassila, O.: A new form of web content that is meaningful to computers will unleash a revolution of new possibilities. Scientific American (May 2001)
5. Gruber, T.: A translation approach to portable ontology specifications. Knowledge Acquisition, 199–220 (1993)
6. Gómez-Pérez, A., Fernández-López, M., Corcho, O.: Ontological Engineering. Springer, Heidelberg (2003)
7. W3C: Web ontology language (2004), http://www.w3.org/2004/OWL/
8. Fernández-López, M., Gómez-Pérez, A., Juristo, N.: Methontology: From ontological art towards ontological engineering. In: Symposium on Ontological Art Towards Ontological Engineering of AAAI, pp. 33–40 (1997)
9. Fundation for Intelligent Physical Agents: Fipa (2005), http://www.fipa.org/
10. The foaf project (2005), http://www.foaf-project.org/
11. CEN: Dublin Core eGovernment Application Profiles (2004), http://www.cenorm.be/cenorm/businessdomains/businessdomains/isss/cwa cwa14860.asp
12. CEN: CEN home page (2005), http://www.cenorm.be/cenorm/index.htm
13. W3C: SPARQL Query Language for RDF (2006), http://www.w3.org/TR/rdf-sparql-query/
14. RSS Advisory Board: RSS 2.0 Specification (2007), http://www.rssboard.org/rss-specification
15. Rosenholtz, R., Twarog, N., Schinkel-Bielefeld, N., Wattenberg, M.: An intuitive model of perceptual groupping for hci design. In: CHI 2009 - Cognitive Modeling and Ssessment, p. 10 (2009); ACM 978-1-60558-246-7/09/04
16. Consortium, W.W.W.: Web services activity (2005), http://www.w3c.org/2002/ws/

Towards the Selection of Testable Use Cases and a Real Experience

Andreia Rodrigues, Plácido Rogério Pinheiro, Maikol Magalhães Rodrigues, Adriano Bessa Albuquerque, and Francisca Márcia Gonçalves

University of Fortaleza (UNIFOR) – Graduate Program in Applied Computer Sciences
Av. Washington Soares, 1321 - Bl J Sl 30 - 60.811-341 - Fortaleza – Ce – Brazil

Abstract. Nowadays, software organizations have needed to develop excellent and reliable products. This scenario has helped to increase the relevance of quality assurance activities, especially the testing discipline. However, sometimes the time and resources are limited and not all tests can be executed, demanding organizations to decide what use cases should be tested, to guarantee the predictability of the project's time and budget. The multiple criteria methodologies support decisions, considering many factors, not only professional experience. This paper presents a multiple criteria model to support the decision of selecting the use cases that should be tested.

Keywords: Tests, Decision and Analysis Resolution, Multiple Criteria Decision Analysis.

1 Introduction

Software testing is one of the disciplines that have the capability of providing assistance to improve the quality of an organization's products, because its goal is to evaluate how the product meets the clients' specified requirements through a controlled execution of the software. In some cases, when there is not enough time and resources to guarantee complete test coverage, software organizations should reduce their scope.

This work has as main objective the implementation of an approach based on a methodology that provides a structured support to multicriteria decisions. This methodology assists the process of deciding which use cases should be selected to be tested by removing subjective decisions in a structured way. The multicriteria methodology helps to generate knowledge about the decision context and thus increases confidence of those making decisions on the outcome results [15]. Other applications which have multicriteria involving software engineering were defined in [9], [10] and [13].

This research was carried out to define and execute an approach to support software organizations at selecting use cases that should be tested. The approach was based on the model defined in [7].

2 Software Test

Testing consists of verifying, dynamically, the software behavior to determine if it adheres to the specifications and executes correctly in the projected environment. As the main

M.D. Lytras et al. (Eds.): WSKS 2009, CCIS 49, pp. 513–521, 2009.

objective is to find software failures, this activity has a destructive nature [5]. Software testing focuses on the product's quality and can not be considered elementary, because many factors can compromise the success of this activity's execution: (i) time limitations; (ii) resource limitations; (iii) lack of skilled professionals; (iv) insufficient knowledge of test procedures and techniques and adequate test planning; (v) subjectivity of requirement and test specifications; and (vi) increase of the systems' complexity.

Moreover, difficulties related to the test activity are also due to the great variety of combinations of input and output data and the large quantity of paths that make it infeasible to execute all possible test cases [2]. Test case is the definition of a specific condition to be tested. Its structure is based on input values, restrictions to its execution and expected results or behaviors [3]. Test cases are designed based on use cases, which represent interactions that users, external systems and hardware have with the software aiming to achieve an objective.

The amount of test cases executed is one of the main factors which may influence the cost of testing [12]. Therefore it is fundamental to define a test scope considering acceptance criteria and business risks.

3 Multiple Criteria Decision Analysis (MCDA)

Decision-making should be exploited when deciding to execute or not some activities or to perform them applying some methods [7]. The multiple criteria decision analysis proposes to reduce the subjectivity on the decision-making. Nevertheless, the subjectivity will be always present, because the mathematically analyzed items are always results of human beings' opinions. These multiple criteria models allow the decision-maker to analyze possible consequences of each action to obtain the best understanding of the relationships between actions and their goals [6]. Objective criteria should be considered, as well as subjective criteria, even being generally disperse and diffuse in a decision context, but they are extremely important to assess actions.

3.1 MACBETH Approach

The MACBETH methodology contemplates the understanding and learning of the problem content and is divided into three phases: structure, evaluation and recommendation, as we can see in Figure 1 [1].

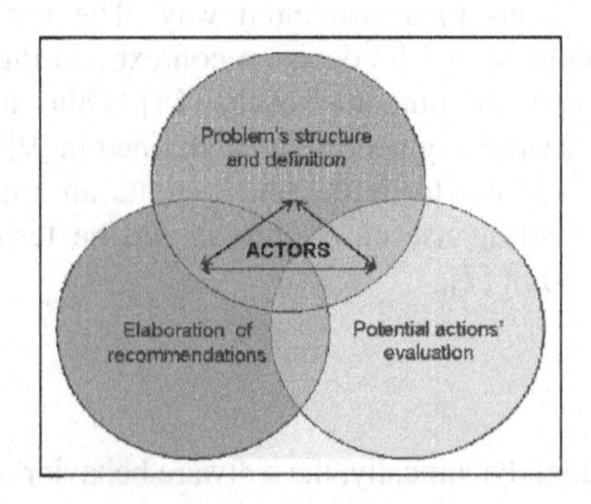

Fig. 1. Phases of the Multiple Criteria Decision Aid

The structure phase focuses on constructing a formal model, capable of being accepted by actors as a structure to represent and organize an entire group of evaluation criteria. It consists of analyzing a specific system and making potential alternatives of decision explicit. The evaluation phase produces matrixes of judgments and provides scales of cardinal value for every criterion. The tasks are implemented with the MACBETH methodology. In the recommendation phase, the results generated by MACBETH are analyzed using scales of values generated by the matrixes of judgments, which are composed of various actions that must be examined according to the decision-maker evaluation [7].

Structuring activities include: (i) definition of a family of fundamental points of view (FPV), (ii) construction of descriptors; and (iii) estimation of the impacts profiles of each action [1]. The construction of descriptors comprehends three stages: (i) description of each descriptor for each fundamental point of view; (ii) access impacts for each fundamental point of view, and (iii) analysis of impacts according to the fundamental points of view [14]. The descriptors are desired to: (i) turn operational the analysis of impacts of the options in a FPV, (ii) describe the impacts with respect to FPVs, (iii) improve the structure of the evaluation model, and (iv) verify the ordinal independence of the corresponding FPVs. The FPV becomes operational if there's a set of impact levels associated with it, defined by Nj, which should be sorted in descending order by the decision makers. Thus, they constitute a range of local preference, limited by the higher level N^{*}_{j}, that has more attractiveness and the lower level N_{*j}, of less attractiveness, should meet the following pre-ordering condition:

$$N_j^* > \ldots > N_{k+1,j} > N_{k,j} > N_{k-1,j} > \ldots > N_{*j}\,.$$

The main difference of MACBETH [11] to other multiple criteria methods is that it requires only qualitative judgments of the difference between the elements' attractiveness aiming to assign values to the options for each criterion and to weigh up the criteria. MACBETH applies the concept of attractiveness to measure the potential actions' values. Therefore, when the decision-maker is demanded to judge the value of a potential action on a specific situation, he should think about his attraction to that action [4].

HIVIEW [8] is a tool to evaluate models defined using multiple criteria methodologies with an aggregation function, like MACBETH. By using HIVIEW, the decision-maker defines, analyzes, evaluates and justifies his preferences, considering existent alternatives. This software facilitates a lot the analysis of complex problems, supporting the elaboration of the problem's structure, specifying the criteria used to choose alternatives and to assign weights to the criteria. The alternatives are evaluated by comparing these criteria and a preference value is assigned to each alternative's criteria. Additionally, it's possible to change judgments and to compare the obtained answers graphically, providing information to the decision-maker for reevaluation. If necessary the decision can be rectified.

4 Model to Select What Use Cases Should Be Tested and the Experience of Use

The proposed model is based on the model presented on [7] and is composed of generic steps, grouped by the phases of the Multiple Criteria Decision Aid (MCDA). These steps are described in Table 1.

Table 1. Model's steps

Phase	Step	Description
Structure	1. Identify criteria	Identify criteria to be used on the use cases evaluation aiming to define their level of priority.
	2. Identify actors and their weights	Identify roles that will expose their point of view, considering also their roles on the decision-making process.
	3. Assign priorities to criteria	Each actor should assign a weight to all criteria, considering the full test process and not only a specific use case.
	4. Execute a partial evaluation	After the execution of the steps listed above, it is necessary to standardize the three sets of values, putting them on the same base (base 1). The goal is to perform a partial evaluation correctly, without any bias. Then, for each actor the three variables should be multiplied, considering each criterion, thus obtaining a specific score.
	5. Calculate the general scores of the criteria	Calculate, for each actor, a score to each criteria of each use case.
Evaluation	6. Apply scores on MACBETH	Construct the matrix of judgments and obtain the cardinal value scales for each defined criteria.
Recommendation	7. Define the level of priorities of the use cases	Prioritize use cases which will be tested, given an analysis of the results obtained from the previous phases.

4.1 Experience of Use

The organization where the experience of use was performed is a government institution with 270 professionals allocated on the Information Technology Area, working on projects that support the organization's businesses.

In 2004, a test team was organized and being external to the projects and responsible for executing systemic tests. Nowadays, the company has high demands of time and resources which make it difficult to satisfy the desired testing coverage in all projects. Therefore, many projects reduce their testing scope to assure the delivery schedule. Priorities have to be applied to use cases and accordingly selected to decrease the testing scope. This is quite relative and varies according to the actors involved and to the criteria they judge relevant. The pilot project selected to apply the proposed approach was a project with a schedule restriction, because the organization agreed on a date with the workers' union, resulting on a fine if it got delayed. The project's life cycle was iterative/incremental and the model was applied on the project's first iteration. The following use cases were part of the test cycle on which the model was applied: UC01_Execute_Sign_In_and_Sign_Out; UC04_Search_Problems; UC05_Demand_Benefit_Permission; UC07_Search_ Demands_Benefit_ Permission; UC08_Approve_Demands_Benefit_Permission; UC10_Manage_Parameters; UC12_Excuse_Sign_in_and_Sign_out.

Bellow, we explain how the steps were performed.

4.1.1 Structure

In step "Identify criteria", we held a meeting with all actors involved at selecting the criteria, given the criteria listed on Table 2. The selected criteria are those highlighted.

A criterion (c) is a tool to evaluate tests that are susceptible to automation in terms of a certain point of view (PV) or concern of the actors responsible for the analysis. The quantity of criteria (n) may vary for each project.

Table 2. List of criteria

Criteria	Reason	Question
Functionality (ISO 9126)	The specific functions and properties of the product should satisfy the user.	Do the specific functions and properties of the product satisfy the user?
Reliability (ISO 9126)	The use case requires maturity on fault tolerance aspects.	Does the use case require maturity on fault tolerance aspects?
Usability (ISO 9126)	**The use case requires a strong interaction with the user and therefore must be usable.**	**Does the use case require a strong interaction with the user, being essential a high level of usability?**
Portability (ISO 9126)	The use case will be executed in different environments.	Will the use case be executed in different environments?
Security (ISO 9126)	**The use case requires a specific control, reducing the access to information.**	**Does the use case require a specific control to reduce the access to informations?**
Availability of time/resource (Gonçalves et al., 2006)	If a use case requires specific resources (people with technological expertise, software and hardware resources, available budget) so that the test can be perfomed.	Are there enough and qualified resources to test the use case?
Repeatability (Gonçalves et al., 2006)	**The use case is being implemented on the first iterations and so will be tested many times.**	**Will the use case's test be repeated many times?**
Complexity (Gonçalves et al., 2006)	**The use case has a large quantity of associated business rules or depends on complex calculations.**	**Does the use case have a lot of associated business rules or depend of complex calculations to function adequately?**
User requirements (Dustin, 2002)	**The use case needs to be implemented because it will satisfy any contractual or legal demand or the organization may have financial loss.**	**Does the use case have to be implemented because it will satisfy any contractual or legal demand or it will prevent the organization against large financial loss?**
Operational characteristics (Dustin, 2002)	**The use case is related to the functions which are frequently used.**	**Is the use case related to frequently used functions?**

In step "Identify actors and their weights", the following actors were selected: (i) Project manager; (ii) Tests coordinator; (iii) Project's system analyst; and (iv) Project's test analyst. The actors answered the questions related to each criterion and the questionnaire applied to obtain the actors' weight, which was defined considering the role performed by the actor on the project e his knowledge of the business for which the software was developed.

A questionnaire was elaborated to obtain the weight of each actor (weight of actor – WA), embracing actor's experience in activities related to tests; roles performed; participation in projects; training and participation in test conferences. Each item had a value and with the measurement of all items, the actor's weight was obtained.

In step "Assign prioritization to the criteria", each actor (a) classified the criteria according to their relevance for the project's test process.

In step "Execute a partial evaluation", we multiplied, for each use case, the values of the actors' point of view (PV), the actors' weights (WA) and the criterion's level of priorities, as can be seen on the formula below:

$$E_x (a, c) = [PV_x (a, c)] * [WA (a, c)] * [\text{Level of Prioritization } (a, c)]$$

It is important to emphasize that the obtained values of the actors' weights and level of priorities were equalized (on the same base – base 1), after they were informed so that a correct evaluation was possible without benefiting a value to the detriment of another.

Finally, in step "Calculate the general scores of the criteria", we calculated the median to obtain the final score of the use cases for each criterion. The value of the median, calculated for each use case, will be used as a basis to prioritize. The following equation illustrates the median calculation.

$$Md(x,c) = \begin{cases} [E_{x,\, j/2} + E_{x,\, ((j/2)+1)}]/2, & \text{if } j \text{ is even,} \\ [E_{x,\, (j+1)/2}], & \text{otherwise.} \end{cases}$$

where x: represents the user case,
j: number of actors and
c: criterion

4.1.2 Evaluation

The evaluation phase was supported by the HIVIEW software. In step "Apply scores on MACBETH", we elaborated the matrixes of judgments in MACBETH for each criterion, having to calculate their subtraction of attractiveness. For this, the values of the median of the criterion are subtracted between the use cases (x_1 to x_k), considering the module of the obtained value. Figure 2 illustrates the matrix of judgments for the "Operational characteristics" criterion.

Figure 3 depicts the level of importance for each analyzed criterion. According to the figure, the criterion with the highest weight was "Operational characteristics" and the one with the lowest weight was "Security".

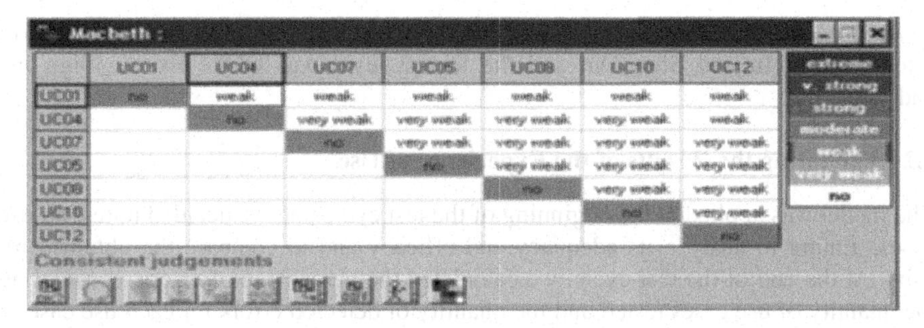

Fig. 2. Matrix of judgements to the criterion "Operational Characteristics"

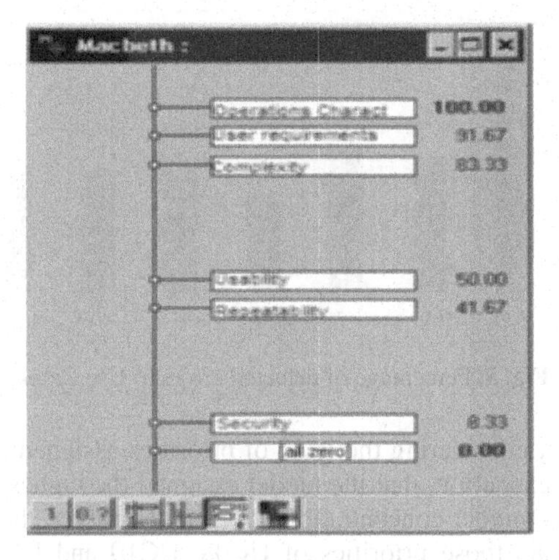

Fig. 3. Criteria's level of prioritization

4.1.3 Recommendation

The results obtained during the evaluation phase generate reports with graphics, and the use cases analyzed are ranked. In this step, we identified the use cases with the highest level of priorities. To represent the general classification of all analyzed criteria we used the graphic showed in Figure 4.

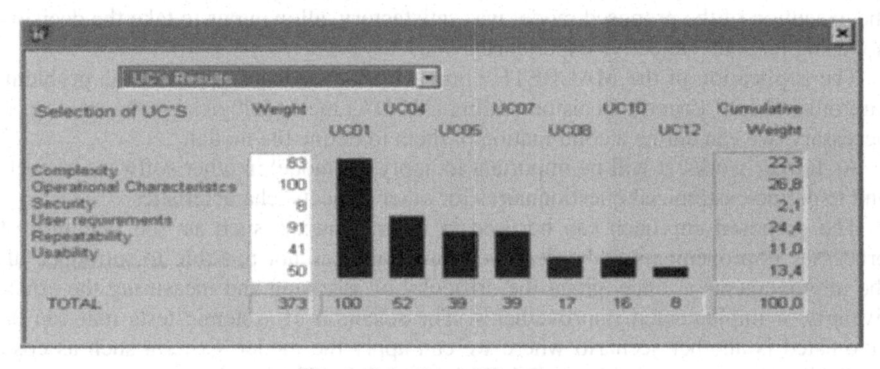

Fig. 4. Selection of Use Cases

As we can see, the use case with the highest level of priorities was UC01_ Execute_Sign_In_and_Sign_Out and the lowest level was UC12_Excuse_Sign_in_ and_Sign_out.

4.2 Validating the Results of the Experience of Use

The model was applied at the beginning of the test cycle and we decided to test all use cases, aiming to analyze its adequacy and efficacy and to compare the obtained results. At the end of the test cycle, we calculated the percentage of errors, considering the quantity of use cases tested and the quantity of detected errors for each use case as displayed in Figure 5.

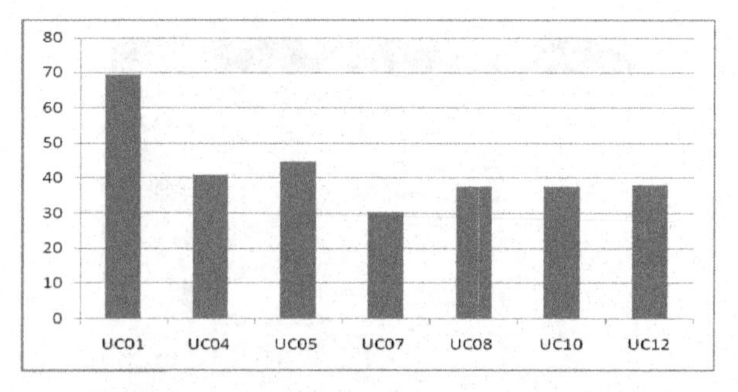

Fig. 5. Percentage of detected errors to Use Cases

We could observe, considering the level of priorities assigned to the use cases obtained by the model execution, that the model assigned the highest level of priority to those use cases with a higher concentration of errors. However, the level of priority of UC07 was higher than those priorities of UC08, UC10 and UC12, but even so in UC07 were detected fewer errors.

5 Conclusion and Further Works

This work helped us to conclude that the restriction of time and resources is a real problem on many organizations and that an approach to support the decision to select use cases to be tested is very relevant. Besides, with the experience, we could see that the execution of the proposed model was satisfactory, allowing us to take the decision not using just our subjective experiences.

The application of the MACBETH approach was adequate to model the problem, but, others Multi Criteria Decision Aiding (MCDA) methodologies can be used or, if necessary, we can define a combination of them to define the model.

As further works, it will be important to apply the model in other software projects and to define customized questionnaires for other projects' characteristics.

The proposed approach can be used in other contexts, such as: the selection of processes improvements to be deployed, as it often is not possible to introduce all the improvements at once, given the difficulty of assessing and measuring the effectiveness of implemented improvements. The selection of systemic tests that can be automated is another scenario where we can apply the model. Criteria such as cost, availability of resources and repeatability of tests should be considered.

References

1. Bana, Costa, C.A., Beinat, E., Vickerman, R.: Model-structuring and impact assessment: qualitative analysis of policy attractiveness. CEG-IST Working Paper n.25 (2001)
2. Bandeira, L.R.P.: Product Quality from Test Automation: a Case Study. Monograph to obtain the B.Sc in Computer Science Title, University of the Ceara State (2005)
3. Craig, R.D., Jaskiel, S.P.: Systematic Software Testing. Artech House Publishers, Boston (2002)
4. Cunha, J.A.O.: Decisius: Um Processo de Apoio à Decisão e sua Aplicação na Definição de um Índice de Produtividade para Projetos de Software, Dissertation of M.Sc., CIN/UFPE, Recife, PE, Brazil (2008)
5. Dias, A.: Introduction to Software Testing. Magazine of Software Engineering, on-line (2007), http://www.devmedia.com.br/esm
6. Flament, M.: Multiple Criteria Glossary. Red Iberoamericana de Evaluación y Decisión Multicritério, Espanha (1999), http://www.unesco.org.uy/redm/glosariom.htm
7. Gonçalves, F.M.G.S., Donegan, P.M., Castro, A.K., Pinheiro, P.R., Carvalho, A., Belchior, A.D.: Multi-criteria Model for Selection of Automated System Tests. In: The IFIP International Conference on Research and Practical Issues of Enterprise Information Systems, Viena. IFIP International Federation for Information Processing (2006)
8. Pinheiro, P.R., Castro, A., Pinheiro, M.: A Multicriteria Model Applied in the Diagnosis of Alzheimer Disease: A Bayesian Network. In: 11th IEEE International Conference on Computational Science and Engineering, July 2008, pp. 15–22 (2008), doi:10.1109/CSE.2008.44
9. Marilia, M., Carvalho, A.L., Furtado, M.E.S., Pinheiro, P.R.: A co-evolutionary interaction design of digital TVapplications based on verbal decision analysis of user experiences. International Journal of Digital Culture and Electronic Tourism (IJDCET) 1, 312–324 (2009)
10. Marilia, M., Carvalho, A.L., Furtado, M.E.S., Pinheiro, P.R.: Co-evolutionary Interaction Design of Digital TV Applications Based on Verbal Decision Analysis of User Experiences. In: WSKS 2008, vol. 19, pp. 702–711 (2008), doi:10.1007/978-3-540-87783-7
11. Meskens, N., Roubens, M. (eds.): Advances in Decision Analysis. Book Series: Mathematical Modelling: Theory and Applications, vol. 4, pp. 131–157. Kluwer Academic Publishers, Dordrecht (1999)
12. Myers, G.J.: The Art of Software Testing. John Wiley & Sons, Inc., New York (1979)
13. Oliveira, J.M.M., Oliveira, K.B., Castro, A.K., Pinheiro, P.R., Belchior, A.D.: Application of Multi-Criteria to Perform an Organizational Measurement Process. In: Elleithy, K. (ed.) Advances and Innovations in Systems, Computing Sciences and Software Engineering, pp. 247–252. Springer, Heidelberg (2007)
14. Pinheiro, P.R., Souza, G.G.C., Castro, A.K.A.: Structuring of the Multicriteria Problem for the Production of Journal. Pesqui. Oper. 28(2), 203–216 (2008)
15. Evangelou, C., Karacapilidis, N., Khaled, O.A.: Interweaving knowledge management, argumentation and decision making in a collaborative setting: the KAD ontology model. Int. J. of Knowledge and Learning 2005 1(1/2), 130–145 (2005)

Softwares
Hiview Software, Catalyze Ltd. www.catalyze.co.uk
M-Macbeth, www.m-macbeth.com

Towards the Neuropsychological Diagnosis of Alzheimer's Disease: A Hybrid Model in Decision Making

Ana Karoline Araujo de Castro, Placido Rogerio Pinheiro,
and Mirian Caliope Dantas Pinheiro

Graduate Program in Applied Informatics, University of Fortaleza, Av. Washington Soares, 1321 - Bloco J sala 30, CEP: 60811-905, Fortaleza, Ceara, Brazil
akcastro@gmail.com, {placido,caliope}@unifor.br

Abstract. Dementias are syndromes described by a decline in memory and other neuropsychological changes especially occurring in the elderly and increasing exponentially in function of age. Due to this fact and the therapeutical limitations in the most advanced stage of the disease, diagnosis of Alzheimer's disease is extremely important and it can provide better life conditions to patients and their families. This work presents a hybrid model, combining Influence Diagrams and the Multicriteria Method, for aiding to discover, from a battery of tests, which are the most attractive questions, in relation to the stages of CDR (Clinical Dementia Rating) in decision making for the diagnosis of Alzheimer's disease. This disease is the most common dementia. Influence Diagram is implemented using GeNie tool. Next, the judgment matrixes are constructed to obtain cardinal value scales which are implemented through MACBETH Multicriteria Methodology. The modeling and evaluation processes were carried out through a battery of standardized assessments for the evaluation of cases with Alzheimer's disease developed by Consortium to Establish a Registry for Alzheimer's disease (CERAD).

Keywords: Alzheimer's disease, Diagnosis, Multicriteria Method, Bayesian Network, Influence Diagram.

1 Introduction

Demographic studies in developed and developing countries have showed a progressive and significant increase in the elderly population in the last years [25].

Alzheimer's disease is the most frequent cause of dementia and is responsible (alone or in association with other diseases) for 50% of the cases in western countries [27]. Dementias are syndromes described by a decline in memory and other neuropsychological changes especially occurring in the elderly and increasing exponentially in function of age. According to [25], despite its high incidence, doctors fail to detect dementia in 21 to 72% of their patients.

One way to identify whether a patient is having a normal aging, or are developing some form of dementia, is through a neuropsychological evaluation [5].

There are several tests available, and one of the major challenges is to find out which test would be more efficient in establishing the diagnosis of dementia. One

M.D. Lytras et al. (Eds.): WSKS 2009, CCIS 49, pp. 522–531, 2009.

factor that must be observed is the brevity of the tests, that is, the shorter the test, it will be more effective.

The main focus of this work is to develop a multicriteria model for aiding in decision making in order to find out which are the most attractive, for a given test or set of tests for the diagnosis of Alzheimer's disease.

The multicriteria methodology helps to generate knowledge about the decision context and thus increases confidence of those making decisions on the outcome results [9]. Other applications which have multicriteria involving with different areas of knowledge were defined in [18], [19] and [20].

Due to be a difficult diagnosis of disease, with initial symptoms subtle, progressing slowly until it is clear and devastating. The battery of tests used in this work is from the Consortium to Establish a Registry for Alzheimer's disease (CERAD) [10]. It was necessary to construct value scales originating from semantic judgments of value with the objective of defining a ranking with the classification of the impact of issues in relation to the stages of the CDR. Finally conclusions and futures works are shown.

2 Diagnosis of Alzheimer's Disease

Diagnosis of Alzheimer's disease is carried out in several steps. Initially, syndromic diagnosis is defined, which informs whether the patient presents the diagnostic criteria for dementia. After Dementia is confirmed, etiological diagnosis follows, which informs which disease is causing the dementia. In this case, we are looking for Alzheimer's disease [15] and [21].

For the global cognitive evaluation, the Mini-Mental State Examination was recommended; memory evaluation: delayed recall [28] or of objects presented as drawings; attention: trail-making or digit-span; language: Boston naming, naming test from ADAS-Cog or NEUROPSI [24]; executive functions: verbal fluency or clock-drawing; conceptualization and abstraction; similarities from CAMDEX [30] or NEUROPSI; construction abilities. For functional evaluation, IQCODE, or the Pfeffer Questionnaire or the Bayer daily activity scale is recommended [15].

Due to conditions inherent to the public health situation in Brazil, some laboratories can be more or less useful for differential diagnosis. In addition to this, health care conditions are not yet ideal, such that the timeframe for evaluation and the cost of procedures should be taken into consideration in order to get maximum accuracy with the lowest cost least amount of time.

Due to these limitations, this study seeks to help in deciding the best manner to define the diagnosis. To do this, we sought to choose the most important questions in the diagnosis of Alzheimer's disease, using a battery of CERAD tests [31]. This battery was chosen because it encompasses all the steps of diagnosis and has been used all over the world.

Therefore, the questions selected through this decision making process will be applied preferentially because, in accordance with the decision maker, these questions play a main role in diagnosis.

3 CERAD – An Overview

The original mandate of the Consortium to Establish a Registry for Alzheimer's Disease (CERAD) in 1986 was to develop a battery of standardized assessments for the evaluation of cases with Alzheimer's disease who were enrolled in NIA-sponsored Alzheimer's Disease Centers (ADCs) or in other dementia research programs [12] Despite the growing interest in clinical investigations of this illness at that time, uniform guidelines were lacking as to diagnostic criteria, testing procedures, and staging of severity. This lack of consistency in diagnosis and classification created confusion in interpreting various research findings. CERAD was designed to create uniformity in enrollment criteria and methods of assessment in clinical studies of AD and to pool information collected from sites joining the Consortium.

The major standardized instruments developed by CERAD are now used by many AD research centers in the US and abroad, by physicians in clinical practice, and in population-based surveys. They have been translated into Bulgarian, Chinese, Dutch, Finnish, French, German, Italian, Japanese, Korean, Portuguese, and Spanish [10].

CERAD developed the following standardized instruments to assess the various manifestations of Alzheimer's disease: Clinical Neuropsychology, Neuropathology, Behavior Rating Scale for Dementia, Family History Interviews and Assessment of Service Needs [22].

4 Model Construction

4.1 Definition of Problem

In studies developed by [7] and [26] the application of the multicriteria model for aiding in diagnosis of Alzheimer's disease was presented. In [26] we analyzed the results of the implementation of a case study conducted with the battery of the CERAD neuropathological assessment.

In [7] we sought to validate the model through of neuropsychological data of patients. The data used in the analysis of the study are part of the battery of neuropsychological CERAD [23].

In the present study, we sought to validate the model in the identification of issues that have greater impact on each stage of the Clinical Dementia Rating (CDR), in deciding the diagnosis of AD, will be held from the combination of the battery of neuropsychological tests of CERAD with the scale functional CDR [14].

We selected six of the eight tests of the battery of neuropsychological CERAD for applying the model of decision support that will assess what are the issues (among all the issues that are implemented in selected tests) that have greater attraction for each stage of CDR, for the definition of the diagnosis of Alzheimer's disease. The tests selected are: Verbal Fluency (J1), Boston Naming Test (J2), Word List Memory (J4), Constructional Praxis (J5), Word List Recall (J6) and Word List Recognition (J7) [23], [32].

The CDR was chosen to be a tool that allows the classification of the prevalence of the various degrees of dementia, on six cognitive-behavioral categories: memory, orientation, judgment and problem solving, community affairs, home and hobbies and personal care [14].

Furthermore, the CDR identifies the questionable cases, or those that are not classified as normal subjects. These cases may correspond to the so-called cognitive decline associated with aging, mild cognitive impairment or, in other epidemiological studies that are part of the group that has a higher rate of conversion to dementia.

Despite of the CDR has only five stages of dementia: none, questionable, mild, moderate and severe, the CERAD implemented a change in scale including two stages: profound and terminal. For the application of the model will be taken into account the scale of the CDR modified by CERAD [29].

Next, the application of the decision model will be presented for solving the problem of choosing the issues considered most attractive in the definition of the neuropsychological diagnosis of AD.

Phase 1: Structuring

Step 1: Identify the decision makers. Individuals classified as cases in the database of CERAD were defined as the decision makers (actors) involved in the process of building the model of defining the issues of greatest impact in defining the neuropsychological diagnosis of AD. That decision was taken, considering that, from the values (responses) issued by the cases, it was the definition of the degree of dementia.

Analyzing the data pertaining to the cases through the database of CERAD was found a negligible quantity of actors to evaluate the attractiveness of the model in multicriteria. The degrees of dementia: none, profound and terminal, are respectively 0, 1 and 2 answers, i.e. between the cases that have been assessed with dementia-type: none, profound and terminal, only 0, 1 and 2 people respectively, answered each of the issues of the CERAD battery. Therefore, these degrees of dementia have not been evaluated by the model.

Table 1. Classification of variables in the problem of decision

Problem: Define the issues of greatest impact in each stage of the CDR in the neuropsychological diagnosis of Alzheimer's disease Heading level	
Objectives	To establish between the various levels of dementia classified by CDR (questionable, mild, moderate, severe), which are the items that have the greatest impact on the decision of a test or set of tests for the definition of the diagnosis of AD
The set of the possible actions	The set of actions (alternative) is defined as (A), where the issues are that are part of the battery of neuropsychological tests of the CERAD were selected for the implementation of the decision model: Verbal Fluency (J1), Boston Naming Test (J2), Word List Memory (J4), Constructional Praxis (J5), Word List Recall (J6) and Word List Recognition (J7).
Criteria	Correspond to the CDR stages of dementia that had a significant amount of actors (questionable, mild, moderate, severe).
Restrictions (properties of criteria that are specified as desirable)	CDR stages: - CDR_QUESTIONABLE > 0 and CDR_QUESTIONABLE <= 0.5; - CDR_MILD > 0.5 and CDR_MILD <= 1; - CDR_MODERATE > 1 and CDR_MODERATE <= 2; - CDR_SEVERE > 2 and CDR_SEVERE <= 3;

Step 2: Identify the alternatives and criteria relevant to the issue of decision. Definition of the hierarchical structure of the problem. This step is related to identify the variables of interest and the determination of the interrelationship between them. The variables can be classified as: objectives, actions, criteria, restrictions and factors, as shown in table 1 [17].

The end result of this step is to define the hierarchical structure of the problem by creating a graphic model represented by a Directed Acyclic Graph (DAG), as shown in figure 1. We have identified 23 of probability, including one decision nodes and 7 utility nodes [11], [16].

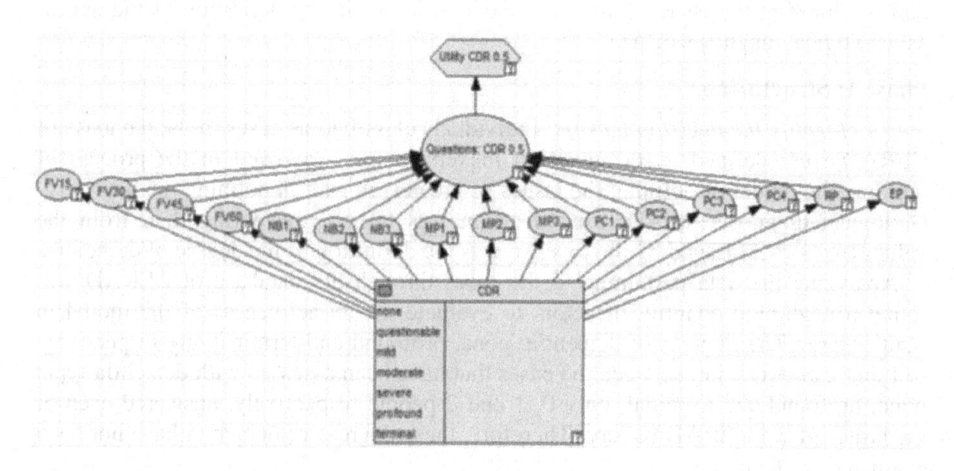

Fig. 1. Hierarchical structure of the problem of the decision to stage the CDR questionable

After defining the structure of the network should be carried out to quantify the probability of us, which was made from the calculation of probabilities in the form of a CPT (Conditional Probability Table) with the data obtained by the database of the CERAD [8], [13].

For to define the issues more attractive, it is necessary to examine the level of impact (or attraction) of the responses of actors in each of the stages of the CDR. This initial assessment is important because the database of the CERAD, you can discover from the responses of a particular actor, which is the stage of CDR.

The attractiveness of the responses is measured by use of the construction of judgments matrixes of value and obtains the scales of global value for each question [1]. Table 2 shows the values of each level of impact for all FPVs in relation FV15.

Table 2. Values of each level of impact for each FPV in relation FV15

Alternatives	FPV1 – CDR: Questionable	FPV2 – CDR: Mild	FPV3 – CDR: Moderate	FPV4 – CDR: Severe
FV15_99	0.50	1.00	2.00	3.00
FV15_0	0.46	0.96	1.92	2.92
…	…	…	…	…
FV15_11	0.00	0.51	1.01	2.01

With this result, it will be possible to apply the model to discover the attractiveness of the issues involved in the battery of neuropsychological CERAD for each stage of the CDR.

Step 3: Definition of descriptors. The construction of descriptors should be made for each point of view of the fundamental problem. Thus, for this problem, have been identified two sets of descriptors with each of three phases: (i) description of each descriptor for each of the fundamental points of view (FPVs), (ii) obtain the impacts according to each key point of view, and (iii) analysis of the impacts each second fundamental point of view [3].

The number of states of each PVF will always be equivalent. It was defined 16 descriptors for each PVF. The states of PVFs are not equivalent, therefore, cannot be the representation of more than one state at a single level of impact. The table 3 shows the descriptors for the FPV1. The levels of impact of each descriptor were ordered based on each issue that has relevance for each stage of the CDR, as regards the issue that has greater influence in defining the diagnosis of AD. This relevance was defined based on the sum of the result obtained in the judgment matrixes of decision in the application of the model on the answers of the questions.

Table 3. Descriptor for FPV1 - CDR: Questionable

NI	Description	Order
N16	FV15: I want you to tell me all the animals you can think (from 0-15 seconds)	1°
N15	MP2: Repeat list of 10 words - attempt 2	2°
...	...	...
N01	PC1: Draw circle	16°

Step 4: Carry out the analysis of impacts. This step is related to the definition of impact assessment according to each FPV. We defined the upper and lower values of each impact and relevant aspects of the distribution of impacts in each of them. In all the FPVs of this model, instead of scoring is attributed to the involvement of dementia in accordance with each stage of the CDR which is being evaluated [2].

The table 4 presents one summary table that shows the descriptors and their values lower and higher to be considered for obtaining the basis of value for each FPV.

Table 4. Summary table of descriptors and impacts seconds each PVF

FPV	Descriptor	Upper Level	Lower Level
FPV1 – CDR: Questionable	Answers a question from the battery of neuropsychological CERAD	0	0.50
FPV2 – CDR: Mild		0.51	1
FPV3 – CDR: Moderate		1.01	2
FPV4 – CDR: Severe		2.01	3

Step 5: Definition of a function of value for each alternative. This function was obtained from the division between the sum of the results obtained through the judgment matrixes in relation to the responses to a question, by the sum of the results obtained

through the judgment matrixes in relation to the issue or set of issues that are part of a battery of neuropsychological subtest of CERAD, on a stage of the CDR.

$$v(b_j) = \frac{\sum_{n}^{1} a_i}{\sum_{n}^{1} b_j}. \quad (1)$$

Where: $a \in$ A (represents all the alternatives - issues), i.e. $A = \{a_i, a_{i-1},\ldots,a_1\}$; $b \in B$ (represents the subtest), i.e. $B = \{b_j, b_{j-1},\ldots,b_n\}$.

This function of value was applied to all sets of issues relevant to their respective tests. In the next stage, will be the implementation of the function of value for a particular set of options.

Step 6: Construction of the judgment matrixes. In this step were performed the following steps: (i) the calculation of the difference of attractiveness in the judgment matrix, and (ii) the equivalence of the dimensions of attractiveness, and (iii) the way it was maintained the consistency of judgment matrixes [4].

For the evaluation of issues, all the FPVs were worked through a descriptor with 16 reference levels, and a lower limit (which was generated from the lower value, the sum obtained regarding the outcome of the evaluation of the issues), an upper limit (which was generated from the higher value, the sum obtained regarding the outcome of the evaluation of the issues) and 14 intermediate levels of reference. Shown in figure 2 is a matrix of assessment of value and scale of cardinal value obtained with the methodology for the FPV1 MACBETH - CDR: Questionable.

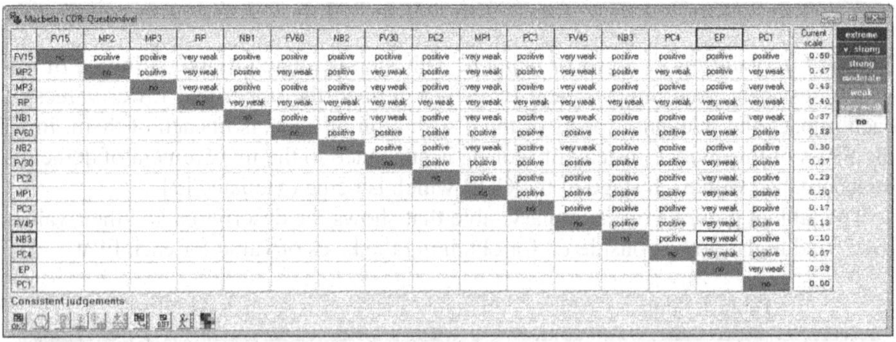

Macbeth : CDR: Questionável

	FV15	MP2	MP3	RP	NB1	FV60	NB2	FV30	PC2	MP1	PC3	FV45	NB3	PC4	EP	PC1	Current scale
FV15	no	positive	positive	very weak	positive	positive	positive	positive	positive	very weak	positive	very weak	positive	positive	positive	positive	0.50
MP2		no	positive	very weak	positive	very weak	positive	very weak	positive	very weak	positive	very weak	positive	very weak	positive	very weak	0.47
MP3			no	very weak	positive	positive	positive	very weak	positive	very weak	positive	very weak	positive	positive	positive	very weak	0.43
RP				no	very weak	very weak	very weak	very weak	very weak	very weak	very weak	very weak	very weak	very weak	very weak	very weak	0.40
NB1					no	positive	positive	very weak	positive	very weak	positive	very weak	positive	positive	positive	very weak	0.37
FV60						no	positive	positive	positive	positive	positive	positive	positive	positive	very weak	positive	0.33
NB2							no	positive	positive	very weak	positive	very weak	positive	positive	positive	positive	0.30
FV30								no	positive	positive	positive	positive	positive	positive	very weak	positive	0.27
PC2									no	positive	positive	positive	positive	positive	very weak	positive	0.23
MP1										no	positive	positive	positive	positive	very weak	positive	0.20
PC3											no	positive	positive	positive	very weak	positive	0.17
FV45												no	positive	positive	very weak	positive	0.13
NB3													no	positive	very weak	positive	0.10
PC4														no	very weak	positive	0.07
EP															no	very weak	0.03
PC1																no	0.00

Consistent judgements

Fig. 2. Matrix of judgment of value and scale for the FPV1 - CDR: Questionable

Following the procedure for construction of judgments matrixes of value and obtain the scales of global value for each of the FPV.

The result of the judgment matrixes shows that the stage of the CDR was questionable that most benefited from the implementation of the model. CDR: questionable obtained the highest value in relation to other criteria, and through the accumulated weights for each option, with the CDR: questionable accumulating 50% of the total weight of the criteria.

This result is very positive, because one of the major goals of medicine in the search for a diagnosis, especially that of Alzheimer's disease, is get it in earlier stages of the disease.

5 Conclusion

The diagnosis of Alzheimer's disease is made up of many steps. The first step is to discover if the patient has dementia and then the patient is assessed to see if he has Alzheimer's [6].

The methodologies applied have been crucial to the analysis of the most attractive in the definition of the diagnosis of Alzheimer's disease, while the methodological design of the model, mapped the possibilities regarding the performance results for the decision.

The model in question, which applies structured assumptions in decision-making problems, provided important impacts for the research and supported in the chain of neuropsychological responses to identify the diagnostic criteria.

As a future project, this model can be extended with the inclusion of new criteria or new models which can be developed using other batteries of assessments.

Acknowledgments. The authors thank the Consortium to Establish a Registry for Alzheimer's Disease (CERAD) for the divulgation of the data utilized in this case study.

References

1. Bana e Costa, C.A., Corte, J.M.D., Vansnick, J.C.: MACBETH. LSE-OR Working Paper, 56 (2003)
2. Bana e Costa, C.A., Corte, J.M.D., Vansnick, J.C.: Software M-MACBETH version 1.1. European Working Group "Multiple Criteria Decision Aiding 3(12) (2005a)
3. Bana e Costa, C.A., Corte, J.M.D., Vansnick, J.C.: On the Mathematical Foundations of MACBETH. In: Figueira, J., Greco, S., Ehrgott, M. (eds.) Multiple Criteria Decision Analysis: State of the Art Surveys Series: International Series in Operations Research & Management Science, vol. 78, XXXVI, pp. 410–442 (2005b)
4. Bana e Costa, C.A., Ensslin, L., Correa, E.C., Vansnick, J.C.: Decision support systems in action: integrates application in a multicriteria decision aid process. European Journal of Operational Research 133, 315–335 (1999)
5. Bassett, S.S.: Attention: Neuropsychological Predictor of Competency in Alzheimer's Disease. Journal of Geriatric Psychiatry and Neurology 12, 200–205 (1999)
6. Castro, A.K.A., Pinheiro, P.R., Pinheiro, M.C.D.A.: A Multicriteria Model Applied in the Diagnosis of Alzheimer's Disease. In: Wang, G., Li, T., Grzymala-Busse, J.W., Miao, D., Skowron, A., Yao, Y. (eds.) RSKT 2008. LNCS (LNAI), vol. 5009, pp. 612–619. Springer, Heidelberg (2008a)
7. Castro, A.K.A., Pinheiro, P.R., Pinheiro, M.C.D.A.: A Hybrid Model for Aiding in Decision Making for the Neuropsychological Diagnosis of Alzheimer's Disease. In: Chan, C.-C., Grzymala-Busse, J.W., Ziarko, W.P. (eds.) RSCTC 2008. LNCS (LNAI), vol. 5306, pp. 495–504. Springer, Heidelberg (2008b)

8. Detwarasiti, A., Shachter, R.D.: Influence Diagrams for Team Decision Analysis. Decision Analysis 2(4), 207–228 (2005)
9. Evangelou, C., Karacapilidis, N., Khaled, O.A.: Interweaving knowledge management, argumentation and decision making in a collaborative setting: the KAD ontology model. Int. J. of Knowledge and Learning 2005 1(1/2), 130–145 (2005)
10. Fillenbaum, G.G., van Bellec, G., Morris, J.C., Mohs, R.C., Mirraf, S.S., Davis, P.C., Tariot, P.N., Silvermani, J.M., Clarkj, C.M., Welsh-Bohmer, K.A., Heymanl, A.: Consortium to Establish a Registry for Alzheimer's Disease (CERAD): The first twenty years. Alzheimer's & Dementia 4, 96–109 (2008)
11. Genie. Graphical Network Interface. Decision Systems Laboratory, University of Pittsburgh (2008)
12. Heyman, A., Peterson, B., Fillenbaum, G., Pieper, C.: The Consortium to Establish a Registry for Alzheimer's Disease (CERAD). Part XIV: Demographic and clinical predictors of survival in patients with Alzheimer's disease. Neurology 46(3), 656–660 (1996)
13. Howard, R.A., Matheson, J.E.: Influence Diagrams. Decision Analysis 2(3), 127–143 (2005)
14. Hughes, C.P., Berg, L., Danzinger, W.L., Coben, L.A., Martin, R.L.: A New Clinical Scale for the Staging of Dementia. British Journal of Psychiatry 140, 566–572 (1982)
15. Jorm, A.F., Jacomb, P.A.: The Informant Questionnaire on Cognitive Decline in the Elderly (IQCODE): Socio-Demographic Correlates, Reliability, Validity and Some Norms. Psychol. Med. 19, 1015–1022 (1989)
16. Kjærulff, U.B., Madsen, A.L.: Bayesian Networks and Influence Diagrams: A Guide to Construction and Analysis. Series: Information Science and Statistics, XVIII, 318 p. Springer, Heidelberg (2007)
17. Korb, K.B., Nicholson, A.E.: Bayesian Artificial Intelligence. Chapman & Hall/CRC (2003)
18. Korhonen, P.: Interactive Methods. In: Figueira, J., Greco, S., Ehrgott, M. (eds.) Multiple Criteria Decision Analysis: State of the Art Surveys. Series: International Series in Operations Research & Management Science, vol. 78, XXXVI, pp. 641–665 (2005)
19. Marilia, M., Carvalho, A.L., Furtado, M.E.S., Pinheiro, P.R.: Co-evolutionary Interaction Design of Digital TV Applications Based on Verbal Decision Analysis of User Experiences. In: WSKS 2008. Communications in Computer and Information Science, vol. 19, pp. 702–711 (2008), doi:0.1007/978-3-540-87783-7
20. Marilia, M., Carvalho, A.L., Furtado, M.E.S., Pinheiro, P.R.: A co-evolutionary interaction design of digital TVapplications based on verbal decision analysis of user experiences. International Journal of Digital Culture and Electronic Tourism (IJDCET) 1, 312–324 (2009)
21. Mohr, E., Dastoor, D., Claus, J.J.: Neuropsychological Assessment. In: Gauthier, S. (ed.) Clinical Diagnosis and Management of Alzheimer's Disease, pp. 93–106. Martin Dunitz, London (1999)
22. Moms, J.C., Heyman, A., Mohs, R.C., Hughes, J.P., van Belle, G., Fillenbaum, G., Mellits, E.D., Clark, C.: Part I. Clinical and neuropsychological assesment of Alzheimer's disease The Consortium to Establish a Registry for Alzheimer's Disease (CERAD). Neurology 39, 1159–1165 (1989)
23. Morris, J.C., Heyman, A., Mohs, R.C., et al.: The Consortium to Establish a Registry for Alzheimer's Disease (CERAD): Part 1. Clinical and Neuropsychological Assessment of Alzheimer's Disease. Neurology 39, 1159–1165 (1989)
24. Ostrosky-Solis, A.-R.M.: NEUROPSI: A Brief Neuropsychological Test Battery in Spanish with Norms by Age and Educational Level. J. Int. Neuropsychol.Soc. 5, 413–433 (1999)

25. Petersen, R.C., Smith, G.E., Waring, S.C., et al.: Mild Cognitive Impairment: Clinical Characterization and Outcome. Archives Neurology 56, 303–308 (1999)
26. Pinheiro, P.R., Castro, A.K.A., Pinheiro, M.C.D.: Multicriteria Model Applied in the Diagnosis of Alzheimer's Disease: A Bayesian Network. In: Proceedings of the 11th IEEE International Conference on Computational Science and Engineering, São Paulo, vol. 1, pp. 15–22. IEEE Computer Society, Danvers (2008)
27. Porto, C.S., Fichman, H.C., Caramelli, P., Bahia, V.S., Nitrini, R.: Brazilian Version of the Mattis Dementia Rating Scale Diagnosis of Mild Dementia in Alzheimer.s Disease. Arq. Neuropsiquiatr. 61(2-B), 339–345 (2003)
28. Prince, M.J.: Predicting the Onset of Alzheimer's Disease Using Bayes' Therorem. American Journal of Epidemiology 43, 301–308 (1996)
29. Rosen, W.G., Mohs, R.C., Davis, K.L.: A New Rating Scale for Alzheimer' s Disease. Am. J. Psychiatry 141, 1356–1364 (1984)
30. Roth, M., Tym, E., Mountjoy, C.Q., et al.: CAMDEX - A Standardized Instrument for the Diagnosis of Mental Disorder in the Elderly with Special Reference to the Early Detection of Dementia. Br. J. Psychiatry. 149, 698–709 (1986)
31. Welsh, K., Butters, N., Hughes, J., Mohs, R., Heyman, A.: Detection of Abnormal Memory Decline in Mild Cases of Alzheimer's Disease Using CERAD Neuropsychological Measures. Arch. Neurol. 48, 278–281 (1991)
32. Welsh, K.A., Butters, N., Mohs, R.C., Beekly, D., Edland, S., Fillenbaum, G., et al.: The Consortium to Establish a Registry for Alzheimer's Disease (CERAD): part V—a normative study of the neuropsychological battery. Neurology 44, 609–614 (1994)

Applied Neuroimaging to the Diagnosis of Alzheimer's Disease: A Multicriteria Model

Isabelle Tamanini, Ana Karoline de Castro, Plácido Rogério Pinheiro, and Mirian Calíope Dantas Pinheiro

University of Fortaleza (UNIFOR) - Graduate Program in Applied Computer Sciences
Av. Washington Soares, 1321 - Bl J Sl 30 - 60.811-905 - Fortaleza - Brazil
{isabelle.tamanini,akcastro}@gmail.com, {placido,caliope}@unifor.br
http://www.unifor.br/mia

Abstract. In the last few years, Alzheimer's disease has been the most frequent cause of dementia and it is responsible, alone or in association with other diseases, for 50% of the cases in western countries. Dementias are syndromes characterized by a decline in memory and other neuropsychological changes, especially occurring in elderly people and increasing exponentially along the aging process. The main focus of this work is to develop a multicriteria model for aiding in decision making on the diagnosis of Alzheimer's disease by using the Aranau Tool, structured on the Verbal Decision Analysis. In this work, the modeling and evaluation processes were conducted with the aid of a medical expert, bibliographic sources and questionnaires. The questionnaires taken into account were based mainly on patients' neuroimaging tests, and we analyzed wheter or not there were problems in the patients' brain that could be relevant to the diagnosis of Alzheimer's disease.

Keywords: Verbal Decision Analysis, Diagnosis of Alzheimer's Disease, Neuroimaging, ZAPROS.

1 Introduction

Demographic studies in developed and developing countries have shown a progressive and significant increase in the elderly population in the last few years [12]. Alzheimer's disease is the most frequent cause of dementia and it is responsible (alone or in association with other diseases) for 50% of the cases in western countries [12]. Dementias are syndromes characterized by a decline in memory and other neuropsychological changes, especially occurring in elderly people and increasing exponentially as one gets older. According to [1], despite its high incidence, doctors fail to detect dementia in 21 to 72% of their patients.

Alzheimer's disease is characterized by the presence of senis plaques and tangled neurofibrillaries in the regions of the hippocampus and cerebral cortex, and the neurons appear atrophied in a large area of the brain, as shown in fig. 1. There's a major importance in identifying the cases in which the risks of developing a dementia are higher, considering the few alternative therapies and the greater effectiveness of treatments when an early diagnosis is possible [3].

M.D. Lytras et al. (Eds.): WSKS 2009, CCIS 49, pp. 532–541, 2009.

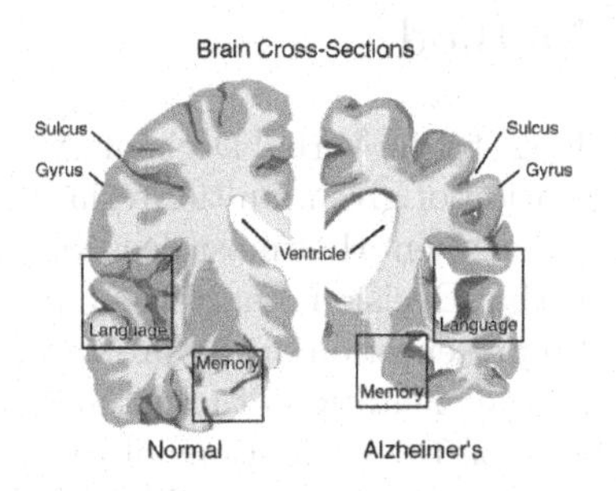

Fig. 1. Normal brain and brain with Alzheimer's disease, respectively

Alzheimer's disease is difficult to be diagnosed, since the initial symptoms are subtle and they progress slowly until they are clear and devastating.

One way to identify whether a patient is having a normal process of aging, or if s/he is developing any form of dementia, is based on a neuropathological evaluation. There are several tests available, and one of the major challenges is to find out which test, or which characteristics of a test, would be more efficient in establishing the diagnosis of dementia.

In this work, we used the battery of tests from the Consortium to Establish a Registry for Alzheimer's Disease (CERAD). The neuropathological battery of CERAD, which is the center of interest of this work, is mainly structured on neuroimaging exams of the patients, with the aim of verifying whether or not there are any other problems in the patients' brain, such as vascular lesions, hemorrhages, microinfarcts, etc, that could be relevant to the diagnosis of Alzheimer's disease.

A multicriteria model based on ZAPROS was developed for aiding in decision making in order to find out which clinical tests are more attractive, for a given set of tests, for the diagnosis of Alzheimer's disease. Multicriteria methodologies help to generate knowledge about the decision context, thus, increasing the confidence of those who make decisions on the results [2]. The ZAPROS method is a verbal multicriteria method and it has been chosen among other available methods because it fits the characteristics of the questioned context, considering the evaluation of the problem, the decision objects and the available information. Besides, the method has already been applied to problems of considerable importance, namely the choice of a prototype for mobile digital television [7].

Our purpose is to determine which of the exams are relevant and would detect faster if the patient is developing Alzheimer's disease. As it has been said, the sooner this information is obtained, the greater the chances of treating and retarding the disease. This way, the data given on the battery was analyzed and, considering the experience of the decision maker, who was the medical expert, a multicriteria problem was structured. The criteria were defined considering the characteristics of each questionnaire of the battery. The preferences were given through the analysis of the questionnaires results and the postmortem diagnosis of each patient. At the end, we had a ranking of the questionnaires, from the one that had the greatest importance on the diagnosis, to the one that had the least.

2 The ZAPROS Method

The ZAPROS method belongs to the Verbal Decision Analysis (VDA) framework and it aims at the classification of given multicriteria alternatives. The method is structured on the acknowledgment that most decision making problems can be verbally described. The Verbal Decision Analysis supports the decision making process the by the verbal representation of the problem [4].

The method is based on the elicitation of preferences around the values that represent the distances between the evaluations of two criteria. A scale of preferences can be structured, enabling the comparison of alternatives.

Before the alternatives comparison process, one should consider:

- The preferences must be elicited in such a way that a decision rule can be formed before the presentation of the alternatives;
- The comparisons between the criteria will be made by human beings, symbolizing the decision maker (DM);
- The quality graduations of the criteria values are verbal and defined by the DM.

Among the advantages of the ZAPROS method utilization, we can say that [15]:

- It presents questions on the elicitation of the preferences process understandable to the decision maker, based on the criteria values. This procedure is psychologically valid (because it respects the limitations of the human information processing system) and represents the method's greatest feature;
- It presents considerable resistance to the decision maker's contradictory inputs, being capable of detecting and requesting a solution to these problems;
- It specifies all the information of the qualitative comparison in a language that is understandable to the decision maker.

A disadvantage of the method is that the number of criteria and values of the criteria supported are limited, since they are responsible for the exponential growth of the problem alternatives and of the information required on the process of preferences elicitation.

The scale of preferences is essentially qualitative, defined with verbal variables, causing losses on the comparison power, because these symbols are not assigned of exact values (which implies in the inexistence of overall values - best or worst in any kind of situation) and can't be recognized computationally. So, there are a lot of incomparable alternatives, which can lead to an absence of an acceptable result.

According to [6], an estimate of the number of incomparable alternatives (and, consequently, of the method's decision power) can be made by calculating the number of pairs of alternatives ($Q = 0.5n^N(n^N - 1)$, where N represents the number of criteria and n is the number of criteria values) and the subset that will be related by Pareto's dominance (D). From the difference between Q and D, we have the set of alternatives that depends directly of the preference scale obtained by the decision maker's answers. This is the set with the greatest probability of presenting contradictory pairs of alternatives. After that, the index of the decision power of the method can be obtained as follows: $P = \frac{1-S}{B}$, where B is the difference between Q and D, and S is the number of alternatives that can't be compared based on the DM's scale of preferences (incomparable alternatives).

3 A New Approach Methodology

A methodology structured basically on the ZAPROS method [5] is proposed. It presents three main stages: *Problem Formulation*, *Elicitation of Preferences* and *Comparison of Alternatives*, as proposed on the original version of the ZAPROS method. These stages are described as follows.

3.1 Formal Statement of the Problem

The methodology follows the same problem formulation proposed in [5]:

Given:

1) K = 1, 2,..., N, representing a set of N criteria;
2) n_q represents the number of possible values on the scale of q-th criterion, $(q \in K)$; for the ill-structured problems, as in this case, usually $n_q \leq 4$;
3) $X_q = x_{iq}$ represents a set of values to the q-th criterion, which is this criterion scale; $|X_q| = n_q (q \in K)$; where the values of the scale are ranked from best to worst, and this order does not depend on the values of other scales;
4) $Y = X_1 * X_2 * ... * X_n$ represents a set of vectors y_i, in such a way that: $y_i = (y_{i1}, y_{i2}, ..., y_{iN})$, and $y_i \in Y$, $y_{iq} \in X_q$ and $P = |Y|$, where $|Y| = \prod_{i=1}^{i=N} n_i$.
5) $A = \{a_i\} \in Y$, i=1,2,...,t, where the set of t vectors represents the description of the real alternatives.

Required: The multicriteria alternatives classification based on the decision maker's preferences.

3.2 Elicitation of Preferences

In this stage, the scale of preferences for quality variations (Joint Scale of Quality Variations - JSQV) is constructed. The elicitation of preferences follows the order of steps shown in fig. 2 [14]. This structure is the same proposed in [5], however, substages 2 and 3 (numbered on the left side of the figure) were put together in just one substage.

Instead of setting the decision maker's preferences based on the first reference situation and, then, establishing another scale of preferences using the second reference situation, we propose that the two substages be transformed in one. The questions made considering the first reference situation are the same as the ones made considering the second reference situation. So, both situations will be presented and must be considered in the answer to the question, in order not to cause dependence of criteria. The alteration reflects on an optimization of the process: instead of making 2n questions, only n will be made. The questions to

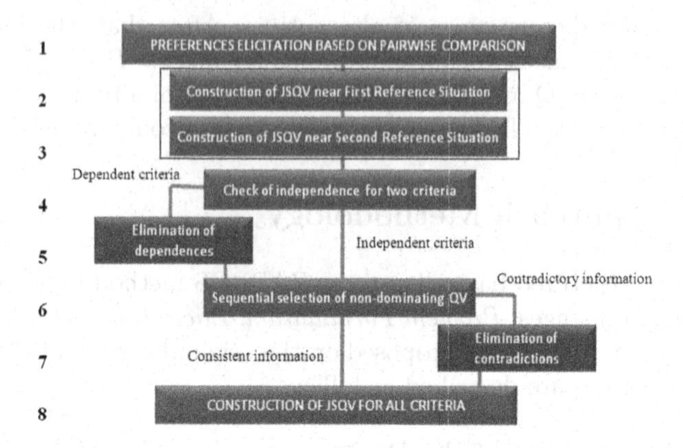

Fig. 2. Elicitation of preferences process

Quality Variations (QV) belonging to just one criteria will be made as follows: supposing a criterion A having $X_A = A_1, A_2, A_3$, the decision maker will be asked about his preferences between the QV $a_1 - a_2$, $a_1 - a_3$ and $a_2 - a_3$. Thus, there is a maximum of three questions to a criterion with three values ($n_q = 3$).

The question will be formulated in a different way on the preferences elicitation for two criteria, because there were difficulties in understanding and delay in the decision maker's answers when exposing the QV of different criteria.

The question will be made dividing the QV into two items. For example, having the set of criteria k = A, B, C, where $n_q = 3$ and $X_q = q_1, q_2, q_3$, considering the pair of criteria A, B and the QV a_1 and b_1, the decision maker should analyze which imaginary alternative would be preferable: A_1, B_2, C_1 or A_2, B_1, C_1. However, this answer must be the same to alternatives A_1, B_2, C_3 and A_2, B_1, C_3. If the decision maker answers that the first option is better, then b_1 is preferable to a_1, because it is preferable to have B_2 on the alternative instead of A_2.

3.3 Comparison of Alternatives

With the aim of reducing the number of incomparability cases, we apply the same structure proposed in [5], but modifying the comparison of pairs of the alternatives' substage according to the one proposed in [10]. Fig. 3 shows the structure of the comparison of the alternatives' process.

Each alternative has a function of quality - V(y) [5], depending on the evaluations of the criteria that it represents. In [10], it is proposed that the vectors of ranks of the criteria values, which represent the function of quality, are rearranged in an ascending order. Then, the values will be compared to the corresponding position of another alternative's vector of values based on Pareto's dominance rule. Meanwhile, this procedure was modified for implementation because it was originally proposed to scales of preferences of criteria values, not for quality variation scales.

So, supposing the comparison between alternatives $Alt1 = A_2, B_2, C_1$ and $Alt2 = A_3, B_1, C_2$, considering a scale of preferences: $a_1 \prec b_1 \prec c_1 \prec a_2 \prec b_2 \prec$

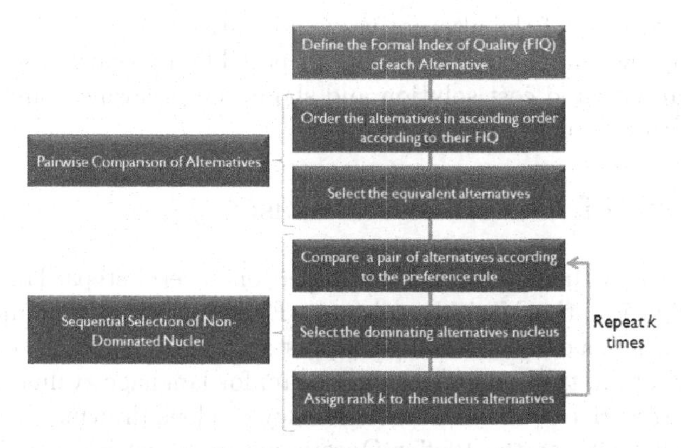

Fig. 3. Alternatives comparison process

$c_2 \prec a_3 \prec b_3 \prec c_3$, we have the following functions of quality: V(Alt1) = (0, 0, 2) and V(Alt2) = (0, 3, 4), which represents the ranks of, respectively, b_1 and c_1, a_2. Comparing the ranks presented, we can say that Alt1 is preferable to Alt2.

However, there are cases in which the incomparability of real alternatives will not allow the presentation of a complete result. These problems require further comparison. In such cases, we can evaluate all possible alternatives to the problem in order to rank the real alternatives indirectly. The possible alternatives should be rearranged in an ascending order according to their Formal Index of Quality (FIQ) and only the significant part will be selected for the comparison process (the set of alternatives presenting FIQ between the greatest and the smallest real alternatives' FIQ). After that, the ranks obtained will be passed on to the corresponding real alternatives.

3.4 Proposed Tool to the New Methodological Approach

In order to facilitate the decision process and perform it consistently, observing its complexity and with the aim of making it accessible, we present a tool implemented in Java, structured on the Verbal Decision Analysis, considering the new approach methodology.

The tool is presented by the sequence of actions that follows:

- *Criteria Definition:* First of all, the user should define the criteria presented by the problem. At this stage the problem formulation occurs.
- *Preferences Elicitation:* This process occurs in two stages: the elicitation of preferences for quality variation on the same criteria and the elicitation of preferences between pairs of criteria.
- *Alternatives Definition:* The alternatives can be defined only after the construction of the scale of preferences.
- *Alternatives Classification:* After the problem formulation, the user can verify the solution obtained to the problem. The result is presented to the decision maker so that it can be evaluated.

If the classification of the alternatives is not satisfactory, the decision maker can request a new comparison based on all possible alternatives for the problem. This is an elevated cost solution and should be performed only when it is necessary for the problem resolution.

4 Diagnosis of Alzheimer's Disease

The diagnosis of Alzheimer's disease is based on several steps. For the global cognitive evaluation, the Mini-Mental State Examination is recommended; for memory evaluation, a delayed recall [8] or presentation of objects as drawings; for attention evaluation, trail-making or digit-span; for language evaluation: Boston naming; and executive functions, verbal fluency or clock-drawing; for functional evaluation, IQCODE, or the Pfeffer Questionnaire or the Bayer daily activity scale is recommended. And then, we have the clinical tests based on laboratory and neuroimaging tests.

This study aims at helping decide which would be the best manner to define the diagnosis. To do this, we sought to choose the most important questions in the diagnosis of Alzheimer's disease, using a battery of CERAD tests. This battery has been chosen because it encompasses all the steps of the diagnosis and it is used all over the world.

Therefore, the questionnaires selected through this decision making process will be applied preferentially because, in accordance with the decision maker, these questions play a main role in the diagnosis.

5 CERAD - An Overview

The original mandate of the CERAD, in 1986, was to develop a battery of standardized assessments for the evaluation of patients with possible Alzheimer's disease who enrolled in NIA-sponsored Alzheimer's Disease Centers (ADCs) or in other dementia research programs [9]. Despite the growing interest in clinical investigations of this illness at that time, uniform guidelines were lacking as to diagnostic criteria, testing procedures, and staging of severity. CERAD developed the following standardized instruments to assess the various manifestations of the disease: Clinical Neuropsychology, Neuropathology, Behavior Rating Scale for Dementia, Family History Interviews and Assessment of Service Needs.

6 Multicriteria Model for Alzheimer's Disease Diagnosis

A multicriterion model was structured based on three tests of the CERAD's battery Neuropathology: Clinical History, Cerebral Vascular Disease Gross Findings and Microscopic Vascular Findings - Major Non-vascular Microscopic Findings; and on a case study presented on [11], that considered the ventricular enlargement after six months based on MRI tests as a measure of the disease's progression. Considering the CERAD data, only the results of the tests of patients that had already died and on which the necropsy has been done were selected, because it is known that necropsy is essential for validating the clinical diagnosis of dementing diseases. So, the model was based on 122 cases.

The criteria are related to the characteristics of the CERAD's questionnaires and the MRI analysis, and the values identified for them are exposed in table 1.

Table 1. Criteria involved on the early diagnosis of Alzheimer's disease

Criteria	Values of Criteria
A - Verification of Existence of Other Diseases	A1. There are questions about the severity of other brain problems A2. There are questions about the existence of brain problems A3. There are no questions about the existence brain problems
B - Type of Tests Required to Answer the Questionnaries	B1. Neuroimaging tests specific for the patient's problems were done to him/her B2. Laboratory tests were done to the patient B3. General neuroimaging tests were done to the patient
C - Measure of Cerebral Ventricular Enlargement After Six Months	C1. Ventricular enlargement was considered on the patient's evaluation C2.No measure of ventricular enlargement was done

The elicitation of preferences was done by means of analysis with the medical expert of the answers obtained on each questionnaire and the patients' final diagnosis, considering also the results of the disease progression measure obtained on [11]. The preferences scale is given as follows: $b_2 \prec b_3 \prec a_1 \prec b_1 \prec a_2 \prec a_3 \equiv c_1$.

After the elicitation of preferences, the alternatives will be defined. As the aim of this paper is to give a preference order of application of the tests, the alternatives will be formulated according to the characteristic of each test, so, we'll have the tests described as criterion values. The model was applied to the Aranaú Tool [13] and the results obtained are shown in table 2.

The order of preference to apply the tests is given by: T_1-weighted MRI, Cerebral Vascular Disease Gross Findings, Microscopic Vascular Findings - Major Non-vascular Microscopic Findings, and Clinical History; which means that the test T_1-weighted MRI applied according to [11] is more likely to detect a

Table 2. Classification of the neuropathological tests considered on the modeling

Alternatives	Criteria Evaluations	Rank
Clinical History	A3B2C2	4
Cerebral Vascular Disease Gross Findings	A1B1C2	2
Microscopic Vascular Findings - Major Non-vasc. Microscopic Findings	A2B3C2	3
T_1-weighted MRI	A3B1C1	1

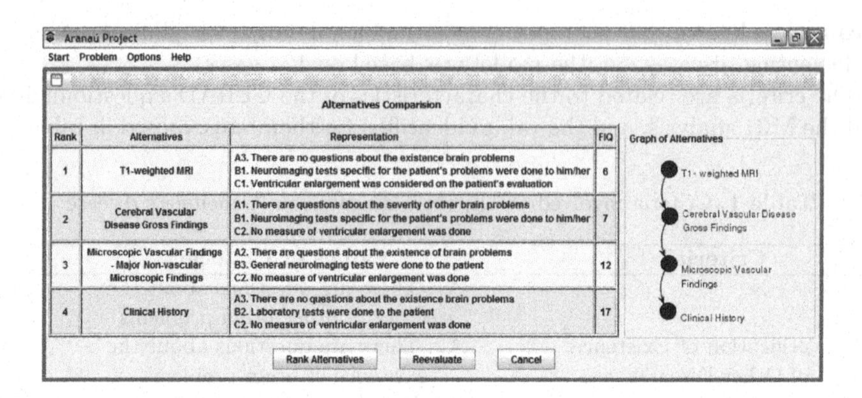

Fig. 4. Presentation of results

possible case of Alzheimer's disease than the others, so, it should be applied first. Fig. 4 shows the results presentation screen of the tool for this problem.

7 Conclusions

A multicriteria model was formulated based on the data available on CERAD and on the case study presented on [11]. For the CERAD data, the criteria were formulated based on the characteristics of neurophatological battery questionnaires, on which, most of the times, a neuroimaging exam of the patient was required. The analysis of the patient's ventricular enlargement after six months was also considered important to the diagnosis, as it is a reliable measure of the Alzheimer's disease development [11]. The preferences were elicitated based on information given by a medical expert that analyzed the results of each questionnaire and the ventricular enlargement measure and, then, the alternatives were structured according to the characteristics of these tests.

This paper contribution is a ranking of the tests, from the most likely to lead to a diagnosis of Alzheimer's disease, to the least. This would enable a faster detection of patients that would develop the disease, increasing their chances of treatment. To do so, a multicriteria model was submitted to a tool, structured basically on the ZAPROS method and with some modifications to improve the alternatives comparison process, and the result obtained was the tests order.

As future works, we intend to improve the model proposed, involving more tests and questionnaires, so that a complete ranking will be done. This would decrease the number of tests that patients have to go through and the number of questionnaires they would have to answer in order to get to the diagnosis.

Acknowledgments. The authors are thankful to the National Counsel of Technological and Scientific Development (CNPq) for the support received on this project and to the Consortium to Establish a Registry for Alzheimer's Disease (CERAD) for making the data used in this case study available.

References

1. Bassett, S.S.: Attention: Neuropsychological Predictor of Competency in Alzheimer's Disease. Journal of Geriatric Psychiatry and Neurology 12(4), 200–205 (1999)
2. Evangelou, C., Karacapilidis, N., Khaled, O.A.: Interweaving knowledge management, argumentation and decision making in a collaborative setting: the KAD ontology model. International Journal of Knowledge and Learning 1(1/2), 130–145 (2005)
3. Hughes, C.P., Berg, L., Danzinger, W.L., et al.: A New Clinical Scale for the Staging of Dementia. British Journal of Psychiatry 140(6), 566–572 (1982)
4. Larichev, O., Moshkovich, H.M.: Verbal decision analysis for unstructured problems. Kluwer Academic Publishers, Boston (1997)
5. Larichev, O.: Ranking Multicriteria Alternatives: The Method ZAPROS III. European Journal of Operational Research 131(3), 550–558 (2001)
6. Larichev, O.: Method ZAPROS for Multicriteria Alternatives Ranking and the Problem of Incomparability. Informatica 12(1), 89–100 (2001)
7. Mendes, M., Carvalho, A.L., Furtado, M.E.S., Pinheiro, P.R.: Co-evolutionary Interaction Design of Digital TV Applications Based on Verbal Decision Analysis of User Experiences. In: WSKS 2008. Communications in Computer and Information Science, vol. 19(1), pp. 702–711 (2008), doi:0.1007/978-3-540-87783-7
8. Mohr, E., Dastoor, D., Claus, J.J.: Neuropsychological Assessment. In: Gauthier, S. (ed.) Clinical Diagnosis and Management of Alzheimer's Disease, pp. 93–106. Martin Dunitz, London (1999)
9. Morris, J.C., Heyman, A., Mohs, R.C., et al.: The Consortium to Establish a Registry for Alzheimer's Disease (CERAD): Part 1. Clinical and Neuropsychological Assessment of Alzheimer's Disease. Neurology 39(9), 1159–1165 (1989)
10. Moshkovich, H., Mechitov, A., Olson, D.: Ordinal Judgments in Multiattribute Decision Analysis. European Journal of Operational Research 137(3), 625–641 (2002)
11. Nestor, S., Rupsingh, R., Borrie, M., et al.: Ventricular Enlargement as a Possible Measure of Alzheimer's Disease Validated Using the Alzheimer's Disease Neuroimaging Initiative Database. Brain 131(9), 243–2454 (2008)
12. Prince, M.J.: Predicting the Onset of Alzheimer's Disease Using Bayes' Theorem. American Journal of Epidemiology 143(3), 301–308 (1996)
13. Tamanini, I., Pinheiro, P.R.: Towards the New Approach Methodology with ZAPROS. Pesquisa Operacional (to appear, 2009)
14. Tamanini, I., Pinheiro, P.R.: Challenging the Incomparability Problem: An Approach Methodology Based on ZAPROS. In: Modeling, Computation and Optimization in Information Systems and Management Sciences, Communications in Computer and Information Science, vol. 14(1), pp. 344–353. Springer, Heidelberg (2008)
15. Ustinovich, L., Kochin, D.: Verbal Decision Analysis Methods for Determining the Efficiency of Investments in Construction. Foundations of Civil and Environmental Engineering 5(1), 35–46 (2004)

An Expedient Study on Back-Propagation (BPN) Neural Networks for Modeling Automated Evaluation of the Answers and Progress of Deaf Students' That Possess Basic Knowledge of the English Language and Computer Skills

John Vrettaros[1], George Vouros[2], and Athanasios S. Drigas[1]

[1] NCSR DEMOKRITOS,
Institute of Informatics and Telecommunications
Net Media Lab
Ag. Paraskevi, 15310, Athens, Greece
[2] Aegean University, Info and Communication Systems Eng,
83200, Karlovassi, Samos, Greece
{dr,jvr}@iit.demokritos.gr, georgev@aegean.gr

Abstract. This article studies the expediency of using neural networks technology and the development of back-propagation networks (BPN) models for modeling automated evaluation of the answers and progress of deaf students' that possess basic knowledge of the English language and computer skills, within a virtual e-learning environment. The performance of the developed neural models is evaluated with the correlation factor between the neural networks' response values and the real value data as well as the percentage measurement of the error between the neural networks' estimate values and the real value data during its training process and afterwards with unknown data that weren't used in the training process.

Keywords: Neural Networks, Back-propagation Network models, diagnosis.

1 Introduction

One of the most promising branches of current educational research is student diagnosis which according to VanLehn ([8]) is the process of inferring students' internal characteristics from their recognizable behavior. Student diagnosis is considered more than necessary as according to learning theories adaptive learning is a fruitful one if one wishes to obtain effective tutoring. That is the reason why one of the latest trends in educational software development is attempting to implement a simulation of the way that a human teacher adapts his tutoring to the individualization of each student [5]. In order for this kind of intelligence to be achieved, researchers have adopted many Artificial Intelligence methods. The most famous among them are neural networks, fuzzy logic as well as several search methods such as genetic algorithms.

M.D. Lytras et al. (Eds.): WSKS 2009, CCIS 49, pp. 542–551, 2009.

Neural networks are on the top of the researchers' choice since they provide a system the ability to recognize patterns, to derive meaning from vague data and to identify matching in similar cases ([6]). Fuzzy set theory is widely used since it can deal in a reliable way with human uncertainty and it obtains smooth modelling of human decision – making. Genetic algorithms are ideal for efficient expert knowledge representation. Finally, Neuro – Fuzzy synergism is getting more and more popular in this area since it seems to overcome obstacles that come up when each of the methods involved is solely applied ([7]). Below we present several typical examples of the application of these methods in student's diagnosis.

Weon and Kim ([3]), implemented a system aiming at individual learning evaluation by making use of fuzzy linguistic variables representing each question. Finally, assessment came up taking into consideration the membership degree of uncertainty factors.

Vrettaros et al. ([4]) introduced a diagnostic system of taxonomies using fuzzy logic. Filled up questionnaires were processed in order for students' classification into one of the predefined knowledge levels to be achieved according to the given answers.

Stathakopoulou et al. ([5]) attempted to infer students' individual characteristics and use them in order not only to create but also to update the student model. This neural network - based fuzzy model is consisted of a fuzzy component and a neural network trained through actual students' profiles.

Hommsi et al. ([2]), presented an Adaptive and Intelligent Web – Based Educational System (AIWBES) based on Fuzzy - ART2 neural network and Hidden Markov Model (HMM), which is a stochastic method. The goal of the system described by the authors was to assess the learners' knowledge level and to obtain students' classification in one of six different levels taking into account several parameters.

In general, the aim of the e-learning environment that is studied is the realization of an e-educational system for the evaluation of the teaching of the English language to deaf people. The crucial point in the definition of the theoretical and technical aspects of the correlation between the e-learning models' subsystems is the definition of the configuration and the characteristics of their corresponding interconnections. According to the structural and operational details of the e-learning procedure the performance factor of the expert system and the collaboration of the e-learning model is the encoding and the contents of the inputs and outputs of the expert system as well as the structure, the standardization and the content of the database questions that require further attention and skilful handling.

The evaluation procedure for the teaching of English to deaf students relies on the attainment of the ESOL (English for Speakers of Other Languages) models (levels 1 and 2). These levels comprise five sections, which in ascending order are [A], [B], [C], [D] and [E]. Section [A] represents the Letter Recognition and their Alphabetical Order, section [B] represents the Spelling and the Vocabulary, section [C] represents the Grammar and the Sentence Structure, section [D] represents the Reading and section [E] represents the Writing.

According to the ESOL e-learning environment specifications the input / output parameters of an expert system can be determined without doubt, while simultaneously their translation is simple and quite straightforward.

As far as the input is concerned, there are five pairs of parameters in total and per question, which are: $a=\{a_{val},a_{rel}\}$, $b=\{b_{val},b_{rel}\}$, $c=\{c_{val},c_{rel}\}$, $d=\{d_{val},d_{rel}\}$ and $e=\{e_{val},e_{rel}\}$. In other words, every pair corresponds to a language section of a certain level. Parameter a describes the recognition of the letters and the alphabetical order of section $[A]$, parameter b relates to the spelling/vocabulary of section $[B]$, parameter c represents the grammar/sentence structure for section $[C]$, the corresponding parameter for reading in section $[D]$ is d, while the writing skill of section $[E]$ is quantified with parameter e. The abbreviation index val (value) represents the evaluation of the specific section based on a particular response, while the abbreviation index rel (relevance) recognizes the level of relevance /weight of a specific question with respect to the contents of a section.

The evaluation values of the input parameters a_{val}, b_{val}, c_{val}, d_{val} and e_{val} originate from the range $S=\{-1\}\cup[0,\ 1]$. When a section is not examined by a question of the corresponding parameter, then the range is characterized by the value -1. A wrong answer according to a certain section leads to a corresponding value of zero (0), while the value of a sections' parameter is (1) if the chosen answer is correct according to that section. Accordingly, answers that are partially correct based on some sections have their values placed in between.

On the other hand, someone can argue that the relevance parameters a_{rel}, b_{rel}, c_{rel}, d_{rel} and e_{rel} characterize the question itself rather than its possible answers. Although this is true, the negotiation with relative parameters as part of a given answer is handy from an evaluation viewpoint. As a result, the relevance/weight is considered range between $[0,\ 1]$, where the value zero (0) or values close to zero mean low relevance, while the value one (1) or values close to one mean high relevance. Accordingly, the other weight values range somewhere in between. It must be pointed out, that the relevance parameters are the same for all the answers of a specific question.

The most skilful method for the provision of information in an expert system in relation with these sections is the sequence in array format of the ten values for the parameters of the input pairs:

a_{val}	a_{rel}	b_{val}	b_{rel}	c_{val}	c_{rel}	d_{val}	d_{rel}	e_{val}	e_{rel}

As an example, let's consider that a question shows low relevance in section A, high relevance in section C and medium relevance in section B. Moreover, let's assume that the question under consideration does not include information about sections D and E. Now let's consider an answer to the previous question, which is correct according to section A, partially correct according to section C, wrong according to section B. Obviously, it does not contain information relating to sections D and E. Such an answer ends in a ten value sequence, which comprises elements that are defined in the range $S = \{-1\} \cup [0, 1]$. In addition, it is obvious that the aforementioned ten value sequence can be directly encoded as an arithmetic string similar to the one below:

a_{val}	a_{rel}	b_{val}	b_{rel}	c_{val}	c_{rel}	d_{val}	d_{rel}	e_{val}	e_{rel}
1	0.1	0	0.5	0.7	0.9	-1	0	-1	0

This way, this particular arithmetic string can easily be used as an input to an expert system.

As far as the output is concerned, from observations and/or monitoring of the operational and correlating characteristics of the expert system, the conclusion that is derived is that that output parameters of the system are six (6) namely y_1, y_2, y_3, y_4, y_5 and y_6. The first five parameters are the evaluation/estimation of the lingual skills per section, while the sixth parameter represents the total estimation of the user for the sum of all of the lingual skills, as follows:

y_1 = letter recognition and alphabetical order skills

y_2 = spelling / vocabulary skills

y_3 = grammar / sentence structure skills

y_4 = reading skills

y_5 = writing skills

y_6 = total lingual skills **(essentially, it is a weighted average of y_1 - y_5, *representing a general estimation of the linguistic level of the student, as a professional pedagogic would have determined in a real case scenario*).**

It is obvious that the output parameters are continuous. Due to the fact that the expert system outputs represent a certain estimation in connection with a specific lingual section, the evaluation is considered to be normalized in the range $[0, 1]$. The translation of the final arithmetic values is simple: zero means no lingual skills, one means perfect lingual skills, while all the other lingual skills levels may be estimated with similar arithmetic insertions. The output values, which are already arithmetically encoded, may be inserted in the e-learning environment as a six value arrayed sequence:

y_1	y_2	y_3	y_4	y_5	y_6

Let's assume that the next estimation is real for a particular student:

0.6 = letter recognition and alphabetical order skills
0.4 = spelling / vocabulary skills
0.2 = grammar / sentence structure skills
0.5 = reading skills
0.3 = writing skills
0.4 = total lingual skills

This particular six value arrayed sequence which comprises continuous elements could be directly encoded as an arithmetic string similar to the one below:

y_1	y_2	y_3	y_4	y_5	y_6
0.6	*0.4*	*0.2*	*0.5*	*0.3*	*0.4*

This way the final outputs are available straight away to rest of the e-learning environment.

The above (according to the ESOL specifications) skilful encoding and inputs and outputs content as well as the structure, standardization and content of the database questions is based on the use of neural networks technology for the e-learning expert system modeling for the automated estimation of the evaluation values of the teaching of English to deaf people.

The neural networks technology has been applied with success to many estimation problems with similar input/output characteristics. This study examines the expediency of applying this technology for modeling the automated evaluation of the deaf students' answers in the form of questions divided in five sections within the e-learning environment of the expert system.

In this study, section 2 provides a description of the neural networks characteristics and details about the development of a typical BPN model. In section 3, various Back-Propagation Networks (BPN) type neural networks models are developed for the automated evaluation of the deaf students' education and progress and their results are presented, determining hence, the expediency of the use of neural networks technology in such problems. In section 4, general neural networks technological issues are presented about generalization and over-fitting for the successful development and realization of effective models. Finally, in section 5 conclusions are presented and ideas are suggested for further research.

2 Development of Back-Propagation Network (BPN) Type Neural Networks for the Automated Evaluation of Deaf Students' Progress

In this study, the general schema of input/output data for supervised learning and their connection with neural networks results from the definition of ten (10) variables namely, v, v', w, w', x, x', y, y', z and z' , which belong in the range $S = \{-1\} \cup [0,\ 1]$ for the input and with the definition of six (6) variables, namely,

h, i, j, k, l and n that belong in the range $[0,\ 1]$, for the output. Hence, each input/output model from the total education data could be similar to the next model (where the symbol ∅ denotes the blank value):

Ten (10) Input Values	a	b	c	d	e	
	(v, v')	(w, w')	(x, x')	(y, y')	(z, z')	
Six (6) Output Values	y_1	y_2	y_3	y_4	y_5	y_6
	h *if* $v \neq -1$ *else* ∅	i *if* $w \neq -1$ *else* ∅	j *if* $x \neq -1$ *else* ∅	k *if* $y \neq -1$ *else* ∅	l *if* $z \neq -1$ *else* ∅	n

A sample of real values for seven (7) models from the sum of the education data is depicted in the following Table 1.

Table 1. A sample of supervised learning education data

		pattern (#)	1	2	3	4	5	6	7
Input Variables	a	a_{val}	0.6	0.1	1.0	0.9	0.6	0.1	0.1
		a_{rel}	0.5	0.5	0.5	0.7	0.7	0.7	0.2
	b	b_{val}	0.8	0.7	0.1	-1	-1	0.8	-1
		b_{rel}	0.8	0.8	0.8	0	0	0.6	0
	c	c_{val}	0.6	0.8	0.2	-1	-1	0.9	0,1
		c_{rel}	0.1	0.1	0.1	0	0	0.5	0.8
	d	d_{val}	-1	-1	-1	0.9	0.1	0.6	-1
		d_{rel}	0	0	0	0.5	0.5	0.4	0
	e	e_{val}	-1	-1	-1	0.6	0.8	0.2	0.9
		e_{rel}	0	0	0	0.3	0.3	0.1	0.3
Output Variables		y_1	0.9	0.2	0.3	0.5	0.8	0.4	0.7
		y_2	0.7	0.8	0.2	∅	∅	0.1	∅
		y_3	0.4	0.5	0.2	∅	∅	0.1	0.5
		y_4	∅	∅	∅	0.9	0.0	0.1	∅
		y_5	∅	∅	∅	0.6	0.2	0.6	0.8
		y_6	0.1	0.5	0.7	0.1	0.2	0.9	0.7

It is obvious that the information that is contained in such education data, should be gathered and processed by a professional pedagogist, due to the fact that such a person seems to be the most suitable and competent person to create the aforementioned content.

For the development of typical back-propagation network (BPN) neural networks models, this study used the modeling and simulation software package Predict by NeuralWare, which is described in reference [11]. This software package provides the necessary mechanisms for the automated modifications and executes the scaling and limiting procedures to the users' input/output data. The basic idea for the modification of data, for the better identification of the statistical distribution of the input variables for the neural networks' training and for the maximization of its performance, is described in references [12, 13].

The software package Predict by NeuralWare demands that the presentation of the variables of the input vector, x, to be of the following pattern:

$$x=\begin{bmatrix} a_{val} & a_{rel} & b_{val} & b_{rel} & c_{val} & c_{rel} & d_{val} & d_{rel} & e_{val} & e_{rel} \end{bmatrix}$$

with values for the evaluation parameters a_{val}, b_{val}, c_{val}, d_{val} and e_{val} that originate from the aforementioned range $S=\{-1\}\cup[0,\ 1]$ and with values for the relevance parameters a_{rel}, b_{rel}, c_{rel}, d_{rel} and e_{rel} that originate from the aforementioned range $[0,\ 1]$. For the variables of output vector, y, the software package Predict by NeuralWare demands that the presentation is of the following pattern:

$$y=\begin{bmatrix} y_1 & y_2 & y_3 & y_4 & y_5 & y_6 \end{bmatrix}$$

with values that originate from the aforementioned normalized range $[0,\ 1]$.

The developed BPN neural network used the first 28 of the 36 available data from Table 7 for its training and the remaining 8 were used for its evaluation trials. Its training was based on the maximization of the correlation with the adaptive gradient-descent technique and its final architecture is of this pattern: 26-2-22/0.9622. This pattern means that the final trained neural network comprises 26 neurons in the input (added inputs due to the modifications), 2 neurons in the hidden layer and 22 neurons in the output (added outputs due to the modifications) with a correlation factor of 0.9622.

The performance of the BPN network Architecture #3 is evaluated by comparing the results of the trained neural network for all of the input/output data, using the recall phase. Table 9 presents the real values of the 36 input/output data, the corresponding values of the BPN#3 neural network and the corresponding indication of the correct or wrong answer (estimation) of the BPN#3 neural network.

From the above Table 2 one can derive that the BPN network Architecture #3 (with verbally encoded outputs) truly learned 27 out of a total of 28 model inputs/outputs, hence a percentage of 27/28 = 96.43%, which is comparable with the correlation factor percentage during its training, which is 96.22%. In addition, from the above Table 2 one can derive that the BPN network Architecture #3 estimates correctly almost 8 of the 8 output values, hence, a 8/8 = 100.0% percentage.

From the performances of the three aforementioned BPN architectures namely, BPN network Architecture #1, BPN network Architecture #2 και BPN network Architecture #3, it is obvious that the neural networks technology can be used in the development of a model for the automated evaluation of the answers and progress of deaf students that possess basic knowledge of the English language as well as computer skills within a virtual e-learning environment. However, this study focused on the development of typical BPN models based only on a small sum of data. It is regarded as worthwhile to develop more BPN models, which are trained for a larger sum of data and to determine the level of their best performance using all the performance improvement techniques of the neural networks, which are presented in the next section. In general though, the acceptance of a model depends on its performance, which may be evaluated based on statistical methods with the only reserve being that the remaining errors between the models' output values and the real values follow approximately the normal distribution.

Table 2. BPN network Architecture #3 – Recall Phase Results - Single Select Questions

Table 9: BPN network Architecture #3 – Recall Phase Results - Single Select Questions														
	Test#,	**Real Input Values (verbally encoded)**						**Output Values BPN#3 (verbally encoded)**						**BPN#3 Correct or Wrong Indication**
		Y1	**Y2**	**Y3**	**Y4**	**Y5**	**Y6**	**Y1**	**Y2**	**Y3**	**Y4**	**Y5**	**Y6**	
1	t1,q2	zero	N	N	zero	N	zero	zero	n	n	zero	n	zero	**Correct**
2	t1,q2	one	N	N	one	N	one	one	n	n	one	n	one	**Correct**
3	t1,q2	zero	N	N	zero	N	zero	zero	n	n	zero	n	zero	**Correct**
4	t1,q2	three	N	N	three	N	three	three	n	n	three	n	three	**Correct**
5	t1,q3	zero	zero	zero	zero	zero	zero	zero	zero	zero	zero	zero	zero	**Correct**
6	t1,q3	zero	zero	five	zero	zero	two	zero	zero	zero	zero	zero	zero	**Correct**
7	t1,q3	zero	zero	zero	zero	zero	zero	zero	zero	zero	zero	zero	zero	**Correct**
8	t1,q3	one	one	one	one	one	one	one	one	one	one	one	one	**Correct**
9	t1,q4	one	one	one	one	one	one	one	one	one	one	one	one	**Correct**
10	t1,q4	zero	zero	zero	zero	zero	zero	zero	zero	zero	zero	zero	zero	**Correct**
11	t1,q4	zero	zero	five	zero	zero	two	zero	zero	five	zero	zero	two	**Correct**
12	t1,q4	zero	zero	five	zero	zero	two	zero	zero	five	zero	zero	two	**Correct**
13	t1,q5	zero	zero	zero	zero	zero	zero	zero	zero	zero	zero	zero	zero	**Correct**
14	t1,q5	one	one	one	one	one	one	one	one	one	one	one	one	**Correct**
15	t1,q5	zero	zero	zero	zero	zero	zero	zero	zero	zero	zero	zero	zero	**Correct**
16	t1,q5	zero	zero	zero	zero	zero	zero	zero	zero	zero	zero	zero	zero	**Correct**
17	t3,q1	N	zero	zero	zero	N	zero	n	zero	zero	zero	n	zero	**Correct**
18	t3,q1	N	zero	zero	zero	N	zero	n	zero	zero	zero	n	zero	**Correct**
19	t3,q1	N	zero	zero	zero	N	zero	n	zero	zero	zero	n	zero	**Correct**
20	t3,q1	N	one	one	one	N	one	n	zero	zero	zero	n	zero	**Wrong**
21	t3,q2	N	five	zero	zero	N	two	n	five	zero	zero	n	two	**Correct**
22	t3,q2	N	zero	zero	zero	N	zero	n	zero	zero	zero	n	zero	**Correct**
23	t3,q2	N	one	one	one	N	one	n	one	one	one	n	one	**Correct**
24	t3,q2	N	zero	zero	zero	N	zero	n	zero	zero	zero	n	zero	**Correct**
25	t3,q3	N	zero	five	zero	N	two	n	zero	five	zero	n	two	**Correct**
26	t3,q3	N	one	one	one	N	one	n	one	one	one	n	one	**Correct**
27	t3,q3	N	five	zero	zero	N	two	n	five	zero	zero	n	two	**Correct**
28	t3,q3	N	zero	zero	zero	N	zero	n	zero	zero	zero	n	zero	**Correct**
29	t3,q4	N	one	one	one	N	one	n	one	one	one	n	one	**Correct**
30	t3,q4	N	zero	zero	zero	N	zero	n	zero	zero	zero	n	zero	**Correct**
31	t3,q4	N	five	zero	zero	N	two	n	five	zero	zero	n	two	**Correct**
32	t3,q4	N	zero	zero	zero	N	zero	n	zero	zero	zero	n	zero	**Correct**
33	t3,q5	N	zero	five	zero	N	two	n	zero	five	zero	n	two	**Correct**
34	t3,q5	N	one	one	one	N	one	n	one	one	one	n	one	**Correct**
35	t3,q5	N	zero	zero	zero	N	zero	n	zero	zero	zero	n	zero	**Correct**
36	t3,q5	N	zero	zero	zero	N	zero	n	zero	zero	zero	n	zero	**Correct**

3 Conclusions

In this study, using the ESOL specifications for the skilful encoding and the inputs and outputs content and also the structure, standardization and content of the database questions of an e-learning expert system, three BPN type models were developed for the estimation of the expediency of applying neural network technology towards the automated estimation of the values, for the evaluation and progress of teaching English to deaf people.

Based on the performances of the three type BPN architectures that were developed, namely, BPN network Architecture #1, BPN network Architecture #2 and BPN network Architecture #3, it was derived that the neural networks technology can be used towards the development of such an automated model for the evaluation of the answers and the progress of deaf students that possess basic knowledge of the English language and also basic computer skills, within a virtual e-learning environment.

However, it is regarded that this estimate would be more accurate and would have more certitude if the available input/output data were more. Additionally, in this study only BPN type neural network models were tested and it is thought of as necessary to develop additional models for this particular problem and to determine the level for their best performance using a larger sum of available input/output data as well as all the performance improvement techniques for the neural networks.

References

1. Rumelhart, D.E., McClelland, J.L. (eds.): Parallel Distributed Processing: Explorations in the Microstructure of Cognition. Foundations, vol. I. MIT Press, Cambridge (1986)
2. Rumelhart, D.E., McClelland, J.L., The PDP Research Group: Learning and Relearning in Boltzman Machines. In: Hinton, G.E., Sejnowski, T.J. (eds.) Parallel Distributed Processing. Foundations, ch. 7, vol. 1. MIT Press, Cambridge (1985)
3. Carpenter, G.A., Grossberg, S.: The ART of Adaptive Patern Recognition by a Self-Organizing Neural Network. Computer, 77–88 (March 1988)
4. Carpenter, G.A., Grossberg, S.: A Massively Parallel Architecture for a Self-Organizing Neural Pattern Recognition Machine. Computer Vision, Graphics and Image Processing 37, 54–115 (1987)
5. Carpenter, G.A., Grossberg, S.: ART 2: Self-Organization of Stable Category Recognition Codes for Analog Input Patterns. Applied Optics 26, 4919–4930 (1987)
6. Kohonen, T., et al.: Statistical Pattern Recognition with Neural Networks: Benchmark Studies. In: Proceedings of the Second Annual IEEE International Conference on Neural Networks, vol. 1 (1988)
7. Kohonen, T.: Self-Organization and Associative Memory, 3rd edn. Springer, New York (1989)
8. Murray, S.: Neural Networks for Statistical Modeling. Springer, Van Nostrand Reinhold (1993)
9. NeuralWorks manual, Neural Computing: A Technology Handbook for Professional II/Plus and NeuralWorks Explorer, and NeuralWorks Advanced Reference Guide: Software reference for Professional II/Plus and NeuralWorks Explorer, NeuralWare, Inc., USA (1995)

10. Rumelhart, D.E., Hinton, G.E., Williams, R.J.: Learning Internal Representations by Error Propagation. In: Parallel Distributed Processing, vol. I, pp. 318–364. MIT Press, Cambridge (1987)
11. NeuralWorks, Predict: Computer Software package for applications of Neural Networks, NeuralWare, Inc., USA
12. Tukey, J.W.: Exploratory Data Analysis. Addison-Wesley, Reading (1977)
13. Chatterjee, S., Price, B.: Regression Analysis by Example. John Wiley & Sons, Chichester (1991)

[1] Stathakopoulou, R., Magoulas, G., Grigoriadou, M., Samarakou, M.: Neuro –Fuzzy knowledge processing in intelligent learning environments for improved student diagnosis (2005)
[2] Homsi, M., Lutfi, R., Carro, R.M., Ghias, B.: Student modeling using NN – HMM for EFL course
[3] Weon, S., Kim, J.: Learning achievement evaluation strategy using fuzzy membership function (2001)
[4] Vrettaros, J., Vouros, G., Drigas, A.: Development of a diagnostic system of taxonomies using fuzzy logic – case SOLO (useful for e-learning system) (2004)
[5] Vrettaros, J., Pavlopoulos, J., Vouros, G., Drigas, S.: The development of a self-assessment system for the learners answers with the use of GPNN (2008)

Decade Review (1999-2009): Artificial Intelligence Techniques in Student Modeling

Athanasios S. Drigas, Katerina Argyri, and John Vrettaros

NCSR DEMOKRITOS,
Institute of Informatics and Telecommunications
Net Media Lab
Ag. Paraskevi, 15310, Athens, Greece
{dr,jvr}@iit.demokritos.gr, akate82@gmail.com

Abstract. Artificial Intelligence applications in educational field are getting more and more popular during the last decade (1999-2009) and that is why much relevant research has been conducted. In this paper, we present the most interesting attempts to apply artificial intelligence methods such as fuzzy logic, neural networks, genetic programming and hybrid approaches such as neuro – fuzzy systems and genetic programming neural networks (GPNN) in student modeling. This latest research trend is a part of every Intelligent Tutoring System and aims at generating and updating a student model in order to modify learning content to fit individual needs or to provide reliable assessment and feedback to student's answers. In this paper, we make a brief presentation of methods used to point out their qualities and then we attempt a navigation to the most representative studies sought in the decade of our interest after classifying them according to the principal aim they attempted to serve.

Keywords: student modeling, student diagnosis, fuzzy logic, neural networks, genetic programming, student assessment, student evaluation, adaptive learning.

1 Introduction

Traditionally artificial intelligence aims at simulating filtering information, handling constraints, recognizing patterns and making logical inferences as well as other activities necessary in order to deal with real – life problems in an automated way. In the last decades, each and every one of the activities mentioned above, have been a real challenge, as the researchers' community realized that they are of significant practical value for learning sciences especially when seeking to obtain in an automatic way active and reflective learning just like all learning theories suggest.

Artificial intelligence has a wide range of applications in the educational field and new directions are constantly given in educational research. Among the most typical of them are several areas such as intelligent information retrieval [11], natural language processing in order to accomplish tasks such as evaluation of student's work [13], intelligent agents [25] (in the wider field of expert systems [40]), robotics [26], intelligent virtual environments [10], voice and image recognition [31] especially used for data input in cases of learning or physical impairments and e-tutoring [1].

M.D. Lytras et al. (Eds.): WSKS 2009, CCIS 49, pp. 552–564, 2009.

In this paper, we will focus on the techniques incorporated during last decade (1999-2009) and on student modeling as it seems to be one of the latest research trends and in the same time one of the most significant and challenging tasks for an instructor, let alone for an intelligent tutoring system.

According to Stathakopoulou et al. (2005), student modeling is consisted of two components: the student model and the diagnostic module [2]. The student model is one of the components of an Intelligent Tutoring System which provides a description of student related information such as his knowledge level, skills or even preferences while diagnosis is the inference process which in the end updates the student model.

Eklund and Brusilovsky (1998) suggest that the student model keeps track of students' state of mind on the basis of their responses [27]. Consequently, reliable student modeling comes up via careful student assessment, the process that allows the expert to diagnose the learner's mental state and knowledge status in order to check on the efficiency of teaching and to detect possible learning deficiencies.

The idea of the creation of a student model via an automated way is a realistic and hopeful one since according to Frias – Martinez et al. [21], a user (in the more generalized user modelling problem) leaves trail while using a hypermedia system and these data can be stored and used in order to generate further information, patterns and predictions.

The popularity of this task is more than justified if one takes into consideration the fact that "student model enables these systems (adaptive and intelligent web-based Educational Systems) provide individualized course content and study support to help students with different background and knowledge status to achieve their learning objectives" [15]. After all, high adaptability in teaching, a goal strongly connected with student diagnosis, has proved to be a fruitful way to maximize learning results [22]. Indeed, studies sought have shown that students process knowledge and learn in different ways and that is why researchers strongly believe that students learn in an efficient way when teaching is tailored according to their personal learning style and other individual characteristics [40].

In this context many artificial intelligence techniques keep on triggering researchers' attention and their incorporation in more and more aspects of the educational field is being adopted.

To name just a few, neural networks, Bayesian networks, Markov models, control theory, fuzzy logic as well as search methods such as genetic and evolutionary algorithms are used to serve this purpose, alone, in all possible combinations among them, or in combination with different techniques, such as other machine learning algorithms for more efficiency.

2 Artificial Intelligence in Student Modeling

As it was mentioned above, student modeling has been a real challenge for researchers in the educational field especially over the last decade. But in order for that to be achieved, the knowledge as well as the experience of the educational expert has to be represented in a precise way and most important, the inference process used by the expert must be modelled and simulated [4].

The most useful tools for approaching those goals and finally obtaining reliable student diagnosis will be described in a few words and the qualities of each technique will be pointed out.

2.1 Artificial Intelligence Methods Used

Artificial Neural Networks (ANN) are computational models, inspired by the way biological neurons process information. These decision – making tools are weighted interconnected networks of artificial neurons and have the ability of "learning" through experience via algorithms. Currently, neural networks become more and more popular in various scientific fields including the educational one. Just a few of their significant abilities are the fact that they recognize patterns, they derive meaning from vague data and not only they learn in an adaptive way, but they also identify matching in similar cases Stath [21]. In education the qualities of the neural networks are used in order to simulate and monitor a learner's cognitive progress [2], [9] and to obtain classification of students sharing common characteristics.

Fuzzy logic is a multi-valued version of Boolean logic based on fuzzy set theory. It was introduced (formally by Zadeh, who axiomatized fuzzy set theory in 1965 [10]) in order to handle uncertainty in everyday problems caused by imprecise and incomplete data, as well as human subjectivity. A *fuzzy set* is defined as an ordered set $(x,\ u_A(x))$, όπου $x \in X$ και $u_A(x) \in [0,1]$, equipped with a *membership function*

$$\mu_A(x): X \to [0,1], \quad \text{where} \quad \mu_A(x) = \begin{cases} 1, & x \quad absolutely\ in\ A \\ 0, & x \quad absolutely\ not\ in\ A\ . \\ (0,1), & x \quad partially\ in\ A \end{cases}$$

Value $u_A(x)$ is called *degree of membership* or *membership value*. Consequently, one could say that Boolean logic is nothing but a generalization of fuzzy logic since both versions coincide when membership function is allowed to have only two possible values (0/1). Fuzzy logic is gaining more and more popularity (even in business world since already by middle nineties more than 6.1 billion dollars were gained in a year just in Japanese fuzzy logic industry) and according to Stathakopoulou et al. (2005) the application of fuzzy logic principles in student modeling is more than attractive since it overcomes computational complexity issues and thanks to its human-like nature is user and designer friendly [2]. In education, researchers try to take advantage of the way that fuzzy logic model human decision – making aiming to achieve diagnosis of a student's knowledge level [10], [11]. Hawkes et al. were among the first to adopt fuzzy student modeling [32].

Genetic programming is a search algorithm used to identify the most efficient candidates for a specific task via the creation of a computer program. It is based on biological evolution principles, namely the Darwinian survival of the fittest. The steps that a genetic algorithm follows are presented below:

- assess the initial population according to specified criteria,
- if a possible solution satisfies the criteria, stop, else,
- select the most satisfying candidate and reproduce (via crossover and mutation),
- return to the first step

In education, genetic algorithms are a valuable tool since they can be used to derive a reliable expert knowledge representation.

Last but not least, many hybrid approaches are among the most popular choices of educational researchers since through research conducted so far, they have proved to deal better with complexity and vagueness issues. Among the most popular synergies are neuro – fuzzy systems and GPNN methodology. In the first case many problems coming up during separate use of the two technologies are being overcome since their combination offers a fuzzy inference system which uses a neural network learning process [24]. The latter approach, takes advantage of the fact that genetic algorithms can provide optimal network architectures [33]. In other words, genetic programming is applied on an initial population of neural networks in order to obtain an ideal one via reproduction, crossover and mutation.

2.2 Literature Sources and Filtering Formulas

During the last decades and especially in the last one, research of artificial intelligence in education has expanded in so many levels that it would be impossible for someone to study in detail all kinds of research achievements. Consequently, we chose to analyze a subset of all the papers examined, the one that covers issues useful in obtaining accurate and reliable student modeling. The papers we studied were included in research platforms such as ISI Web of Knowledge, database bibliography sites such as Digital Bibliography & Library Project (DBLP) and Scopus as well as web search engines such as Google Scholar. The research was conducted combining the following keywords: student model, student modeling, student diagnosis, intelligent learning systems, intelligent tutoring systems, fuzzy logic, neural networks, genetic programming, neuro-fuzzy systems, learning styles, adaptive learning and assessment. So, after completing a detailed literature review and a filtering with respect to the publication year, the artificial intelligence methods used and the main relevant conferences and journals, we came up with the final selection of papers. In the end, we came to the conclusion that the most meaningful way to organize the research in the particular field is to categorize the studies conducted according to the purpose they attempted to serve. Equations should be punctuated in the same way as ordinary text but with a small space before the end punctuation mark.

3 State of the Art

First of all we need to observe that one common goal that the majority of papers under study shared was classification of the students, namely the attempt for every student to be mapped in one of several predefined groups. The classification purpose though was different from study to study. There were researchers that chose to achieve this categorization giving emphasis to the individual characteristics (such as learning style) of a student and so the enhancement of individualized learning was the actual goal. There where also researchers that gave emphasis to coming up with a reliable assessment method of students cognitive state some of them seeking to the traditional students' assessment and ordering and others to predict students' future performance or even to fix in time serious learning deficiencies.

3.1 Studies Classifying Students According to Their Knowledge Level / Modeling Their Future Performance

Ma and Zhou (2000) implemented a fuzzy set approach in order to assess the outcomes of learning process. In this paper fuzzy set principles were applied to the determination of the assessment criteria and the corresponding weights and finally students' performance was evaluated on a fuzzy grading scale according to the selected criteria. The application of this method as well as the use of a group decision support system (GDSS [34]) was attempted in the Department of Information Systems at the City University of Hong Kong [35], [36]. With the help of a control group, experimental results have proved that this method empowers a deeper approach of learning [28].

Olds et al. (2000) have implemented a software package called Cogito aiming at accurate and inexpensive measurement of the students' ability to think critically acquired so far. Cogito makes use of a neural network in order to obtain pattern recognition in learners' data and finally matching with models of intellectual development that have come up via traditional interviews. Cogito keeps being developed, retraining the neural network with new data extracted from interviews on 60 additional subjects, updating its interface and forming extra partnerships with several institutions [8].

Weon & Kim (2001) introduced a learning evaluation system aiming at providing more flexible results comparing to numerical results which come up applying traditional assessment methods. This approach makes use of fuzzy linguistic variables representing each question and then evaluates a score taking into account the membership degree of uncertainty factors. This method has been tested on 4th grade students of an elementary school with promising results [12].

In 2004, Hammadi and Milne designed a student classification method using neuro-fuzzy techniques in order to obtain learners' performance prediction before admission to college. The learner's secondary school marks and his college entry test performance are the inputs of the neuro-fuzzy system. The system outputs the student's performance in the first semester of his college studies. This system, NEFCLASS, was proposed for pattern recognition and fuzzy data analysis. The system has been tested using actual student exam results and participants were classified into three categories according to the expected performance with encouraging results in most cases [14].

In the same year Vrettaros et al., developed a diagnostic system of taxonomies based on fuzzy logic which is mostly ideal for e – learning systems [41], [42]. The procedure described by the authors consists of an initial processing of the filled up questionnaires, students' categorization into level in five separate theme sections and students' categorization according to the given answers into one of the predefined knowledge levels. This system has been tested on 100 high school and senior high school students in mathematics and results were in significant accordance with cognitive science expert's results [9].

Nykanen (2006) introduced a fuzzy system aiming at predicting students' performance and finally preventing students from failing, dealing in time with serious learning deficiencies and giving extra assistance if needed. Experiments have been conducted using two data sets which describe students' performance in a university mathematics course in 2003 and 2004 at the Tampere University of Technology, Finland [20].

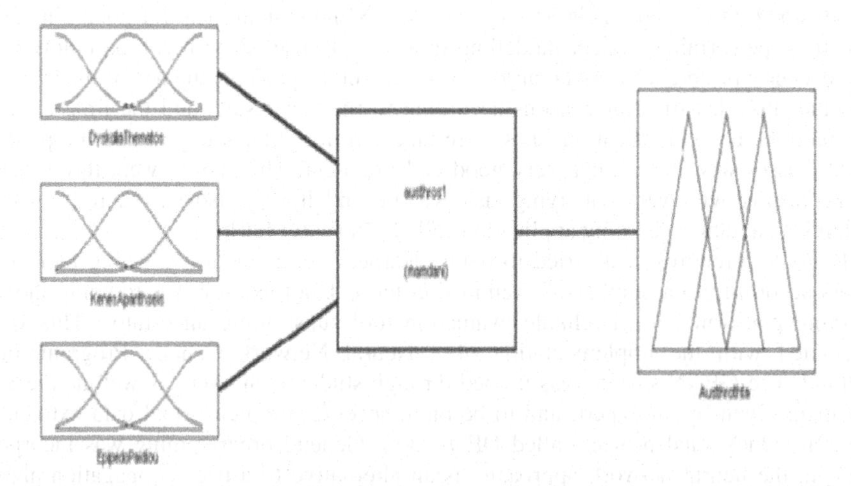

Fig. 1. Inputs and output of the system [9]

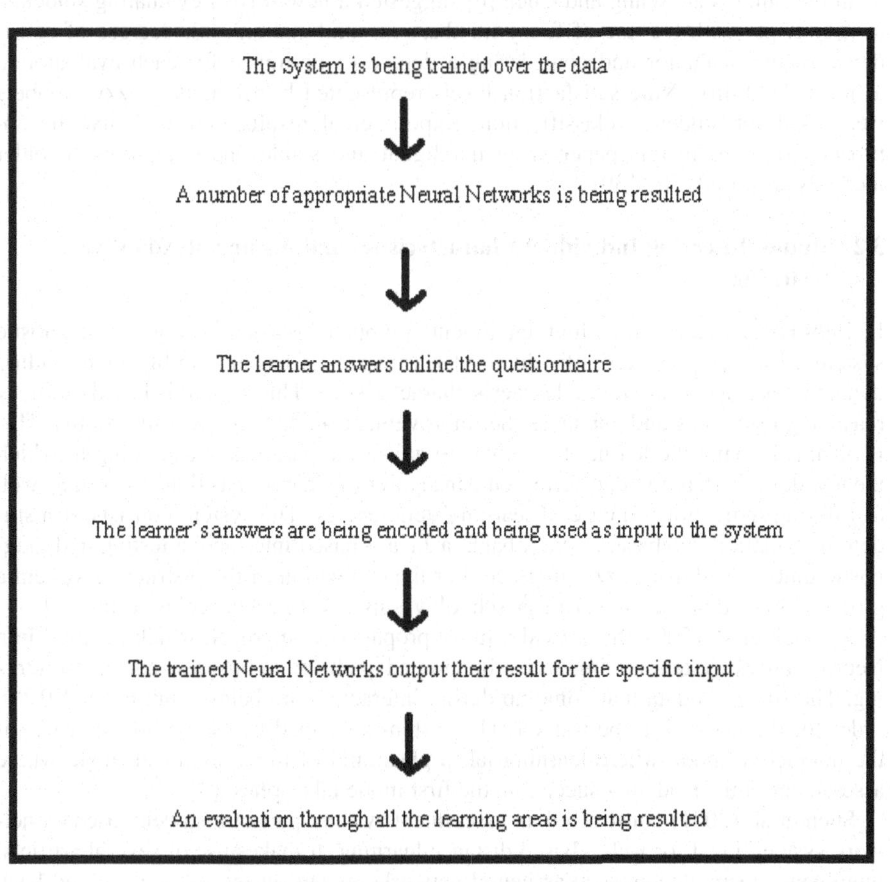

Fig. 2. The steps of the assessment system [1]

Sevarac (2006) used a neuro-fuzzy reasoner (NFR) system ideal for e-learning applications performing student modelling in a flexible way since it can be tailored to the teacher's preferences. As an application, the author presents the use of the system for students' classification reasons according to the test results and the time of test completion. The classification rules were taken from a human expert and the predefined classes were bad, good, very good and excellent. The system went through an initial testing with very satisfying success rate and it was also successfully tested through an actual e- learning application called Multitutor [30].

In 2008, Vrettaros et al. tried to obtain learners' assessment evaluating their answers according to a number of examined criteria. The final aim was to equip the e-learning platform with a reliable evaluation tool substituting an e-tutor. This was attempted with the implementation of a Neural Network Genetic Programming method. The GPNN system was trained through students' answers as well as the assessment given by an expert, and to be more specific, through actual data extracted from an educational project called DEDALOS. Genetic programming was incorporated in the neural network approach (as an alternative to a back propagation algorithm) for optimization reasons such as quick convergence to the solution [1].

In the same year, Wang and Chen [6] suggested a new way of evaluating students' answerscripts with the use of fuzzy numbers in combination with degrees of confidence of the evaluator and considering a degree of optimism for each evaluator to achieve flexibility. Nine satisfaction levels represented by triangular fuzzy numbers were used for students' classification. Experimental results indicated that the approach proposed in this paper is an intelligent and stable one comparing to older methods proposed [38], [39].

3.2 Studies Inferring Individual Characteristics and Aiming at Adaptive Learning

In 1999 Hwang came up with an intelligent computer- assisted testing and diagnostic system which empowered individualized learning as it had the ability of providing dynamic tests according to the learner's characteristics. This system is based on fuzzy reasoning principles and enhances the improvement of students' learning status. This is obtained giving the learner the choice to adjust the parameters according to which the test items differentiate, offering guidance after the learner has done the test as well as offering indications in case of learning deficiencies. The system consists of a student information database, an item bank, a Java – based interface, a testing and diagnostic unit as well as a fuzzy interface. For the evaluation of the instructing system a group of 30 students from a primary school was used with encouraging results [7].

Kinshuk et al. (2001) have used a fuzzy propagation approach which is based in a Neuro – Fuzzy system aiming at maximizing adaptability in business education tutoring. The student's data that come up during interaction are being input to the FBP in order for the network to be trained. The system described by the authors consists of the interactive mode where learning takes place and of the assignment mode where assessment of the student's success in the first mode takes place [3].

Shen et al. (2001) constructed an assessment system, part of a student oriented network system, ideal for web – based distance learning. It makes use of several artificial intelligence technologies such as neural networks, fuzzy set theory, genetic and back propagation algorithms in order to succeed not only in assessing and classifying students but also in constructing types of student profiles so as to empower the adaptability of the learning content to each individual's needs. Via those technologies the results are impressive even in cases where data is extremely vague and subjective [16].

Grigoriadou et al. (2002) have attempted to achieve precise student diagnosis incorporating fuzzy logic and multicriteria decision-making in INSPIRE (Intelligent System for Personalized Instruction in a Remote Environment), a web – based AEHS (Adaptive Educational Hypermedia System - Brusilovsky, [43]). Via this combination, enhancement of adaptability of INSPIRE is being achieved. The authors focus on student's diagnostic module which aims at student's classification according to his knowledge level and learning style. The method described has been tested over 20 postgraduate students of the Department of Informatics and Telecommunications of the University of Athens and proved to offer students' assessment as much reliable as the one performed by a human teacher [4].

In 2003, De Arriaga et al. created an ILS (Intelligent Learning System) based on multi – agent architectures. Via this system, using a fuzzy set for each individual and for each one of the basic learning styles, the determination of the learning style is obtained. Secondly, the student's knowledge level as well as the learning ability is being assessed and finally, the overall student grade comes up. This technique has already been used in TUTOR3 and according to the results, the assessment performed is reliable and the individualization of the learning process is enhanced [5].

Stathakopoulou et al. (2004) attempted to create and update a student model identifying students' individual characteristics. For this purpose, a neural network - based fuzzy model was implemented where the fuzzy component aimed at simulating human decision making and the neural networks were incorporated to provide it learning and generalization abilities. Neural networks were trained through actual students' profiles. The diagnosis that came up with the help of the model was compared to the one based on the judgement of a group of five teachers and the model proved to handle uncertainty in a smooth way even if we are dealing with marginal cases [2].

Mir Sadique & Ghatol (2004), attempted to use the architecture of the Adaptive Neuro Fuzzy Inference System (ANFIS) in the field of Intelligent Tutoring Systems aiming at reliable and precise student modelling. According to the authors' description, learners are tested in concept understanding, memorizing skills and possible misconceptions and results of these tests are given as input in the inference system. Finally, learners' performance is categorized as poor, fair, good or excellent. This method has proved to give results matching to those that come up from human teachers' evaluation [29].

According to De Arriaga et al. (2005), FINANCE, a system for learning Financial Accounting and Business Analysis based on a fuzzy intelligent research system called NEOCAMPUS2, offers a full and reliable student model. According to experimental results with the necessary use of control groups, FINANCE turned out to be a fruitful way of learners' assessment especially comparing to the traditional evaluation methods. Most important of all, this method achieved significant acceleration of the novices' turning into experts (more than 25% time needed for the control groups) [18].

Homsi et al. (2008) introduced an Adaptive and Intelligent Web – Based Educational System (AIWBES) which makes use of an algorithm based on Fuzzy – ART2 neural network and Hidden Markov Model (HMM), which is a stochastic method.

Table 1. Papers' Categorization

title	year	individual characteristics /adaptive learning	students' classification based on their performance	prediction of students' performance
Development of an intelligent testing and diagnostic system on computer networks	1999	yes	-	-
Fuzzy set approach to the assessment of student-centered learning	2000	-	yes	-
Measuring the intellectual development of students using intelligent assessment software	2000	-	yes	-
Learning achievement evaluation strategy using fuzzy membership function	2001	-	yes	-
The intelligent assessment system in web_based distance learning education	2001	yes	yes	-
Adaptive tutoring in business education using fuzzy backpropagation approach	2001	yes	-	-
Fuzzy Inference for student diagnosis in Adaptive Educational Hypermedia	2002	-	yes	-
Fuzzy logic applications to students' evaluation in intelligent learning systems.	2003	yes	yes	yes
A neuro – fuzzy classification approach to the assessment of student performance	2004	-	-	yes
A neuro – fuzzy classification approach to the assessment of student performance.	2004	-	yes	yes
Development of a diagnostic system of taxonomies using fuzzy logic – case SOLO (useful for e-learning system)	2004	-	yes	-

Table 1. *(continued)*

A Neuro Fuzzy Inference System For Student Modeling In Web-Based Intelligent Tutoring Systems	2004	yes	yes	-
Neuro –Fuzzy knowledge processing in intelligent learning environments for improved student diagnosis	2005	yes	-	-
Evaluation of fuzzy Intelligent Learning Systems	2005	yes	yes	yes
Neuro Fuzzy Reasoner for Student Modeling	2006	-	yes	-
Inducing Fuzzy models for student classification	2006	-	yes	yes
The development of a self-assessment system for the learners answers with the use of GPNN	2008	-	yes	-
Student modeling using NN – HMM for EFL course	2008	yes	yes	-
Evaluating Students' Answerscripts Using Fuzzy Numbers Associated With Degrees of Confidence	2008	-	yes	-

Reliable student modelling is attempted through two stages, the initializing where personal and educational information is gathered and the updating one which updates the knowledge status. The goal of this system is assessing learners taking into consideration five parameters and determining their knowledge level in order to categorize them in six levels. This study attempted a hybrid approach combining a machine learning algorithm with a soft computing synergy and according to experimental results, the combination which was used achieved a precise learners' categorization [15].

The online version of the volume will be available in LNCS Online. Members of institutes subscribing to the Lecture Notes in Computer Science series have access to all the pdfs of all the online publications. Non-subscribers can only read as far as the abstracts. If they try to go beyond this point, they are automatically asked, whether they would like to order the pdf, and are given instructions as to how to do so.

Please note that, if your email address is given in your paper, it will also be included in the metadata of the online version.

The correct BibTeX entries for the Lecture Notes in Computer Science volumes can be found at the following Website shortly after the publication of the book: http://www.informatik.uni-trier.de/~ley/db/journals/lncs.html

4 Conclusions

In order to permit cross referencing within LNCS-Online, and eventually between different publishers and their online databases, LNCS will, from now on, be standardizing the format of the references. This new feature will increase the visibility of publications and facilitate academic research considerably. Please base your references on the examples below. References that don't adhere to this style will be reformatted by Springer. You should therefore check your references thoroughly when you receive the final pdf of your paper. The reference section must be complete. You may not omit references. Instructions as to where to find a fuller version of the references are not permissible.

We only accept references written using the latin alphabet. If the title of the book you are referring to is in Russian or Chinese, then please write (in Russian) or (in Chinese) at the end of the transcript or translation of the title.

The following section shows a sample reference list with entries for journal articles [1], an LNCS chapter [2], a book [3], proceedings without editors [4] and [5], as well as a URL [6]. Please note that proceedings published in LNCS are not cited with their full titles, but with their acronyms!

Acknowledgments. The heading should be treated as a 3rd level heading and should not be assigned a number.

References

1. Vrettaros, J., Pavlopoulos, J., Vouros, G.A., Drigas, A.: The Development of a Self-assessment System for the Learners Answers with the Use of GPNN. In: Lytras, M.D., Carroll, J.M., Damiani, E., Tennyson, R.D. (eds.) WSKS 2008. LNCS (LNAI), vol. 5288, pp. 332–340. Springer, Heidelberg (2008)
2. Stathakopoulou, R., Magoulas, G., Grigoriadou, M., Samarakou, M.: Neuro –Fuzzy knowledge processing in intelligent learning environments for improved student diagnosis. Information Sciences 170, 273–307 (2005)
3. Kinshuk, A., Nikov, A., Patel, A.: Adaptive Tutoring in Business Education Using Fuzzy Backpropagation Approach. In: Proceedings of the Ninth International Conference on Human-Computer Interaction, pp. 465–468 (2001)
4. Grigoriadou, M., Kornilakis, H., Papanikolaou, K., Magoulas, G.: Fuzzy Inference for Student Diagnosis in Adaptive Educational Systems. In: Vlahavas, I.P., Spyropoulos, C.D. (eds.) SETN 2002. LNCS (LNAI), vol. 2308, pp. 191–202. Springer, Heidelberg (2002)
5. de Arriaga, F., Arriaga, A., El Alami, M., Laureano-Cruces, A., Ramírez-Rodríguez, J.: Fuzzy Logic Applications To Students' Evaluation in Intelligent Learning Systems. In: En Memorias XVI Congreso Nacional y II Congreso Internacional de Informática y Computación de la ANIEI, Zacatecas, 22-24 de octubre del, vol. I, pp. 161–166 (2003)
6. Hui-Yu, W., Shyi-Ming, C.: Evaluating Students' Answerscripts Using Fuzzy Numbers Associated With Degrees of Confidence. IEEE Transactions on Fuzzy Systems 16(2) (April 2008)
7. Hwang, G.J.: Development of an intelligent testing and diagnostic system on computer networks. In: Proceedings of the National Science Council of ROC, vol. 9(1), pp. 1–9 (1999)

8. Olds, B., Miller, R., Pavelich, M.: Measuring the Intellectual Development of Engineering Students Using Intelligent Assessment Software. In: Proceedings of the International Conference on Engineering Education, Taipei, Taiwan, August 14-18 (2000)
9. Vrettaros, J., Vouros, G., Drigas, A.: Development of a Diagnostic System of Taxonomies Using Fuzzy Logic - Case SOLO (useful for e-learning system). In: Proceedings of 5th WSEAS International Conference on Automation & Information (ICAI 2004), Venice, Italy, November 15-17 (2004); Selected and is included also in WSEAS Transactions on Information Science and Applications 1(6) (December 2004)
10. Zadeh, L.A.: Fuzzy Sets. Information and Control, 338–353 (1965)
11. Drigas, A.S., Vrettaros, J.: An Intelligent Tool for Building E-Learning Contend-Material Using Natural Language in Digital Libraries. In: Proceedings of WSEAS Int. Conf. on E-AVTIVITIES (E-AVTIVITIES 2004), Rethymno, Crete, Greece, October 24-26 (2004); Selected and is included also in WSEAS Transactions on Information Science and Applications 5(1) (November 2004)
12. Weon, S., Kim, J.: Learning achievement evaluation strategy using fuzzy membership function. In: Proceedings of the 31st ASEE/IEEE frontiers in education conference, Reno, NV, vol. 1, pp. 19–24 (2001)
13. McAlister, M., Wermter, S.: Rule Generation from Neural Networks for Student Assessment. In: Proceedings of the International Joint Conference on Neural Networks, Washington, USA (July 1999)
14. Al-Hammadi, A.S., Milne, R.H.: A Neuro-Fuzzy Classification Approach To the Assessment of Student Performance. In: IEEE International Conference on Fuzzy Systems, July 2004, vol. 2, pp. 837–841 (2004)
15. Homsi, M., Lutfi, R., Rosa, M.C., Barakat, G.: Student modeling using NN – HMM for EFL course. In: 3rd International Conference on Information and Communication Technologies: From Theory to Applications, ICTTA 2008, April 7-11, pp. 1–6 (2008)
16. Shen, R., Tang, Y., Zhang, T.: The Intelligent Assessment System in Web_based Distance Learning Education. In: 31st ASEE/IEEE Frontiers in Education Conference (2001)
17. Othman, M., Ku-Mahamud, K.R., Abu Bakar, A.: Fuzzy evaluation method using fuzzy rule approach in multicriteria analysis. Yugoslav Journal of Operations Research 18(1), 95–107 (2008)
18. de Arriaga, F., El Alami, M., Arriaga, A.: Evaluation of Fuzzy Intelligent Learning Systems. In: Méndez-Vilas, A., et al. (eds.) Recent Research Develoments in Learning Technologies, Formatex (2005)
19. Kasabov, N.K., Kim, J.S., Gray, A.R., Watts, M.J.: FuNN – A Fuzzy Neural Network Architecture for Adaptive Learning and Knowledge Acquisition, Technical Report, Department of Information Science, University of Otago, Dunedin, New Zealand (1997)
20. Nykänen, O.: Inducing Fuzzy Models for Student Classification. Educational Technology & Society 9(2), 223–234 (2006)
21. Frias-Martinez, E., Magoulas, G.D., Chen, S.Y., Macredie, R.D.: Modeling Human Behavior in User-Adaptive Systems: Recent Advances Using Soft Computing Technique. Expert Systems with Applications 29(2) (2005)
22. Lane, H.C.: Intelligent Tutoring Systems: Prospects for Guided Practice and Efficient Learning. Whitepaper for the Army's Science of Learning Workshop, Hampton, VA, August 1-3 (2006)
23. Brusilovsky, P., Peylo, C.: Adaptive and Intelligent Web – Based Educational Systems. International Journal of Artificial Intelligence in Education 13, 156–169 (2003)

24. Al Hamadi, A.S., Milne, R.H.: A neuro – fuzzy approach for student performance modeling. In: Proceedings of the 2003 10th IEEE International Conference on Electronics, Circuits and Systems, ICECS 2003, vol. 3, pp. 1078–1081 (2003)
25. Baylor, A.L.: Intelligent agents as cognitive tools for education. Educational Technology 39(2), 36–40 (1999)
26. Alimisis, D., Moro, M., Arlegui, J., Pina, A., Frangou, S., Papanikolaou, K.: Robotics & Constructivism in Education: the TERECoP project (2007)
27. Brusilovsky, P., Eklund, J.: A Study of User Model Based Link Annotation in Educational Hypermedia. Journal of Universal Computer Science 4(4), 429–448 (1998)
28. Ma, J., Zhou, D.: Fuzzy set approach to the assessment of student-centered learning. IEEE Trans. Educ. 43, 237–241 (2000)
29. Ali, M.S., Ghatol, A.A.: An Adaptive Neuro Fuzzy Inference System For Student Modeling. In: Web-Based Intelligent Tutoring Systems (2004)
30. Sevarac, Z.: Neuro Fuzzy Reasoner for Student Modeling. In: Proceedings of the Sixth International Conference on Advanced Learning Technologies, pp. 740–744 (2006)
31. Salleh, N.S.M., Jais, J., Mazalan, L., Ismail, R., Yussof, S., Ahmad, A., Anuar, A., Mohamad, D.: Sign Language to Voice Recognition: Hand Detection Techniques for Vision-Based Approach. In: Uniten Student Conference on Research and Development (SCOReD), May 14-15 (2007)
32. Hawkes, L.W., Derry, S.J., Rundensteiner, E.A.: Individualized tutoring using an intelligent fuzzy temporal relational database. International Journal of Man–Machines Studies 33, 409–429 (1990)
33. Siddique, M.N.H., Tokhi, M.: Training Neural Networks: Backpropagation vs. Genetic Algorithms. In: Proceedings of the IEEE International Joint Conference on Neural Networks, vol. 4, pp. 2673–2678 (2001)
34. Ma, J.: Group decision support system for assessment of problem-based learning. IEEE Trans. Educ. 39, 388–393 (1996)
35. Zhou, D., Ma, J., Kwok, R.C.W., Tian, Q.: Group decision support system for project assessment based on fuzzy set theory. Presented at the Proc. 32nd Hawaii Int. Conf. System Sciences (HICSS-32), Honolulu, HI (January 1999)
36. Kwok, R., Ma, J.: Use of group support system for collaborative assessment. Comput. Educ. Int. J. 32, 109–125 (1999)
37. VanLhen, K.: Student Modeling. In: Polson, M.C., Richardson, J.J., Lea (eds.) Foundations of Intelligent Tutoring Systems, Hove & London (1988)
38. Biswas, R.: An application of fuzzy sets in students' evaluation. Fuzzy Sets Syst. 74(2), 187–194 (1995)
39. Chen, S.M., Lee, C.H.: New methods for students' evaluating using fuzzy sets. Fuzzy Sets Syst. 104(2), 209–218 (1999)
40. Liao, S.-H.: Expert system methodologies and applications - a decade review from 1995 to 2004. Expert Syst. Appl. 28(1), 93–103 (2005)
41. Drigas, A., Vrettaros, J., Kouremenos, D.: Tele education and e-learning services for teaching English as a second language to Deaf People, whose first language is the Sign Language. WSEAS Transactions on Information Science and Applications 1(3) (September 2004)
42. Drigas, A.S., Vrettaros, J., Stavrou, L., Kouremenos, D.: Elearning Environment for Deaf people in the E-Commerce and New Technologies Sector. In: 6th WSEAS International Conference on EACTIVITIES, Rethymno (October 20, 2004)
43. Brusilovsky, P.: Methods and Techniques of Adaptive Hypermedia. User Modeling and User-Adapted Interaction, vol. 6, pp. 87–129. Kluwer Academic Publ., Netherlands (1996)

Digital Dividend Aware Business Models for the Creative Industries: Challenges and Opportunities in EU Markets

Vassiliki Cossiavelou

University of the Aegean, SOC/CI, Greece
Communications Secretary A' at State
v.cossiavelou@ct.aegean.gr

Abstract. EU counties have a historically unique opportunity to enable their creative industries to promote the knowledge societies, applying new business models to their media content and networks markets, that are digital dividend (DD) aware. This new extra-media gatekeeping factor could shape new alliances and co operations among the member states and the global media markets, as well.

Keywords: Knowledge Society, e/m-governance, e/m-learning, e/m-health, e/m-education, e/m- communications, e/m-entertainment. digital dividend, radio spectrum, creative industries, infotainment. FTTH, CEPT, extra-media gatek-eeping factor.

1 Introduction

The Knowledge Society is starving for information and IMS to gather, to edit, to disseminate, to store and to retrieve it. Furthermore, to gather knowledge compiling, summarizing, sharing and interactively communicating information.

The sources for these two similar but of increasing complexity procedures are getting rarer and rarer. The knowledge society is starving from sources to implement its targets. E/m-governance, e/m-learning, e/m-health, e/m-education, e/m-communications, e/m-entertainment new technologies awards the society of these targets with a "rare resource". This is the Digital Dividend (DD) in Radio Spectrum.

The location and allocation of digital dividend is a new challenge and opportunity first of all for creative industries, the industries of infotainment. Their business models that will exploit or not this new resource will define the have and have not of the knowledge society, in the new digital era.

1.1 Creative Industries Business before and After Digital Dividend

The National Telecommunications Commissions throughout the EU countries are the Regulatory Authorities which has the role of supervising and regulator in the areas of electronic communications. Curators from industry and academia are examining, in national and European level, the new resource, the "digital dividend" (DD) in the sectors of telecommunications and the mass media. The subject of their research is to

M.D. Lytras et al. (Eds.): WSKS 2009, CCIS 49, pp. 565–569, 2009.

present new opportunities for the economy and communication in the knowledge society through the use of «digital dividend».

As a «digital dividend» indicated the important part of the digital spectrum, which is expected to be released (1:4 approximately) given the decision for the termination of analogue broadcasting transmission by 2012 at European level. This spectrum may be available for public uses, for society needs. As economy is going towards intangible New Economy, which is the Knowledge-based Economy and the societies are asking for universal services the released spectrum is recognized as «the rare natural resource» and of statutory public goods with high added value for the communication and the economy [1]. As indicative uses include mobile TV services (Mobile TV broadcast networks) [3], high-definition television (HDTV), [14], universal broadband access via wireless broadband, [15] extending the coverage of mobile telephony the 3G/4G applications in low-cost, e.g. by using WiMAX, etc., and developed universal social services such as e-health, e-learning and e-government.

Especially, in these circumstances the international economic crisis uses digital dividend associated with the development of broadband networks as a major contribution to national GDP growth of around billion, and create new jobs in the range of tens of thousands. [3], [4], given that in Europe, the ICT industry produces 25% of total development. [5].

1.2 The Situation in the EU

The use of the digital dividend (considered to be the band 790-862 MHz) is among the strategies of the Treaty of Lisbon [7] and approved by resolution of the European Parliament (24/9/2008) which calls on Member States to develop common methodology in the national strategies of dividends by the end of 2009. The transition is scheduled to be completed by 2012 and is expected to be finalized in 2015, both at European and global level. The EU Member States are of the key managers of the new radio spectrum itself but of the new partnerships and allies to be built up at European and global level. Providers should take the risk of investment to achieve them. The majority of providers are public, and in this case, by their institutional role are responsible for the proper functioning of public services, social and regional cohesion, together with ensuring multilingualism and cultural diversity, [8] necessary components of Knowledge Society.

Most of the European Union is moving steadily to total elimination of analogue, terrestrial television and its replacement by digital. The specifics of the institutional frameworks of each country, however, make it difficult to quantify the digital dividend. Also, the limits of the digital dividend that is not commonly available in all countries of Europe [13]. The European Commission is preparing a common, as far as possible, use of radio spectrum to be released by digital switchover, in order to maximize benefits [3], [12] especially given that the borders of national territory does not coincide with telecommunications «boundaries» [11].

For example, the UK intends to follow a market approach, with the DD auction to be placed in 2010, in agreement with CEPT (Comité Européen des Postes et Télécommunications) and with release of intermediate spectrum range packages (shared with DTT) available for programs production and Special events coverage (PMSE-Programme Making and Special Events) [4].

Finally, in economic terms, the role of radio spectrum DD to offer services and develop their respective markets is currently at 2.2% of GDP of the EU [8], while the estimated number of subscriptions by 2012 are about 2 billion €, of which 2 / 3 will derive by mobile broadband.[9].

1.3 The Situation in Greece

Greece is struggling to eliminate the «digital divide» into her territory, and among big cities and periphery. The proper use of the digital dividend is a political objective of significant national priority. [6].

The established Greek analogue television networks can be served by a few digital television multiplexes and to release hundreds of spectrum MHz, resulting to development growth and jobs creation. [3].

The first stage for the digital television adoption is the transition period. During this time the analogue signal is going to coexist to digital signal, while is already completed technical study, for 23 digital signal broadcast centres. [6] The Greek government has already announced that it intends to promote a plan for the massive development of the access network with optical fiber to the home (FTTH-Fiber-To-The-Home). The public contribution will be given to joint ventures in the form of public-private partnerships (PPPs). [6].

1.4 The Institutional Framework for the Digital Dividend in Greece

The precondition of use the digital dividend is the completion of digital transition. The frequencies management, into the frame of digital transition and the facing of institutional and legal issues related to the network provider are issues that should be answered soon, in terms of global market, at a national level. During the exploitation of the digital dividend period the strategic options for the frequencies allocation should be set, and the responsibility of its management, as well. A key process should be the procedure of assignment for parts of DD to the beneficiaries and the legal frame according to witch are entitled to participate in the exploitation of the DD. [10], [11].

The issues raised at the level of international agreements and the fermentation of which Greece is going to be asked as member state of EU to declare her policy on exploitation scenarios of the digital dividend is mainly the followings: a. the interference issues, b. the protection of existing uses of radio spectrum, c. the existing antenna system, d. the preferred frequencies for individual uses, e. the policy about related to DD economies of scale, f. the cost of existing services redeploying, g. the content intellectual property rights and the right to have/use the infrastructure (networks, antennas, etc), i. the clarification of intellectual property management, k. the standards to be reached to obtain the related permits, l. the protection of social value, etc. [3].

The intensity of the claim of DD from the audiovisual and telecommunications providers and strategic investors, [7] shows the need to regulate the operation of a public good, to meet the challenges of a globalized market. The market of media content, [7] information, entertainment, advertising educational media, user-produced content in communications 2.0.

2 Conclusions

EU member states are committed to the end of 2009 to meet the challenge and opportunity of the digital transition and use the digital dividend, which will derive. In the coming years, and by 2012, as a credible contribution to the European organs for the creation of the relevant international agreements and the information of global audience, about the new investment opportunities emerging in EU countries can be a important element promoting contemporary image of EU in the field of competitiveness.

The early establishment of institutional framework that ensures the distribution and management of the digital dividend is evident. EU has a unique opportunity to promote its media content and networks markets as part of the global knowledge-based social policy.

The targets of Knowledge society, e/m-services could be reached using the new "rare resource" of DD in its creative industries of information and entertainment, as well.

The new business models should be seen as gatekeepers of avoiding the social exclusion and facilitate the DD management to be used as the new extra-media gatekeeping factor, in favour of alliances among states and markets to achieve well established knowledge societies.

Acknowledgments. To my advisor Philemon Bantimaroudis, Assistant Professor in the Department of Cultural Informatics at the University of the Aegean, Greece, for the inspiration, for his great help, and the great guidance during the supervision in my Ph.D. research, based on his international academic experience.

To George Carayannis, General Director of Research and Innovation Centre in Information, Communication and Knowledge Technologies "Athena" and Professor NTUA, Greece with my deep respect as this research could not be conducted at the same level, and be noticed in public without the ILSP scholarship that award me and his kind personal help in some critical turning points of this academic work.

References

1. Cossiavelou, V., Bantimaroudis, P.: Mediation of the Message in a Wireless Global Environment: Revisiting the Media Gatekeeping Model. In: WTS, April 22-24, p. 8. IEEE/Springer (2009)
2. Reding, V.: Television is going Mobile - and needs a pan European policy approach. International, CeBIT Summit, Hannover, Germany, March 8 (2006)
3. Abecassis, D.: Understanding the European dimension of the Digital Dividend, Analysis Mason study (due to be completed by September 2009), NTPC, Athens (2009)
4. Conway, M.: Awarding the UK's digital dividend (2009), `http://www.ofcom.org.uk/radiocomms/ddr/events/geoseminar.pdf`
5. Staple, G., Kevin Werbach, K.: The End of Spectrum Scarcity, New technologies and regulatory reform will bring a bandwidth bonanza 41(3), 48–52 (2004)
6. Stefanopoulos, G.: Mobile Communications and its contribution to the overall Economy, Financial Times, Telecommunications, Media & Technology Conference (2009), `http://www.eekt.gr/Files/Presentations/131.3.ppt?PHPSESSID`

7. Dominguez, E.: Pan-European Digital Strategy: Collaboration, Not Competition! NTPC, Athens (2009)
8. McGonagle, T.: The Promotion of Cultural Diversity via New Media Technologies: An Introduction to the Challenges of Operationalisation, European Audiovisual Observatory (2008)
9. Ercole, R.: Technology-Evolution and Applications: How mobile broadband can boost broadband penetration in Greece, NTPC, Athens (2009)
10. BBC's Response to Ofcom's Consultation on the Digital Dividend (2009), http://www.bbc.co.uk/info/policies/pdf/bbcresponse_digital_dividend.pdf
11. Kapsalis, H.: Challenges and Opportunities in the New Digital Age, EETT, Athens (2009)
12. Niepold, R.: The Digital Dividend-an EU perspective, NTPC, Athens (2009)
13. Fournier, E.: Digital Dividend in France, NTPC, Athens (2009)
14. Crawford, D.I.: Spectrum for Mobile Multimedia Services. In: IEEE Tenth International Symposium on ISCE (2006)
15. Ratkaj, D.: Digital Dividend: Challenges and Opportunities in the New Digital Era' Broadcasters' perspective, EETT, Athens (2009)

Integrating Adults' Characteristics and the Requirements for Their Effective Learning in an e-Learning Environment

Maria Pavlis Korres[1], Thanassis Karalis[2], Piera Leftheriotou[3], and Elena García Barriocanal[1]

[1] Computer Science Department
University of Alcalá, Spain
eumarcor@otenet.gr
[2] Department of Educational Science and Early Childhood Education, University of Patras, Greece
[3] School of Humanities, Hellenic University of Greece

Abstract. Learning technology, through e-learning, allows adults to adapt learning to their own time, place and pace. On the other hand, the adults' specific characteristics as learners and the requirements for their effective learning must be integrated in the design and the development of any learning environment addressed to them. Adults in an online environment have also to deal with new barriers related to access to the courses, the sense of isolation and the sense of immediacy with educator and other learners. This paper is dealing with the way through which an online environment can overcome these barriers and can integrate adults' characteristics and requirements for effective learning. The use of the appropriate communication tools by designers, developers and educators seem to provide the answers as these tools promote immediacy and interaction, both considered very important factors in online educational environments and affect the nature and the quality of communication and learning.

Keywords: Adult education, E-learning, Communication tools, Immediacy, Interaction.

1 Introduction

In modern societies adults' education is a constantly growing field as adults are more oriented towards lifelong learning in order to upgrade their knowledge and skills so that they can meet with the increased demands of personal, social and professional life. Today, learning technology, especially through e-learning, is providing adults with the tools which enable them to adapt learning to their own place, time and pace. On the other hand, adults as learners have considerably different characteristics from children, and these characteristics, as well as the requirements for effective adult learning must be taken into consideration and be integrated in any e-learning environment. This paper is dealing with the adults' characteristics as learners, the requirements for effective adults' education, the barriers adults are facing in an e-learning environment, in relation to immediacy, interaction, active participation and collaboration and investigating the most effective on-line environment for adult education.

M.D. Lytras et al. (Eds.): WSKS 2009, CCIS 49, pp. 570–584, 2009.

2 Adult Education

It was in 1968 when in the United States Malcom Knowles[1] used the term androgogy[2] in an article in Adult Leadership and he has become known as the principle expert of andragogy which is a set of assumptions about how adults learn.

Malcolm Knowles [1] who is considered the father of Andragogy worked in terms of identifying the characteristics of adult learners as opposed to children as learners. His five assumptions are listed below:

1. Self-concept: As a person matures his self concept moves from one of being adependent personality toward one of being a self-directed human being
2. Experience: As a person matures he accumulates a growing reservoir of experience that becomes an increasing resource for learning.
3. Readiness to learn. As a person matures his readiness to learn becomes oriented increasingly to the developmental tasks of his social roles.
4. Orientation to learning. As a person matures his time perspective changes fromone of postponed application of knowledge to immediacy of application, and accordingly his orientation toward learning shifts from one of subject centeredness to one of problem centeredness.
5. Motivation to learn: As a person matures the motivation to learn is internal

2.1 Characteristics of Adults as Learners

Studying the literature on adults' learning brings up the fact that adults as learners have specific characteristics that set them apart from children.

Although the characteristics pointed out by authors and researchers are various, there seems to be a general consensus in the literature on some common characteristics that have an impact on adults' learning efficacy and the overall classroom experience [2], [1], [3], [4], [5], [6], [7], [8], [9], [10], [11].

Table 1. Characteristics of adults as learners

1.	Adults participate in the learning process with concrete intents, goals and expectations
2.	Adults already have certain knowledge and experience as well as established perspectives
3.	Adults have already developed personal styles of learning
4.	Adults prefer self-directed learning and active involvement in the educational endeavor
5.	Adults have to deal with certain barriers on their learning process

1. They participate in the learning process with concrete intents, goals and expectations.

Adult learners are coming in the educational process with concrete and immediate goals (e.g. professional, social, personal development). Learners have specific

[1] Knowles,M. S. (1968). Androgogy, not pedagogy, Adult Leadership, 16, 350-386.

[2] Andragogy ("andr"-meaning "man") could be contrasted with paedagogy (from greek paedagogos, from "pais"-meaning child+ "agogos" meaning leader).

expectations from the learning process and when this process meets their expectations then their motivation for learning is empowered, their positive attitudes are enhanced and their negative attitudes are transformed to positive ones, contributing to the achievement of the educational goals.

2. They already possess certain knowledge and experience as well as established perspectives.

Adults enter a learning situation having a specific spectrum of prior knowledge and a variety of life experiences – different for each individual. Adults would prefer that these knowledge and experience are both considered and exploited during their current educational process. Learning is facilitated when the instruction is related to these experiences. The rejection of learners' experience is often taken in as a personal rejection which leads to negative reactions and attitudes in the context of the educational process.

3. They have already developed personal styles of learning

Each adult has already developed his own learning 'model'. In order to have an effective learning process, the learning style and personal pace of each learner has to be taken into consideration, leading to the adoption of the appropriate learning methods and techniques.

4. Adults prefer self-directed learning and active involvement in the educational endeavor

Adult desire and strive for self-directedness, emancipation and active participation in every situation in life in which they are involved. This fact affects their attitude towards active participation in the educational level. Usually, adults prefer to be self-directed learners.

5. They have to deal with certain barriers on their learning process

The educational process of adults may face barriers which could render the whole procedure ineffective or cause its termination if not dealt with appropriately. There are three main categories of barriers which adults face as learners [5], [12].

- □ Barriers related to the organization of educational programs (goals, coordination, infrastructure etc.)
- □ Barriers arising from the situation in which the learners are i.e. physical or contextual factors, concerns or troubles related to personal problems, bad relation between teacher and learners, or between learners, lack of communication inside the learning group.
- □ Internal barriers which the adults may possess towards themselves.

By studying the above characteristics we realize that in the context of adult education the learning process is influenced by many interrelated factors with unpredictable results. All the characteristics of adults could have controversial effects in the learning process, operating either as catalysts for effective learning or as hindering factors [5], [13], while variety is the main characteristic of adult learners and adult learning processes [6].

2.2 Requirements for Effective Adult Learning

The main requirements for effective adult learning, as these are defined by taking into consideration the above mentioned characteristics and in the relative literature [14], [15], [16], [17], [6], [8], [11], are depicted in the following table.

Table 2. Requirements for effective adult learning

a. education is centered on the learners	The education meets the needs, the interests and the expectations of learners
	The ways of learning which the learners prefer are seriously taken into consideration when instruction is organized
	The knowledge and the experiences of the adult learners should be used as much as possible in the educational process
	The barriers which the learners usually deal with are defined and ways to overcome these are sought after
b. the active participation of the learners is both encouraged and intended	The learners are participating actively in the transformation of the learning process (curriculum, choice of educational material and methods as well as the arrangement of many practical issues which rise during the education e.g. time schedule of the meetings, use of audiovisual media etc)
	In the learning process active educational practices are used. Through these practices the development of a critical way of thinking is promoted as well as the "learning to learn" strategy in order for the learners to be able to continue their learning progress after the end of educational process.
c. the creation of a learning environment based on communication, cooperation and mutual respect	Bidirectional relations between educators and learners are cultivated, governed by sincerity, respect and acceptance.
	Relations of collaboration, mutual respect and trust are cultivated between learners, functioning as a group.

3 The e-Learning Environment for Adults

Taking into consideration the factors which affect adults' learning, as they have been analyzed in the previous section of this paper, the online learning environment within which an adult learner could learn effectively has to meet one's needs, expectations and interests, to take advantage of one's existing knowledge and learning styles as well as to promote the multiple intelligences, active learning and participation, to respect each learner's own pace in time, space and momentum and to help overcome the barriers which an adult has to face during the educational procedure (personal, social, vocational).

Besides the adult related barriers adult learners are already facing, in an online environment they have to deal with new barriers related to access to the courses, the sense of isolation and the sense of immediacy with educator and other learners [18], [12]. An effective educational environment has to overcome all these barriers. In the next sections we shall analyze the notion of immediacy and interaction as both are considered very important in online educational environments and affect the nature and the quality of communication and learning. In the same direction we shall elaborate on the preferable communication mode and the importance of communication tools in an online environment.

3.1 Immediacy and Interaction in Online Educational Environments

The two very important intertwined issues are those of interaction and immediacy and their role in online education. Immediacy and Interaction are considered vitally important for online educational environments and affect the nature and the quality of communication [19], [20].

3.1.1 Immediacy in On-Line Learning Environment

Mehrabian [21] defined immediacy as the extent to which selected communicative behaviors enhance physical or psychological closeness in interpersonal communication or in other words lessen the psychological distance between communicators. Immediacy can have verbal and non-verbal forms. Non-verbal immediacy would therefore be understood as a sense of psychological closeness produced by physical communicative behaviors such as facial expression, eye contact, posture, proximity, and touch. Verbal immediacy would thus be a sense of psychological closeness produced by word selection. For example, the use of the word "we" fosters increased relational closeness and is considered more immediate than the comparable statement "you and I" [22].

Merlose and Bergeron [20] mention that immediacy behaviors are believed to enhance instructional effectiveness in online classrooms, although the non verbal cues are absent and the construct is not easy to articulate. The experience of liking and feeling close to instructors can lead to positive effects in online classrooms and there are correlations between immediacy and affective learning [22], [23].

LaRose and Whitten [24] created a model concerning the interaction and instructional immediacy for Web-based courses. Their model incorporated not only teacher and student immediacy, but also computer immediacy, which they proposed as a result of an ethnographic content analysis of three Web courses. Within this social cognitive framework, they concluded: "There are three possible sources of immediacy in the virtual classrooms of the Web that may create feelings of closeness: 1) the interactions between teacher and students (teacher immediacy); 2) interactions between students (student immediacy); and 3) interactions with the computer system that delivers the course (computer immediacy). Collectively, these sources constitute instructional immediacy. In each case, learning is motivated either through social incentives (e.g., approval for good behavior, expressions of interest in the student) or status incentives that recognize or enhance the status of the learner. The immediacy mechanism is enactive if it results from the interaction between a specific individual learner and one of the other agencies present in the classroom. Immediacy is vicarious if it operates through the observation of other learners as they interact "(p. 336).

Butland and Beebe [25] find evidence that instructor immediacy in a synchronous e-learning environment, such as immediate verbal and nonverbal communications, including timely feedback and use of emoting in text (such as using a word or phrase enclosed in angle-brackets to express emotion, e.g., <sigh>, <grin>), promote increased learning. Grooms [19] in her study concerning communication immediacy in an online doctoral level course points out the importance of the nature and medium of response as well as the frequency of response as variables of interaction. Thus, online educators need to manifest immediate behaviors when providing feedback to distant learners. Immediacy concerning time is very crucial in order to overcome time and space as barriers and promote interaction in online learning.

3.1.2 Interaction in Online Learning Environment

Interaction is at the heart of the online learning experience. Moore's transactional distance theory considers interaction [26] a defining characteristic of education and regards it as vitally important in the design of distance education.

Researchers have shown [27], [28], [29]) that interaction is a significant component in promoting learners' positive attitudes towards distance education and when learners perceive a high level of interaction, they will be more satisfied, but when they perceive low interaction, they are dissatisfied and their academic achievement is harmed.

Moore [26] identified three kinds of interactions that support learning: learner-content, learner-instructor, and learner-learner interactions. Learner-content interaction is the process in which students examine, consider, and process the course information presented during the educational experience (learners' interaction with the knowledge, skills and attitudes being studied). According to Moore and Kearsley [30], "Every learner has to construct knowledge through a process of personally accommodating information into previously existing cognitive structures. It is interacting with content that results in these changes in the learner's understanding" (p. 128). Learner-instructor interaction is communication between the instructor – educator and the learner in a course. In the case of online learning, such interaction usually occurs via computer-mediated communication and is not strictly limited to instructional communication that occurs during the educational experience, but may include advising, offline communication, and personal dialogue. Interaction with instructors includes the myriad ways instructors motivate, enhance and maintain the learners' interest, present information, demonstration of skill, or modeling of certain attitudes and values, organize students' application of what is being learned, evaluate, counsel, support and encourage learners.

According to Rovai [31] in Asynchronous Learning Networks (ALN) learner-instructor interaction takes the form of intellectual discussion or stimulating exchanges of ideas. He stresses that facilitating productive interactions is probably the most important responsibility of the online educator.

Finally, learner-learner interaction is communication between two or more learners, alone or in group settings, with or without the real-time presence of an instructor. Such interaction often occurs via asynchronous computer-mediated communication, although it may include other forms of interpersonal and small group communication, online and offline, that occurs during the duration of a course. Learner-learner interaction among members of a class or other group is sometimes extremely valuable resource for learning, and is sometimes even essential.

Moore at the end of the 80's supported that it was needed to organize programs to ensure maximum effectiveness of each type of interaction, and ensure they provide the type of interaction that is most suitable for the various teaching tasks of different subject areas and for learners at different stages of development. In our days with the possibilities which the learning technology offers we can plan for all three kinds of interaction selecting the appropriate media and tools from a rich array.

This threefold interaction construct has been extended and adapted by subsequent researchers in the area of distance and Web-based learning. Other types of interaction has been added to the initial model as "learner-interface interaction" which occurs when learners use technologies to communicate with the content, ideas, and information about course content with the educator and their classmates [32]. According to Hillman et al. [32] learners need to be fully literate with the interfaces which are used in communications technologies in an e-learning course or a program. "The learner must be skilled in using the delivery system in order to interact fully with the content, instructor and other learners" (p.40). Furthermore, Brunham and Walden [33] have defined learner-environment interaction which is "a reciprocal action or mutual influence between a learner and the learner's surroundings that either assists or hinders learning".

Anderson and Garrison [34] added three more types of interaction: teacher-teacher, teacher-content and content-content.

Anderson [34] has developed an equivalency theorem concerning interaction and its educational effectiveness as follows:

> *Deep and meaningful formal learning is supported as long as one of the three forms of interaction (student–teacher; student-student; student-content) is at a high level. The other two may be offered at minimal levels, or even eliminated, without degrading the educational experience.*
>
> *High levels of more than one of these three modes will likely provide a more satisfying educational experience, though these experiences may not be as cost or time effective as less interactive learning sequences.*

This theorem implies that an instructional designer can substitute one type of interaction for one of the others (at the same level) with little loss in educational effectiveness – thus the label of an equivalency theory. (p. 4).

As Anderson [34] very accurately states, "Efforts at enhancing teacher-student interaction through an increase in teacher immediacy (McCrosky and Richmond, 1992), or through use of theatrical or multimedia presentation techniques, can also be expected to increase the quality of student-teacher interaction. Further efforts at enhancing student-student interaction in the classroom through case or problem based learning activities, have long been shown to increase not only student achievement, but also student completion and enjoyment rates (Slavin, 1995). In these types of activities, increased student-student interaction is substituting for student-teacher interaction"(p.6).

3.2 Online Communication Modes

A closer examination of the barriers adults may face in an e-learning environment reveals that some of them are related with e-learning itself, i.e. computer literacy, while others are related with the selected mode of online communication (the two basic modes being synchronous or asynchronous) and the way each mode affects isolation, interaction, immediacy etc. The mode of online communication is also affecting some other important adults' requirements for effective learning such as time management, active participation and the development of the appropriate learning environment.

Both communication modes have their strong and weak points. As in many other areas there is no magic recipe. We have considered the asynchronous learning environment to be better suited to the characteristics and requirements for effective adults' learning, as analyzed in the next section.

3.2.1 Asynchronous Learning Environment in On-Line Learning Environment

The asynchronous online learning environment seems to be preferable over the synchronous online learning environment for adults' education, as it allows learners to follow their own pace, overcoming the constrains of time and to harmonize their personal, vocational and social life with education [35],[36]. Furthermore, asynchronous online education appears as one of the most appealing instructional methods for adults' education as it combines flexible access to teaching material with time to reflect and self-study techniques with collaborative learning, while involving the use of low-cost technology.

The greatest benefit of asynchronous learning environment is its flexibility, as it gives the freedom to learners to access the course and participate at any time and from any location they choose through an Internet connection [37], [38], [39], [40], [41].

The advocates of asynchronous learning support that this kind of a learning environment provides a "high degree of interactivity" between participants who are separated both geographically and temporally [35]. Since learners have an equal opportunity to participate in an asynchronous communication from where and when it suits them, they can express their thoughts without interruption, they have more time to reflect on and respond to class materials and their classmates than in a traditional classroom [42]. In communication which takes place in synchronous mode the learners who have a language barrier or those lacking enough confidence they do not dare to speak up. Even such learners are seen vehemently participating in electronic discussions through asynchronous mode [43].

Many studies have also highlighted that on-going asynchronous interaction -such as forums-are preferable to synchronous computer-mediated communication groups in that they help participants to build a better context in which learning can take place [44].

Besides the argumentation against distance learning as stated by Kochery [45], who found that students learning over a distance often feel alone and separated from not only the teacher, but also from the socialization with other students, the adversaries to asynchronous learning environment support that students experience isolation and social disconnectedness which, correlated with students' difficulties with the course, result in failing grades, noncompletion, or withdrawal [18].

The answer to the above criticism comes from recent technological innovations, which have reduced significantly the barriers in communication and interaction and have allowed new forms of personal and group interaction as well as course delivery [46], [47].

It is broadly accepted that learning technology has changed the teaching and learning process. Multimedia, communication tools and Internet navigations are becoming more widely used in different educational levels, influencing education, motivating students, promoting learning, and changing classroom interaction.

Another advantage of the asynchronous environment is that all materials and all interactions that occur within this environment, such as e-mails, discussions etc are archived, so that learners and educators can go back and review course materials, assignments, presentations as well as correspondence and discussions between participants [48].

In order to benefit from an asynchronous online learning environment, learners have to overcome the barriers inherent to any online learning environment which are related to the learners' access to computers and the Internet. Although personal computers and web access are becoming more and more pervasive every day, this requirement can be a barrier to entry for many learners.

Once the access barrier is overcome, the acquisition of skills needed to participate in the electronic environment by learners and educators in the beginning of each educational program are considered essential as they influence directly all forms of interaction which are taking place in an e-learning environment [32]. The institutions must also provide a computer network infrastructure and the technical support needed to develop and maintain asynchronous learning environments.

3.3 The Role of Communication Tools in an Asynchronous Online Learning Environment in Reducing Isolation and Promoting Immediacy, Active Participation, Interaction and Collaboration Communication Modes

In order to promote immediacy, active participation, interaction and collaboration, as well as fight isolation in an asynchronous online learning environment, communication tools such as email, forums, threaded discussions, conferencing systems, online discussion boards, wikis and blogs, video-conferences etc. become of paramount importance. Course management systems such as Blackboard, WebCT, Moodle, Dokeos and Sakai, have been developed to support online interaction and collaborative learning, providing tools that allow users to organize discussions, post messages and replies, upload or download and access multimedia and working in smaller or larger groups.

Some of the above communication tools are specifically connected to certain barriers or requirements for effective adult learning, therefore their importance is further elaborated.

3.3.1 The Importance of Forum/Electronic Discussions Groups in an Asynchronous Online Environment

As Freire [49] supports learning is itself a reflective process and it is dialogue that is central to this reflection.

Discussions are especially important when we are working with the middle and higher level of the cognitive domain (application, analysis, synthesis and evaluation) as well as with all levels of the affective domain [10].

Brookfield [50] also says that discussion supports both cognitive and affective ends, such as problem solving, concept exploration, and attitude change, as well as the kind of active participatory learning that results in engaged learning within the classroom.

By using asynchronous communications tools, learners actively construct their own learning by engaging themselves and others in reflective explorations of ideas, drawing conclusions based on their explorations and synthesizing those conclusions with previous knowledge.

Therefore, forums/electronic discussions groups seem to have taken the lead among asynchronous communication tools [44], promoting collaborative learning and reflection and improving the quality and quantity of education in online learning environments [48], [51], [52].

Discussions help learners explore different perspectives, recognize their own values and assumptions, develop their ability to defend ideas and learn to respect others' opinions and viewpoints. Discussion topics should be interesting, meaningful and relevant to everyone in the group.

In forums and electronic discussion groups, people work together to form ideas, argue points, and solve problems. All learners have a voice and no one can dominate the conversation. The asynchronous nature of the discussion also makes it impossible for even an instructor to control. Accordingly, many educators note that students perceive online discussion as more equitable and more democratic than traditional classroom discussions [53].Whereas in face-to-face meetings learners must make their statements one after the other synchronously within a limited timeframe, in forums they can take their time and write their messages asynchronously when it suits them, or within a larger timeframe. Since learners can express their thoughts without interruption and in time convenient for them, they have the opportunity to reflect on their classmates' contributions while creating their own, and on their own writing before posting them [42], [54]. It is possible for learners to "rewind" a conversation and thus they have time to carefully consider their own and other learners' responses leading to deeper discussion ([48]. This tends to create a certain mindfulness among learners, encourages deeper level of thinking, discourse and a culture of reflection in an online course [55], [53],[56]). Many researchers suggest that asynchronous threaded discussion boards are a viable instructional method for sustained written interaction that promotes critical thinking [18].

Despite the fact that forums/electronic discussion groups are text-based and so lacking in visual and verbal cues, most participants find them strangely personal [57] and J. Walter has called them "hyperpersonal" [53].

MacNamara and Brown [48] support that discussion forums need to be carefully structured and managed to ensure that they result in the deep level of collaborative reflection that is desired. They propose three factors which should be considered in planning an online discussion: the organization of the forum, the motivation of students to participate and the ability of students to participate effectively.

At this point the role of the educator must be stressed. According to the literature the most appropriate role for the educator using threaded discussions is that of facilitator [18]. The educator's tasks with regard to the facilitation of discussion boards are: a) setting the scene, b) monitoring participation, c) facilitating critical thinking and d) promoting student collaboration [58].

Educators may lead or facilitate discussion by asking for clarification, summarizing major points, and focusing on the issue, or they may participate as a member of the group while learners take on the roles of keeping things on track and summarizing [10].

Summing up, it is evident that the use of forums/electronic discussion groups in the e-education of adults could play an important role promoting reflection, critical thinking, collaborative learning and interaction between learners and educator as well as interaction between learners. The asynchronous mode seems to provide a more equitable and democratic environment and better time management for learners enhancing the role of the educator as facilitator.

4 Conclusions

Adults as learners have specific characteristics which must be observed in the design and development of an e-learning environment. Adults in an online environment have to deal also with new barriers related to access to the courses, the sense of isolation and the sense of immediacy with educator and other learners. An effective educational environment has to overcome all these barriers. Designers, developers and educators using effectively the appropriate communication tools have to promote immediacy and interaction as they are both considered very important in online educational environments and affect the nature and the quality of communication and learning.

More precisely in order to enhance adults' learning the following points must be observed in the design and development of e-learning processes:

- ☐ Asynchronous learning mode seems to suit better to adults' characteristics and requirements for effective learning.
- ☐ Learners must actively participate as much as possible through all the phases of the educational process (needs analysis, design, development, implementation, evaluation)
- ☐ The content of the education must be formed and adapted according to the needs analysis results. That means that the content will meet the expectations, needs and interests of the learners.
- ☐ The appropriate communication tools which learning technology offers in our days must be integrated in the e-learning environment, in order to

promote immediacy and interaction between educator and learner, between the learners, between learner and content and between learner and interface. Forums and electronic discussions groups have a central role in the educational process.

- Activities that promote and support
  - higher-order thinking (analysis, synthesis, evaluation),
  - critical thinking
  - collaborative work
  - self-directed learning and learners' explorations

  must be designed.
- Design for supporting personalized learning must be implemented. The content and the presentation of the educational material must correspond to the different needs and learning profiles of the learners.
- The course must be structured in a way that educational material will be adapted to the pace of learning and the specific needs of the learners.
- The counseling and facilitating dimensions of the educator's role must be strengthened. Less emphasis should be placed on transmitting information, and more emphasis should be placed on developing students' skills.
- The learners must participate in the formative and summative evaluation of the course.
- The learners' experience in using the Web as a learning environment should be considered in structuring the content and the presentation of the learning material.
- In the planning of a course of e-learning instruction and exercises must be included that will provide learners with the appropriate skills needed to participate in the electronic classroom.

References

1. Knowles, M.: Andragogy in Action. Applying modern principles of adult education. Jossey Bass, San Francisco (1984)
2. Brookfield, S.: Understanding and Facilitating Adult Learning. Open University Press, Stony Stratford (1986)
3. Knowles, M.: The modern practice of adult education. Follett, Chicago (1980)
4. Knowles, M.: The Adult Learner. Gulf Publishing Company, Houston (1998)
5. Rogers, A.: Teaching Adults, 3rd edn. Open University Press, Stony Stratford (2007)
6. Rogers, A.: Adult learners: characteristics, need, learning styles. In: Kokkos, A. (ed.) International conference for adults' learning. Metexmio, Athens (2002) (in Greek)
7. Jarvis, P.: Adult and continuing education. Theory and practice. Routledge, London (1995)
8. Cross, K.P.: Adults as Learners. Jossey-Bass, San Francisco (1981)
9. Jackson, L., Caffarella, R.: Experiential learning: A new approach. Jossey-Bass, San Francisco (1994)
10. Cranton, P.: Planning Instruction for Adult Learners, 2nd edn. Wall & Emerson, Inc., Toronto (2000)
11. Leftheriotou, P.: *Διερεύνηση των εκπαιδευτικών αναγκών των εκπαιδευτών ενηλίκων* (Needs assessment of adults' educators) Master Thesis, Hellenic Open University (2005)

12. Karalis, T., Koutsonikos, G.: Issues and Challenges in Organising and Evaluating Web-based Courses for Adults. Themes in education 4(2), 177–188 (2003)
13. Kokkos, A.: Adult Education: tracing the field.Metaixmio, Athens (2005) (in Greek)
14. Noye, D., Piveteau, J.: Guide pratique du formateur. INSEP Editions (1997)
15. Courau, S.: Les outils d' excellence du formateur, 2nd edn. ESF editeur, Paris (1994)
16. Brookfield, S.D.: Developing Critical Thinkers: Challenging Adults to Explore Alternate Ways of Thinking and Acting. Jossey-Bass, San Francisco (1991)
17. Jaques, D.: Learning in groups. Kogan Page (2000)
18. Waltonen-Moore, S., Stuart, D., Newton, E., Oswald, R., Varonis, E.: From Virtual Strangers to a Cohesive Online Learning Community: The Evolution of Online Group Development in a Professional Development Course. J. of Technology and Teacher Education 14(2), 287–311 (2006)
19. Grooms, L.: Computer-Mediated Communication: A vehicle for learning. International Review of Research in Open and Distance Learning 4(2) (2003), http://www.irrodl.org/index.php/irrodl/article/view/148/709 (retrieved in January 12, 2009)
20. Merlose, S., Bergeron, K.: Instructor immediacy strategies to facilitate group work in online graduate study. Australasian Journal of Educational Technology 23(1), 132–148 (2007)
21. Mehrabian, A.: Orientation behaviors and nonverbal attitude communication. Journal of Communication 17, 324–332 (1967)
22. Woods, R., Baker, J.: Interaction and immediacy in online learning. The International Review of Research in Open and Distance Learning 5(2) (2004), http://www.irrodl.org/index.php/irrodl/article/view/186/268 (retrieved in February 20, 2009)
23. Russo, T., Benson, S.: Learning with invisible others: Perceptions of online presence and their relationship to cognitive and affective learning. Educational Technology & Society 8(1), 54–62 (2005), http://www.ifets.info/journals/8_1/8.pdf
24. LaRose, R., Whitten, P.: Re-thinking Instructional Immediacy for Web Courses: A social cognitive exploration. Communication Education 49, 320–338 (2000)
25. Butland, M.J., Beebe, S.A.: A Study of the Application of Implicit Communication Theory to Teacher Immediacy and Student Learning. Paper presented at the Annual Meeting of the International Communication Association, Miami (ERIC Document Reproduction Service No. ED 346 532) (1992)
26. Moore, M.G.: Three types of interaction. The American Journal of Distance Education 3(2), 1–6 (1989)
27. Booher, R.K., Seiler, W.J.: Speech communication anxiety: An impediment to academic achievement in the university classroom. Journal of Classroom Interaction 18(1), 23–27 (1982)
28. Thompson, G.: How can correspondence-based distance education be improved? A survey of attitudes of students who are not well disposed toward correspondence study. Journal of Distance Education 5(1), 53–65 (1990)
29. Fulford, C., Zhang, S.: Perception of interaction: the critical predictor in distance learning. American Journal of Distance Education 7(3), 8–12 (1993)
30. Moore, M.G., Kearsley, G.: Distance Education: A systems view. Wadsworth, Belmont (1996)
31. Rovai, F.: A preliminary look at the structural differences of higher education classroom communities in traditional and ALN courses. JALN 6(1) (2002), http://www.aln.org/publications/jaln/v6n1/pdf/v6n1_rovai.pdf

32. Hillman, D., Willis, D., Gunawardena, C.: Learner-Interface Interaction in Distance Education: An Extension of Contemporary Models and Strategies for Practitioners. The American Journal of Distance Education 8(2), 30–42 (1994)
33. Burnham, B.R., Walden, B.: Interactions in Distance Education: A report from the other side. Paper presented at the 1997 Adult Education Research Conference, Stillwater, Oklahoma (1997), http://www.edst.educ.ubc.ca/aerc/1997/97burnham.html (retrieved May 30, 2005)
34. Anderson, T.: Getting the Mix Right Again: An updated and theoretical rationale for interaction. International Review of Research in Open and Distance Learning 4(2) (2003)
35. Mayadas, F.: Asynchronous learning networks: a sloan foundation perspective. Journal of Asynchronous Learning Networks 1 (1997)
36. Harsh, O.K.: World Wide Web (WWW) and Global Learning Environment for adults. Learning Technology newsletter 4(1) (2002)
37. Tsinakos, A.: Distance Teaching using SYIM educational environment. Learning Technology newsletter 4(4), 2–5 (2002)
38. Kalin, S.: Collaboration: A key to Internet training. American Society for Information Science 20(3), 20–21 (1994)
39. Khan, B.H.: Web-based instruction. Educational Technology Publications, New Jerssey (1997)
40. Dillon, A., Zhu, E.: Design Web-based instruction: A human-computer interaction perspective. In: Khan, B.H. (ed.) Web-Based Instruction, pp. 221–224. Educational Technology Publications, New Jersey (1997)
41. Bostock, S.J.: Designing Web-based instruction for active learning. In: Khan, B.H. (ed.) Web-based instruction, pp. 225–230. Educational Technology Publications, New Jersey (1997)
42. Shea, P.J., Pickett, A.M., Pelz, W.E.: A follow-up investigation of "teaching presence" in the SUNY Learning Network. Journal for Asynchronous Learning Networks 7, 61–80 (2003)
43. Shankar, V.: A Discourse on Synchronous and Asynchronous E-Learning (2007), http://www.articlealley.com/article_142663_22.html (retrieved in February 11, 2009)
44. Karsenti, T.: Teacher Education and Technology: Strengths and Weaknesses of Two Communication Tools. In: Proceedings of the 2007 Computer Science and IT Education Conference (2007), http://csited.org/2007/83KarsCSITEd.pdf (retrieved in January 2009)
45. Kochery, T.S.: Distance education: A delivery system in need of cooperative learning. In: Proceedings of selected research and development presentations at the 1997 National Convention of the Association for Educational Communications and Technology, Albuquerque, NM (ERIC Document Reproduction Service No. ED 409 847) (1997)
46. Pantelidis, V., Auld, L.: Teaching virtual reality using distance education. Themes in Education 3(1), 15–38 (2002)
47. Galusha, J.M.: Barriers to Learning in Distance Education. University of Southern Mississippi (2009), http://www.infrastruction.com/barriers.htm (retrieved in May 2009)
48. McNamara, J., Brown, C.: Assessment of collaborative learning in online discussions. In: Proceedings ATN Assessment Conference 2008, Engaging Students in Assessment. University of South Australia, Adelaide (2008)
49. Freire, P.: Pedagogy of the Oppressed. Herder and Herder, New York (1970)

50. Brookfield, S.D.: Discussion. In: Galbraith, M.W. (ed.) Adult learning methods: A guide to effective instruction, pp. 187–204. Robert E. Krieger, Malabar (1990)
51. Clark, J.: Collaboration tools in online environments (2009), http://www.aln.org/publications/magazine/v4n1/clark.asp (retrieved in February 12, 2009)
52. Hiltz, S.: Collaborative Learning in Asynchronous Learning Networks: Building Learning Communities. In: Web 1998 Symposium, Orlando, Florida (1998), http://eies.njit.edu/~hiltz/collaborative_learning_in_asynch.htm
53. Swan, K.: Threaded Discussion (2005), http://www.oln.org/conferences/ODCE2006/papers/Swan_Threaded_Discussion.pdf (retrieved in February 2009)
54. Pincas, A.: Features of online discourse for education. Learning Technology Newsletter 2(1) (2000)
55. Hiltz, S.R.: The Virtual Classroom: Learning without Limits via Computer Networks, Norwood, NJ (1994)
56. Rheingold, H.: The virtual community. Minerva, London (1994)
57. Gunawardena, C., Zittle, F.: Social presence as a predictor of satisfaction within a computer mediated conferencing environment. American Journal of Distance Education 11(3), 8–26 (1997)
58. Youngblood, P., Trede, F., DiCorpo, S.: Facilitating online learning: A descriptive study. Distance Education 22(2), 264–284 (2001)

Author Index